IN VITRO HAPLOID PRODUCTION IN HIGHER PLANTS

Current Plant Science and Biotechnology in Agriculture

VOLUME 25

Aims and Scope
The book series is intended for readers ranging from advanced students to senior research scientists and corporate directors interested in acquiring in-depth, state-of-the-art knowledge about research findings and techniques related to all aspects of agricultural biotechnology. Although the previous volumes in the series dealt with plant science and biotechnology, the aim is now to also include volumes dealing with animals science, food science and microbiology. While the subject matter will relate more particularly to agricultural applications, timely topics in basic science and biotechnology will also be explored. Some volumes will report progress in rapidly advancing disciplines through proceedings of symposia and workshops while others will detail fundamental information of an enduring nature that will be referenced repeatedly.

The titles published in this series are listed at the end of this volume.

In Vitro Haploid Production in Higher Plants

Volume 3 – Important Selected Plants

Edited by

S. MOHAN JAIN

Plant Production Department, University of Helsinki, Helsinki, Finland

S.K. SOPORY

School of Life Science, Jawaharlal Nehru University, New Delhi, India

and

R.E. VEILLEUX

Department of Horticulture, Virginia Polytechnic Institute and State University, Blacksburg, Virginia, U.S.A.

KLUWER ACADEMIC PUBLISHERS

Dordrecht/Boston/London

Library of Congress Cataloging-in-Publication Data

In vitro haploid production in higher plants / editors, S. Mohan Jain,
S.K. Sopory, R.E. Veilleux.
p. cm. -- (Current plant science and biotechnology in
agriculture ; v. 23)
Includes index.
Contents: v. 1. Fundamental aspects
ISBN 0-7923-3580-5 (set : alk. paper). -- ISBN 0-7923-3577-5
(hardback : alk. paper)
1. Micropropagation. 2. Haploidy. 3. Crops--Genetic engineering.
4. Plant breeding. I. Jain, S. Mohan. II. Sopory, S. K.
III. Veilleux, R. E. IV. Series: Current plant science and
biotechnology in agriculture ; 23.
SB123.6.I45 1996
631.5'23--dc20 95-304

ISBN 0-7923-3579-1 (Volume 3)
SET: ISBN 0.7923-3580-5

Published by Kluwer Academic Publishers,
P.O. Box 17, 3300 AA Dordrecht, The Netherlands.

Kluwer Academic Publishers incorporates
the publishing programmes of
D. Reidel, Martinus Nijhoff, Dr W. Junk and MTP Press.

Sold and distributed in the U.S.A. and Canada
by Kluwer Academic Publishers,
101 Philip Drive, Norwell, MA 02061, U.S.A.

In all other countries, sold and distributed
by Kluwer Academic Publishers Group,
P.O. Box 322, 3300 AH Dordrecht, The Netherlands.

Printed on acid-free paper

Printed in the Netherlands

Table of Contents

Haploid production in higher plants
A dedication

INDRA K. VASIL

The value of haploids in genetic analysis and plant breeding has been known for a long time. Natural haploid embryos and plants, derived from gametophytic cells, have been described in about 100 species of angiosperms. However, haploids occur only rarely in nature. To be useful, they must be produced in large numbers. Therefore, many attempts have been made over the years to increase the efficiency of *in ovulo* haploid production, but none of these has proven to be of wide practical utility. The early attempts to obtain haploid plants from the male gametophyte of gymnosperms (Tulecke, 1953) and angiosperms (Yamada *et al.*, 1963) resulted only in the production of haploid callus tissues (Vasil, 1980).

Embryo-like structures formed in cultured anthers of *Datura innoxia* were first described by Guha and Maheshwari (1964). They were considered to have originated from the somatic tissues of the anther. In a subsequent study, it was determined that the somatic embryos and the resulting plantlets were indeed derived from the developing microspores and were haploid in nature (Guha and Maheshwari, 1966). As is true of most pioneering studies, these first androgenic haploids were neither grown to maturity, nor were the experimental conditions for their production clearly defined. Their real value was in demonstrating the feasibility of the experimental production of haploids. Haploid plants were soon obtained from cultured anthers of *Nicotiana sylvestris* and *N. tabacum* by Bourgin and Nitsch (1967). These and subsequent studies by Nitsch and Nitsch (1969) clearly established that the culture of excised anthers at a precise stage of development was the most important requirement for switching the development of pollen from a gametophytic to a sporophytic phase, resulting in the formation of haploid embryos and/or plants. They also described a simple nutrient medium for the culture of anthers, and an easy procedure for obtaining dihaploid homozygous plants. The elegant, simple and reliable method of haploid production invented by Jean Pierre Nitsch and his associates provided much stimulus for future studies by many others.

During the past three decades many improved methods as well as nutrient media have been developed to increase the efficiency of production of andro-

Prof. S.C. Maheshwari

Dr. J.P. Nitsch

genic haploids, from cultured anthers as well as isolated microspores, in a wide variety of species. Success has also been achieved in obtaining gynogenic haploids from cultured ovaries or ovules. As a result, haploids are being used increasingly and profitably in breeding programmes for the development of new and improved cultivars. The various chapters in this and the companion volumes describe in detail the basic as well as many applied aspects of haploid production and utilization.

It has been my pleasure and privilege to have known the late Jean Pierre Bourgin, Sipra Guha-Mukherjee, Satish C. Maheshwari, Colette Nitsch and the late Jean Pierre Nitsch, all pioneers in haploid research. These volumes are dedicated to them for their seminal contributions to the experimental production of haploids and for creating a whole new field of basic and applied plant research.

References

Bourgin, J.P. and J.P. Nitsch, 1967. Obtention de *Nicotiana* haploïdes à partir d'étamines cultivées *in vitro*. Ann. Physiol. Veg. 9: 377–382.

Guha, S. and S.C. Maheshwari, 1964. *In vitro* production of embryos from anthers of *Datura*. Nature 204: 497.

Guha, S. and S.C. Maheshwari, 1966. Cell division and differentiation of embryos in the pollen grains of *Datura in vitro*. Nature 212: 97–98.

Nitsch, J.P. and C. Nitsch, 1969. Haploid plants from pollen grains. Science 163: 85--87.

Tulecke W., 1953. A tissue derived from the pollen of *Ginkgo biloba*. Science 117: 599–600.

Yamada, T., T. Shoji and Y. Sinoto. 1963. Formation of calli and free cells in the tissue culture of *Tradescantia reflexa*. Bot. Mag. Tokyo 76: 332–339.

Vasil, I.K., 1980. Androgenetic haploids. Int. Rev. Cytol. Suppl. 11A: 195–223.

General Preface

Since the beginning of agricultural production, there has been a continuous effort to grow more and better quality food to feed ever increasing populations. Both improved cultural practices and improved crop plants have allowed us to divert more human resources to non-agricultural activities while still increasing agricultural production. Malthusian population predictions continue to alarm agricultural researchers, especially plant breeders, to seek new technologies that will continue to allow us to produce more and better food by fewer people on less land. Both improvement of existing cultivars and development of new high-yielding cultivars are common goals for breeders of all crops. *In vitro* haploid production is among the new technologies that show great promise toward the goal of increasing crop yields by making similar germplasm available for many crops that was used to implement one of the greatest plant breeding success stories of this century, i.e., the development of hybrid maize by crosses of inbred lines. One of the main applications of anther culture has been to produce diploid homozygous pure lines in a single generation, thus saving many generations of backcrossing to reach homozygosity by traditional means or in crops where self-pollination is not possible.

Because doubled haploids are equivalent to inbred lines, their value has been appreciated by plant breeders for decades. The search for natural haploids and methods to induce them has been ongoing since the beginning of the 20th century. Blakeslee (1921) first identified naturally occurring haploids of *Datura stramonium* and subsequently, natural haploids of many other plants were reported by various researchers. However, naturally occurring haploids could not be produced in sufficient numbers by reliable techniques for their extensive use in breeding programmes. In 1964, the research group headed by Prof. S.C. Maheshwari, Department of Botany, Delhi University, India, reported haploid production in *Datura innoxia* for the first time by anther culture. Since this discovery, many of the limitations of the technique have been overcome such that it is currently employed for the production of haploids and doubled haploids of many crop species throughout the world. The early contributions of Drs. C. Nitsch and J.-P. Nitsch (France), G. Melchers (Germany), M.S. Swaminathan (India), I.K. Vasil (USA), N.

Sunderland (UK), and Hu Han (China) towards overcoming these limitations and adapting the technology to a variety of crops must be acknowledged. Their realization of the potential of anther culture and tireless pursuit of reliable techniques that would facilitate its success has led to its implementation in breeding programmes.

In addition to its practical applications, *in vitro* haploid extraction has changed our understanding of developmental processes in plants. Androgenesis can be thought of as a type of somatic embryogenesis that involves cells, i.e., microspores, that at first thought would not have been expected to be embryogenically competent. How can the natural course of microsporogenesis be diverted onto an embryogenic pathway? Why are some microspores competent for androgenic development and not others? How can the process of anther culture be a heritable trait in crosses between competent and non-competent parents? Does the process of anther culture impose some selection pressure on the population of microspores or otherwise result in some undesirable change expressed in the population of regenerated plants? Why are albinos so common among the anther-derived regenerants of some species when it is obvious that microspores must contain proplastids in order for green plants to be regenerated at all? We have only begun to answer some of these questions.

This book project was submitted with the consent of co-editors to the Kluwer Academic Publishers, Dordrecht, The Netherlands. The publisher had this project reviewed by anonymous reviewers. Finally, on the basis of the positive comments of the reviewers, the publisher gave us the contract to proceed with this book project. We have not followed any conservative format of chapters and gave all the liberty to the authors to write the way they felt appropriate. Most of the chapters are reviews of work done. However, in some cases, where a lot of work has not been done in the past, the authors have been encouraged to give their own research findings in details.

In this set of volumes, we have made an attempt to assimilate detailed descriptions of various aspects of anther culture and related *in vitro* procedures. Many chapters have been written by experts in the various applications of anther culture to specific crops. In addition to crop-by-crop discussions on the progress of anther culture, we have also included chapters on other topics concerning the utilization of *in vitro* haploids in plants. Embryogenic microspores have recently been regarded as ideal targets for genetic transformation. Molecular markers such as RFLPs, RAPDs, or SSRs can be used to determine disturbed segregation ratios in haploid populations or to tag traits of interest to plant breeders. The potential of pollen protoplasts is discussed. *In vitro* selection during androgenesis, both imposed and inadvertent, is also considered.

The series is divided into five volumes. Volume 1 contains 18 chapters and primarily covers fundamental aspects of haploidy and various methods of haploid extraction, e.g., anther culture, microspore culture, ovary culture, etc. The second volume comprises 21 chapters and describes applications of haploid breeding in protoplast manipulations, mutation breeding, RFLP mapping, identification of quantitative trait loci (QTLs), cryopreservation, chromosome engineering by anther culture, molecular biology of pollen rejection, transformation of pollen/microspores, etc. The third volume has 20 chapters focussed on haploid breeding in selected important crops including vegetables (*Allium* spp., *Brassica* spp., *Capsicum*, *Cichorium*, *Cucumis*, *Solanum melongena*, *Solanum tuberosum*); fruit crops (*Malus*, *Fragaria*, *Vitis*); and other miscellaneous crops (*Beta*, *Coffea*, *Ginkgo*, *Glycine*, *Medicago*, *Saccharum*, *Sinocalamus latiflora*). We have included 11 chapters in the fourth volume on haploid breeding in cereals (wheat, rice, barley, oats, sorghum, maize, triticale, rye, pearl millet, buckwheat). The fifth volume has 13 chapters, mainly dealing with ornamentals, tobacco, tomato, cotton, linseed, sunflower, asparagus, niger, gynogenic haploids in angiosperms and haploids in potato interspecific somatic hybrids.

While preparing these volumes, we were overwhelmed by the enthusiastic response and timely cooperation of invited authors and the many research scientists who gave freely of their time to review the manuscripts. The reviewers were: R.I.S. Brettel (Australia); B.S. Ahloowalia (Austria); J.M. Bonga, K.N. Kao, K. Kasha, L.K. Kott, K.P. Pauls, R. Sadashivaiah (Canada); Hu Han (China); H. Ahokas, V. Kauppinen, J. Peltonen, S. Sarvori, L. Simola, P.M.A. Tigerstedt (Finland); C. Dóre, R.S. Sangwan (France); B. Foroughi-Wehr, D. Hess, S. Deimling, H. Uhrig, G. Wenzel (Germany); S.S. Bhojwani, P.B. Kirti, S.C. Maheshwari, A.F. Mascarenhas, P.S. Nadgauda, D. Pental, S.K. Raina, P.S. Rao, N. Sarin, G. Lakhshmi Sita (India); A. Mujeeb-Kazi (Mexico); H. Dons, K. Sree Ramulu (The Netherlands); D.S. Brar, G.S. Khush (Philippines); M. Zenkteler (Poland); A.M. Vieitez (Spain); C. Bornman, K. Glimelius, A. Wallin (Sweden); W. Chang (Taiwan); J. Dunwell, V.E. Franklin-Tong, W. Powell (UK); P.S. Baenziger, E. Earle, G.J. Galletta, D.J. Gray, P.K. Gupta, J.P. Helgeson, J. Janick, S.M. Reed, H.S. Rines, G. Schaeffer, T.L. Sims, I.K. Vasil, J.M. Widholm (USA).

It is still too early to write the last chapter on *in vitro* haploids. There are those who argue that its great potential will result in improved cultivars of many of our major crops. On the other hand, there are those who think its potential has been overrated, that the severe inbreeding depression observed among primary doubled haploids and the lack of selection pressure for functional sexual flower parts during the process of androgenesis will result only in useless, fruitless plants. The first cultivars employing anther-derived doubled haploids in their pedigree have already been released for a number of crops including wheat, rice, maize, and asparagus. Whether these cultivars and future such releases will endure remains to be seen.

S. Mohan Jain
S.K. Sopory
R.E. Veilleux

Preface to Volume 3

This third volume on *In Vitro* Haploid Production in Higher Plants comprises 20 chapters that review the current status of haploid breeding in a variety of plants. Each chapter contains some introductory material about the selected plant, the techniques used most successfully for haploid production (anther culture, microspore culture, and ovary culture), the factors influencing the success of these techniques, the identification and genetic characterization of haploid regenerants, the application of haploids in breeding, and a brief conclusion on the potential of haploid breeding for the crop. The plants include vegetable crops (*Allium* spp., *Brassica* spp., *Capsicum*, *Cichorium*, *Cucumis*, *Solanum melongena*, *Solanum tuberosum*); fruit crops (*Malus*, *Fragaria*, *Vitis*); and other miscellaneous crops (*Beta*, *Coffea*, *Ginkgo*, *Glycine*, *Medicago*, *Saccharum*, *Sinocalamus latiflora*, forest trees). In some chapters, protoplast manipulation and genetic transformation have also been covered.

We are grateful to authors and reviewers for working to improve the quality of the manuscripts.

S. Mohan Jain
S.K. Sopory
R.E. Veilleux

Acknowledgements

Ever since I finished my Ph.D. on *In vitro* haploids in higher plants, I have wanted to edit a book on this important subject and tucked this thought in the back of my mind. On separate opportunities, I mentioned it to Profs. S.K. Sopory and Richard E. Veilleux, inviting them to co-edit such a book. They graciously accepted my invitation to become co-editors. Their critical review of manuscripts and valuable suggestions have substantially improved the quality of these volumes. I am thankful to both Sudhir and Richard for helping me on this ambitious project and it has been a great pleasure working with them.

I appreciate the invited authors for their punctuality in meeting deadlines for submission of their contributions and all the reviewers (named in the General Preface) for constructive and timely critical reviews of the manuscripts. Their comments have been extremely useful for improving the quality of these volumes.

I am thankful to my colleagues Prof. Eija Pehu, Mr. Tapio Poutala, and Mr. Matti Teittinen for their assistance.

While editing this book, I had the opportunity to visit the University of Tuscia, Italy, as a visiting professor fellow. I am thankful to Prof. Eddo Rugini for his warm hospitality. During my short stay, I managed to find the time to edit several manuscripts.

Also, with great love and affection, I want to thank my daughters Sarita and Sonia, and my wife, Marja Liisa, for their unceasing patience and understanding while I was working on this time-consuming project.

Finally, I express my deepest sense of appreciation to Adrian Plaizier of Kluwer Academic Publishers, The Netherlands, for giving us the opportunity to work on this project. Adrian has always been cooperative and helpful, encouraging me with intelligent advice.

Book Project Leader
S. Mohan Jain

1. Haploids of sugarcane

MAUREEN M.M. FITCH and PAUL H. MOORE

Contents

1. Introduction

Sugarcane is one of the world's largest cultivated grasses and belongs to the family *Poaceae*. Sugarcane as a crop is grown in 106 tropical and subtropical countries (FAO, 1992). In 1992, the crop was grown on 1.8×10^7 hectares yielding 1.1×10^9 metric tons (MT) of cane giving an average yield of 61 MT/ha. The crop is grown mainly for sucrose and for fiber and ethanol production. Sugarcane ranks second in production after cereals (2.0×10^9 MT), the cane tonnage being equivalent to that of wheat plus rice (0.5 and 0.6×10^9 MT, respectively).

The interspecific hybrid is believed to have evolved in New Guinea where prehistoric settlers selected clones probably for sweet-tasting juice (Daniels and Roach, 1987). The genus *Saccharum* includes six species, *S. officinarum* L., *S. spontaneum* L., *S. robustum* Brandes and Jeswiet ex Grassl, *S. barberi* Jesw., *S. sinense* Roxb., and *S. edule* Hassk. Other important related genera in this tropical and subtropical tribe Agropogoneae are *Erianthus*, *Miscanthus*, *Narenga*, and *Sclerostachya*. Interspecific and intergeneric hybrids are possible in this group and account for the genetic complexity of sugarcane cultivars that are grown today, many clones of *S. officinarum* exhibit thick stalks and high concentrations of sucrose which accumulate in stalk storage parenchyma cells. These plants are called the "noble" canes and are responsible for high concentrations of sucrose in hybrids. These plants along with *S. barberi* and *S. sinense* store sucrose in the vacuole. The sucrose may amount to as much as 20% of a cell's fresh weight (Hawker, 1985).

Saccharum species have high chromosome numbers, ranging from $2n = 40$ to $2n = 128$ (Daniels and Roach, 1987). These species are considered to

S.M. Jain, S.K. Sopory & R.E. Veilleux (eds.), In Vitro *Haploid Production in Higher Plants, Vol. 3*, 1–16.

be octoploids with base chromosome numbers of 8, 10, or 12 (Sreenivasan *et al.*, 1987). The hybrids between *S. officinarum* ($2n = 80$) and *S. spontaneum* ($2n = 40$–128), however, often exhibit chromosome numbers around $2n = 120$, indicative of a $2n + n$ transmission. Hybrids inherit the full chromosome complement of the *S. officinarum* parent and the gametic chromosome complement of the *S. spontaneum* parent, apparently due to a doubling of the egg complement in the "noble" parent. Most of the sugarcane cultivars have germplasm from the two species mentioned. In addition, *S. robustum*, *S. barberi*, and *S. sinense* have been used for introducing disease resistance in the modern cultivars (Sreenivasan *et al.*, 1987). In the recurrent backcrossing program, aimed at the introgression of desirable genetic components from the other species, *S. officinarum* serves as the female parent. Since the chromosome number is large and the parentage of these interspecific hybrids is complex, haploids were sought as a simpler genome to study.

2. Review of the literature

Most of the recent research work on sugarcane haploids was reviewed recently by Moore *et al.* (1989) and Moore and Fitch (1990). The previous reviews did not include unpublished results from India. The present review summarizes the earlier work, including that on anther culture in India which was started before 1978 (Jalaja, 1978) and continues until now (Table 1). In addition, our work on haploids since 1990 is described.

3. Production of haploids

3.1. *Interspecific hybridization and anther culture*

Sugarcane haploids were first reported by Roach (1966) who obtained five haploid seedlings ($2n = 40$) out of 886 progeny from crosses between several *S. officinarum* ($2n = 80$) and *S. spontaneum* clones. The plants from Korpi × Mandalay ($2n = 96$), Korpi × Molokai 1032 ($2n = 64$), and Badila × 51NG2 ($2n = 80$) apparently arose from failure of the normal $2n + n$ transmission of gametes typical of those crosses. Instead, seed development attributed to apomixis resulted in haploid *S. officinarum* plants. Panicles of the haploids failed to emerge. Apparently, the haploid lines were not maintained nor subjected to additional studies. The haploid plants may have arisen in an analogous manner to "*bulbosum*" haploids in barley where a cross between *Hordeum vulgare* and *H. bulbosum* results in the elimination of the latter chromosomes in the formation of *H. vulgare* haploids (Kasha, 1974).

Anther culture-derived haploid sugarcane plants were reported by groups in China, Hawaii, and India using a range of clones and methods (Table 1). The first haploid sugarcane plants were produced from a Chinese cultivar

Table 1. Summary of anther culture in China, Hawaii, and India

Author	Cultivar	Type	2N	Stage	Step	Medium	Sucrose (%)	Growth Regulator (mg/l)	Duration	Result	2N
Chen	Y31	Commercial	120	diad,	1) high	MS	20	2,4-D 2	nm	callus	
et al.,		hybrid		tetrad,	sugar	or		KIN 2			
1979,				early		N6					
1983				uninu-							
				cleate	2) grow	MS	5	NAA 2	nm	green	50-90
					out			BAP 2		plants	
Fitch	SES 208,	*S. spon-*									
&	SES 205B,	*taneum*	64	late	1) cold	MS	3	2,4-D 1	3	pollen	
Moore,	SES 365,		64	uninu-				IAA 1	wk	divisions	
1983	US 56-15-8		80	cleate,				BAP 1			
			80	early				10°C			
				binu-							
				cleate	2) float	MS	3	2,4-D 1	7-	multi-	
								IAA 1	10 d	cellular	
								BAP 1			
								27°C			
					3) nurse	MS	2	2,4-D 3	3-4	callus:	
								CW 20%	wk	208	32,
								+/-			32+64
								charcoal+			
								2,4-D 10			
					4) grow	MS	2	CW 20%	3-4	green	
					out				wk	plants:	
										208	32
										205A	64
Fitch	SES 208,	*S. spon-*									
&	SES 365,	*taneum*	64	late	1) cold	MS	3	2,4-D 3	4-10	multi-	
Moore,,	US 56-15-8		80	uninu-				CW 20%	wk	cellular	
1984			80	cleate		or					
				early		0.5MS	2	2,4-D 2			
				binu-							
				cleate		or					
						H	2	2,4-D 2			
								BAP 1			
						or					
						N6	5	2,4-D 2			
								BAP 1			
								KIN 0.5			
					2) float	0.5MS	2	2,4-D 2	1-8	callus	
									wk		
						or					
						B5	2	2,4-D 2			
								BAP 1			
					3) nurse	MS	2	PIC 0.25-0.5	2-8	green,	
									wk	albino	
										plants:	
										208	n,2n
										365	2n
										15-8	n,2n
Fitch	SES 205B,	*S. spont-*		late	1) cold	0.5MS	2	2,4-D 1	4-8	multi-	
et al.,	SES 208,	*aneum*	64	uninu-					wk	cellular	
1989,	SES 365		64	cleate,							
Nagai			80	early	2) float	B5	2	2,4-D 2	2-4	callus	
et al.,				binu-				BAP 1	wk		
1989,				cleate							
Nagai	US 56-15-8,		80		3) nurse	MS	2	PIC 0.25-0.5	2-8	green	
&	US 56-20-1		80						wk	and	
Ching,										albino	
1990,										plants:	
C.	US 66-56-9	Hybrid	98							205B	n,2n
Nagai,		(1/4								208	n,2n
(per-		"spont")								365	n,2n
sonal										15-8	n,2n
commu-	NG 77-154	*S. offi-*	80							20-1	nd
nication)		*cinarum*								56-9	nd
										154	nd

Table 1. Continued

Author	Cultivar	Type	2N	Stage	Step	Medium	Sucrose (%)	Growth Regulator (mg/1)		Duration	Result	2N
Sreenivasan & Jalaja, 1979,	IMP 1533	*S. spontaneum*	nm	late uninucleate	1) cold	MS	6	2,4-D KIN 14°C	8 4	7–10 d	synchronize pollen divisions	
Jalaja & Sreenivasan, 1988, 1991, 1992, T.V. Sreenivasan (personal communication)	SES 24, SES 161B, SES 274, SES 305	*S. spontaneum* *E. elephantinus*	80 nm 64 20		2) grow out	MS	6	2,4-D KIN 25°C	8 4		plants	2n, all
	Isiolo	*S. spontaneum*	124–128									
	CoC 671	Commercial hybrid	110									88-114
	Co 7219, Co 7717	Commercial hybrids	nm nm									
	MC 132	Commercial hybrid	nm									
		E. arundinaceus	nm									

Abbreviations: BAP = benzylaminopurine; CW = coconut water; 2,4-D = 2,4,-dichlorophenoxyacetic acid; H = Nitsch medium (Nitsch and Nitsch, 1969); IAA = indole-3-acetic acid; KIN = kinetin; MS = Murashige and Skoog medium; N6 = Chu N6 medium (Chu *et al.*, 1975); NAA = 1-naphthaleneacetic acid; nd = no data; nm = not mentioned; PIC = picloram.

(Chen *et al.*, 1979) primarily of *S. sinense* origin (Hu and Zeng, 1984). In Hawaii, *S. spontaneum* haploids were produced from three clones (Fitch and Moore, 1983, 1984). Additional haploid plants were regenerated from several more *S. spontaneum* clones, one *S. officinarum* clone, and an interspecific hybrid that was one-quarter *S. spontaneum*, but only the "spont" clones survived greenhouse culture (Nagai *et al.*, 1989; Nagai and Ching, 1990; Nagai, unpublished data). Jalaja and Sreenivasan obtained several plants from anther culture of *S. spontaneum*, *Erianthus*, and Indian commercial sugarcane cultivars; however, these plants were all doubled haploids (Sreenivasan and Jalaja, 1987; Jalaja and Sreenivasan, 1992; T.V. Sreenivasan, personal communication).

The protocols used by the three groups are summarized in Table 1 and have certain elements in common. Each group used modified MS medium (Murashige and Skoog, 1962) containing 2,4-dichlorophenoxyacetic acid (2,4-D) and produced calli after multiple steps following osmotic (Chen *et al.*, 1979) or temperature stress applied to panicles prior to culturing of anthers (Fitch and Moore, 1983, 1984; Sreenivasan and Jalaja, 1984). Microspores were at diad, tetrad, or uninucleate stages in the Chinese protocol, while late uninucleate to early binucleate pollen was used in Hawaii and India. Fitch and Moore (1983) determined that cold treatment was critical for inducing pollen divisions and incubated panicles in cold treatment media (Table 1) at 10 °C for 21 days while Jalaja and Sreenivasan (1991) incubated panicle branches at 14 °C for 7–10 days. Both groups observed that in some cases pollen calli were released into the medium from the cultured anthers and subsequently formed plants. Albino as well as green plants were obtained in Hawaii (Fitch and Moore, 1984; Nagai *et al.*, 1989) and India (Jalaja and Sreenivasan, 1992). Haploid plants were produced in China and Hawaii.

3.2. *Improvements in anther culture*

The research group in Hawaii found that anther culture of *S. officinarum* was difficult. Since *S. spontaneum* flowered one to two months earlier than *S. officinarum* and the two species had different flowering characteristics, studies were conducted to determine if these differences could be manipulated to improve androgenesis in *S. officinarum*. Minimum temperatures in Hawaii are low between October–November (20–21.5 °C) when *S. officinarum* flowers, and may have been too low for optimal androgenesis in this species. To test this hypothesis, five *S. officinarum* genotypes with high pollen viability (55–78%) were planted in 60-liter pots in a greenhouse with supplemental heat (22–23 °C). This treatment did not advance flowering or increase pollen viability or culturability for haploids (Nagai *et al.*, 1989). Another possible problem with *S. officinarum* was that its panicle maturation rate was slower than its rate of emergence compared to that of *S. spontaneum*. In *S. officinarum*, panicles emerged from the leaf sheath before the pollen reached the late uninucleate stage which normally occurs within the sheath in *S. spontaneum*. Therefore, surface sterilization of *S. officinarum* panicle branches was required prior to culture. This was accomplished by soaking panicles in 1.05% sodium hypochlorite for 20 min, rinsing three times with sterile water, and placing branches into cold treatment medium (0.5 MS, Table 1). Multicellular pollen, calli, and plants subsequently developed. Thus, the major problem with culturing *S. officinarum*, which results from delayed development of the anthers as compared with time of panicle emergence, was resolved.

Since cultured anthers turn brown and phenolic compounds inhibit cell cultures, it appeared that death of the anther wall produced compounds like phenolics which inhibited microspore development. To test this hypothesis, we cultured isolated microspores. Cultured microspores did not brown but continued development to become multicellular; however, no macroscopic callus developed (Hinchee and Fitch, 1984).

Hinchee *et al.* (1984) identified biochemical characteristics of cultured anthers that might aid in selecting anthers suitable for culture. Pollen did not divide from anthers containing higher concentrations of asparagine relative to glutamine. Anthers initially high in reducing sugars, free amino acids, and amylase activity yielded more dividing microspores than those low in one or more of the biochemical components. Glutamine enrichment of the media has potential for stimulating haploid development.

Media modifications that led to improved efficiency in anther culture of wheat and barley included increased glutamine (Henry and De Buyser, 1981), inositol (Sunderland and Xu, 1982), maltose (Orshinsky *et al.*, 1990), or

Table 2. Effect of different disaccharides or glutamine supplement on anther culture of SES 208 (from Fitch *et al.*, 1989)

Medium Supplement	# Panicles	# Anthers	# Calluses	%Calluses	Range of %Calluses	%Panicles w/ calluses
Melibiose	33	34,950	32	0.09	0-1.1	30
Maltose	35	40,050	547	1.37	0-6.3	74
Glutamine	40	43,050	1,476	3.43	0-19.2	87
Sucrose	41	47,100	2,706	5.75	0-33.4	95
Total		165,150	4,761			

melibiose (Sorvari and Schieder, 1987). Therefore, experiments were conducted to determine the effects of these media changes on sugarcane anther culture. None of the additives increased callus yield above that of sucrose (Table 2). This is in contrast to results of an earlier study indicating that increased inositol, 1–10 g/l in half-strength MS (0.5 MS) but not full strength MS medium, resulted in higher callus yields compared to media containing 0.1–1.5 g/l sucrose (Fitch and Moore, 1983b). Since inositol was beneficial in a low salt medium only, the beneficial effect might have been osmotic. The osmotic potential of the medium increased as inositol concentration increased. More studies based on varying the concentrations of these additives in the cold and/or float media are needed to determine if such treatments consistently increase anther culture efficiency. For example, in barley pollen culture, a 9:1 ratio of $NO_3:NH_4$ in the medium had no effect on the frequency of mitotic divisions, increased plating efficiency moderately, but resulted in a pronounced increase on embryogenesis and plant regeneration (Mordhorst and Lörz, 1993). After cold treatment of sugarcane panicles, we observed twice as many calli on a B5 medium (4:1 ratio, $NO_3:NH_4$) than on 0.5 MS medium (2:1 ratio, $NO_3:NH_4$; Fitch and Moore, 1984). Therefore, the higher ratio of $NO_3:NH_4$ had a stimulatory effect on sugarcane anther culture. We have not determined if an increase in the $NO_3:NH_4$ ratio improves regeneration from sugarcane pollen callus.

3.3. *Difficult genotypes*

In our hands, haploids were most readily obtained from *S. spontaneum* genotypes (Table 1). A majority of the haploid plants was produced from a *S. spontaneum* clone from India, SES 208. Haploid and doubled haploid plants were also obtained from SES 205A, SES 205B, SES 365, US 56–15–8, and US 56–20–1. Additional clones from which callus was obtained were SES 106B ($2n = 48$), AP83–0141 (doubled haploid), and AP85–0068 (doubled haploid), but no plants were regenerated. The low chromosome number of the former and the doubled haploid genome of the latter two may not have the necessary genomic elements for completing embryogenesis. Besides the *S. spontaneum* group, we have succeeded in producing anther-derived callus from clones of *S. robustum*, *S. officinarum*, and an interspecific hybrid. A single *S. robustum*, IJ76–459, produced numerous multicellular pollen grains, but only one developed into callus and that one callus failed to develop beyond 2–3 mm. Similar problems were encountered with most of the *S. officinarum* calli. A single callus from NG77–154 developed into a plant that died. Plants from the "quarter spont" genotype US66–56–9 were weak and also died. Since multicellular masses and calli were initiated in these two species, an increase in the number of anthers cultured could conceivably lead to the induction of larger number of calli and success in plant regeneration. In contrast, researchers in China and India produced anther culture-derived plants from hybrid cultivars (Table 1). The haploid work completed in Hawaii has been summarized in Table 3. Efficiency of callus induction in *S. spontaneum* clones SES 208, SES 365, and US 56–15–8 increased from 1983 to 1989, but regeneration in SES 208 did not. Anther-derived plants were considerably more difficult to produce from the other clones.

4. Characterization of anther culture-derived plants

4.1. *Morphological characteristics*

The haploids and doubled haploids were generally smaller in stature and weaker than the parent clone SES 208. Ploidy of plants derived from anther culture of SES 208 was estimated by measuring guard cell length (Fitch and Moore, 1984); lengths smaller than 22.5 μm were characteristic of haploids, those larger than 27.5 μm were characteristic of doubled haploids (Fitch *et al.*, 1987).

Flowering percentage of mature stalks of the parent SES 208 was 90% while that of the haploid and doubled haploid lines ranged from 0–90% (Nagai and Schnell, 1987). Anthers of SES 208 were 1.7 mm long, haploids were 0.8–1.5 mm, and doubled haploids were 1.2–1.8 mm. Pollen stainability was 85.2% for SES 208, 1.4–49.7% for haploids and 0–84% for doubled haploids. Despite the moderately good fertility of some haploids, 10 different crosses failed to produce progeny. Although differences were observed between morphological characteristics of haploid and doubled haploid plants from SES 208, no significant differences at the 5% confidence level were observed in stalk number (n = 10.8 vs. 2n = 5.3), plant height (n = 82.3 cm vs. 2n = 91.5 cm), leaf width (n = 9.9 mm vs. 2n = 10.5 mm), leaf length (n = 69.9 cm vs. 2n = 71.3 cm), and leaf area (n = 71.3 cm^2 vs. 2n = 75.1 cm^2) (Nagai, 1990).

Haploids exhibited characteristics not found in their donor clones. Leaves

Table 3. Efficiency of anther culture of sugar cane in Hawaii (from Fitch *et al.*, 1989; Moore *et al.*, 1989; Nagai *et al.*, 1989; Nagai and Ching, 1990, C. Nagai, personal communication)

Year	Clone	2N	#Anthers	#Calli	Callus%	#Plants	Plant%	Pl./Cal.%	%H
1983	SES 205A	64	5,000	10	0.2	10	0.20	100.0	0
1983	SES 208	64	6,000	6	0.1	2	0.04	33.3	50
1984	SES 208	64	23,205	218	1.0	59	0.25	27.1	30
1985	SES 208	64	30,000	500	1.7	500	1.70	100.0	nd
1989	SES 208	64	165,150	4,761	2.9	60	0.01	0.6	20
1983	SES 365	80	4,000	2	0.1	0	0.00	0.0	0
1984	SES 365	80	1,200	3	0.3	2	0.17	66.7	0
1989	SES 365	80	5,460	240	4.4	12	0.22	5.0	30.8
1989	SES 106B	48	9,000	15	0.2	0	0.00	0.0	0
1989	SES 205B	64	70,070	180	0.3	7	0.01	2.2	71.4
1983	US56-15-8	80	4000	5	0.10	0	0.00	0.0	0
1984	US56-15-8	80	25,000	55	0.20	39	0.17	70.9	nd
1989	US56-15-8	80	41,335	1,430	3.50	300	0.73	21.0	~30
1989	US56-20-1	80	24,540	6	0.02	3	0.01	50.0	nd
S. robustum									
1985	IJ76-459	80	2,000	1	0.05	0	0.00	0.0	0
S. spontaneum X *S. officinarum* (1/4 "spont")									
1989	US66-56-9	98	9,315	7	0.08	3	0.03	42.9	33.3
S. officinarum									
1989	NG77-154	80	21,833	4	0.02	1	0.01	25.0	nd
1989	NG77-151	80	8,950	mult.	0.00	0	0.00	0.0	0
1989	H9811	80	6,700	mult.	0.00	0	0.00	0.0	0
S. spontaneum doubled haploids									
1985	AP83-0103	64	1,000	mult.	0.00	0	0.00	0.0	0
1985	AP83-0141	64	2,000	50	2.50	0	0.00	0.0	0
1990	AP85-0068	64	11,400	40	0.13	0	0.00	0.0	0

Table 3. Continued

Year	Clone	2N	#Anthers	#Calli	Callus%	#Plants	Plant%	Pl./Cal.%	%H
S. spontaneum haploids									
1984	AP83-0106	32	500	mult.	0.00	0	0.00	0.0	0
1985	AP83-0106	32	500	mult.	0.00	0	0.00	0.0	0
1990	AP85-0441	32	30,000	0	0.00	0	0.00	0.0	0

Attempted culture of eight other *S. officinarum* clones, IJ76-325, IJ76-361, IJ76-321, IM76-325, Manjiri Red, Fiji24, 28NG21, and Korpi; either viability was poor or contamination or other problems resulted.
Abbreviations: %H = percentage of haploids, mult. = multinucleate or multicellular, nd = no data, Pl./Cal.% = number of plants obtained from number of calli ×100.

of some haploids and doubled haploids of SES 208 exhibited red blotches, particularly when the plants were grown in the field (Moore *et al.*, 1989). Pathogens were not associated with the blotch, so the trait appears to be a mineral or herbicide toxicity response. Leaf variegation, consisting of alternating light and dark transverse bands across the leaf blade which resembled "zebra stripes", was observed in all four haploid lines generated from SES 205B (Nagai and Ching, 1990). A photo-induced lesion may have been responsible for the striping.

4.2. *Cytological observations*

Cytological observations on the chromosomes of the SES 208 haploids were made at meiosis and mitosis (young leaves and root tips). Meiosis appeared normal with 16 pairs of bivalents, even in clones having low fertility (10–15%) and vigor, e.g., AP83–106 (Fitch, 1985). Laggards or univalents were not observed. In one line of SES 205A-derived doubled haploids, however, 4–10 laggards were observed at anaphase, but telophase appeared normal with $2n = 32$ chromosomes.

Mitotic studies were conducted by applying an enzymatic maceration method to root tips. Giemsa staining revealed the presence of chromosome satellites (S. Ha, unpublished data). Variation in the number of satellite chromosomes was found among the haploids. Satellites are chromosome fragments that are separated from the rest of the chromosome by a secondary constriction (Fig. 1A, B). Satellite chromosomes usually have a region called the nucleolus organizer that is responsible for the formation of the nucleolus and ribosomal RNA synthesis. Satellite chromosomes are sometimes called nucleolus organizer chromosomes and each species usually possesses at least one homologous pair of these chromosomes. Very often, the basic genome

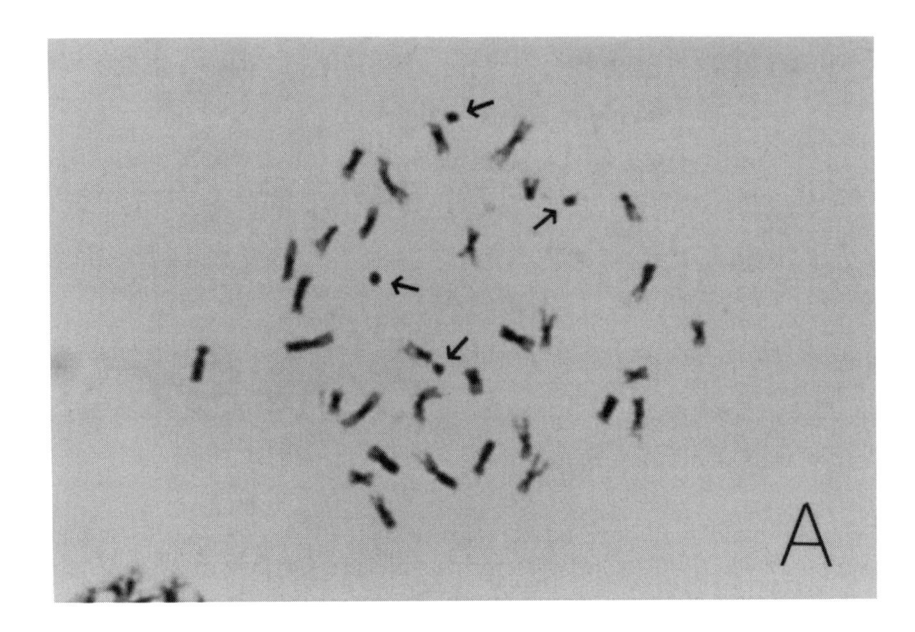

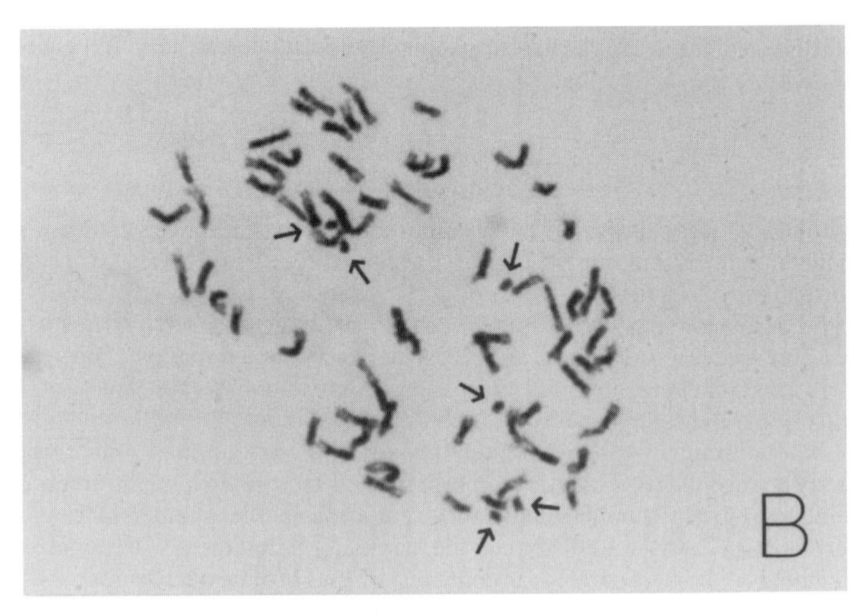

Figure 1. Giemsa-stained root tip chromosomes of *S. spontaneum* clones. (A) Haploid clone AP85–0361 ($2n = 32$) showing 4 satellite chromosomes (arrows). (B) Parent clone SES 208 ($2n = 64$) showing 6 satellite chromosomes (arrows) (photographs by S. Ha).

number x is equal to the number of nucleolus organizer chromosomes. Thus, satellite chromosomes can serve as marker chromosomes for quantifying the basic genome size.

Some haploid lines contained three satellite chromosomes, others contained four (Fig. 1A), while the parent contained at least six satellites (Fig. 1B). In addition, two of the haploids examined apparently were mixoploid, $2n = 32$ and $2n = 64$ (S. Ha, unpublished data). Further cytological studies are in progress on the SES 208 haploids to karyotype the 32 chromosomes.

4.3. *Field studies on anther-derived plants*

In China, hundreds of green plants regenerated from 30 callus cultures grew normally in the field and showed segregation for agronomic characters (Chen *et al.*, 1983). One of the anther-derived plants showed a 3% higher brix (percentage by weight of the soluble solids in sugarcane juice) than the donor plant and was clonally propagated.

In India, the anther culture-derived plants were examined for field performance and yield characteristics (Jalaja and Sreenivasan, 1992). Plants from the commercial cultivar CoC 671 ($2n = 110$), possessing *S. officinarum* and *S. spontaneum* germplasm, were chromosomal mosaics with a range in chromosome number from $2n = 88$ to 114. Somatic tissue culture-derived populations of this cultivar did not produce such an extensive variation in chromosome number. Field-grown doubled haploids from CoC 671-TC 1251 (a plant line regenerated from tissue culture) were analyzed for stalk weight, Brix, sucrose, purity (percentage of juice solids that is sucrose), CCS (ratio of weight of sucrose per unit fresh weight of cane), and NMC/clump (number of mature canes per plant). Yield and quality of clones were equal or superior to that of the donor plant CoC 671. Pollen fertility ranged from 12–56% in anther-derived plants while it was 47–59% in the donor plants. These fairly high fertility numbers probably reflected the doubled haploid nature of the plants.

Fifteen doubled haploid CoC 671 plants were selfed producing about 700 S_1 plants (Jalaja and Sreenivasan, 1992). The number of seedlings per cross ranged from 2–214 and averaged nearly 50 seedlings per selfed cross. This represents a surprisingly high number of seedlings considering the low fertility usually observed on selfing the donor clone CoC 671. In addition, five of the dihaploid plants were crossed as females to two *Erianthus* species which produced 50 seedlings. Sugarcane-*Erianthus procerus* ($2n = 40$) progeny were tall and vigorous with few characteristics of the wild species (Sreenivasan and Jalaja, 1991). In addition, the group intends to study the role *Erianthus* plays in the origin of *S. officinarum* and *S. robustum*. Hybrids of *S. officinarum* × *E. arundinaceus* expressed characteristics intermediate between the two parents and also yielded androgenic lines resembling *Erianthus* with $2n = 60$ chromosomes.

The attempts to produce progenies from the SES 208 haploids for five years were not successful; no plants were recovered either from selfing or hybridizing in 30 crosses (C. Nagai, unpublished data). However, 200 progenies were produced by backcrossing AP85–0068, a doubled haploid of SES 208, with SES 208 in 1987. A haploid clone of US 56–15–8 (*S. spontaneum*) from Thailand was crossed with L.A. Purple (*S. officinarum*), and 150 plants were obtained in 1993.

5. Haploids in genome analysis

5.1. *RFLPs*

Haploids and doubled haploids both have a reduced number of alleles and could potentially substitute for inbred lines which do not exist in sugarcane (Moore *et al.*, 1989). Therefore, haploids were evaluated for use in genome mapping using restriction fragment length polymorphisms (RFLPs). DNA fragments from a *Pst*I-digested genomic DNA library of the hybrid cultivar NCo 310, three cDNA libraries of the *S. officinarum* cultivar Badila, and selected cDNAs from other grasses were mapped. Segregation in a population of 80 haploid lines from SES 208 was skewed for one of the three batches of anther-derived plants (Da Silva *et al.*, 1993). Therefore, mapping was conducted on an F_1 population derived from a cross between a doubled haploid and the donor parent, SES 208. A low density RFLP map derived from single-dose markers (Wu *et al.*, 1992) was generated that contains markers about 25 cM apart (Da Silva *et al.*, 1993). The development of this RFLP map demonstrated that single dose traits could be mapped in *Saccharum* and served as a model for sugarcane scientists to produce a map of the important sugarcane species *S. officinarum* and its hybrids.

5.2. *Isozymes*

Isozyme analysis of the mapped population and of other anther culture-derived plants was carried out on starch gel electrophoresis. Out of eight enzyme systems, peroxidase (PX), phosphoglucosemutase (PGM), phosphoglucoseisomerase (PGI), and esterase (EST) showed isozyme polymorphisms in SES 208-derived haploids (Nagai and Schnell, 1986a,b). Polymorphisms in five band regions for PX, PGI, and EST were also established for SES 205B, SES 365, and US 56–15–8 plants from anther culture (Nagai and Ching, 1991). Variations in band patterns were observed for PX and EST in 70 haploids. Selfed SES 208 progeny showed the same band segregations while somaclones of SES 208 from tissue culture did not. Upon further study, diaphorase showed polymorphism in SES 208 haploids (T. Wenslaff and C. Nagai, unpublished data). A single polymorphic diaphorase band was found that segregated 1:1 in the AP85–0068 × SES 208 progeny, providing a single dose marker for segregation analysis. Compared to the large number of polymorphic markers obtained through RFLP analysis, isozyme analysis was less efficient but was useful for determining the doubled haploid nature of some $2n = 64$ anther culture-derived plants (Nagai and Schnell, 1987).

6. Conclusions

Plants of sugarcane from anther culture have been produced by groups in China, Hawaii, and India, but an efficient protocol is still far from routine. The most important treatment to induce sugarcane haploids was some form of stress, either osmotic or cold. The media for multicellular microspore formation appear to be adequate; however, regeneration from callus needs to be improved. To increase the numbers of plants regenerated from anther-derived callus, nitrogen availability in 9:1 NO_3-NH_4 ratio should be attempted. Among the five sugarcane species, *S. spontaneum* is still the most easily cultured. Cytological analysis is relatively simple with haploids and may eventually give insights into chromosome base numbers, pairing behaviour, and evolutionary relationships among species of the *Saccharum* complex. Haploids and doubled haploids of hybrid cultivars in field studies in China and India segregated for traits associated with yield, but no improved cultivars have yet been identified. On the other hand, haploids have been useful in molecular genetic analysis of the genus. A population from progeny of a doubled haploid of *S. spontaneum* crossed with its donor parent was used to produce the first low density map of the genus. That effort helped in developing protocols for a project to produce an *S. officinarum* map using progeny from *S. officinarum* cultivar L.A. Purple × *S. robustum* cultivar Molokai 5829 (K.K. Wu, P.H. Moore, M.M.M. Fitch, and T. Wenslaff, unpublished data).

7. Ackowledgements

We wish to thank the following individuals for sharing their unpublished data: Dr. T.V. Sreenivasan, Head, Tissue Culture Section, Sugarcane Breeding Station, Coimbatore, India; Dr. Sen Ha, Research Associate, Hawaiian Sugar Planters' Association; Dr. Chifumi Nagai, Plant Breeder, Hawaiian Sugar Planters' Association; Dr. Timothy Wenslaff, Research Associate, Hawaiian Sugar Planters' Association.

8. References

Chen, Z.-H., Z.-T. Deng, N.-S. Huang and S.-L. Wu, 1983. Induction of sugarcane haploid plants. Cell and Tissue Culture Techniques for Cereal Crop Improvement, Proceedings of a workshop co-sponsored by the Institute of Genetics, Academia Sinica and the International Rice Research Institute, p. 437. Science Press, Beijing.

Chen, Z., C. Qian, M. Qin, C. Wang, C. Suo, F. Chen and Z. Deng, 1979. The induction of pollen plants of sugarcane. Annu. Rep. Inst. Genet., Academia Sinica 79: 91–93.

Chu, C.-C., C.-C. Wang, C.-S. Sun, C. Hsu, K.-C. Yin, C.-Y. Chu and F.-Y. Bi, 1975. Establishment of an efficient medium for anther culture of rice through comparative experiments on the nitrogen sources. Sci. Sin. 18: 659–668.

Daniels, J. and B.T. Roach, 1987. Taxonomy and evolution. In: D. Heinz (Ed.), Sugarcane Improvement through Breeding, pp. 7–84. Elsevier, Amsterdam.

Da Silva, J.A.G., M.E. Sorrells, W.L. Burnquist and S.D. Tanksley, 1993. RFLP linkage map and genome analysis of *Saccharum spontaneum*. Genome 36: 782–791.

FAO, 1992. Production 1992, FAO Yearbook, Volume 46. FAO Statistics Series #112, pp. 71–77, 168, 169. Food and Agriculture Organization of the United Nations, Rome.

Fitch, M., 1985. Meiosis in haploid and diploid plants derived from anther cultures of *S. spontaneum*. In: D. Heinz (Ed.), Annu. Rep. Hawaiian Sugar Plant. Assoc., pp. 8–9. Hawaiian Sugar Planters' Association, Aiea.

Fitch, M.M. and P.H. Moore, 1983a. Haploid production from anther culture of *Saccharum spontaneum* L. Z. Pflanzenphysiol. 109: 197–206.

Fitch, M.M. and P.H. Moore, 1983b. Haploid progress. In: D. Heinz (Ed.), Annu. Rep. Hawaiian Sugar Plant. Assoc., pp. 10–11. Hawaiian Sugar Planters' Association, Aiea.

Fitch, M.M. and P.H. Moore, 1984. Production of haploid *Saccharum spontaneum* L. – Comparison of media for cold incubation of panicle branches and for float culture of anthers. J. Plant Physiol. 117: 169–178.

Fitch, M.M.M., C. Nagai, L. Clemente, R. Ching and H. Ginoza, 1989. Anther culture for haploid plants. In: D. Heinz (Ed.), Annu. Rep. Hawaiian Sugar Plant. Assoc., pp. 6–7. Hawaiian Sugar Planters' Association, Aiea.

Fitch, M.M., C. Nagai and R.J. Schnell, 1987. Relationship of chromosome number to stomate guard cell length in *S. spontaneum* anther culture-derived plants. In: D. Heinz (Ed.), Annu. Rep. Hawaiian Sugar Plant. Assoc., p. 8. Hawaiian Sugar Planters' Association, Aiea.

Hawker, J.S., 1985. Sucrose. In: P.M. Dey and R.A. Dixon (Eds.), Biochemistry of Storage Carbohydrates in Green Plants, pp. 1–51. Academic Press, London.

Henry, H. and J. De Buyser, 1981. Float culture of wheat anthers. Theor. Appl. Genet. 60: 77–79.

Hinchee, M.A.W., A. Dela Cruz and A. Maretzki, 1984. Developmental and biochemical characteristics of cold-treated anthers of *Saccharum spontaneum*. J. Plant Physiol. 115: 271–284.

Hinchee, M.A.W. and M.M.M. Fitch, 1984. Culture of isolated microspores of *Saccharum spontaneum*. Z. Pflanzenphysiol. 113: 305–314.

Hu, H. and J.Z. Zeng, 1984. Development of new varieties via anther culture. In: P.V. Ammirato, D.A. Evans, W.R. Sharp and Y. Yamada (Eds.), Handbook of Plant Cell Culture, Vol. 3. Crop Species, pp. 65–91. Macmillan, New York.

Jalaja, N.C., 1978. Preliminary studies on tissue culture in sugarcane: (b) Anther culture. In: K. Mohan Naidu and S. Arulraj (Eds.), Annu. Rep. Sugarcane Breed. Inst. Coimbatore, p. 33. Sugarcane Breeding Institute, Coimbatore.

Jalaja, N.C. and T.V. Sreenivasan, 1988. Utilization of tissue culture techniques in sugarcane improvement: (b) Anther culture. In: K. Mohan Naidu and S. Arulraj (Eds.), Annu. Rep. Sugar. Breed. Inst. Coimbatore, p. 33. Sugarcane Breeding Institute, Coimbatore.

Jalaja, N.C. and T.V. Sreenivasan, 1991. Genetics and cytogenetics: Anther culture. In: K. Mohan Naidu and S. Arulraj (Eds.), Annu. Rep. Sugarcane Breed. Inst. Coimbatore, p. 29. Sugarcane Breeding Institute, Coimbatore.

Jalaja, N.C. and T.V. Sreenivasan, 1992. Genetics and cytogenetics: Anther culture. In: K. Mohan Naidu and S. Arulraj (Eds.), Annu. Rep. Sugarcane Breed. Inst. Coimbatore, pp. 34–35. Sugarcane Breeding Institute, Coimbatore.

Kasha, K.J., 1974. Haploids from somatic cells. In: K.J. Kasha (Ed.), Haploids in Higher

Plants, Advances and Potential, Proceedings of the First International Symposium, June 10–14, 1974, pp. 67–87. University of Guelph, Guelph.
Moore, P.H., C. Nagai and M.M.M. Fitch, 1989. Production and evaluation of sugarcane haploids. Proc. Int. Soc. Sugar Cane Technol. 20: 599–607.
Moore, P.H. and M.M.M. Fitch, 1990. Sugarcane (*Saccharum* spp.): Anther culture studies. In: Y.P.S. Bajaj (Ed.), Biotechnology in Agriculture and Forestry, Vol. 12, Haploids in Crop Improvement I, pp. 480–497. Springer-Verlag, Berlin.
Mordhorst, A.P. and H. Lörz, 1993. Embryogenesis and development of isolated barley (*Hordeum vulgare* L.) microspores are influenced by the amount and composition of nitrogen sources in culture media. J. Plant Physiol. 142: 485–492.
Murashige, T. and F. Skoog, 1962. A revised medium for rapid growth and bioassays with tobacco tissue cultures. Physiol Plant. 15: 473–497.
Nagai, C., 1990. Morphological evaluation of haploid/doubled haploid and selfed populations of *Saccharum spontaneum* clone SES208. In: D. Heinz (Ed.), Annu. Rep. Hawaiian Sugar Plant. Assoc., pp. 14–15. Hawaiian Sugar Planters' Association, Aiea.
Nagai, C. and R. Ching, 1990. Anther culture of *S. officinarum* and *S. spontaneum* clones. In: D. Heinz (Ed.), Annu. Rep. Hawaiian Sugar Plant. Assoc., pp. 12–13. Hawaiian Sugar Planters' Association, Aiea.
Nagai, C. and R. Ching, 1991. Continuing studies of haploids from two *Saccharum* species. In: D. Heinz (Ed.), Annu. Rep. Hawaiian Sugar Plant. Assoc., pp. 3–4. Hawaiian Sugar Planters' Association, Aiea.
Nagai, C., R. Ching and M.M.M. Fitch, 1989. Anther culture of various genotypes of sugarcane. In: D. Heinz (Ed.), Annu. Rep. Hawaiian Sugar Plant. Assoc., pp. 7–9. Hawaiian Sugar Planters' Association, Aiea.
Nagai, C. and R.J. Schnell, 1986a. Genetic study of *Saccharum* haploids. In: D. Heinz (Ed.), Annu. Rep. Hawaiian Sugar Plant. Assoc., pp. 3–4. Hawaiian Sugar Planters' Association, Aiea.
Nagai, C. and R.J. Schnell, 1986b. Isozyme analysis of anther-derived, selfed, and tissue culture-derived plants of *Saccharum spontaneum*. In: Proceedings, 78th Annu. Mtg. Agron. Soc. Am., New Orleans, LA, Nov. 30–Dec. 5, 1986. Agron. Abstr., pp. 74–75. Am. Agron. Soc., Madison.
Nagai, C. and R.J. Schnell, 1987. Flowering of haploid and doubled haploid plants. In: D. Heinz (Ed.), Annu. Rep. Hawaiian Sug. Plant. Assoc., p. 5. Hawaiian Sugar Planters' Association, Aiea.
Nitsch, J.P. and C. Nitsch, 1969. Haploid plants from pollen grains. Science 163: 85–87.
Orshinsky, B.R., L.J. McGregor, G.I.E. Johnson, P. Hucl and K.K. Kartha, 1990. Improved embryoid induction and green shoot regneration from wheat anthers cultured in medium with maltose. Plant Cell Rep. 9: 365–369.
Roach, B.T., 1966. Parthenogenesis and n+n inheritance in *Saccharum*. Sugarcane Breeders' Newslett. 17: 6–10.
Sorvari, S. and O. Schieder, 1987. Influence of sucrose and melibiose on barley anther cultures in starch media. Plant Breed. 99: 164–171.
Sreenivasan, T.V., B.S. Ahloowalia and D. Heinz, 1987. Cytogenetics. In: D. Heinz (Ed.), Sugarcane Improvement through Breeding, pp. 211–254. Elsevier, Amsterdam.
Sreenivasan, T.V. and N.C. Jalaja, 1979. Studies on tissue culture in sugarcane. In: K. Mohan Naidu and S. Arulraj (Eds.), Annu. Rep. Sugarcane Breed. Inst. Coimbatore, p. 19. Sugarcane Breeding Institute, Coimbatore.
Sreenivasan, T.V. and N.C. Jalaja, 1984. Utilization of tissue culture techniques in sugarcane improvement: (b) Anther culture. In: K. Mohan Naidu and S. Arulraj (Eds.), Annu. Rep. Sugarcane Breed. Inst. Coimbatore, p. 62. Sugarcane Breeding Institute, Coimbatore.
Sreenivasan, T.V. and N.C Jalaja, 1987. Utilization of tissue culture techniques in sugarcane improvement: (b) Anther culture. In: K. Mohan Naidu and S. Arulraj (Eds.), Annu. Rep. Sugarcane Breed. Inst. Coimbatore, p. 39. Sugarcane Breeding Institute, Coimbatore.
Sreenivasan, T.V. and N.C. Jalaja, 1991. Origin of sugarcane. In: K. Mohan Naidu and S.

Arulraj (Eds.), Annu. Rep. Sugarcane Breed. Inst. Coimbatore, pp. 25–26. Sugarcane Breeding Institute, Coimbatore.

Sunderland, N. and Z.H. Xu, 1982. Shed pollen culture in *Hordeum vulgare*. J. Exp. Bot. 33: 1086–1095.

Wu, K.K., W. Burnquist, M.E. Sorrells, T.L. Tew, P.H. Moore and S.D. Tanksley, 1992. The detection and estimation of linkage in polyploids using single-dose restriction fragments. Theor. Appl. Genet. 83: 294–300.

2. Haploidy in sugar beet (*Beta vulgaris* L.)

HANS C. PEDERSEN and BIRGIT KEIMER

Contents

1. Introduction

The cultivated beet is a biennial member of the Chenopodiaceae. It is bred for high yield of extractable sugar (sugar beet) or for high yield of roots with good feed value (fodderbeet). At the end of the first growing season, the thickened fleshy roots can be harvested and utilized. In the second year, after cold treatment (vernalization) in the field or in cold storage, the compressed stem will elongate into a 1–2 m tall inflorescense with numerous small petalless flowers in an open panicle.

While old beet cultivars were multigerm, having 3–5 seeds per cluster, modern cultivars are usually monogerm, since commercial seed is mainly harvested on genetically monogerm plants. Furthermore, these varieties are 100% hybrids due to the introduction of cytoplasmic male sterility (CMS). This makes sugar beet breeding rather complex. The breeder has to maintain two main programmes. Firstly, the monogerm programme, where monogerm, male-sterile, diploid CMS lines and their corresponding maintainer lines (OT) are inbred through 3–4 generations to create more or less homozygous lines with fixed traits. Due to inbreeding depression inbred CMS lines are crossed to unrelated OT lines to make single-cross lines (SX) with good seed-plant characteristics. Secondly, the selected single-crosses are then crossed to selected diploid or tetraploid, multigerm lines from the pollinator programmes. Diploid or triploid hybrids are harvested on the single-crosses as potential cultivars.

Because sugar beet is one of the major crops in the world, the effort

S.M. Jain, S.K. Sopory & R.E. Veilleux (eds.), In Vitro *Haploid Production in Higher Plants, Vol. 3,* 17–36.

on biotechnological developments in sugar beet has been rather intensive. Although it has been regarded as a relatively recalcitrant species in *in vitro* culture, regeneration of whole plants from callus, suspension culture and protoplasts has been reported, but the ability to regenerate seems to be genotype dependent. Genetic manipulation, using the well-established *Agrobacterium tumefaciens* technique for transfer of genes into sugar beet, is now routine in many laboratories (Fry *et al.*, 1990; Lindsay and Gallois, 1990; D'Halluin *et al.*, 1992). Thus, many *in vitro* techniques are presently in use successfully on sugar beet. However, one technique has failed: the haploidization of sugar beet via anther or microspore culture.

To obtain haploid ($2n = x = 9$) or doubled haploid sugar beets in quantities attractive for breeding purposes, it has been necessary to use the more labour intensive ovule culture techniques. The relatively high cost per doubled haploid (DH) sugar beet line produced by this technique is one of the main reasons for the rather low demand so far for doubled haploid techniques in breeding of sugar beet. But as the technique becomes more efficient and reliable it is foreseen that use of doubled haploid plants will be a natural choice in future breeding of sugar beet, either directly as 100% inbred components in the breeding programme or as an efficient method for fixing of desirable genes. Furthermore (doubled) haploids are valuable in RFLP mapping, as tools for diverse genetic studies, and in the study of reproductive biology and in biotechnological research.

Haploid sugar beets can be obtained by *in vivo* or *in vitro* techniques. Although *in vivo* induction of haploid sugar beets has been known for a long time (Levan, 1945; Kruse, 1963; Bosemark, 1971; Yüce, 1973; Seman, 1983), the relatively new haploid induction technique *in vitro* has become the method of choice for routine production. If chromosome doubling is needed, this can be done either by treatment of the haploid plantlets with a suitable chromosome doubling agent such as colchicine or alternatively by treatment in the early stage of *in vitro* culture (Hansen *et al.*, 1994).

In this paper, we intend to review the current status, problems, and prospects of utilizing haploids in improving sugar beet. We have paid most attention to the ovule culture technique for haploid production. Due to the inter-company competitive nature of the methods for (doubled) haploid production in sugar beet, results from the often intensive work done or sponsored by breeding companies has usually only been presented as posters or as oral presentations and the actual protocols used by the companies are not always available.

2. Haploid production by *in vivo* techniques

Haploids obtained spontaneously during sexual reproduction in sugar beet are rare (Levan, 1945; Kruse, 1963; Bosemark, 1971; Yüce, 1973; Seman, 1983). The first haploid sugar beet was found after colchicine treatment of seeds (Levan, 1945). Seman (1983) obtained 0.013% haploids after pollination of pollen-sterile sugar beets with red table beets, having the dominant red hypocotyl as a selectable marker. A maximum of 0.26% haploids was reported by Bosemark (1971) among seeds harvested on diploid multigerm pollen-sterile plants pollinated with pollen from tetraploid pollinators. In general, pollen-sterile plants seem to give higher number of haploid progenies than fertile plants. Yüce (1973) found that seeds harvested on pollen sterile plants resulted in more than 10 times as many haploids than seeds harvested on pollen fertile plants. However, the number of haploids was extremely low. Since the most efficient selection has been done on male-sterile plants, self-pollination was impossible, thus requiring vegetative propagation *in vitro* of the maternal parents of haploids using this method. The selection of spontaneous haploids has been superseded by the new efficient *in vitro* haploidization techniques.

Attempts to increase the number of haploids by different treatments such as use of gamma-irradiated pollen or pollination with pollen from distantly related species have been rather unsuccessful. After pollination of male sterile plants with 60 Kr or 100 Kr gamma irradiated pollen, Yüce (1973) obtained 11 seedlings of which 8 haploids survived. Use of pollen from the distantly related *Beta* species did not increase the average frequency of haploid induction: 8 haploids were obtained after pollination with *B. trigyna* (6X) pollen. Seman (1983) obtained a single haploid plant after pollination with low dose-irradiated pollen. Attempts to stimulate the stigma chemically were unsuccessful (Seman, 1983).

By combining pollen irradiation and embryo rescue techniques Buchter-Larsen (1986) obtained one haploid and possibly one doubled haploid plant after pollination with 50 Kr irradiated pollen. In this case a rather high number of seedlings survived, but almost all of them were heterozygous for one or more of the isozymes tested. Embryo rescue following pollination with pollen from other *Beta* species resulted in a relatively high number of hybrids without haploids (Buchter-Larsen, 1986).

While using irradiated pollen or pollen from other *Beta* species, it is possible that a fragment(s) from the male genome can stably integrate into the otherwise female haploid genome in the developing embryo or fragment(s) can exist as an additional – more or less unstable – extrachromosomal fragment(s). These special techniques have been used for the transfer of desired traits from other beets into sugar beet (Pedersen, unpublished results).

3. Haploid production by *in vitro* techniques

3.1. *Anther culture*

Although haploid plant production has proved to be successful from anther or isolated microspore cultures in a number of species, e.g., within Brassicaceae (Beversdorf and Kott, 1987; Hu and Huang, 1987; Siebel and Pauls, 1989; Huang, 1992), induction of haploid tissue leading to plant formation from isolated anthers or microspores of sugar beet has until now been rather unsuccessful. Numerous attempts to induce haploid plant formation from isolated anthers or microspores have only led to induction of proembryoid structures, occasionally ending in callus and/or root formation. Banba and Tanabe (1972) obtained one plantlet. Herrman and Lux (1988) reported regeneration of 17 plants from a total of 141 isolated proembryoid structures (average yield of proembryoid structures 0.12%), but none of the regenerated plants was haploid. Goska and Rogozinska (1981) obtained numerous plants from one anther-derived callus but in this case also, the regenerated plants seemed to be non-androgenic.

The embryo induction response seems to be highly genotype dependent (Van Geyt *et al.*, 1985; Atanassov and Butenko, 1980; Herrman and Lux, 1988), with up to 15% of genotypes producing embryo structures. Recent attempts to induce androgenesis from sugar beet and from 7 different wild *Beta* species confirmed that the ability to produce embryo structures is genotype dependent. Contrary to earlier attempts (Rogozinska and Goska, 1982) some of the wild relatives were more amenable to embryo induction than cultivated sugar beet. However, no plants were regenerated from the embryo structures (H. Krogaard and S.B. Andersen, pers. comm.).

In general, cytological analysis has demonstrated that multicellular structures can be induced up to about two weeks after anther culture. Heat shock treatment or cold pretreatment of buds has only weak, if any, effect and should not exceed 8 days of treatment (Herrman and Lux, 1988). The microspore state of development also seems to be critical. Only microspores in late meiosis to early mitosis are able to undergo division (Van Geyt *et al.*, 1985; Herrman and Lux, 1988).

The reason for the lack of success is unknown. In general plants having amyloplasts in the pollen grains are recalcitrant to *in vitro* androgenesis (Sangwan and Sangwan-Norreel, 1987). Sugar beet belongs to this group of plants having uninucleate pollen containing starch grains in their plastids.

3.2. *Ovule culture*

3.2.1. *Introduction*

The first sugar beet plants obtained from *in vitro* cultured flower buds and ovules were diploid (Goska, 1982). Hosemans and Bossoutrot (1983) were first to report on the induction of haploid sugar beet plants through *in vitro* culture of unpollinated ovules. Seventeen haploid plants were raised from a total of 7237 ovules isolated from male-sterile donor plants. Although the frequency was low (0.23%), it was higher than that previously reported by any of the classic techniques.

For breeding purposes only doubled haploid plants from self-fertile plants are of interest. Therefore, the two researchers attempted to extend the technique to male fertile plants (Bossoutrot and Hosemans, 1985; Hosemans and Bossoutrot, 1985). The two best basal media found in the first experiments were used and resulted in the development of only 3 haploid plants, which was about 5 times less than from ovules isolated from male sterile plants in the same experiment.

These first encouraging reports on successful *in vitro* gynogenesis in sugar beet were soon followed by several other reports (Goska, 1985; Bornman, 1985; Magnusson *et al.*, 1985; Keimer and D'Halluin, 1985; D'Halluin and Keimer, 1986; Barocka *et al.*, 1986; Van Geyt *et al.*, 1987; Doctrinal *et al.*, 1989, 1990; Lux *et al.*, 1990; Dubois *et al.*, 1990; Galatowitsch and Smith, 1990). Since the first report in 1983, the embryo induction frequency has increased dramatically. Methods have been improved enabling almost all donor plant genotypes to respond and in some laboratories, the doubling of chromosomes is now included as a part of the *in vitro* culture protocol. The actual percentage of doubled haploid plants that can be transferred to soil now exceeds 5% of cultured ovules. This frequency, although on paper seemingly rather low, is sufficient for routine production of doubled haploid sugar beets for different breeding, selection, and research purposes.

Some main problems yet to be solved for the improvement of the number of doubled haploid plants per ovule can be identified as:

1. the overall ovule response and variation in ovule response between donor plants (intra- as well as intervarietal);
2. reduction of the, often high, embryo mortality in the step between induced embryos and rooted plantlets;
3) reduction of number of haploid plants after chromosome doubling treatment.

In the following pages, a more detailed survey of the problems in haploidy formation is given.

3.2.2. *Donor plant growth conditions*

There is a clear correlation between origin of harvested ovules on donor plants and ovule response: comparison of embryo induction from ovules harvested at the stem apex and ovules harvested on lateral branches from different side stems revealed that lateral branches gave the highest response (80% higher embryo induction, Fig. 1). Likewise, embryo induction from different lateral branches showed a clear gradient of responding ovules (Table 1, Fig. 1). Ovules harvested from the first-formed lateral branches (at the base of the plant) gave the highest response, almost 100% higher than from the sixth-formed lateral branch (D'Halluin and Keimer, 1986). By comparing

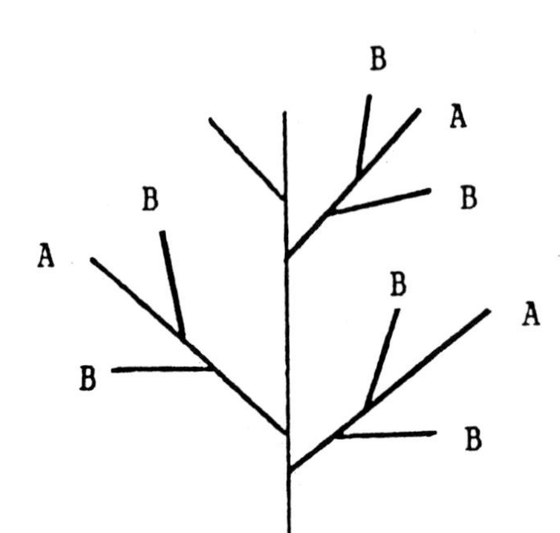

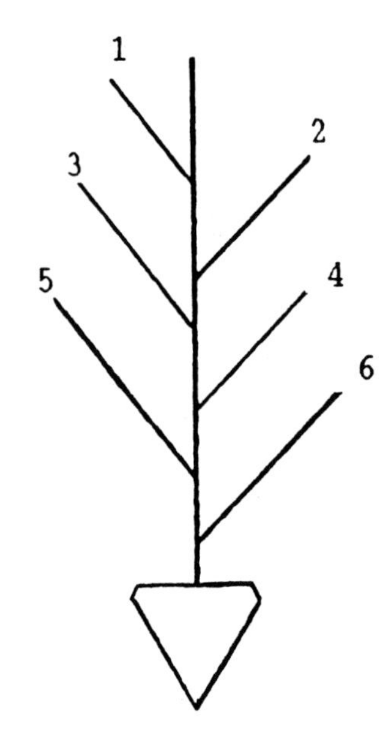

Figure 1. Ovule response in relation to position on plant. A – 3.6%; B – 6.6% (from D'Halluin and Keimer, 1986).

Table 1. Embryo induction frequency from ovules isolated from lateral branches (from D'Halluin and Keimer, 1986)

Lateral branches (stem no.)	% embryo induction (avg. of 4 genotypes)
1 top	5.2
2	7.2
3	7.4
4	9.1
5	9.3
6 bottom	10.8

these data with a sugar beet seed germination cartogram (Alcaraz, 1986) there seems to be a reverse correlation between ability of seed germination and gynogenesis.

The ovule response seems to be influenced by the light regime during plant development. The maximal response in embryo induction has been found in donor plants grown under natural daylight and the normal life cycle of sugar beet (Doctrinal *et al.*, 1989, 1990; Lux *et al.*, 1990).

Embryo induction frequency from donor plants grown under artificial light and out of season has been investigated by Keimer and Pedersen (1988): to fit a sugar beet breeding program it may be necessary to start the ovule

Table 2. Embryo induction from ovules harvested from greenhouse and growth chamber grown donor plants. (*Cursive*): Low-high embryo induction response from harvested ovules (%) from different donor plants

Donorplants grown in:	Greenhouse		Growth Chamber 1*		Growth Chamber 2*	
Year	Embryos/ovules	%	Embryos/ovules	%	Embryos/ovules	%
1985	165/575	28.7 *(4–68)*	67/375	17.9 *(0–35)*		
1986	983/3844	25.6 *(7–33)*	888/3070	28.9 *(8–49)*	941/3271	28.8 *(6–43)*
1986/87 Small stecklings	91/310	29.4 *(21–39)*	24/210	11.4 *(4–28)*	91/200	45.5 *(4–67)*
1986/87 Big stecklings	580/3060	19.0 *(12–29)*	453/2780	16.3 *(8–23)*	538/2770	19.4 *(5–39)*
Average	1819/7789	23.4	1432/6435	22.3	1570/6241	25.2

* Greenhouse and growth chambers conditions: *Greenhouse*: Shaded and cooled by ventilation. 160 m^2. Additional light: 48 mercury halogen lamps, type Osram Power Star HQI-T 400W, 10 h. Photoperiod min. 16 h. *Growth Chamber* 1: 12 m^2. Height from floor to lamps = 2.4 m. Light source: 10 Osram SON/T 400W + 4 Osram HPI/T 400W + 6 Philips Argenta 150W E27. Photoperiod 16 h/8 h. Temp. 20 °C/16 °C. *Growth Chamber* 2: 12 m^2. Height from floor to lamps = 2.4 m. Light source: 12 Osram Power Star HQI-/ 250W/D + 12 Osram Vialox NAV-T 250W + 7 Argenta 150W E27. Photoperiod and temp. as Growth Chamber 1.

culture during the winter. Donor plants which flowered during this period were raised from normal field-grown plants, which after vernalization were cultivated in a greenhouse or in growth chambers under artificial light. The conclusion from these experiments was that there is no significant difference in ovule response whether the donor plants were grown in the greenhouse or in growth chambers, provided the growth chamber light source was supplemented with incandescent light. The average embryo induction over three years' experiments is listed in Table 2.

3.2.3. *Harvest of ovules*

The embryo sac of sugar beet is of the *Polygonum* type (Maheswari, 1950). The structure of the mature embryo sac was described by Artschwager (1927) using light microscopy and, recently, in more detail with the aid of electron microscope by Bruun (1987). Artschwager and Starrett (1933) reported that the embryo sacs in sugar beet are mature at the time of anthesis. Moreover, the sugar beet pollen is viable and the stigma receptive before anthers dehisce. This fact might cause some problems in ovule culture of sugar beet. To avoid culture of a high proportion of fertilized ovules care has to be taken during excision of the ovules from pollen-producing plants. Excision of ovules at an early developmental stage where no self-pollination can occur or use of isolated highly self-incompatible donor plants helps to minimize this problem.

Bossoutrot and Hosemans (1985) emasculated flower buds from the male fertile donor plants. Most self-pollination can be avoided by excision of ovules by cutting through the bud slightly above the junction of the receptacle and the ovary and thus excising the ovule at the distal end of the bud (Galatowitsch and Smith, 1990). With this method more than 90% of 536 plants regenerated from ovules derived from self-fertile plants (OT and diploid pollinator lines) were scored as haploids in an experiment where the ovules were excised from flower buds (bud size 0.3–0.6 mm) before or even shortly after anthesis (Keimer and D'Halluin, 1985; D'Halluin and Keimer, 1986). Most of the remaining plants were diploid and appeared to be spontaneously doubled haploids as determined by isozyme analysis at six loci. Less than 1% were tetraploids.

3.2.4. *Embryo sac stage*

There is no clear correlation between size of flower buds and ovule response. Neither could a correlation between ovule size and ovule response be found (D'Halluin and Keimer, 1986). In contrast to *in vitro* androgenesis, it is possible to induce a gynogenic response from ovules over a broad range of developmental stages. By cytological analysis, Dubois *et al.* (1990) confirmed that complete maturation of the ovule is not necessary for induction of gynogenesis. Culture of pistils having 0.3 mm long ovules, also resulted in embryo induction (D'Halluin and Keimer, 1986).

3.2.5. *Haploid embryo development*

Histological sectioning of ovules cultured on Maribo's B11-medium showed that shortly after initiation of *in vitro* culture, the ovules responded by initiating haploid embryo formation in most of the ovules (L. Bruun, pers. comm.). This high primary haploid embryo induction frequency facilitated the attempts to trace the exact cellular origin of the haploid embryos (Olesen *et al.*, 1988). Following our B11 protocol, all of the haploid embryos seem to induce from the unfertilized egg cells. This is consistent with the results obtained by Goska *et al.* (1990), who also found that in most cases the embryo developed from the egg cells and rarely from the synergids. This was later on confirmed by Ferrant and Bouharmont (1994), who excluded regeneration from the synergids because of their early degeneration in sugar beet. Sporadic regeneration of diploid adventitious embryos from the nucellar tissue does occur (Goska and Jassem, 1988).

The primary stages in haploid embryo development are morphologically similar to the comparable stages of zygotic embryogenesis (Olesen *et al.*, 1988; Goska *et al.*, 1990). The induced haploid embryos develop smaller cotyledons than zygotic embryos, but apart from this, they look like zygotic embryos and develop through precocious germination without a dormancy stage into complete embryos and plantlets. At the cellular level, *in vitro* cultured embryos have a higher degree of vacuolization than the corresponding embryos developed *in situ* (Bruun, 1991).

3.2.6. *Culture conditions*
Standard basal media solidified with agarose are usually used. Sucrose concentrations have been between 3% and 10%. Inclusion of the amino acid glutamine increases the frequency of haploid embryo induction (Lux *et al.*, 1991). Barocka *et al.* (1986) added complex vitamin solutions to the medium.

The carbohydrate source seems to play an important role in embryo induction frequency and embryo quality. A high quality of the induced embryos results in a better establishment of plants. Maltose enhances the average embryo induction compared with sucrose, probably because no free fructose is generated (Lux *et al.*, 1991).

Media which induce high ovule response from most of the donor plants seem to result in poorer quality of the induced embryos and vice versa. This is the case for the two media N61 and N62 (Doctrinal *et al.*, 1989) – both containing 6% sucrose – the former eliciting response from most of the donor plants used whereas the latter results in a high quality of the induced embryos (Ferrant and Bouharmont, 1994).

The breeding company Maribo has reported the embryo induction frequency using their B11 protocol (Keimer and Pedersen, 1988). This medium resulted in more than 15% embryo induction and a response from almost all donor plants used, but the quality of the induced embryos was not always good. The B11 medium is based on the PGoB-basal medium (De Greef and Jacobs, 1979) and contains 3% sucrose and 0.5% activated charcoal. Successful embryo induction on media containing activated charcoal indicates that addition of low levels of exogenous growth regulators that can be inactivated by charcoal only plays a minor role in the induction and culture of haploid embryos. In general, increased cytokinin concentrations in the induction medium result in an increase in embryo yield, whereas auxins seem to have the reverse effect or no effect at all (Lux *et al.*, 1990).

In our laboratory, culturing ovules without light for the first 3–6 weeks until embryos have developed gives the best quality of embryos, but good results have also been obtained by illumination with cool white fluorescent light (about 25 $\mu Em^{-2}s^{-1}$) 16 h/day (Doctrinal *et al.*, 1989; Lux *et al.*, 1990). Ovules are often cultured at approximately 25 °C, but at 27 °C embryo induction is increased and embryos develop more rapidly compared with cultures at 24 °C (Doctrinal *et al.*, 1989).

Pretreatment of flower buds at 4 °C has resulted in a slight increase in yield of embryos, with a maximum after a 5 day treatment (Lux *et al.*, 1990). These findings allow more flexible planning in the laboratory since harvested floral stems can be stored at 4 °C for up to a week.

In our experience, culture of the ovules with the funiculus facing downwards on the solid medium has resulted in a better response than random orientation of the ovules. This is probably due to the need for an alternative nutritional pathway after ovules have been transferred to culture *in vitro*. A detailed discussion on this aspect and a histological description of the cellular responses within sugar beet ovules as a consequence of *in vitro* culture have been given by Bruun (1991).

Normally a two step protocol has been used. After embryos have appeared on the induction medium, they have been transferred to developmental medium, usually with lower osmotic strength and a low exogenous level of growth regulators. In many cases, the embryo developed directly into a normal plantlet but, depending on the quality of the induced embryos, additional subcultures may be needed for normal shoot formation. Multiplication and rooting of shoots have been done on standard multiplication and rooting media, respectively.

3.2.7. *Ovule response*

Direct haploid embryogenesis from the cultured ovules is the most common in sugar beet. Haploid embryo formation through a primary haploid callus induction from isolated ovules has been reported by Barocka *et al.* (1986) and Galatowitsch and Smith (1990), using a complex vitamin solution and 2iP in the ovule culture medium. In addition, both embryogenic and organogenic calli were produced.

As mentioned earlier, the B11 protocol resulted in more than 15% haploid embryo induction. The variation in embryo induction from ovules isolated from different donor plants was between 1% and 68% and almost all of the donor plants responded. On this medium, callus formation accounts for less than 5% of the structures formed. By careful histological analysis of the cultured ovules on this medium, Olesen *et al.* (1988) and Bruun (1991, pers. comm.) observed that most callus formation within the embryo sac derives from the suspensor region in the developing embryo. Callus derived from cells within the embryo sac itself has rarely been observed. Thus, it can be expected that a haploid embryogenic callus induction pathway proceeds through a callusing phase in the suspensor region of the primary embryo originated from the egg-cell.

Callus from nucellus tissues accounts for most of the callus obtained, unless activated charcoal has been added to the embryo induction medium. Inclusion of activated charcoal in the medium prevents callus formation from the mother tissue, which hampers the development of haploid plantlets (Speckmann *et al.*, 1986; Van Geyt *et al.*, 1987).

The occurence of polyembryony and secondary embryogenesis is a common phenomenon in sugar beet ovule culture (Keimer and D'Halluin, 1985; D'Halluin and Keimer, 1986; Goska and Jassem, 1988). In a typical experiment we have found that 105 of 3628 responding ovules showed polyembryony, producing an additional 146 embryos (Table 3). Most of the polyembryogenic response resulted in twins, but up to 9 embryos have been isolated from a single ovule (Fig. 2).

Table 3. Polyembryony. Experiment including 6 cultivars. A total of 17714 ovules (2.682–3.481 ovules/cultivar) was cultured and average ovule response was 19.7% (cultivar range 5.7–27.6%). The average survival of induced embryos to plants: 61.3%

Cultivar	Responding ovules	Number of induced embryos	Number of twins and secondary embryos	Number of ovules with polyembryony	Average number of embryos if polyembryony
205	588	610	22	19	2.2
206	773	801	28	22	2.3
228	662	730	68	38	2.8
229	159	164	5	4	2.3
230	752	768	16	15	2.1
231	548	555	7	7	2.0

3.2.8. *Plant regeneration*

A relatively large proportion of induced embryos from unfertilized ovules does not regenerate. Hosemans and Bossoutrot (1983) observed that only 11% of the induced structures (calli and embryos) regenerated into plantlets. Lux *et al.* (1990) reported that an average of 25% of the induced embryos produced plants. In this experiment, the embryo induction medium contained 10% sucrose.

In our own experiments, we found that both the induction frequency and the survival rate of the induced embryos were donor plant dependent and it was not unusual to lose more than 50% of the induced embryos before they were ready for transplanting into soil. Some of the embryos died at an early stage of embryogenesis, whereas others were lost during subculture of the harvested embryos. Besides the effect of inadequate culture conditions, expression of lethal genes or failure in precocious germination may have affected embryo survival.

Lethal genes result in a skewed segregation and can act at both the gametophytic and zygotic levels. To rule out the influence of lethal genes or other genetic factors on the success rate of ovule culture, we included culture of ovules from a doubled haploid plant in our routine production with our standard protocol. This doubled haploid was produced by ovule culture two years earlier. Since the egg cell survived and no recessive lethal genes can be masked allelically in a haploid genome, we expected this doubled haploid donor plant to be devoid of lethal genes acting on gametophytic and zygotic levels.

Table 4 shows the results of this experiment. Although the ovule response from the doubled haploid donor plant was higher than in the experiment in which the donor plant itself was produced, it was not higher than average for ovules isolated from heterozygous donor plants in this experiment. However, the percentage of induced embryos that died on the induction medium was significantly lower for haploid embryos isolated from the "pre-rinsed"

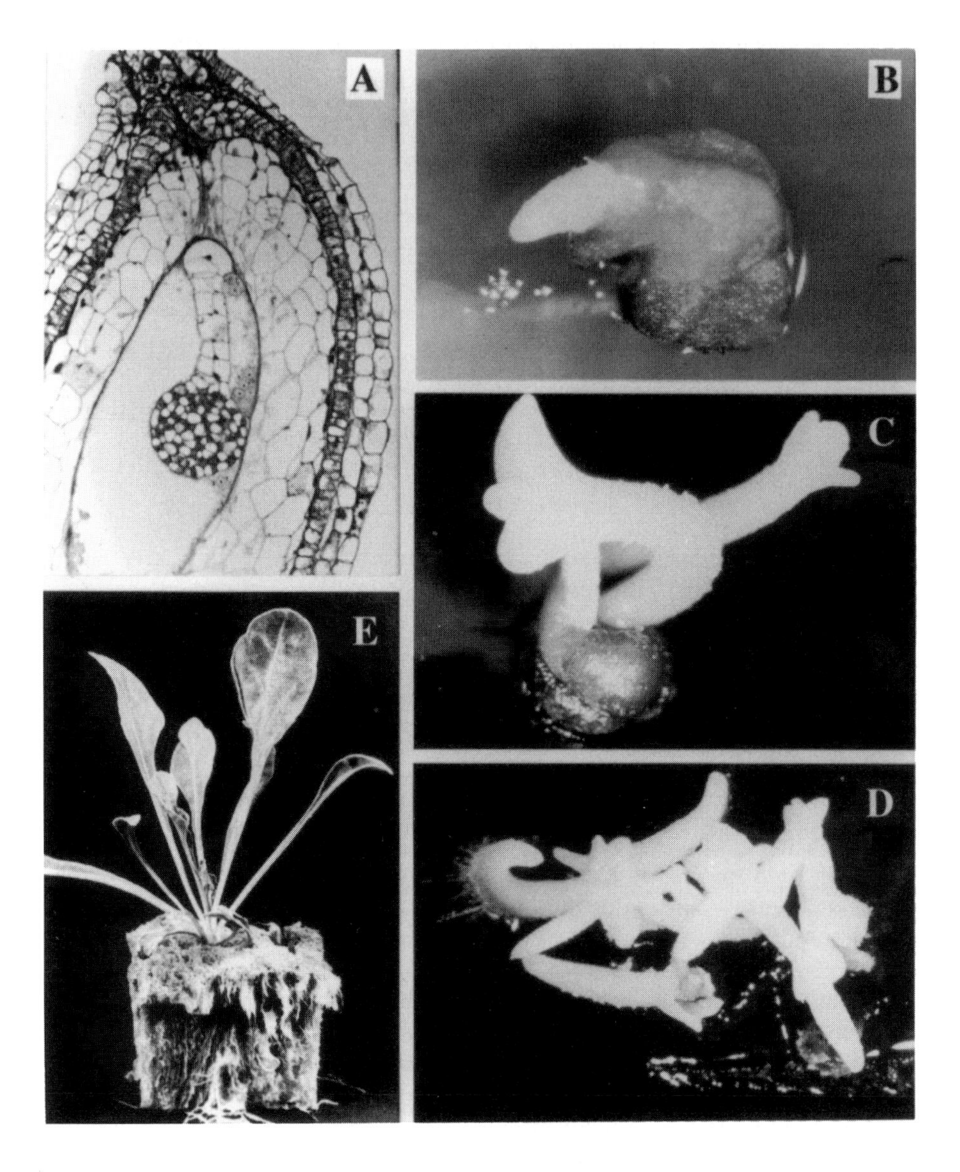

Figure 2. Ovule culture in sugar beet. (A) Globular haploid embryo with long suspensor, ×220; (B) Induction of haploid embryo from ovule; (C) Induction of twin embryos; (D) Polyembryony and secondary embryogenesis; (E) Haploid plant established in Grodan.

doubled haploid donor plant than for those isolated from heterozygous donor plants. After subculture of the embryos, the mortality rate was equal. Table 4 shows that, from embryos to surviving haploid plants, the success rate was a little higher for ovules excised from doubled haploid donor plants than for those taken from heterozygous donor plants.

Table 4. Results from culture *in vitro* of unfertilized sugar beet ovules, summer, 1987. The cultivar 251DH is a doubled haploid donor plant, produced by previous ovule culture. All other donor plants are heterozygous OT plants or 2n pollinators. Figures are corrected for polyembryony (2.3% on average). Ovule culture was done on a medium containing 3% sucrose and 0.8% agarose

Cultivar	Ovules cultured	Ovules responding[1]	Embryos lost during induction[2]	Embryos lost during subculture[3]	Embryos surviving to plantlets
101	2061	428 (20.8%)	54 (12.6%)	141 (37.5%)	233 (54.2%)
102	1968	467 (23.7%)	57 (12.1%)	154 (37.2%)	257 (54.6%)
103	2048	547 (26.7%)	101 (17.3%)	213 (44.0%)	243 (41.5%)
104	2059	233 (11.3%)	45 (18.9%)	75 (38.9%)	116 (48.7%)
106	494	62 (12.6%)	13 (21.0%)	23 (46.9%)	26 (41.9%)
107	485	11 (2.3%)	4 (36.4%)	5 (71.4%)	2 (18.2%)
108	472	181 (38.3%)	51 (27.7%)	44 (33.1%)	86 (46.7%)
109	468	54 (11.5%)	23 (41.1%)	10 (30.3%)	21 (37.5%)
110	477	102 (21.4%)	25 (23.8%)	29 (36.3%)	49 (46.7%)
111	1922	227 (11.8%)	53 (22.2%)	88 (47.3%)	89 (37.2%)
112	1791	177 (9.9%)	57 (31.7%)	68 (55.3%)	52 (28.9%)
114	1785	258 (14.5%)	72 (27.8%)	102 (54.5%)	83 (32.0%)
115	1715	211 (12.3%)	32 (15.1%)	85 (47.2%)	95 (44.8%)
259	120	28 (23.3%)	10 (35.7%)	15 (83.3%)	3 (10.7%)
261	491	55 (11.2%)	22 (38.6%)	23 (65.7%)	11 (19.3%)
Total/average	18356	3041 (16.6%)	619 (19.9%)	1075 (43.0%)	1366 (43.8%)
251DH	990	143 (14.4%)	5 (3.5%)	59 (42.8%)	79 (55.2%)

[1] Figures represent ovules with embryo induction. The average callus induction frequency from ovules was 1.8%.
[2] Induced embryos lost before transfer to first subculture medium (expressed as percentage of induced embryos).
[3] Embryos lost after the transfer from induction medium to regeneration medium (as percentage of alive embryos transferred).

The improvement can be interpreted as an example of genetic adaptation to tissue culture conditions or a result of the elimination of lethal factors, such as expression of lethal genes during embryo development. It could also be a combination of a number of factors. A selective advantage for embryo survival on the artificial ovule culture medium cannot be excluded from this experiment, but the mortality rate after the harvest of embryos indicates that, if so, it is not a general selection for vigour in tissue culture. In fact, the mortality rates in the following subcultures was likely due to the effect of inadequate chemical or physical growth conditions of the weak haploid embryos and shoots, rather than caused by the haploid genotype.

By cytological analysis of chimeric embryos (mixoploids) obtained after colchicine treatment of ovules at an early stage of embryo induction, we observed that haploid tissue is more sensitive to culture conditions than diploid tissue, and that prolonged culture of chimeric tissues or embryos results in the selection of the more vigorous doubled haploid plantlets.

4. Chromosome doubling

Haploid plants are valuable for RFLP mapping, mutation research, genetic engineering, as a protoplast source for somatic hybridization and for *in vitro* selection. But for direct use in plant breeding, haploid plants have no value, unless their chromosomes can be doubled.

Spontanoeus chromosome doubling in sugar beet occurs at a rather low frequency. Only between 2 and 10% of the established plants have spontaneously doubled (Bossoutrot and Hosemans, 1985; D'Halluin and Keimer, 1986; Lux *et al.*, 1990). Thus an efficient chromosome doubling technique is needed in sugar beet doubled haploid production.

Treatment of sugar beet plantlets with colchicine has been effective for chromosome doubling (Jensen, 1974; Speckmann, 1975). Either the whole plantlet has been dipped in a colchicine solution for several hours or the basal meristem has been treated with colchicine using a cotton plug technique (Ragot and Steen, 1992). To obtain a high proportion of doubled plantlets, it is necessary to ensure that the treatment for efficient chromosome doubling has affected the part of the basal meristem that will develop the inflorescence. Insufficient colchicine treatment results in doubling of only the external cells in the basal meristem.

Having an effective routine production of haploid plantlets, hundreds or thousands of haploid plantlets may require chromosome doubling within a short period of time in order to be used in a breeding programme. This makes it almost impossible to nurse carefully individual colchicine treated plantlets. As a consequence many of the haploid plantlets will have been insufficiently treated and may revert to the haploid level. To avoid loss of this valuable material in the final stage, the development of an effective large scale chromosome doubling technique for sugar beets has been a top priority.

Plever and later Hansen (Hansen *et al.*, 1994) have shown that it is possible to double chromosomes at an early stage of ovule culture. The treatment of ovules with colchicine 4–7 days after the initiation of ovule culture resulted in a high frequency of chromosome doubling. Moreover, this doubling seemed to be stable. Treatment with chromosome doubling agents at this early stage of embryo development resulted in doubled haploid or chimeric embryos. Preliminary results from our laboratory indicate that doubled areas may be positively selected during culture *in vitro* due to greater vigour of doubled cells. The toxic effect of colchicine treatment has resulted in a minor loss of developing embryos, but this decrease in efficiency of embryo formation was more than compensated by the efficiency of chromosome doubling.

5. The use of doubled haploids in breeding

The importance of using doubled haploid plants in sugar beet breeding can be reduced to two main fields:

Table 5. Percent yield relative to control of pure homozygous OT lines derived from haploid plants from family 483 (from Steen *et al.*, 1988)

Family/line	Root yield	Sugar yield	POL*	IV**
Selfed progeny of family 483	102	99	97	112
Homozygous line, no. 1	47	47	100	100
no. 2	35	36	102	101
no. 3	85	86	100	101
no. 4	97	100	104	94
no. 5	6	–	–	–
no. 6	40	38	96	123
no. 7	34	35	102	108
no. 8	33	33	99	108

* POL: Polarization value; **IV: Impurity value.

1. The creation of homozygous monogerm lines.
2. The fixing (stabilization) of specific traits.

In spite of the obvious advantage of having a 100% homozygous monogerm OT line in a single step, instead of using 3–4 generations in developing sufficiently inbred lines, the use of doubled haploid plants so far has been limited for sugar beet breeding purposes. Reasons for this may be: 1. Production of homozygous seed is restricted to OTs, whereas doubled haploid CMS plants must be propagated vegetatively; 2. The doubled haploid lines need to be evaluated whereas in traditional breeding programmes, evaluation and selection can partly take place parallel to the inbreeding process; 3. The generally low number of doubled haploid lines produced within a programme may reflect only a fraction of the potential variation within the germplasm; to be competitive with traditional breeding techniques the efficiency of the doubled haploid technique must be improved; and 4. Inadvertent selection in the process of their production can hamper the use of double haploids.

Due to the inbred nature of DH lines, the evaluation of pure lines – suffering more or less from inbreeding depression – can be difficult. Some results, however, have been reported on the performance of doubled haploid OT lines (Steen *et al.*, 1988). Table 5 shows an example of the variation found among 8 DH lines generated from the same moderately inbred OT line. The highest yielding progeny after self-pollination within the same OT line are shown for comparison. All entries have been compared to a standard set at 100% for all characters. The pronounced variation in root yield was not caused by seed quality, but can be explained by the variation in plant vigour and morphology. Line no. 4 equalled the standard in sugar yield and had better results in the quality parameters polarization (pol) and juice impurity value (IV). Also as a seed plant, line no. 4 was better than the average. These results do not predict combining ability, but they demonstrate the potential of DHs evaluated on the same terms as normal inbreds.

By pollinating CMS-populations with pollen from doubled haploid sugar beet lines, Zahariev *et al.* (1992) found that some of the doubled haploid lines had better combining ability than the initial parental population and gave rise to hybrid progeny with high yield and good quality.

In sugar beet haploid techniques can be used for the production of homozygous maintainer or 2n-pollinator lines with useful resistance genes.

At the former IfR, Dr. Lux and co-workers isolated a haploid line that was resistant to three different viruses. In field tests, this line resisted attacks of virus yellows, whereas the surrounding controls were severely attacked. Unfortunately, this line could only be maintained by vegetative propagation due to self-incompatiblility (H. Lux, pers. comm.).

We have selected double haploids tolerant to rhizomania. In field tests conducted in soil infested with rhizomania, populations of selfed doubled haploid lines segregated clearly into suceptible and tolerant lines. Selected DH lines are now used in breeding programmes for the introduction of rhizomania tolerance.

The doubled haploid technique can also be used as a rapid method for fixing transgenes. Thus, doubled haploid transgenic OT lines have been produced from elite breeding material. If, however, scorable markers for homozygosity are present among the transgenes that have been transformed into self-fertile material, the benefit of using the DH technique is questionable. In sugar beet, the widely used β-glucuronidase marker gene – GUS (Jefferson *et al.*, 1987) – is usually expressed in pollen (Steen and Pedersen, 1994). Since transformants regarding the introduced transgene(s) behave as heterozygotes, GUS analysis of pollen from S1 plants in late bud stage can be used for the selection of plants having marker-linked transgenes as homozygous traits (close to 100% of pollen expressing the GUS-gene). Thus these S_1 plants, homozygous for the specific transgene, can be utilized in the breeding programme in the same season that DH lines would be generated from the same transformant.

To fix a specific trait (disease resistance, transgene) as homozygous, but at the same time to retain most of the variation harboured within the genotype or small population of plants used for ovule culture, it is necessary to extract many haploids from the specific genotype or selected population. Our standard has been to produce at least 100 doubled haploid genotypes, select those expressing the desired trait (resistance, transgenes) and intercross these selections directly or after one self-pollination. In theory, starting with a diploid heterozygous population with the desired trait in a single allele, 50% of the doubled haploids should be found usable for restoration of the genotype or population. Of course, if the desired traits are due to more than one allele, a much smaller percentage of the doubled haploids can be used in the restoration. So far, we have not found any unexpected allelic segregation or for transgenes that could have developed as a result of ovule culture.

6. Conclusion

At present culture of unfertilized ovules is the method of choice for obtaining homozygous sugar beet plants. When compared with anther or free microspore culture, ovule culture is far more labour intensive, but as long as the androgenic response fails in beets, the gynogenic pathway is the only realistic alternative to traditional inbreeding.

Ovule culture is now used routinely by most sugar beet breeding companies for obtaining doubled haploid plants. The most straightforward use of doubled haploid lines is for fixing qualitative traits such as disease resistance or transgenes. If the production of doubled haploid sugar beets can be elicited by a high ovule response from all donor plants, it can be integrated into most breeding programmes.

Haploids and doubled haploids are excellent material for unmasking the undesired recessive traits in beet populations. Within one generation, completely homozygous lines can be established as opposed to 3–4 generations (of two years) in traditional breeding for sufficiently inbred material to utilize hybrid vigor. Breeders can select desirable traits and discard undesirable material during this inbreeding process, leaving only a small proportion of the original material. Provided a doubled haploid is male fertile and self-compatible, the breeder can retain it as a non-segregating line indefinitely and in this way store well-characterized germplasm in a gene bank for future use.

7. Acknowledgements

The authors would like to thank Drs. Per Steen and Trevor Morris for their enthusiastic engagement in tissue culture work in relation to breeding and for their critical reading of the manuscript. We also gratefully acknowledge the work of our co-workers during the development of our haploid protocol. In this connection we give special thanks to Drs. Kathleen D'Halluin and A. Lisbeth Hansen. Finally, our warmest thanks to Jan E. Andersen and his team in the greenhouses for taking care of motherplants as well as all the plantlets resulting from this work.

8. References

Alcaraz, G., 1986. Sémences de betterave à sucre: cartographie du porte-graine. Bulletin fnams sémences 97: 47–49.

Artschwager, E. and R.C. Starrett, 1933. The time factor in fertilization and embryo development in the sugar beet. J. Agr. Res. 47: 823–843.

Artschwager, E., 1927. Development of flowers and seed in the sugar beet. J. Agr. Res. 34: 1–25.

Atanassov, A.I. and R.G. Butenko, 1980. Culture of sugar beet isolated anthers. Fiziol. Biokhim. Kult. Rast. 12(1): 49–56.

Banba, H. and H. Tanabe, 1972. A study of anther culture in sugar beets. Bull. Sugar Beet Res. (Tensai Kenkyu Hokaku) Suppl. 14: 9–16.

Barocka, K.H., F. Sievert and R. Nehls, 1986. Morphogenesis in ovule-derived sugar beet tissues. In: D.A. Somers, B.G. Gengenbach, D.D. Biesboer, W.P. Hackett and C.E. Green (Eds.), VIth Int. Congress Plant Tissue and Cell Culture, Univ. Minnesota, p. 403.

Beversdorf, W.D. and L.S. Kott, 1987. An *in vitro* mutagenesis/selection system for *Brassica napus*. Iowa State J. Res. 61: 435–443.

Bornman, C.H., 1985. Haploidization of sugar beet (*Beta vulgaris*) via gynogenesis. In vitro 21(3,II): 36A (Abstr.).

Bosemark, N.O., 1971. Haploids and homozygous diploids, triploids and tetraploids in sugar beet. Hereditas 69: 193–204.

Bossoutrot, D. and D. Hosemans, 1985. Gynogenesis in *Beta vulgaris* L.: From *in vitro* culture of unpollinated ovules to the production of doubled haploid plants in soil. Plant Cell Rep. 4: 300–303.

Bruun, L., 1987. The mature embryo sac of the sugar beet, *Beta vulgaris*: A structural investigation. Nord. J. Bot. 7(5): 543–551.

Bruun, L., 1991. Histological and semiquantitative approaches to *in vitro* cellular responses of ovules, embryo and endosperm in sugar beet, *Beta vulgaris* L. Sex. Plant Reprod. 4: 64–72.

Buchter-Larsen, A., 1986. *In situ* Induktion af Haploide Embryoer på Sukkerroer efterfulgt af *In vitro* Embryokultur. Thesis, Acad. Technical Sciences, Copenhagen.

De Greef, W. and M. Jacobs, 1979. *In vitro* culture of the sugar beet: Description of a cell line with high regeneration capacity. Plant Sci. Lett. 17: 55–61.

D'Halluin, K. and B. Keimer, 1986. Production of haploid sugar beets (*Beta vulgaris* L.) by ovule culture. In: W. Horn, C.J. Jensen, W. Odenbach and O. Schieder (Eds.), Genetic Manipulation in Plant Breeding, pp. 307–309. De Gruyter, Berlin.

D'Halluin, K., M. Bossut, E. Bonne, B. Mazur, J. Leemans and J. Botterman, 1992. Transformation of sugarbeet (*Beta vulgaris* L.) and evaluation of herbicide resistance in transgenic plants. Bio/Technology 10: 309–314.

Doctrinal, M., R.S. Sangwan and B.S. Sangwan-Norreel, 1989. *In vitro* gynogenesis in *Beta vulgaris* L.: Effect of plant growth regulators, temperature, genotypes and season. Plant Cell Tiss. Org. Cult. 17: 1–12.

Doctrinal, M., R.S. Sangwan and B.S. Sangwan-Norreel, 1990. Sugar beet (*Beta vulgaris* L.): *In Vitro* Induction of Haploids. In: Y.P.S. Bajaj (Ed.), Biotechnology in Agriculture and Forestry, Vol. 12. Haploids in Crop Improvement, pp. 346–357. Springer Verlag, Berlin.

Dubois, F., P. Lenee, R.S. Sangwan and B.S. Sangwan-Norreel, 1990. Do developmental stages of ovules influence *in vitro* induction of gynogenetic embryos in sugar beet? In: Seeds: Genesis of natural and artificial forms, p. 249. Le Biopôle Végétal, Amiens (Abstr.).

Ferrant, V. and J. Bouharmont, 1994. Origin of gynogenetic embryos of *Beta vulgaris* L. Sex. Plant Reprod. 7: 12–16.

Fry, J.E., A.R. Barnason and M.A. Hinchee, 1990. Genotype-independent transformation of sugarbeet using *Agrobacterium tumefaciens*. In: Abstracts VIIth Int. Congress on Plant Tissue and Cell Culture, Amsterdam, June 24–29, 1990, p. 57.

Galatowitsch, M.W. and G.A. Smith, 1990. Regeneration from unfertilized ovule callus of sugar beet (*Beta vulgaris* L.). Can. J. Plant Sci. 70: 83–89.

Goska, M., 1982. Tissue Culture in Sugar Beet Breeding. IIRB Breeding and Genetics Study Group, Zaragossa.

Goska, M., 1985. Sugar beet haploids obtained in the *in vitro* culture. Bull. Polish Acad. Sci. Biol. Sci. 33: 31–33.

Goska, M. and B. Jassem, 1988. Histological observations of sugar beet ovules in *in vitro* cultures. Bull. Polish Acad. Sci. Biol. Sci. 36: 171–175.

Goska, M., B. Jassem and E. Jazdzewska, 1990. Embryo development from sugar beet ovules

in *in vitro* culture. In: Int. Inst. Sugar Beet Research, 53th Winter Congress, Brussels, Feb. 1990, pp. 145–154.
Goska, M. and J.H. Rogozinska, 1981. Recent results on obtaining beet haploids through *in vitro* culture of anthers. Biuletyn Inst. Hodowli I Aklimatyzacji Roslin 145: 141–143.
Hansen, A.L., C. Plever, H.C. Pedersen, B. Keimer and S.B. Andersen, 1994. Efficient chromosome doubling in *Beta vulgaris* L. ovule culture. Plant Breed. 112: 89–95.
Herrman, L. and H. Lux, 1988. Antherenkultur bei Zuckerrüben, *Beta vulgaris* var. *altissima* DÖLL. Arch Züchtungsforsch. Berlin 18(6): 375–383.
Hosemans, D. and D. Bossoutrot, 1983. Induction of haploid plants from *in vitro* culture of unpollinated beet ovules (*Beta vulgaris* L.). Z. Pflanzenzüchtg. 91: 74–77.
Hosemans, D. and D. Bossoutrot, 1985. *In vitro* culture of unpollinated beet (*Beta vulgaris* L.) ovules of male sterile and male fertile plants and induction of haploid plants. In: G.P. Chapman, S.H. Mantell and R.W. Daniels (Eds.), The Experimental Manipulation of Ovule Tissues, pp. 79–88. Longman Inc., New York.
Hu, H. and B. Huang, 1987. Application of pollen-derived plants to crop improvement. Int. Rev. Cytol. 107: 293–313.
Huang, B., 1992. Genetic manipulation of microspores and microspore-derived embryos. In Vitro Cell. Dev. Biol. 28: 53–58.
Jefferson, R.A., T.A. Kavanagh and M.W. Bevan, 1987. GUS-fusions: Beta-glucuronidase as sensitive and versatile gene fusion marker in higher plants. EMBO J. 6: 3901–3907.
Jensen, C.J., 1974. Chromosome doubling techniques in haploids. In: C.J. Jensen (Ed.), Haploids in Higher Plants: Advances and Potential, pp. 153–190. University of Guelph Press, Guelph.
Keimer, B. and K. D'Halluin, 1985. Production of haploid sugar beets (*Beta vulgaris* L.) by ovule culture: preliminary results. Hereditas 3 (Suppl.): 145–146.
Keimer, B. and H.C. Pedersen, 1988. Induction of haploid sugar beet embryos from ovule culture. Effect of donor plant growth conditions. In: 2nd Nordic Symposium on Cell and Tissue Culture, Elsinore, Denmark, p 27.
Kruse, A., 1963. Pure lines and their hybrids in beets, *Beta vulgaris* L. Royal Vet. and Agric. Coll., Copenhagen, Yearbook 1963: 42–53.
Levan, A., 1945. A haploid sugar beet after colchicine treatment. Hereditas 31: 193–204.
Lindsay, K. and P. Gallois, 1990. Transformation of sugarbeet (*Beta vulgaris*) by *Agrobacterium tumefaciens*. J. Exp. Bot. 41: 529–536.
Lux, H., L. Herrmann and C. Wetzel, 1990. Production of haploid sugar beet (*Beta vulgaris* L.) by culturing unpollinated ovules. Plant Breed. 104: 177–183.
Lux, H., C. Frahnert, S. Kotter, C. Wetzel, L. Herrmann and K.-H. Weege, 1991. *In Vitro* Präparetezubereitung für die Erzeugung Haploider Beta-Rübenpflanzen. Patent Application DD 289425.
Magnusson, I., E. Wremerth-Weich and C.H. Bornman, 1985. Haploidization of *Beta vulgaris* through ovule culture. Hereditas 3: 147–148 (abstract).
Maheswari, P., 1950. An Introduction to the Embryology of Angiosperms. McGraw-Hill Book Co., New York.
Olesen, P., E. Buck and B. Keimer, 1988. Structure and variability of embryos, endosperm and perisperm during *in vitro* culture of sugar beet *Beta vulgaris* ovules. In: M. Cristi *et al.* (Eds.), Sexual Reproduction in Higher Plants, pp. 107–112. Springer-Verlag, Berlin.
Ragot, M. and P. Steen (1992) Genetic and environmental effects on chromosome doubling of sugar beet (*Beta vulgaris* L.) haploids. Euphytica 63: 233–237.
Rogozinska, J.H. and M. Goska, 1982. Attempts to induce haploids in anther cultures of sugar, fodder and wild species of beet. Acta Soc. Bot. Poloniae 51(1): 91–105.
Sangwan, R.S. and B.S. Sangwan-Norreel, 1987. Ultrastructural cytology of plastids in pollen grains of certain androgenic and nonandrogenic plants. Protoplasma 138: 11–22.
Seman, I., 1983. Possibilities of detection and induction of haploids in *Beta vulgaris* L. Biologia (Bratislava) 38(11): 1113–1122.

Siebel, J. and K.P. Pauls, 1989. A comparison of anther and microspore culture as a breeding tool in *Brassica napus*. Theor. Appl. Genet. 78: 473–479.

Speckmann, G.J., 1975. Methods for chromosome doubling in fodder crops. In: B. Nuesch (Ed.), Ploidy in Fodder Crops. Report of the Meeting of the Eucarpia Fodder Crop Section, April 23–25, Zürich-Reckenholz, 1975, pp. 90–95.

Speckmann, G.J., J.P.C. Van Geyt and M. Jacobs, 1986. The induction of haploids of sugarbeet (*Beta vulgaris* L.) using anther and free pollen culture or ovule and ovary culture. In: W. Horn, C.J. Jensen, W. Odenbach and O. Schieder (Eds.), Genetic Manipulation in Plant Breeding, pp. 351–353. De Gruyter, Berlin.

Steen, P., B. Keimer and E. Smed, 1988. Homozygous lines in sugar beet (*Beta vulgaris* L.). In: 2nd Nordic Symposium on Cell and Tissue Culture, Elsinore, Denmark, p. 28.

Steen, P. and H.C. Pedersen, 1994. Gene transfer for herbicide resistance. J. Sugar Beet Res. (in press).

Van Geyt, J., K. D'Halluin and M. Jacobs, 1985. Induction of nuclear and cell divisions in microspores of sugar beet (*Beta vulgaris* L.). Z. Pflanzenzüchtg. 95: 325–335.

Van Geyt, J., G.J. Speckman Jr., K. D'Halluin and M. Jacobs, 1987. *In vitro* induction of haploid plants from unpollinated ovules and ovaries of the sugar beet (*Beta vulgaris* L.) Theor. Appl. Genet. 73: 920–925.

Yüce, S., 1973. Haploidie bei der Zuckerrübe. Thesis, Justus Liebig-Universität, Giesen.

Zahariev, A., I. Slavova, G. Kikindonov and Zh. Ivanov, 1992. First results of the production of dihaploid multigerm lines of sugarbeet and their use in heterosis breeding. Genetika i Selektsiya 25(5): 375–378.

3. Haploidy in important crop plants – potato

RICHARD E. VEILLEUX

Contents

1. Introduction

Solanum tuberosum L. ($2n = 4x = 48$), the cultivated white or Irish potato, is the world's fourth major economic crop. It is one of four cultivated *Solanum* species found within subsection Tuberarium that contains 150–200 wild tuber-bearing species. The genus forms a polyploid series from $2x$ to $6x$. The most widely cultivated forms, *S. tuberosum* and *S. andigena*, are tetraploid, whereas the diploid *S. phureja* and triploid *S.* × *chaucha* are cultivated to a limited extent in South America. Hybridization between cultivated tetraploids is unrestricted by breeding barriers. It has also been possible to bring about hybridization between diploids and tetraploids by utilizing the occurrence of functional $2n$ gametes in diploids such that $4x$-$2x$ interploid hybridization often results in the production of tetraploid hybrids (Veilleux, 1985).

Parthenogenetically derived dihaploids of *S. tuberosum* have been available for several years through pollination with haploid-inducing pollinators. This was possible by the selection of *S. phureja* that induced embryo development without fertilization or elimination of *S. phureja* chromosomes in hybrid embryos (Van Breukelen, 1981). Such dihaploids ($2n = 2x = 24$) cross readily with cultivated diploids and many of the wild diploids, although crossing barriers exist between some diploid groups. Modern potato breeding efforts generally take advantage of the wealth of germplasm available to incorporate necessary disease or insect resistances or improve characters of economic importance, e.g., specific gravity, reducing sugars, processing characters, etc. (Ross, 1986).

Potato is a highly heterozygous species that suffers severe inbreeding depression on selfing. Crosses between potato cultivars result in progeny that are generally inferior to either parent necessitating the evaluation of a large number of seedlings from which to select a few potentially valuable clones.

S.M. Jain, S.K. Sopory & R.E. Veilleux (eds.), In Vitro *Haploid Production in Higher Plants, Vol. 3,* 37–49.

Various schemes have been envisioned to develop potato hybrids comprising different genomic compositions to optimize heterotic allelic interactions (Chase, 1963; Mendiburu *et al.*, 1974; Wenzel *et al.*, 1979; Meyer *et al.*, 1992).

Recent progress has been made to develop potato cultivars that can be propagated from true potato seed (TPS) rather than vegetatively. This would circumvent the problem of transmission of many virus diseases through vegetative parts and reduce the expense of transporting and storing bulky seed tubers to nations that cannot produce their own. Currently available TPS cultivars are selections of $4x$ clones that suffer less inbreeding depression than most cultivars so that open-pollinated seed, which generally consist of a large proportion of first generation inbreds by self-pollination, can be used. The derivation of inbred lines could facilitate the development of hybrid TPS cultivars that are expected to exhibit considerably better performance. One such scheme would employ four independently selected inbred diploids. Two inbreds could be selected for the ability to generate a high frequency of $2n$ female gametes (diplogynoids), then intercrossed to generate a diplogynous F_1. The other two inbreds would be selected for the ability to produce high frequencies of $2n$ pollen (diplandroids) by the genetic equivalent of first division restitution (FDR) (Veilleux, 1985), then crossed to generate a diplandrous F_1. These two F_1s could then be synthesized into heterozygous but genetically homogeneous tetraploids by sexual polyploidization. Hence, the development of anther culture techniques to generate a diverse array of inbred lines would be invaluable to enable our realization of such a scheme.

2. Anther culture – development of the techniques for potato species

In nearly all anther culture studies of potato, anthers have been selected when microspores were at the late uninucleate stage of development (Irikura, 1975; Mix, 1983; Sopory, 1977, Sopory *et al.*, 1978; Batty and Dunwell, 1989; Veilleux *et al.*, 1985). Occasionally late tetrad (Johansson, 1986) or early uninucleate (Calleberg and Johansson, 1993) stage microspores were preferred. Cappadocia (1990) found that late uninucleate or mitotic stage microspores were more responsive than either early uninucleate or binucleate stage microspores of *S. chacoense*. Veilleux (1990) demonstrated that anther length and anther culture response (number of embryos or plantlets per anther) were not well correlated. Typically, buds with anthers between 2.5 and 4.0 mm have been selected for culture.

Non-distribution of the anthers from a single bud to various treatments in anther culture experiments may be an important source of experimental error. Powell *et al.* (1990) found that the response of anthers from the same flower was positively associated. This suggested that if one anther from a single bud responded positively to anther culture, additional anthers from the same bud would be more likely to respond positively than anthers taken from a different bud. In fact, Snider and Veilleux (1994) found significant variation due to distribution of anthers to flasks in an anther culture experiment, i.e., when anthers from a single flower were distributed to five treatments, the likelihood of finding significant treatment effect was greater than when all five anthers from a single bud were placed in the same treatment. Therefore, some randomization of anthers to treatments is highly recommended in order to prevent erroneous conclusions about apparent treatment effects that are due simply to a single treatment receiving an inordinate number of "good" anthers.

Initial success in the regeneration of haploid plants from anther culture of tuber-bearing *Solanum* species was reported by Irikura and Sakaguchi (1972). They obtained two documented haploids ($2n = 1x = 12$) of *S. verrucosum* after anther culture of 24 clones representing a wide array of potato germplasm including several cultivated types. In continued studies, Irikura (1975) examined a wider range of species on various media and obtained haploid plants from 11 species and two interspecific hybrids. Although haploids were obtained on a variety of media, Irikura concluded that Nitsch (1969) or Murashige and Skoog (1962) basal media (MS) with 0.1 mg/L indole acetic acid (IAA), 0.2 mg/L cytokinin [benzyladenine (BA) or kinetin (KIN)] and 3g/L activated charcoal (AC) were most suitable. Dunwell and Sunderland (1973) first reported successful regeneration of dihaploid plants ($2n = 2x = 24$) from cultured anthers of a cultivar (Pentland Crown) of *S. tuberosum*. Pollen embryos were observed on various media but plantlets were too rare to permit conclusive results about medium effect on dihaploid plant formation.

In initial studies of parthenogenically obtained dihaploids ($2n = 2x = 24$) of *S. tuberosum* in anther culture, Sopory (1977) and Sopory and Rogan (1976) reported primarily callus and embryo formation with little or no plant regeneration. Foroughi-Wehr *et al.* (1977) first reported the regeneration of monoploid[1] plants of *S. tuberosum* on Linsmaier and Skoog (1965) basal medium (LS) without $CoCl_2{\cdot}6H_2O$ or $FeSo_4{\cdot}7H_2O$ and with 40 mg/L FeEDTA, 40g/L sucrose, 1.0 mg/L IAA and 1.0 mg/L KIN. Sopory *et al.* (1978) and Sopory (1979) further refined the techniques for a single dihaploid clone by increasing sucrose to 60 g/L, addition of 0.5% AC and optimizing BA concentration at 4×10^{-6} M. Uhrig (1985) was able to obtain as many as 30 regenerated plants per cultured anther by selecting a highly responsive diploid clone and using liquid shake culture.

Powell and Uhrig (1987) compared the reaction of five genetically divergent clones on various media: liquid vs. solid, presence or absence of AC, and sucrose concentrations ranging from 3 to 15%. Three different pretreatments (6 °C for 2 days, 30 °C for 2 days, and direct culture) were also

[1] Throughout this article, the term monoploid, which is synonymous with monohaploid, will be used to describe plants with 12 chromosomes so as to avoid confusion with the 24-chromosome dihaploids of *S. tuberosum*.

Table 1. Potato cultivars from which anther-derived dihaploids have been extracted

Cultivar	Reference
Pito	Tiainen, 1992a,b
Pentland Crown	Dunwell and Sunderland, 1973
Maria	Johansson, 1986
Stina	Johansson, 1986
Elin	Johansson, 1986

examined. Of the three genotypes that produced embryos, all preferred liquid medium with AC confirming Uhrig's (1985) previous report. The optimal sucrose concentration of 6%, first employed by Sopory *et al.* (1978) was also reconfirmed, although genotype and sucrose interaction was noted. Genotype and pre-treatment interaction was highly significant – pretreatment at 30 °C inhibited androgenic embryo production for one genotype but enhanced it for another.

Owen *et al.* (1988a) found that anthers taken from buds of a single clone of *S. phureja* growing under a 14 h photoperiod were more responsive than those taken from the same clone growing under an 18 h photoperiod. Cappadocia and Ahmim (1988) found that field-grown plants of *S. chacoense* produced more calli in anther culture than greenhouse-grown plants but the percent anthers forming plants did not differ.

Production of dihaploids of *S. tuberosum* cultivars by anther culture has been less successful than for less adapted potato germplasm despite considerable effort. Although there have been several reports of plant regeneration from anther culture of unnamed 4*x* clones (Uhrig and Salamini, 1987; Mix, 1983; Johansson, 1986; Calleberg and Johansson, 1993), the list of named potato cultivars from which androgenic dihaploids have been derived remains short (Table 1). Johansson (1983) obtained embryos from nine of 16 tetraploid clones examined and regenerated dihaploids from five of them. Mix (1982) obtained dihaploids from 11 of 48 tetraploid genotypes. The plantlets were obtained through indirect embryogenesis, i.e., first by formation of microspore-derived callus and then regeneration of plantlets from callus, and varied for ploidy level (Mix, 1983). Johansson (1986) was able to increase the androgenic response in 19 of 20 tetraploid clones by using a bilayer medium of liquid over solid, cold pretreatment of the flower buds and treatment of the anther donor plants with Alar. Addition of 15 or 45 mg/L L-cysteine·HCl to the medium also increased androgenesis. Calleberg and Johansson (1993) found that embryo production was dependent on the incubation temperature. Tiainen (1992a) found that seasonal variation for embryo production exceeded all the manipulation in tissue culture but that reducing agents, ascorbic acid and L-cysteine, stimulated embryogenesis (Tiainen, 1992b). Uhrig and Salamini (1987) have proposed that the androgenic competence of advanced breeding lines of 4*x* potato be improved by hybridization with "efficient anther plants producing tetraploid strains," i.e., 4*x* clones that respond well to anther culture but which remain unselected for agronomic traits. Most of the 4*x* clones thus obtained were derived by spontaneous doubling of diploid clones obtained from anther culture of dihaploid potato. Plants combining the androgenic competence of one parent and desirable agronomic attributes of the other could then be selected among the progenies. Repeated cycles of anther culture and field selection could be used to generate 4*x* breeding lines with the potential for cultivar release.

3. Embryo conversion

Although occasional development of androgenic plantlets directly from cultured anthers *in vitro* has been noted in potato, generally an embryo conversion step involving subculture of androgenic embryos to a regeneration medium has been required. Conditions under which anthers have been cultured may affect the conversion frequency of embryos. Batty and Dunwell (1989) found a higher conversion frequency after anthers had been cultured on medium containing maltose instead of sucrose. Calleberg and Johansson (1993) found that anthers cultured at 20 °C produced more regenerable embryos than those cultured at 30 °C, despite higher embryo production at the higher temperature for some clones.

A variety of media has been used to convert androgenic embryos or "macroscopic structures" of potato into plantlets. A two-step procedure was originally proposed by Sopory (1977) where anther-derived callus was grown on MS medium with 2.5×10^{-5} M naphthalene acetic acid (NAA). After greening, the callus was transferred to MS + 10^{-5} M zeatin (ZEA) for differentiation of plantlets. Modifications of the two-step procedure, involving different growth regulators, reduced sucrose, or addition of coconut milk, were subsequently proposed by Wenzel and Uhrig (1981) and Johansson (1986). Cappadocia *et al.* (1984) and Uhrig (1985) employed one-step embryo regeneration procedures, the former using Nitsch (1969) medium + 2% sucrose + 0.1 mg/L indole butyric acid (IBA) and the latter using 1/2 strength Linsmaier and Skoog (1965) medium + 0.1 mg/L gibberellic acid (GA_3) + 2% sucrose. We compared Johansson's (1986) two-step and Uhrig's (1985) one-step procedures on androgenic embryos of *S. phureja* in our lab (unpublished data) and found no significant difference in regeneration rate. Hence, we adopted the simpler one-step procedure. We routinely subculture the unregenerated embryos to fresh medium three times at three week intervals. Johansson (1988) proposed culture of anthers on cubes of solid medium surrounded by liquid, the composition of which could be changed from embryo induction to embryo conversion as development proceeded. Different germplasm may have specific requirements for regeneration.

4. Chromosome number of anther-derived plants

The ploidy level of anther-derived potato plants has been estimated by counting chromosomes in mitotic cells at metaphase in root tip squashes (Sopory, 1977; Jacobsen and Sopory, 1978; Sopory and Tan, 1979; Mix, 1983; Cappodocia *et al.*, 1984; Johansson, 1986), counting chloroplasts in guard cell pairs (Frandsen, 1968; Singsit and Veilleux, 1991) most often combined with chromosome counts when ambiguous, or by DNA quantification in propidium-iodide stained nuclei (Owen *et al.*, 1988c; Snider and Veilleux, 1994). Monoploid potato plants have also been recognized by a composite of morphological characters, especially reduced anther length and corolla width (Pehu *et al.*, 1987; Veilleux, 1990).

A high frequency of diploid regenerants has generally been found among anther-derived plants of diploid potato clones, occasionally comprising most or all of the androgenic plants (Table 2). These can arise by embryogenesis of 2*n* microspores, spontaneous chromosome doubling during the culture phase or afterwards, or by regeneration of plants from somatic tissue. The genetic composition of plants arising from these three sources would differ. The first would be genetically similar but not identical to the anther donor and would exhibit considerable heterozygosity dependent upon the type of meiotic restitution present (Veilleux, 1985), the second would be homozygous inbred lines, and the third would be similar to the anther donor except for possible gametoclonal variation induced during the culture phase.

Because of the paucity of morphological markers in potato, discrimination among these groups has been difficult. Jacobsen and Sopory (1978) used segregation for a single morphological marker heterozygous in the anther donor to demonstrate homozygosity for at least one locus in anther-derived diploid potato. In addition, they proposed that anther donors be constructed that would be heterozygous for isozyme markers in order to have a more complete picture of homozygosity in anther-derived plants. Veilleux *et al.* (1985) used segregation for tuber protein bands in a second generation of anther-derived plants to demonstrate heterozygosity in the first generation of anther-derived diploids. Both of these methods have the limitation that homozygosity/heterozygosity can be determined for only a small part of the genome. Homozygosity at all loci cannot be assumed simply because one or two heterozygous loci in the anther donor can be shown to be homozygous in an anther-derived plant because formation of unreduced microspores may result in considerable but not complete homozygosity (Veilleux, 1985). This limitation can be overcome by using restriction fragment length polymorphisms (RFLPs). RFLPs have been successfully used to discriminate among anther-derived genotypic classes of diploid *S. chacoense* (Rivard *et al.*, 1989) or interdihaploid clones (Meyer *et al.*, 1992, 1993). These have the advantage that, if enough probe/enzyme combinations were tested, sufficient variation has been found to exist within unselected potato clones that the construction of anther donors that are heterozygous at specific loci has been unnecessary.

Table 2. Frequencies of various ploidy levels of androgenic plants derived from anther culture of 2*x* and 4*x* anther donor clones

No. of anther donor clones	Ploidy of anther donor clones	No. of androgenic plants analyzed	Number of				source
			1*x* (%)	2*x* (%)	4*x* (%)	other	
1	2*x*	40	2 (5)	34 (85)	4 (10)		Jacobsen and Sopory, 1978
2	2*x*	32	0	5 (16)	27 (84)		Sopory, 1977
9	2*x*	1189	10 (4)	1119 (44)	60 (5)		Wenzel *et al.*, 1979
5	2*x*	113	0	15 (13)	71 (63)	27	Sopory and Tan, 1979
2	2*x*	1135	0	1135 (100)	0		Przewozny *et al.*, 1980
11	2*x*	2133	469 (22)	1600 (75)	64 (3)		Wenzel and Uhrig, 1981
2	2*x*	306	160 (52)	142 (46)	4 (1)		Cappadocia *et al.*, 1984
19	2*x*	313	2 (<1)	251 (80)	43 (14)	17	Uhrig, 1985
1	2*x*	125	29 (23)	58 (46)	8 (6)		Veilleux *et al.*, 1985
29	2*x*	1143	124 (11)	870 (46)	73 (6)	15	Meyer *et al.*, 1993
5	4*x*	624	0	468 (75)	156 (25)		Uhrig and Salamini, 1987
31	4*x*	1337	0	1257 (94)	80 (6)		Uhrig and Salamini, 1987
10	4*x*	81	0	~72 (>90)	?		Johansson, 1986
5	4*x*	40	0	34 (85)	3 (7)	1	Mix, 1983

Also, there are virtually an unlimited number of loci. Meyer *et al.* (1993) assumed complete homozygosity of androgenic plants when at least five loci (distributed among different linkage groups) that were heterozygous in the anther donor were found to be homozygous in anther-derived plants. Veilleux *et al.* (1995) used the PCR-based molecular markers (RAPDs: randomly amplified polymorphic DNA; and SSRs: simple sequence repeats) to discriminate heterozygous from homozygous anther-derived diploids.

In contrast to the preponderance of higher than expected ploidy levels found among androgenic plants of diploid anther donors, anther culture of tetraploids has been found to yield mostly the expected dihaploids (Table 2). This may be because there is a lower fraction of unreduced gametes in the tetraploid anther donors. Alternatively, dihaploid embryos, because inbreeding would not be expected to be as severe as in a monoploid, may be able to compete better than completely inbred monoploid embryos compared to heterozygous diploid ones in anther culture of diploid anther donors. Dihaploids may also not be as likely to double spontaneously because their level of endopolyploidization has been found to be considerably lower compared to that in monoploids (Owen *et al.*, 1988c).

5. Genetic control of androgenesis

Since success with anther culture of potato was first reported, a strong genotypic response has been evident. Jacobsen and Sopory (1978) determined that specific genotypes responded to anther culture with one of five classes of response; highly responsive genotypes, however, could be selected among the progeny of weakly responding ones. Wenzel and Uhrig (1981) demonstrated that the capacity for producing "macroscopic structures" from cultured anthers was under genetic control due to its segregation in F_1 families between variously responsive parents. Uhrig (1983) recognized four classes of androgenic response among potato clones: those from which direct plantlet formation from cultured anthers could be obtained; those with high embryo formation but requiring transfer to regeneration medium for plant formation; those that formed callus requiring lengthy regeneration procedures; and those for which only a few poorly regenerative embryos could be obtained. In most crosses between a poorly responding clone and highly responsive ones, segregation occurred such that F_1 genotypes better than the poor parent could be selected. Cappadocia *et al.* (1984) and Singsit and Veilleux (1988) confirmed that the induction phase and regeneration process appeared to be under independent genetic control in *S. chacoense* and *S. phureja*, respectively. Simon and Peloquin (1977) reported a paternal inheritance of callusing ability of cultured stamen of various *Solanum* species. Singsit and Veilleux (1988) did not find evidence of maternal effect on embryo forming capacity of *S. phureja*.

Sonnino *et al.* (1989) suggested that androgenic ability was controlled by more than one recessive gene. Taylor and Veilleux (1992) proposed genetic control of embryo formation in anther culture by a single codominant gene with additive effect. They found no significant correlations between anther culture response and leaf disc regeneration or cell colony formation by cultured protoplasts. Veronneau *et al.* (1992), however, found a positive correlation between anther induction and leaf disc regeneration of *S. chacoense*, suggesting a common system of genetic control. Meyer *et al.* (1993) partitioned anther donor clones into classes depending on the frequency of monohaploid regenerants. A significant negative correlation between production of viable $2n$ pollen and monohaploid-producing ability was found. Snider and Veilleux (1994) found homozygous, anther-derived doubled monoploids that were responsive to anther culture despite their lack of fertile pollen. The response, however, was just as variable as that of a heterozygous anther donor, indicating that the anther culture response was determined by the sporophyte, since the genetically homogeneous array of microspores expected from a doubled monoploid responded no better than the genetically heterogeneous array generated by a heterozygous clone. Differential responses of genotypes to high greenhouse temperatures (Snider and Veilleux, 1994), pretreatments, or sucrose level in the medium (Powell and Uhrig, 1987) have also been observed. Calleberg *et al.* (1989) demonstrated a temperature dependent relationship between androgenesis and pollen germination and suggested a correlation between temperature tolerance of a plant and its anther culture capacity. Wenzel and Uhrig (1981) concluded that introgression of androgenic competence into recalcitrant germplasm by hybridization would be more efficient than attempts to customize tissue culture conditions to unresponsive genotypes.

6. Utilization of anther-derived plants

Various breeding schema have been proposed to capitalize on the passage of a heterozygous crop through the so-called "monoploid sieve" (Wenzel *et al.*, 1979). Any homozygous plant regenerated through anther culture would be free of lethal recessive alleles as well as sublethal or severely deleterious alleles. By combining independently selected homozygous diploid clones, heterotic F_1 hybrids can be expected. If such hybrids have functional $2n$ pollen by the genetic equivalent of FDR, they could be used in crosses to tetraploid clones to generate possibly superior $4x$-$2x$ hybrids (Veilleux *et al.*, 1985). Alternatively, four different homozygous diploids could be combined through a double cross hybridization scheme in which the second cross would restore the tetraploid level either by sexual polyploidization or by protoplast fusion (Wenzel *et al.*, 1979). If sufficient seed set could be obtained after sexual polyploidization, release of a TPS cultivar can be envisioned. If protoplast fusion were required, the products would be more suitable for clonal propagation. Variation can be increased among microspore-derived plants by mutagenic treatment of developing microspores before culture (Przewozny *et al.*, 1980). *In vitro* selection during microspore culture offers still more opportunity. Wenzel and Uhrig (1981) found levels of nematode and virus resistance in anther-derived plants equal to that of the anther donor; although segregation for resistances occurred, the possibility of obtaining homozygous forms permitted intensification of resistance alleles. Recurrent selection schema involving simultaneous selection for both competence in anther culture as well as for agronomic traits at the diploid and tetraploid levels have been proposed by Uhrig and Salamini (1987) and Meyer *et al.* (1993).

7. Performance of anther-derived plants

Because of the difficulties in developing the anther culture techniques for potato, the limitation of the technique to only a relatively few competent genotypes, and the low frequencies of monoploids from anther culture of diploids, completely homozygous potato clones have not been sufficiently abundant or representative of a wide enough diversity of germplasm to realize any of the above objectives. Owen *et al.* (1988b) found significant variation for traits associated with tuberization among monoploid clones selected from a single anther donor of *S. phureja*. This indicates the necessity of sampling a large enough population of monoploids from any heterozygous clone to allow selection of the more vigorous and fertile ones. M'Ribu and Veilleux (1991) found significant correlations between monoploids and their corresponding doubled monoploid clones obtained by leaf disc regeneration of the monoploids (M'Ribu and Veilleux, 1990) for most traits including those contributing to tuber yield. This permits efficient selection at the monoploid level before the doubling process, saving time in the breeding process.

In examination of fertility of anther-derived homozygous lines, Wenzel *et al.* (1979) reported that 20% were self-fertile including some spontaneously doubled tetraploids. M'Ribu and Veilleux (1992) found no male fertility among 13 doubled monoploids from a clone of *S. phureja*. Seed set after crossing the doubled monoploids with heterozygous pollinators, however, ranged from 0 to 210 seeds per fruit. Depending on the specific doubled monoploid clone, the progeny resulting from these crosses equalled or exceeded the yield of progeny of the anther donor crossed to the same pollinators. Cappadocia (1990) and Cappadocia *et al.* (1986) were able to obtain limited seed set in crosses among doubled monoploids of *S. chacoense*. In contrast, Uijtewaal *et al.* (1987) reported low fertility among doubled monoploid clones extracted from heterozygous dihaploid potato clones by pollination with a haploid inducing pollinator. The poor performance of these homozygous lines discouraged their use in breeding. As might be expected, the performance of homozygous potato is highly variable and depends primarily on the tolerance to inbreeding of the specific heterozygous clones from which they have been derived. Until a wide array of genetically divergent homozygous lines becomes available, conclusions about their utility cannot be accurately made. However, their potential cannot be denied. Therefore, future effort is certainly warranted toward their unbiased assessment.

8. References

Batty, N. and J. Dunwell, 1989. Effect of maltose on the response of potato anthers in culture. Plant Cell Tiss. Org. Cult. 18: 221–226.

Calleberg, E.K. and L.B. Johansson, 1993. The effect of starch and incubation temperature in anther culture of potato. Plant Cell Tiss. Org. Cult. 32: 27–34.

Calleberg, E.K., I.S. Kristjansdottir and L.B. Johansson, 1989. Anther cultures of tetraploid *Solanum* genotypes – the influence of gelling agents and correlations between incubation temperature and pollen germination temperature. Plant Cell Tiss. Org. Cult. 19: 189–197.

Cappadocia, M., 1990. Wild potato (*Solanum chacoense* Bitt.): *in vitro* production of haploids. In: Y.P.S. Bajaj (Ed.), Biotechnology in Agriculture and Forestry. Vol. 12, pp. 514–529. Springer-Verlag, Berlin.

Cappadocia, M. and M. Ahmim, 1988. Comparison of two culture methods for the production of haploids by anther culture in *Solanum chacoense*. Can. J. Bot. 66: 1003–1005.

Cappadocia, M., D.S.K. Cheng and R. Ludlum-Simonette, 1984. Plant regeneration from *in vitro* culture of anthers of *Solanum chacoense* Bitt. and interspecific diploid hybrids *S. tuberosum* L. × *S. chacoense* Bitt. Theor. Appl. Genet. 69: 139–143.

Cappadocia, M., D.S.K. Cheng and R. Ludlum-Simonette, 1986. Self-compatibility in doubled haploids and their F_1 hybrids, regenerated via anther culture in self-incompatible *Solanum chacoense* Bitt. Theor. Appl. Genet. 72: 66–69.

Chase, S.C., 1963. Analytic breeding in *Solanum tuberosum* L. – a scheme utilizing parthenotes and other diploid stocks. Can. J. Genet. Cytol. 5: 359–363.

Dunwell, J.M. and N. Sunderland, 1973. Anther culture of *Solanum tuberosum* L. Euphytica 22: 317–323.

Foroughi-Wehr, B., H.M. Wilson, G. Mix and H. Gaul, 1977. Monohaploid plants from anthers of a dihaploid genotype of *Solanum tuberosum* L. Euphytica 26: 361–367.

Frandsen, N.O., 1968. Die Plastidenzahl als Merkmal bei der Kartoffel. Theor. Appl. Genet. 38: 153–167.

Irikura, Y., 1975. Induction of haploid plants by anther culture in tuber-bearing species and interspecific hybrids of *Solanum*. Potato Res. 18: 133–140.

Irikura, Y. and S. Sakaguchi, 1972. Induction of 12-chromosome plants from anther culture in a tuberous *Solanum*. Potato Res. 15: 170–173.

Jacobsen, E. and S.K. Sopory, 1978. The influence and possible recombination of genotypes on the production of microspore embryoids in anther cultures of *Solanum tuberosum* and dihaploid hybrids. Theor. Appl. Genet. 52: 119–123.

Johansson, L.B., 1983. Embryogenesis in anther cultures of some species in the families Solanaceae, Ranunculaceae and Papaveraceae. Ph.D. Diss., Uppsala University, Uppsala.

Johansson, L., 1986. Improved methods for induction of embryogenesis in anther cultures of *Solanum tuberosum*. Potato Res. 29: 179–190.

Johansson, L.B., 1988. Increased induction of embryogenesis and regeneration in anther cultures of *Solanum tuberosum* L. Potato Res. 31: 145–149.

Linsmaier, E.M. and F. Skoog, 1965. Organic growth factor requirements of tobacco tissue cultures. Physiol. Plant. 18: 100–127.

Mendiburu, A.O., S.J. Peloquin and D.W.S. Mok, 1974. Potato breeding with haploids and 2*n* gametes. K.J. Kasha (Ed.), Haploids in Higher Plants: Advances and Potential, pp. 249–258. University Press, Guelph.

Meyer, R., F. Salamini and H. Uhrig, 1992. Biotechnology and plant breeding: relevance of cell genetics in potato improvement. Proc. Roy. Soc. Edinburgh Sect. B 99: 11–21.

Meyer, R., F. Salamini and H. Uhrig, 1993. Isolation and characterization of potato diploid clones generating a high frequency of monohaploid or homozygous diploid androgenetic plants. Theor. Appl. Genet. 85: 905–912.

Mix, G., 1982. Dihaploide Pflanzen aus *Solanum tuberosum* Antheren. Landbauforschung Völkenrode 32: 34–36.

Mix, G., 1983. Production of dihaploid plantlets from anthers of autotetraploid genotypes of *Solanum tuberosum* L. Potato Res. 26: 63–67.

M'Ribu, H.K. and R.E. Veilleux, 1990. Effect of genotype, explant, subculture interval and environmental conditions on regeneration of shoots from *in vitro* monoploids of a diploid potato species, *Solanum phureja* Juz. and Buk. Plant Cell Tiss. Org. Cult. 23: 171–179.

M'Ribu, H.K. and R.E. Veilleux, 1991. Phenotypic variation and correlations between monoploids and doubled monoploids of *Solanum phureja*. Euphytica 54: 279–284.

M'Ribu, H.K. and R.E. Veilleux, 1992. Fertility of doubled monoploids of *Solanum phureja*. Amer. Potato J. 69: 447–459.

Murashige, T. and F. Skoog, 1962. A revised medium for rapid growth and bioassays with tobacco tissue cultures. Physiol. Plant. 15: 473–497.

Nitsch, J.P., 1969. Experimental androgenesis in *Nicotiana*. Phytomorphology 19: 389–404.

Owen, H.R., R.E. Veilleux, F.L. Haynes and K.G. Haynes, 1988a. Photoperiod effects on 2*n* pollen production, response to anther culture, and net photosynthesis of a diplandrous clone of *Solanum phureja*. Amer. Potato J. 65: 131–139.

Owen, H.R., R.E. Veilleux, F.L. Haynes and K.G. Haynes, 1988b. Variability for critical photoperiod for tuberization and tuber yield among monoploid, anther-derived genotypes of *Solanum phureja*. J. Amer. Soc. Hort. Sci. 113: 755–759.

Owen, H.R., R.E. Veilleux, D. Levy and D.L. Ochs, 1988c. Environmental, genotypic, and ploidy effects on endopolyploidization within a genotype of *Solanum phureja* and its derivatives. Genome 30: 506–510.

Pehu, E., R.E. Veilleux and K.W. Hilu, 1987. Cluster analysis of anther-derived plants of *Solanum phureja* (Solanaceae) based on morphological characteristics. Amer. J. Bot. 74: 47–52.

Powell, W. and H. Uhrig, 1987. Anther culture of *Solanum* genotypes. Plant Cell Tiss. Org. Cult. 11: 13–24.

Powell, W., M. Coleman and J. McNicol, 1990. The statistical analysis of potato anther culture data. Plant Cell Tiss. Org. Cult. 23: 159–164.

Przewozny, T., O. Schieder and G. Wenzel, 1980. Induced mutants from dihaploid potatoes after pollen mother cell treatment. Theor. Appl. Genet. 58: 145–148.

Rivard, S.R., M. Cappadocia, G. Vincent, N. Brisson and B.S. Landry, 1989. Restriction fragment length polymorphism (RFLP) analyses of plants produced by *in vitro* anther culture of *Solanum chacoense* Bitt. Theor. Appl. Genet. 78: 49–56.

Ross, H., 1986. Potato Breeding: Problems and Perspectives. Verlag Paul Parey, Berlin/Hamburg.

Simon, P.W. and S.J. Peloquin, 1977. The influence of paternal species on the origin of callus in anther culture of *Solanum* hybrids. Theor. Appl. Genet. 50: 53–56.

Singsit, C. and R.E. Veilleux, 1988. Intra- and interspecific transmission of androgenetic competence in diploid potato species. Euphytica 43: 105–112.

Singsit, C. and R.E. Veilleux, 1991. Chloroplast density in guard cells of leaves of anther-derived potato plants grown *in vitro* and *in vivo*. HortScience 26: 592–594.

Snider, K.T. and R.E. Veilleux, 1994. Factors affecting variability in anther culture and in conversion of androgenic embryos of *Solanum phureja*. Plant Cell Tiss. Org. Cult. 36: 345–354.

Sonnino, A., S. Tanaka, M. Iwanaga and L. Schilde-Rentschler, 1989. Genetic control of embryo formation in anther culture of diploid potatoes. Plant Cell Rep. 8: 105–107.

Sopory, S.K., 1977. Differentiation in callus from cultured anthers of dihaploid clones of *Solanum tuberosum*. Z. Pflanzenphysiol. 82: 88–91.
Sopory, S.K., 1979. Effect of sucrose, hormones, and metabolic inhibitors on the development of pollen embryoids in anther cultures of dihaploid *Solanum tuberosum*. Can. J. Bot. 57: 2691–2694.
Sopory, S.K. and P.G. Rogan, 1976. Induction of pollen divisions and embryoid formation in anther cultures of some dihaploid clones of *Solanum tuberosum*. Z. Pflanzenphysiol. 80: 77–80.
Sopory, S.K. and B.H. Tan, 1979. Regeneration and cytological studies of anther and pollen calli of dihaploid *Solanum tuberosum*. Z. Pflanzenzüchtg. 82: 31–35.
Sopory, S.K., E. Jacobsen and G. Wenzel, 1978. Production of monohaploid embryoids and plantlets in cultured anthers of *Solanum tuberosum*. Plant Sci. Lett. 12: 47–54.
Taylor, T.E. and R.E. Veilleux, 1992. Inheritance of competencies for leaf disc regeneration, anther culture, and protoplast culture in *Solanum phureja* and correlations among them. Plant Cell Tiss. Org. Cult. 31: 95–103.
Tiainen, T., 1992a. The influence of culture conditions on anther culture response of commercial varieties of *Solanum tuberosum* L. Plant Cell Tiss. Org. Cult. 30: 211–219.
Tiainen, T., 1992b. The role of ethylene and reducing agents on anther culture response of tetraploid potato (*Solanum tuberosum* L.). Plant Cell Rep. 10: 604–607.
Uhrig, H., 1983. Breeding for *Globodera pallida* resistance in potatoes. I. Improvement of the androgenetic capacity in some resistant dihaploid clones. Z. Pflanzenzüchtg. 91: 211–218.
Uhrig, H., 1985. Genetic selection and liquid medium conditions improve the yield of androgenetic plants from diploid potatoes. Theor. Appl. Genet. 71: 455–460.
Uhrig, H. and F. Salamini, 1987. Dihaploid plant production from 4*x*-genotypes of potato by the use of efficient anther plants producing tetraploid strains (4*x* EAPP-clones) – proposal of a breeding methodology. Plant Breed. 98: 228–235.
Uijtewaal, B.A., E. Jacobsen and J.G.Th. Hermsen, 1987. Morphology and vigour of monohaploid potato clones, their corresponding homozygous diploids and tetraploids and their heterozygous diploid parent. Euphytica 36: 745–753.
Van Breukelen, E.W.M., 1981. Pseudogamic production of dihaploids and monoploids in *Solanum tuberosum* and some related species. Pudoc, Wageningen.
Veilleux, R.E., 1985. Diploid and polyploid gametes in crop plants: mechanisms of formation and utilization in plant breeding. Plant Breed. Rev. 3: 253–288.
Veilleux, R.E., 1990. *Solanum phureja*: anther culture and the induction of haploids in a cultivated diploid potato species. In: Y.P.S. Bajaj (Ed.), Biotechnology in Agriculture and Forestry. Vol. 12, pp. 530–543. Springer-Verlag, Berlin.
Veilleux, R.E., J. Booze-Daniels and E. Pehu, 1985. Anther culture of a 2*n* pollen producing clone of *Solanum phureja* Juz. and Buk. Can. J. Genet. Cytol. 27: 559–564.
Veilleux, R.E., L.Y. Shen and M.M. Paz, 1995. Analysis of the genetic composition of anther-derived potato by randomly amplified polymorphic DNA and simple sequence repeats. Genome: (in press).
Veronneau, H., G. Lavoie and M. Cappadocia, 1992. Genetic analysis of anther and leaf disc culture in two clones of *Solanum chacoense* Bitt. and their reciprocal hybrids. Plant Cell Tiss. Org. Cult. 30: 199–209.
Wenzel, G. and H. Uhrig, 1981. Breeding for nematode and virus resistance in potato via anther culture. Theor. Appl. Genet. 59: 333–340.
Wenzel, G., O. Schieder, T. Przewozny, S.K. Sopory and G. Melchers, 1979. Comparison of single cell culture derived *Solanum tuberosum* L. plants and a model for their application in breeding programs. Theor. Appl. Genet. 55: 49–55.

4. Haploidy in onion (*Allium cepa* L.) and other *Allium* species

E.R. JOACHIM KELLER and LARISSA KORZUN

Contents

1. Introduction

The genus *Allium* contains some of the most important vegetables and spices used throughout the world including onion, leek, garlic, and shallot. Additional *Allium* species are cultivated more regionally including chives (*A. schoenoprasum* L.), Chinese chives (*A. tuberosum* Rottl. ex Spr. and *A. ramosum* L.), rakkyo (*A. chinense* G. Don), and *A. hookeri* Thw. In 1987, the cultivation of onions and shallots comprised 1.25 million ha in developing and 0.5 million ha in developed countries (Currah and Proctor, 1990).

A. cepa L. is, undoubtedly, the most important species within the genus. According to the type of reproduction and use, the forms of this species can be divided into two groups: the common onions, which are propagated mainly by seed, and the *aggregatum* group (including shallots and potato onions) which are primarily propagated vegetatively (Hanelt, 1990). Due to this difference and the sterilities that often occur among the vegetatively propagated types, breeding is more advanced in common onion than in shallot.

The main breeding objectives for *A. cepa* concern quality (dry matter content, flavour, bulb colour and shape), development (specific daylength dependency, uniformity of ripening, and storability), and pest resistance. The latter includes resistance to several viruses: shallot latent virus and onion

S.M. Jain, S.K. Sopory & R.E. Veilleux (eds.), In Vitro *Haploid Production in Higher Plants, Vol. 3,* 51–75.

yellow dwarf virus (Walkey, 1990); fungi: white rot (*Sclerotium cepivorum* Berk.), downy mildew (*Peronospora destructor* [Berk.] Casp.), pink root rot (*Phoma terrestris* Hansen), *Botrytis* and *Fusarium* diseases; and insects: onion thrips (*Thrips tabaci* Lind.) and onion fly [*Delia antiqua* (Meigen)] (Dowker, 1990). A large gene pool of related wild species, possessing useful characters such as disease, frost or salt resistance, will gain increasing importance as biotechnological methods become available to facilitate gene transfer.

A. cepa is a predominantly outbreeding species due to strong protrandry. Pollination from other flowers of the same inflorescence is, however, possible because of a flowering gradient within the inflorescence. A high degree of heterozygosity exists because of these features. Traditional cultivars are heterogeneous populations providing sufficient background for selection. In less developed countries, onion breeding has been concentrated on selection of superior types from land races. In advanced countries, however, breeding programmes to derive hybrids using male sterility have been underway for some time. Heterosis resulting from crosses of homozygous inbreds is expected to hasten the realization of breeding goals. However, inbreeding depression has been reported to diminish the effect of hybrid breeding by influencing the production of homozygous parental lines. Rapid progress in the production of homozygous material for hybrid breeding programs can be made by using DH (doubled haploid) lines derived from haploid plants.

The main pathways for haploid induction are androgenesis (discovered by Guha and Maheshwari, 1964), gynogenesis (first description by San Noeum, 1976), and chromosome elimination [exemplified by the so-called *bulbosum* technique in *Hordeum* (Symko, 1969)]. Of these methods, only gynogenesis has been used effectively to produce haploid and doubled haploid onion plants, although there have been attempts to employ all of the above techniques.

2. Haploid induction via androgenesis

Despite considerable effort in anther culture of onion, it remains among the androgenetically recalcitrant species. Keller (1990a) cultured a total of 98,027 anthers on 25 different culture media but observed no androgenic development. Campion *et al.* (1984) observed nuclear cleavage in onion anther cultures but obtained no further development. The only development obtained in these studies was callus formation from the filaments (Fig. 1). In *A. altaicum* Pall., *A. fistulosum* L. and its hybrids with onion or leek, a high frequency of callus (up to 41.7% in *A. altaicum*) developed. Most callus occurred in media with equimolar (1 μM) concentrations of 2,4–D and BA (Keller, 1990a). An experiment to determine the effect of sucrose in the more reactive species, *A. porrum* L., revealed an optimum of callus formation at a relatively high (10%) sucrose level (Keller, 1992a).

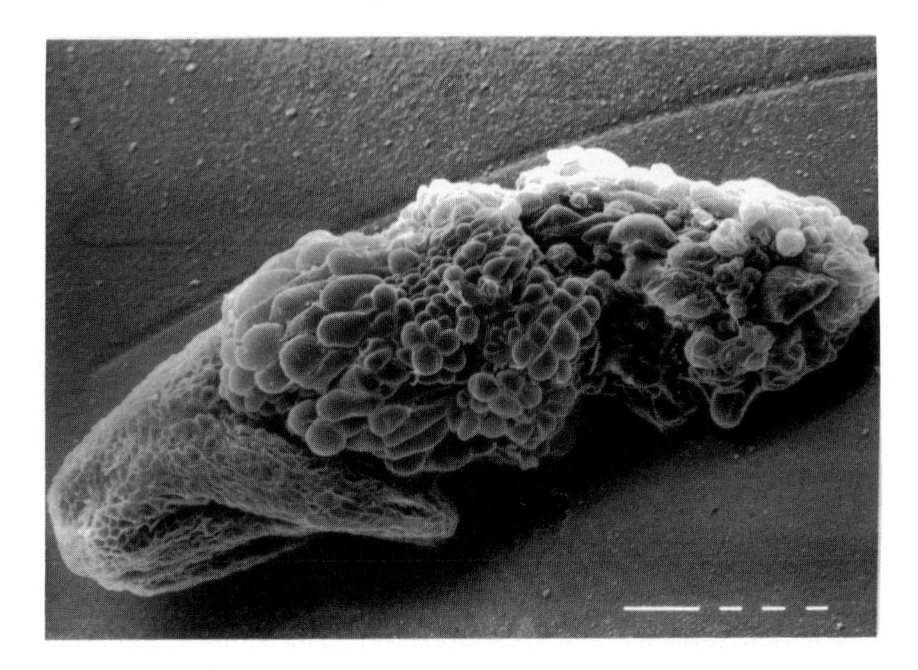

Figure 1. Scanning electron micrograph (SEM) showing an onion anther some weeks after start of the culture. Only the filament has responded by callus formation. Magnification: the left bar is 0.2 mm.

Roy (1989) reported callus formation and differentiation of embryos in anther cultures of onion. Anthers were cultured on MS medium (Murashige and Skoog, 1962) with 10 mg l^{-1} NAA and 0.04 mg l^{-1} kinetin. The anthers became swollen within 4–6 weeks. The author isolated swollen thecae (termed as "anther lobes") and transferred them to several media. Callus was produced from thecae on MS medium with 10 mg l^{-1} NAA, 0.5 mg l^{-1} 2,4–D, and 0.04 mg l^{-1} kinetin. Culture on MS + 10 mg l^{-1} NAA + 0.5 mg l^{-1} 2,4–D + 3 mg l^{-1} BA resulted in callus plus embryo formation. The embryos were of different stages ranging from globular and heart (unexpected for a monocotyledonous embryo) to torpedo stage. However, ploidy analysis was not performed.

Both Flier (1990) and Kanne (1991) failed to obtain haploids from microspore culture of leek. Callus formation has been observed in anther culture of *A. sativum* resulting in the regeneration of plants from adventitious buds, i.e., without embryo formation (Suh and Park, 1986). Ploidy of these plantlets was not reported. Smith *et al.* (1991) were unable to observe androgenic development in anther cultures of *Allium*.

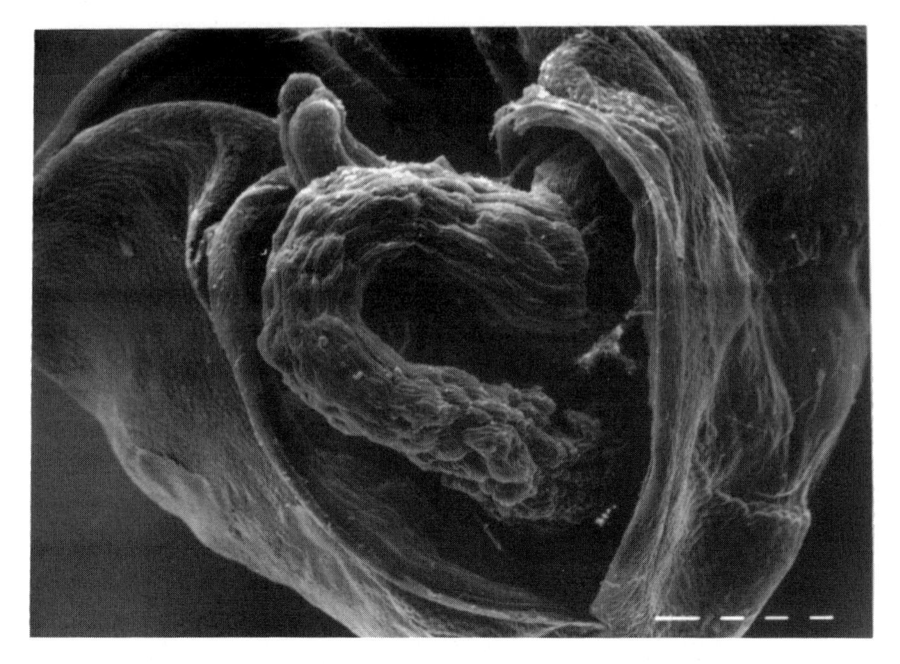

Figure 2. SEM showing a gynogenic embryo just emerging from the ovary. Magnification: the left bar is 0.2 mm.

3. Haploid induction via gynogenesis

The culture of ovaries and ovules of onion species has been more successful than anther culture. Ovary culture of onion was first reported by Guha and Johri (1966) who obtained regenerants only from pollinated ovaries. However, they found that the black seed coat developed even without fertilization, a finding that has since been confirmed in our own investigations (Keller, 1990b).

Other developmental processes have also been observed in ovary cultures. An intensive swelling of the carpels generally occurs within the first few days of culture. Later, these tissues die off, and the ovules or black "seeds" become visible. On some media, the nectarial region forms callus or even adventive plantlets (Keller, 1990b; Cohat, 1994). Such shoots of somatic origin can easily be discarded when the cultures are checked in early stages of development. Gynogenic embryos emerge from carpels, sometimes with the black seed coat still adhering to the plantlets (Figs. 2 and 3). Isolated ovules exhibit similar responses (swelling, black testa) as when cultured inside the entire ovary or bud. Additionally, swelling or callus formation can often be observed in the region of the funiculus.

The frequency of cultured ovaries or ovules that has yielded gynogenic

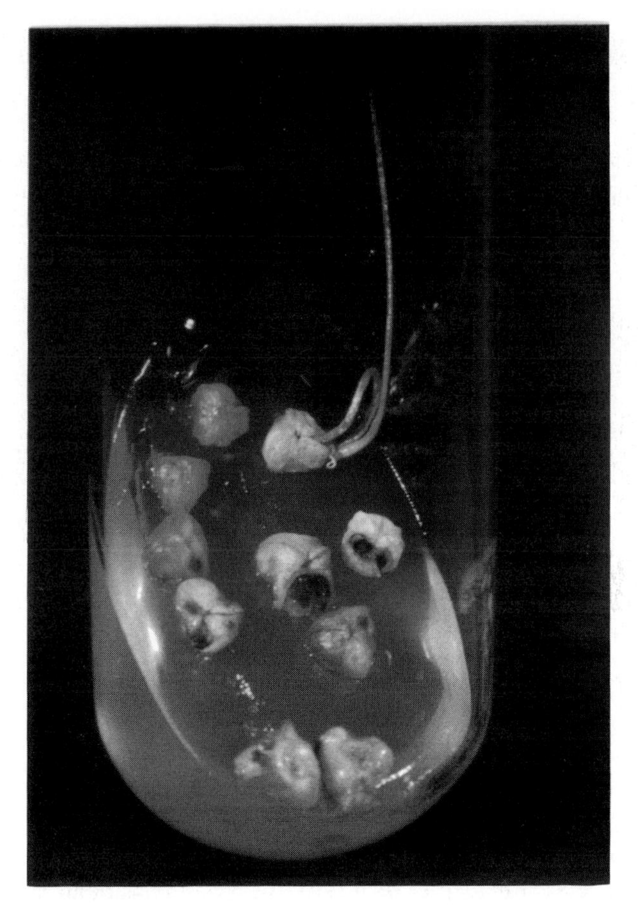

Figure 3. Ovaries of onion about six weeks after the start of culture. Note the gynogenic plantlet in one of the explants and the black seed testae of the ovary at the center.

plantlets has ranged from 0.03 to 21.9% (Table 1). The first report of haploid plantlets derived from *in vitro* culture of unpollinated ovules of *A. cepa* was by Campion and Azzimonti (1988). The authors described the technique of *in vitro* culture of unpollinated ovules, determined ploidy level of regenerated plants, and confirmed the derivation of embryos from reduced cells in the embryo sac. However, the yield of embryos was 0.28% in the best treatment (Campion and Alloni, 1990).

Muren (1989) induced *in vitro* parthenogenesis in unpollinated ovaries of *A. cepa*. Among about 250 regenerants, 70% were haploid. Embryonic structures were produced from unpollinated ovules of onion in whole flower bud culture by Smith *et al.* (1991), but no plants were recovered. Campion *et al.* (1992) obtained haploid plants of *A. cepa* from flower bud and ovary culture on modified medium at 0.7 and 0.64%, respectively. From 2,640

Table 1. Frequencies of developing plants and haploids in culture of isolated ovules (OVU), ovaries (OVA) and flower buds (FLB) of onion (*A. cepa*)

Cultivated type	Organs cultured	Plantlets per cultured organ (%)	Haploid regenerants (%)	Reference
OVU } OVA }	n.i.	7.6	≈50%	Bohanec *et al.* (1995)
OVU	26,000	0.03	42.8%	Campion and Alloni (1990)
OVU	18,193	n.i.	n.i.	Campion *et al.* (1992)
OVA }		1.66* }	88.3%	
FLB }	28,800	0.66* }		
FLB	10,560	21.9	77.6%	Cohat (1994) (shallot)
OVU	12,369	0.07 }		Keller (1990b)
IVA	14,842	0.44 }	n.c.	
FLB	1,821	0.27 }		
OVA	22,869	2.86*	62.2%	Keller (1990c)
OVA	2,640	2.83*	5 plants (19%)**	Martinez *et al.* (1994)
OVA	n.i.	About 250 plants	70%	Muren (1989)
OVA	2,150	3.7*	34.8%**	Schum *et al.* (1993; leek)
FLB	3,087	27 embryo-like structures	0	Smith *et al.* (1991; onion)
FLB	6,024	5.1*	4.4%	(leek)

* In the best variant.
** Figure calculated from the text information.
n.i. No information, n.c. not counted.

cultivated ovaries of *A. cepa*, Martinez *et al.* (1994) obtained only five haploid plantlets. Bohanec *et al.* (1995) proved the haploid origin of their gynogenically derived onion plants both cytologically and by α-esterase isozyme analysis. Regeneration was much higher from ovaries than from ovules for all cultivars.

Ovary and flower bud culture have generally been found to be more efficient than ovule culture because of the less intrusive manipulation (Keller, 1990b; Campion *et al.*, 1992). The latter authors found the percentage of embryos giving haploid plants from flower bud culture to be slightly higher than from ovary culture. However, in our investigations, the yield of haploids was lower in flower bud (0.27%) compared to ovary culture (0.44%). We favored isolated ovary cultures both because of this result and because of the likelihood of regenerating ordinary seedlings from self-pollination in cultures of complete flower buds. However, Cohat (1994) recently reported that onion anthers were unable to dehisce *in vitro*, possibly because of high humidity within the culture vessel. He therefore cultivated only whole flower buds, a simpler procedure that facilitates more efficient production of haploids.

Recently, Campion (1994) reported that the yield of haploid embryos was

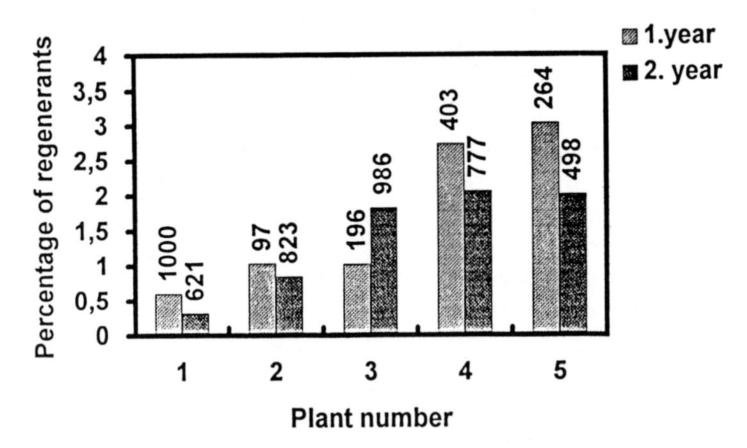

Figure 4. Gynogenic regeneration rate% in ovary culture of five onion plants cv. Stuttgarter Riesen, recorded in two successive years (phytotron conditions identical, conditions of the previous first weeks of culture in greenhouse different). Numbers on the bars represent numbers of ovaries per variant.

genotype-dependent and ranged from 0.1 to 6%. From these embryos, 65% developed into whole plants. Cohat (1994) obtained even higher percentages of haploid embryonic structures (up to 55% in the best genotype) using *in vitro* culture of flower buds in shallot (*A. cepa* L. var. *aggregatum*). Within a year, up to 21.9% of cultured buds had yielded plants. Despite significant genotypic differences, a breakthrough in the efficiency of gynogenesis in *A. cepa* was obtained with this high regeneration frequency; 77.6% of the total number of regenerants were haploid.

4. Factors influencing gynogenesis

4.1. *Genotypic effects*

Like most developmental processes in plants, tissue cultures of onion are dependent on the genotype (Phillips and Luteyn, 1983). This can be an obstacle for the application of *in vitro* methods in plant breeding.

We have found clear genotypic effects for gynogenic frequency among 273 donor plants representing different onion cultivars grown in the same greenhouse (Keller, 1990c). The genotypic effect is highly reproducible, as shown in repeated culture of the same donor plants over several years (Fig. 4). Campion *et al.* (1992) reported that one of four onion genotypes was significantly more responsive in flower bud culture, whereas no significant differences were found in ovary culture. In flower bud culture of 32 genotypes of shallot, Cohat (1994) found extreme differences in gynogenic frequency:

seven showed no embryonic development, two between 0 and 1% induction, thirteen between 1 and 5%, three between 5 and 10%, and seven greater than 10%. In the best genotype, 55% of cultured flower buds yielded embryonic structures. The development of plantlets ranged from 0 to 21.9%. Recalcitrance of some genotypes was attributed to their sterility. In a study of gynogenic induction in six onion cultivars, Muren (1989) grouped two into high (2%), two into medium (1–1.9%), and two into low reactive (0–0.9%) categories. The influence of the genotype on the percentage of developing regenerants was also reported for leek (Schum *et al.*, 1993).

Genotypic effects have been observed not only for frequency of gynogenic embryos but also for the percentage of haploids obtained from such embryos. Schum *et al.* (1993) found that one of three genotypes of leek yielded 89% haploids whereas the other two did not form haploid regenerants. Likewise, we found differences for haploid frequencies among three onion cultivars (in parentheses, percentages haploid/mixoploid/diploid regenerants): "Stuttgarter Riesen" (66/23/11, and 0.6 polyploids), "Zittauer Gelbe" (72/6/22), and "Calbenser Gerlinde" (27/18/55). In five genotypes used by Campion and Azzimonti (1988), two produced only diploid, and the other three haploid regenerants with subsequent spontaneous diploidization. In screening *ex situ* plant material, we even found genotypic differences in polyembryony of *A. tuberosum*. In one accession no plants developed any polyembryony in 100 enzymatically isolated embryo sacs. In another accession, polyembryony was found in 56 of 300 analyzed ovules (18.7%). Genotypic dependencies have been also described for this species by Kojima and Nagato (1992).

4.2. *Influence of culture media constituents*

Various media have been used for induction of gynogenesis in onion species and subsequent maintenance of the gynogenically derived plantlets (Table 2). No conclusions can yet be drawn for the best combination of plant growth regulators (auxins and cytokinins) for induction of gynogenesis. In flower bud culture of onion, Campion *et al.* (1992) emphasized the stimulating effect of 2,4–D which has also been used by Muren (1989), Cohat (1994), Martinez *et al.* (1994), as well as by Smith *et al.* (1991) for leek. We avoid 2,4–D because of its callusogenic influence.

Sucrose concentrations higher than the usual 3% for MS, BDS (Dunstan and Short, 1977) or B5 (Gamborg *et al.*, 1968) media have been found to be important for gynogenic induction in onion. Muren (1989) demonstrated that 10% sucrose was significantly superior to 5 and 15%. Furthermore 10% sucrose has been routinely used by Keller (1990a,b,c), Campion *et al.* (1992), Cohat (1994), and Martinez *et al.* (1994) whereas Smith *et al.* (1991) used 8% and Schum *et al.* (1993) used 6%.

Campion and Alloni (1990) emphasized the favourable influence of adenine sulphate in the culture medium for flower buds and ovules. Bohanec *et*

Table 2. List of the best media used for *in vitro* gynogenesis of onion

Reference	Medium
Part 1: Induction media	
Campion and Azzimonti (1988), Campion and Alloni (1990):	
Two step culture: first flowers, subsequently ovaries:	
– flowers on MS (Murashige and Skoog, 1962) + increased vitamin concentrations: (mg l^{-1}: thiamine-HCl 2, pyridoxine-HCl 1, nicotinic acid 1,C-pantothenate 1, folic acid 1,biotin 0.01), myo-inositol 500, amino acids: glycine 2, L-proline 200, $NaH_2PO_4{\cdot}H_2O$ 250, TIBA 0.2, adenine sulphate 0.3–0.5,	
– ovules on MS + vitamins and myo-inositol as for flowers, amino acids: glycine 2, glutamine 800, with/without L-proline 200, $NaH_2PO_4{\cdot}H_2P$ 250, NAA 1, 2iP 2, adenine sulphate 10.	
Campion *et al.* (1992):	
One step culture; either flowers or ovaries	
– flowers on BDS (Dunstan and Short, 1977) + vitamins as above, myo-inositol 100, 2,4-D 2, BAP 0,12,	
– ovaries on BDS + vitamins as above, myo-inositol 100, NAA1, BAP 0.12.	
Keller (1990a):	BDS + glutamine 500, IBA 2.03, BAP 1.25
Martinez *et al.* (1994):	MS + 2,4-D 2, BAP 2
Muren (1989),	
Cohat (1994),	
Martinez *et al.* (1994):	B5 (Gamborg *et al.*, 1968) + 2,4-D 2, BAP 2
Schum *et al.* (1993):	MS + NAA 1, 2iP 8
Smith *et al.* (1991):	B5 + 2,4-D 1, NAA 0.1
Part 2: Plant development media	
Campion and Azzimonti (1988), Campion *et al.* (1992):	
Reduced MS (1/2 macro-, 4/5 micro-salts) vitamins and myo-inositol as in the induction medium, $NaH_2PO_4{\cdot}H_2O$ 170, IAA 1.5, 2iP 2.	
Cohat (1994):	B5 + IBA 0.1
Keller (1990b):	BDS + IBA 2, BAP 0.12, GA_3 3.5
Muren (1989):	BDS + IBA 0.1
Schum *et al.* (1993):	MS + NAA 1, 2iP 8
Smith *et al.* (1991):	BDS

al. (1995), who followed the protocol of Campion and Alloni (1990), mentioned the favourable effect of 2 mg l^{-1} thidiazuron (TDZ) in the second induction medium on regeneration frequency.

4.3. *Light conditions and temperature*

The entire culture duration has been done under 16 h light (Muren, 1989; Campion and Alloni, 1990; Smith *et al.*, 1991; Campion *et al.*, 1992; Cohat, 1994; Martinez *et al.*, 1994) or in dark for the first four weeks (Keller, 1990a,b,c). Schum *et al.* (1993) used a dark preculture at 10 °C for up to 12 days but did not obtain consistent results for all genotypes. A heat pretreatment of 30 °C led to suppression of regeneration. No or even negative effects of cold pretreatment were reported by Muren (1989); our own results (unpublished) also did not show any clear influence. The temperature for the entire culture has been 25 °C (Muren, 1989), 27 °C for the first four weeks followed by 25 °C during the plant development phase (Keller, 1990a,b), or 27 °C in light and 24 °C in dark (Campion and Alloni, 1990).

5. Gynogenesis induced by irradiated pollen

Doré and Marie (1993) returned to one of the oldest techniques to obtain onion haploids, i.e., pollination with irradiated pollen (Blakeslee *et al.*, 1922). In their approach, no *in vitro* culture was used for embryo rescue. The female parents (male sterile) were crossed with fertile pollen that had been irradiated by a Co_{60} source giving 150 Gy. Seeds that formed after such pollination were sown either *ex vitro* (with very limited success) or *in vitro* with subsequent transfer to soil. A few haploid and mixoploid as well as numerous diploid plantlets were obtained from these seeds. Rates of gynogenic plantlets approached 14.4% in one experiment. In these studies, the use of male sterility for procedural reasons and additionally as a marker in conjunction with other marker characters (skin colour and bulb shape) enabled the authors to select the true gynogenic material visually among other products of pollination. Predominantly diploid plantlets, which may have arisen by spontaneous diploidization, occurred. According to the authors, any possible apomictic origin of the mother-like plants could be excluded. However, homozygosity of such gynogenic diploids was not proven. Nevertheless, the technique presented by Doré and Marie (1993) is relatively simple and offers, alone or in combination with embryo rescue, a chance to broaden the spectrum of usable genotypes for DH breeding if some of its limitations (e.g., use of male sterile material) could be overcome.

6. Haploidy, dihaploidy and apomixis in other *Allium* species

Analogous to reduction of the tetraploid genome to the dihaploid level by means of anther culture in potato (Kohlenbach and Geier, 1972; Sopory and Rogan, 1976), gynogenesis can be used for dihaploid production in tetraploid *Allium* species. Voorrips (1991) was unable to obtain dihaploid plantlets in ovule and ovary cultures of leek, *A. porrum*. He reported, however, the development of two octoploid plantlets from ovary cultures. Smith *et al.* (1991) obtained one dihaploid plant from 6,024 cultivated flower buds. Schum *et al.* (1993) were also successful in ovary culture, obtaining eight dihaploid regenerants (corresponding to 0.16% of cultured ovaries) which is a first step towards a usable system in leek.

Ovule culture is applicable to rescue naturally occurring dihaploid embryos in Chinese chives (*A. ramosum* and *A. tuberosum*; Kojima and Kawaguchi, 1988). From 908 unpollinated ovules, 132 tetraploids and two dihaploids were obtained through preculture of buds, followed by ovule and embryo culture. The occasional dihaploid indicates that some ovules have a normally

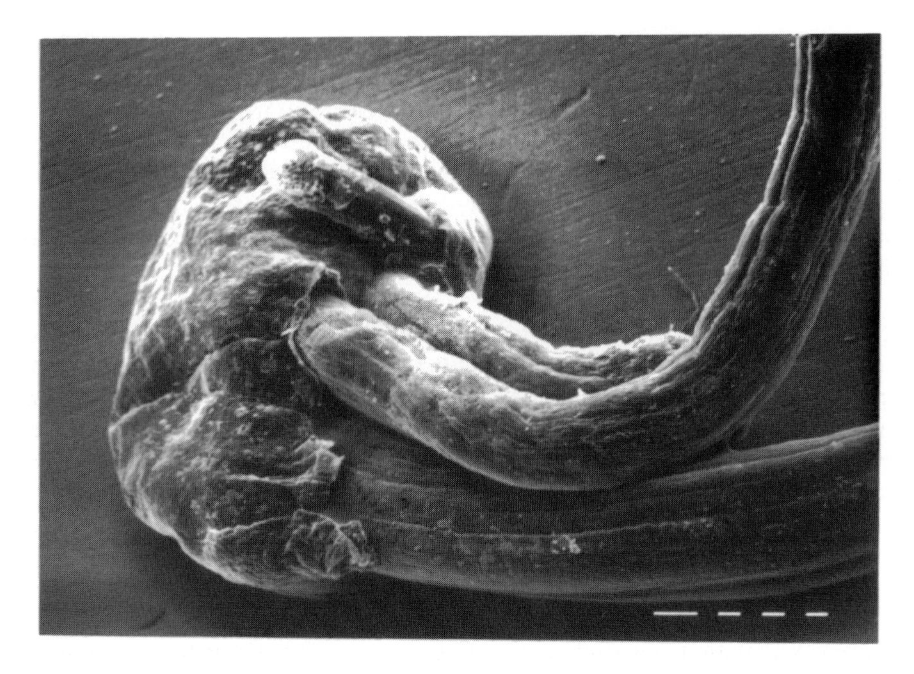

Figure 5. SEM picture showing twin embryos emerging from both ends of an ovule of *A. tuberosum*. Magnification: the left bar is 0.2 mm.

developed (eusporous) embryo sac. One of two dihaploids ($2n = 2x = 16$) had exactly half the tetraploid number of chromosomes. It bolted and flowered 12 months after potting, but its anthers shrivelled before perianth opening and contained no viable pollen. The apomictic development of tetraploid and dihaploid embryos has been reinvestigated in detail by Kojima *et al.* (1991), thus confirming the results of Modilewski (1930) and Hakansson (1951). They used electrophoretic analysis of leaf esterase (EST). The degree of apomixis was greater than 90%.

We have obtained similar results (Keller, 1992b). Twin embryos can be observed to grow from the same ovule (Fig. 5). Dihaploid plants are generally much less vigorous than diploids. In *A. ramosum*, only 1 of 66 regenerants (representing 6.3% regeneration) was a dihaploid/tetraploid chimera; all others were tetraploid. In *A. tuberosum*, 0.25% dihaploid regenerants occurred whereas the percentage of tetraploids was 5.6. The dihaploids were distinctly smaller and entirely sterile. Kojima and Nagato (1992) raised three possibilities for the lower frequency of dihaploid compared to tetraploid regenerants: 1) a lower capacity for parthenogenesis in reduced cells (as discussed by Modilewski, 1930), 2) a retarded cell cycle in the reduced egg cells compared to the tetraploid embryonic structures, and 3) dihaploid embryos might be inferior to tetraploid ones in competitive ability.

Tian and Yang (1989) reported that, in unpollinated ovary cultures of *A. tuberosum*, the egg and antipodal cells were induced to form mature embryos which regenerated into dihaploid plants. Nearly 50% of cultured ovules contained parthenogenetic and/or apogametic proembryos.

Twin embryos, which occurred at a percentage of 0.04–0.32, were investigated in *A. fistulosum* (Turkov *et al.*, 1973). A combination of diploid and haploid twins was found. For *A. ramosum/tuberosum*, which have been formerly described as *A. odorum* L., investigations on apomixis and polyembryony were published by Hegelmaier (1897), Tretjakow (1895), Haberlandt (1923), and others. In analyses of *A. odorum* grown under natural conditions, two levels of ploidy including dihaploidy have been found in polyembryonic complexes of the same embryo sac as early as 1930 by Modilewski. Furthermore, apomixis has been described in *Nothoscordum borbonicum* Kunth (the former *N. fragrans* [Vent.] Kunth) (Hakansson, 1953), *A. albidum* Fisch. ex Bieb., *A. nutans* L., *A. scorodoprasum* L. ssp. *rotundum* (L.) Stearn (the former *A. rotundum* L.), *A. sativum*, *A. schoenoprasum* L., *A. zebdanense* Boiss. et Noe, (Hakansson, 1951; Gvaladze, 1963, 1970) as well as in *A. fistulosum* (here haploid embryos were found; Borisenko *et al.*, 1971). The studies of Kojima and Kawaguchi (1988) and our own results demonstrated that polyembryony has some potential for the derivation of haploids and dihaploids in the more important crops of the genus *Allium*.

Garlic, *A. sativum*, is known to be predominantly seed sterile (Tagaki, 1990). Flower initials generally do not develop to the stage at which haploid gametes would be expected to appear. Thus, chromosomal instability of callus cultures may be more important as a source of haploid cells in this species. Novák (1974) and Novák and Havránek (1974) described callus strains of garlic in which hypoploidy occurred. The callus was induced from leaf segments. True haploid cells accumulated up to 2.03 and 7.75% of the total cell population in two of three lines investigated within 339 days after start of the culture. Plant regeneration from these callus cultures was not reported. In callus of *A. sativum*, 4.1% haploid cells were found among 45.9% diploid, 2.7% triploid, 14.9% tetraploid, 12.3% aneuploid and 20% highly polyploid cells (Maggioni and Marchesi, 1984).

Seo and Kim (1988) found hypodiploid ("hypoamphihaploid") cells in callus cultures of heterozygous shallot ($2n = 16$, but termed "amphihaploid" in this publication), the percentage of which tended to decrease with time. Hypodiploid regenerants were unable to form normal plants. Roy (1980) found dihaploid cells in callus of *A. tuberosum*, but only cells with higher ploidy in *A. cepa*. Haploid cells were found in callus of *A. cepa* (Yamane, 1975), *A. fistulosum* (Yamane, 1983) and *A. schoenoprasum* (Yamane, 1979).

Apart from the problem of chimeric development, a considerable drawback of haploid plants from callus cultures resides in the general genetic instability of callus cultures (somaclonal variation). Thus, undesirable variation may occur in homozygous plants obtained from doubling of callus-derived haploids, rendering them useless.

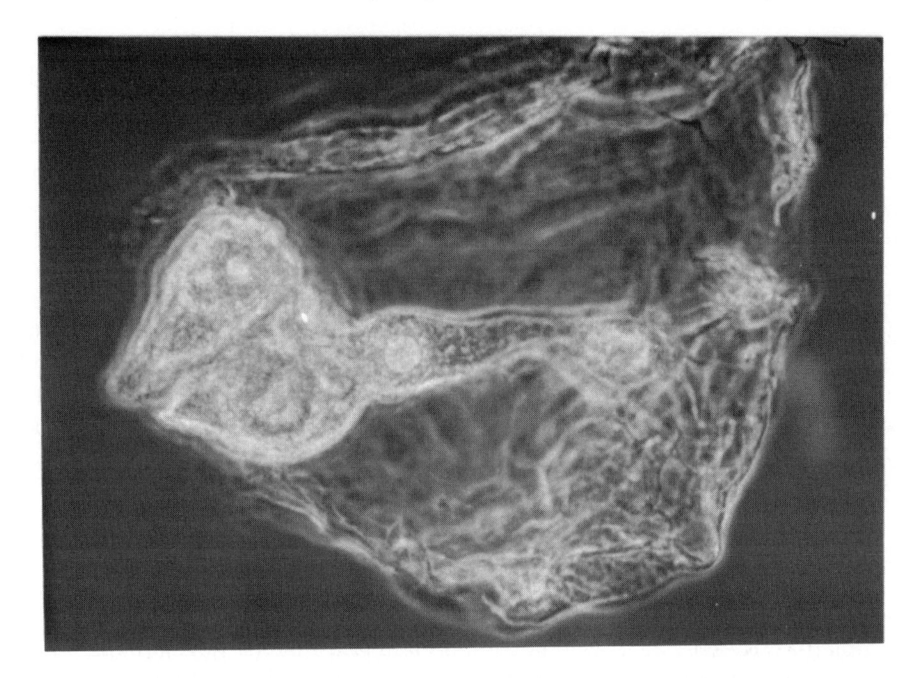

Figure 6. Embryo sac of *A. ericetorum*, isolated by means of enzymatic treatments. The egg cell and one synergid are visible as well as two polar nuclei. Antipodal cells have already degenerated. This is typical also for *A. cepa*. Phase contrast. Length of the embryo sac: 300 μm.

7. Embryo sac isolation as a tool for analysis and culture

Isolation of embryo sacs can be useful both for the study of gynogenesis and as promising targets for biotechnology and genetic engineering (Rangan 1982; Theunis *et al.* 1991). The technique of embryo sac isolation needs a special pretreatment (i.e., enzymatic maceration) followed by delicate and time-consuming manipulations. Using enzymatic maceration of ovules in a mixture of 3% pectinase and 3% cellulase "Onozuka R-10", we have isolated embryo sacs of *A. cepa*, *A. ericetorum* Thore, *A. hookeri*, *A. porrum*, *A. sativum*, *A. senescens* L., *A. sphaerocephalon* L., *A. triquetrum* L., and *A. tuberosum* (Fig. 6). Our purpose was to analyze the organization and development of embryo sac elements.

8. Developmental pathways of gynogenesis

Because the embryo sacs of higher plants are embedded in protective somatic tissues, it is difficult to elucidate gynogenic pathways. Several possibilities have been discussed for various plant species, e.g., embryogenesis of the synergids for rice (Zhou *et al.*, 1986) and of the egg cell for barley and sunflower (Yang *et al.*, 1986). All cells within the embryo sac are potential candidates for embryonic development. Because of additional possibilities to induce embryonic development of the integuments or other somatically derived parts of the ovule, morphogenesis must be investigated for each species separately. Whereas embryogenesis has been reported to occur in nearly all parts of the ovule (integuments, egg apparatus, antipodals) in *A. ramosum* and *A. tuberosum* (the former *A. odorum*), microscopic observations in the main crop species of *Allium* have been limited. Antipodal cells do not persist in many species of *Allium* (Porter, 1936; Glushtchenko, 1957). Thus, in onion, they can be excluded as potential sources for gynogenesis. The occurrence of both twin seeds and heterozygous diploid gynogenic regenerants in onion as well as in many other species of *Allium* and the related genus *Nothoscordum* implies that nucellar embryos, genetically identical to the maternal diploid plant, can develop.

Certain difficulties arise during the processes of haploidization or spontaneous diploidization that result in plants unfit for utilization in breeding programs. Many gynogenic embryos have been found to degenerate immediately after emerging from the carpels (Keller 1990c, 1992b). In onion cv. Stuttgarter Riesen, 33% of regenerants obtained in 1989 degenerated with the following symptoms: irregular swelling, especially in the hypocotyl, and vitrification (hyperhydricity; Fig. 7). Such embryos have been found to contain giant cells with enlarged nuclei (Fig. 8). This phenomenon was only observed at first appearance of the plantlet, not in later stages. Bohanec *et al.* (1995) also found poor vigour and hyperhydricity among about one third of gynogenic onion embryos. Muren (1989) mentioned 25% abnormal plants, with less than 1% albinos. Schum *et al.* (1993) also mentioned recovery of chimera with albino sectors.

Campion and Azzimonti (1988) and Campion and Alloni (1990) analyzed seven gynogenically derived onion clones, three diploid, three haploid, and one tetraploid. After two months of growth, one haploid clone showed haploid-diploid mixoploidy in the roots. For the three initially haploid clones, they found the following ploidy distributions (percentages $n/n + 2n/2n$) at first analysis: 100/0/0; 2 months later: 83/17/0; 5 months later: 23/27/50. The shift towards diploidy was taken as evidence of spontaneous diploidization. These data were obtained by counting chromosomes in mitoses of root tip cells. Cytophotometric measurements have shown that ploidy distributions can differ in other parts of the plant from what would be expected by examination of root tips. Therefore, a conclusion about spontaneous diploidization must be regarded cautiously.

Genetic instability in culture can result in spontaneous chromosome doubling, possibly due to the growth regulators in the induction medium. Bohanec *et al.* (1995) concluded from their flow cytometric investigations that "at least some regenerants have the potential to double spontaneously, while others remain in the haploid stage". In our conditions of slow growth storage, where

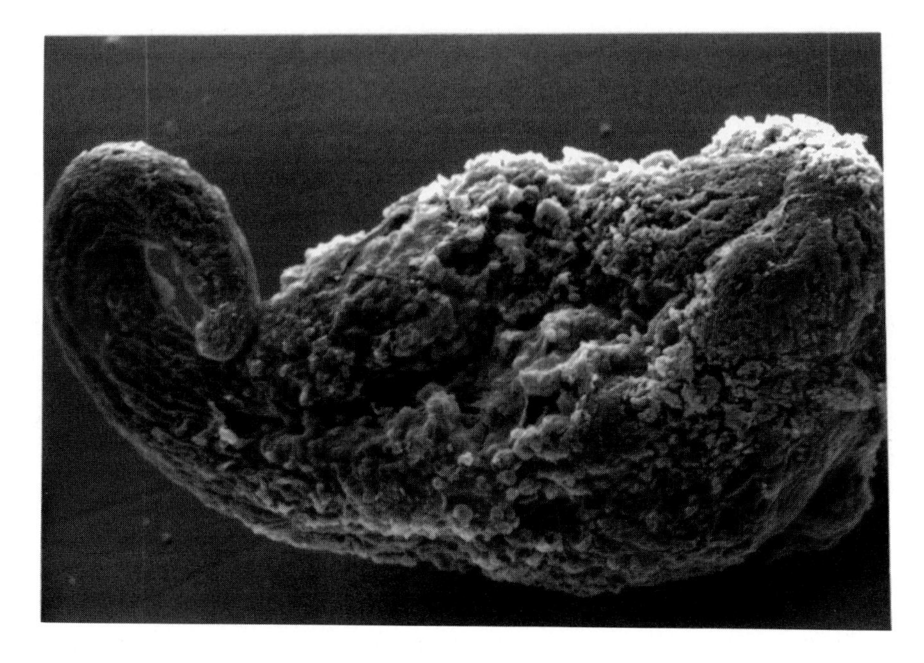

Figure 7. SEM picture of a degenerating gynogenic regenerant of *A. cepa*. Hypocotyl is swollen and shows callus-like disorganization of the superficial cell-layer. Total length of the gynogenic embryo: 4 mm.

growth regulators in the culture medium have been avoided, no chromosome doubling has yet been observed by flow cytometric analyses.

Among gynogenic regenerants from two onion genotypes, Cohat (1994) found several diploid regenerants which differed morphologically from each other and from the mother plant. This could be, indeed, a symptom of spontaneous diploidization. Better characterization is available by using isozyme tests for homozygosity. We investigated the gynogenic offspring derived from different donor plants of onion by means of isozyme tests for MDH (malate dehydrogenase), PGI (phosphogluco-isomerase), PGM (phosphoglucomutase), and GAL (galactosidase) isozymes (Table 3). Of 16 plants, nine were heterozygous for at least one enzyme system (Fig. 9). Thus, 56.2% of the diploid regenerants, i.e., the heterozygous plants, could not have developed via spontaneous diploidization of haploid cultures. On the other hand, we routinely observed the developmental pathway of gynogenic embryos under a dissection microscope. The origin of regenerants from the wall or other parts of the ovary can thus be excluded.

Bohanec *et al.* (1995) used isozyme variation at the α-esterase locus to analyze diploid gynogenic regenerants of onion. Approximately half the regenerants exhibited doubled haploid origin, whereas the others were of maternal type. Provided that somatic regenerants could be excluded, 50%

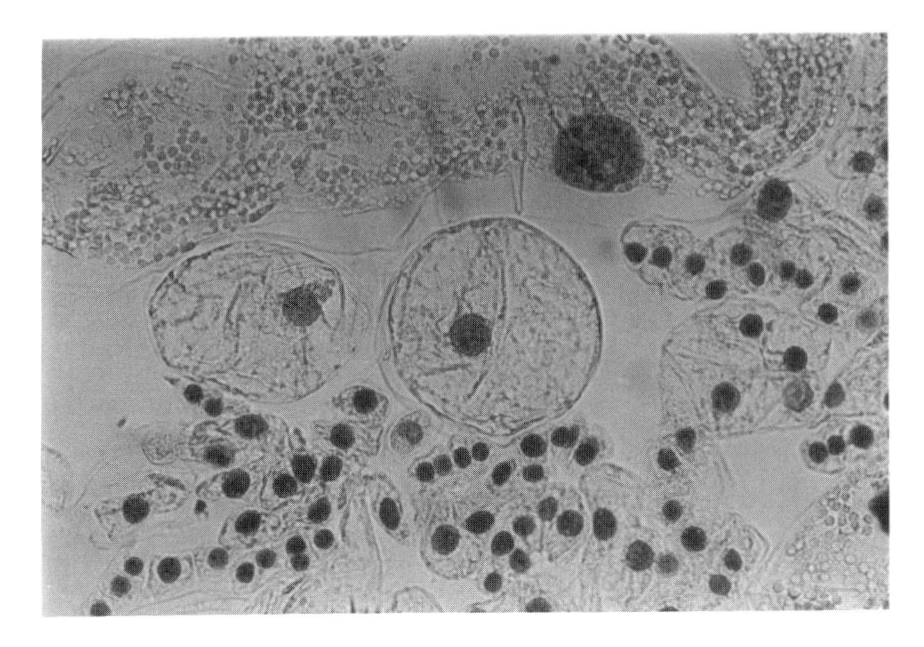

Figure 8. Cells of the basal region of a degenerating gynogenic regenerant. Several ploidy levels are present. In the cluster consisting of smaller cells, haploid and diploid nuclei are mixed. Three tetraploid and one giant nucleus are visible. Length of the giant nucleus: 40 μm.

of the regenerants could not have been from spontaneous diploidization but rather from unreduced gametes or diploid apomixis.

Varshanina (1973) analyzed polyembryony in five cultivars of onion, three of which exceeded 0.1%, with the highest rate amounting to 0.25%; 20% of the twin seedlings were triploid. In a later study, Varshanina (1977) found polyembryony in 64% of 25 onion cultivars, with a mean frequency of 0.06%.

9. Production of diploid homozygous (DH) lines from haploids

For breeding purposes, haploid plants represent only an intermediate stage towards homozygous diploids. Therefore, a diploidization procedure is necessary if no spontaneous diploids occur. *In vitro* polyploidization using colchicine has been obtained in *A. sativum* (Maryakhina *et al.*, 1981; Novák, 1983), *A. cepa* (Maryakhina *et al.*, 1983; Stöldt, 1994), and *A. porrum* (Stöldt, 1994). *In vitro* cultures of onion shoot tips (Novák, 1983; Stöldt, 1994), bulb scale bases prior to formation of adventitious shoots (Maryakhina *et al.*, 1983; Maryakhina *et al.*, 1985; Stöldt, 1994), inflorescences (Maryakhina *et al.*, 1983), and entire basal bulblets (Maryakhina *et al.*, 1990) have been used as targets of colchicine doubling. Lower concentrations (0.05 to 0.3%) of

Table 3. Comparison of homo- and heterozygosity in different regenerants measured by means of isoenzyme patterns

Enzyme Regenerant Cultivar/ donor pl. number/ regenerant number	MDH	PGI	PGM	GAL	Conclusion	Number with + enzymes
CG/4/1	+	–	?	+	+	2
CG/4/7	–	–	–	–	–	0
CG/4/8	–	–	–	–	–	0
CG/4/12	+	–	+	+	+	3
CG/4/19	+	–	–	+	+	2
CG/4/23	+	–	+	+	+	3
SR/5/44	–	–	–	–	–	0
SR/5/48	+	+	+	+	+	4
SR/18/8	–	–	–	–	–	0
SR/27/7	–	–	–	–	–	0
SR/38/2	–	–	–	–	–	0
W/2/5	–	+	–	?	+	1
Z/6/1	–	–	–	?	–	0
Z/6/3	–	+	–	?	+	1
ZG/11/2	–	+	–	?	+	1
ZG/11/5	–	+	–	+	+	2
Sum (+% of 16)	31.2	31.2	18.8	37.5	56.2	

Abbreviations: – homozygous, + heterozygous.
Enzymes: MDH: malate dehydrogenase, PGI: phosphogluco-isomerase, PGM: phosphoglucomutase, GAL: galactosidase; Cultivars: CG: Calbenser Gerlinde, SR: Stuttgarter Riesen, W: Wolska, Z: Zerti, ZG: Zittauer Gelbe.

colchicine can generally be used for *in vitro* cultures compared with *ex vitro* treatments (e.g., 0.4%; Tanaka and Nakata, 1969). Genotypes are variously susceptible to the potential toxicity of colchicine. For leek, Stöldt (1994) found an upper limit for application of *in vitro* colchicine treatments of 0.15% for approximately 15–20 days whereas optimal concentrations for chromosome doubling ranged from 0.05 to 0.1%. In onion, 0.05% colchicine for 10 days appears to be sufficient or occasionally too high.

Colchicine treatments result in a spectrum of changes from no effect to repeated doubling cycles accompanied by chimeric development. Stöldt (1994) obtained about 15–35% diploidization in leek and 18–45% in onion. Chimerism causes complications in subsequent cultivation, because only a part of the plant is dihaploid. *In vitro* culture can be used to induce spontaneous separation of chimeric tissues (Stöldt, 1994). Alternatively, chimeric plants can be grown to maturity. It can be expected that segments of the inflorescence contain diploid tissues that will result in normal flowers capable of self-pollination producing homozygous dihaploid seed.

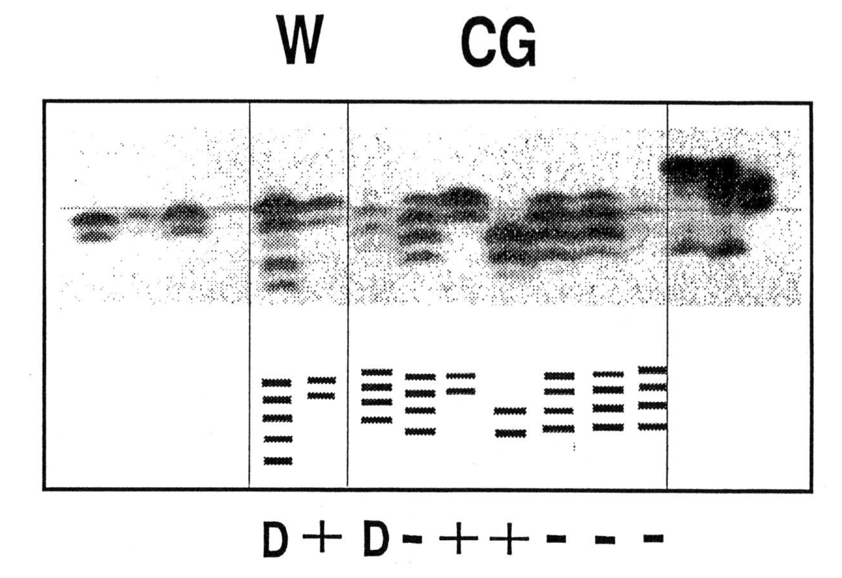

Figure 9. Detail of an isozyme pattern of MDH (malate dehydrogenase) showing heterozygosity in diploid gynogenic regenerants. W – *A. cepa* cv. Wolska; CG – *A. cepa* cv. Calbenser Gerlinde; D – donor plant; – – heterozygous diploid regenerant; + – homozygous diploid regenerant (Keller and Eickmeyer, unpublished).

10. Identification of haploids and doubled haploid plants

Several accounts of haploid ($1n = 1x = 8$) metaphases in root tip squashes of gynogenic onion that unequivocally demonstrate the haploid state have been published (Campion and Azzimonti, 1988; Muren, 1989; Campion and Alloni, 1990; Keller, 1990a; Martinez *et al.*, 1994). However, the presence of countable mitoses is limited to the root and shoot meristems. Such counts provide no information about the ploidy of differentiated tissues. Furthermore, the technique is slow and tedious. Therefore, indirect methods of ploidy estimation have been adopted.

Counting of guard cell chloroplasts has been used successfully for other plants (Mochizuki and Sueoka, 1955; Butterfaß, 1959); however, this method is not feasible in *Allium* because of the small size of the chloroplasts (Allaway and Setterfield, 1972). However, measurements of the size of nuclei or entire cells may be suitable for ploidy determination in plants (Doré, 1986). The size of guard cells and their nuclei was also useful for discrimination between haploid and diploid onion plants (Fig. 10) and even between different tissue sectors in chimera (Fig. 11; Keller, 1990c).

Furthermore, we have also performed ploidy determination by means of scanning and flow cytometry (Keller, 1990b; Keller *et al.*, 1994). For scanning cytometry, epidermal strips were used, in which the guard cells represent a

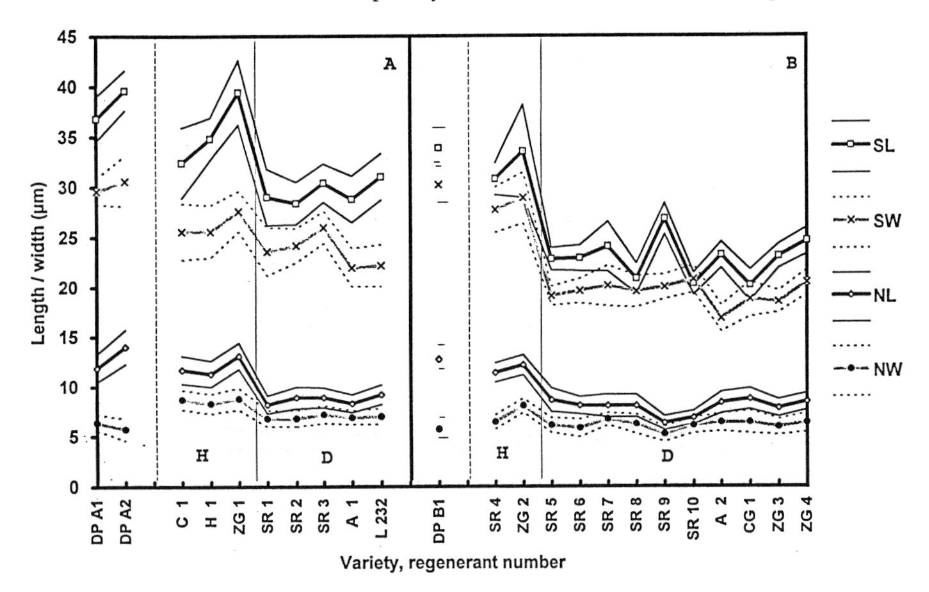

Figure 10. Stoma and guard cell nucleus lengths and widths in different gynogenic plants. DP – diploid donor plants (cv. Stuttgarter Riesen); H – haploid regenerants; D – diploid reg.; SL – stoma lengths; SW – stoma widths; NL – nucleus lengths; NW – nucleus widths; Cultivars: A – Ambros; C – Carmen; CG – Calbenser Gerlinde; H – Hylight; L 232 – breeder's line; SR – Stuttgarter Riesen; ZG – Zittauer Gelbe.

homogeneous population of differentiated cells with nuclei in the G_O phase. Nuclei are stained by the procedure in which Schiff's reagent binds quantitatively to the DNA (Feulgen and Rossenbeck, 1924). Thus, extinction measurements quantify the amount of DNA per cell. We measured 100 nuclei of a reference plant, the ploidy of which had been confirmed by chromosome counts. For routine screening of 312 regenerants, we found that 20 nuclei per preparation were sufficient for the derivation of clear histograms. It was, however, necessary to check several epidermal preparations from different parts of the plant in order to observe chimera. Fig. 11 demonstrates clear differences among haploid, diploid and chimeric preparations.

Flow cytometry, in which absorbance of fluorescent stains that bind quantitatively to DNA in intact nuclei is measured by passing single nuclei in microdrops past a laser, is more efficient for ploidy determinations. Large populations of nuclei representing all the cell types isolated from freshly chopped tissue that has been passed through a mesh are measurable. In diploid onion plants, a pattern of three peaks corresponding to the 2C, 4C, and 8C DNA levels, was observed. Haploid plants exhibited a similar pattern, except that the peaks were shifted to the left, representing instead the 1C, 2C, and 4C DNA contents (Keller *et al.*, 1995). The stability of four gyno-

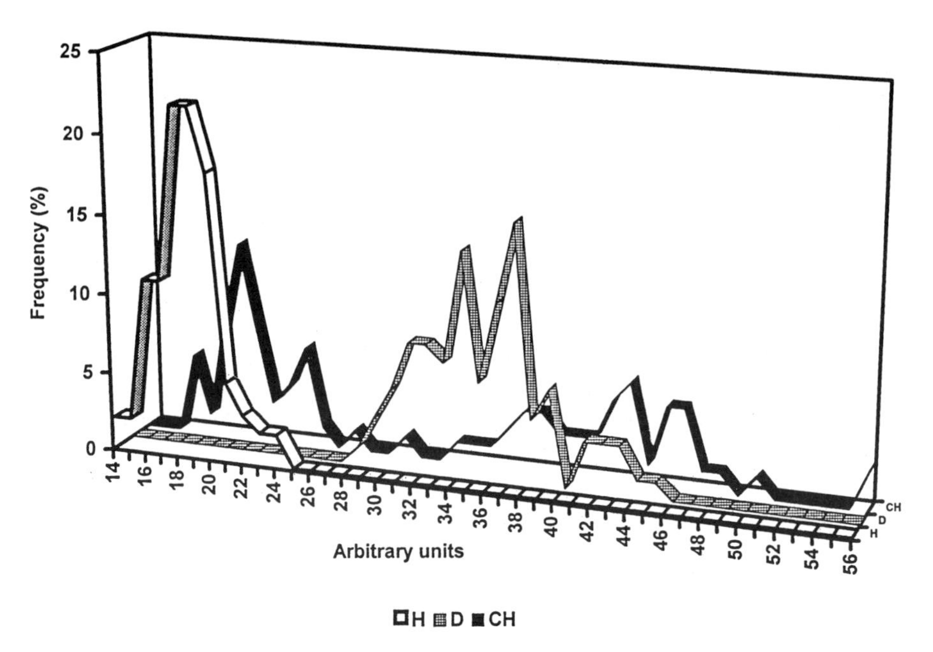

Figure 11. Histograms showing the distribution of quantity of DNA in the nuclei measured as arbitrary units in scanning cytometry. Note the distinct differences between the haploid (H) and diploid (D) distributions as well as the two peak pattern of the chimera (CH). For each plantlet, 100 nuclei were measured.

genic haploids has been checked under slow growth conditions in an *in vitro* gene bank for 5 years.

In scanning cytometrical observations of epidermal explants, a homogenous population of guard cell nuclei can be measured, thus characterizing one tissue layer. Flow cytometric measurements can be made on any part of the plant. Results about mixoploidy must be treated carefully because some cells of higher ploidy (e.g., diploid cells in haploid regenerants) are expected due to the common occurrence of endopolyploidization (D'Amato, 1952). Mixoploid chimeric plants should show a clear change in the peak distribution towards the higher C-values.

11. Utilization of haploids and doubled haploids in onion breeding

Because of the limited availability of clearly defined DH lines, no experience has yet been accumulated about the use of haploids and doubled haploids in breeding of onion or other *Allium* crops. The main expectation, the remarkable acceleration in production of homozygous lines, has not yet been fulfilled. But Campion (1994) promised that "the protocol for the production of DH lines in onion is nearly ready to be utilized for breeding purposes".

The strong genotypic effect on gynogenesis has been cited as a major obstacle for application of haploids in onion breeding (Campion, 1994). If the genetic base for gynogenic ability is simple, this character could be included in onion breeding strategies. In theory, a positive effect of haploidy (or subsequent homozygosity) is that lethal genes, which have been hidden by the heterozygous status of *Allium*, will act directly (Smith *et al.*, 1991). Thus, only vigorous genotypes may survive gynogenesis. It is possible that the abnormal and degenerating embryos observed among the gynogenic regenerants are related to the expression of such lethal genes.

The potential of haploid breeding in *Allium* has justified the considerable effort that has already been expended on development of haploid breeding methods. Rapid inbreeding may result in the development of male sterile lines that could be used for hybrid seed production. Whether DH lines exhibit the same level of inbreeding depression as the corresponding self-pollinated plants remains to be determined. Reduction of tetraploid leek to the dihaploid level may enable breeders to introduce male sterility from onion into leek via interspecific crosses (Schum *et al.*, 1993).

Progress in biotechnology may allow the use of gynogenesis for breeding purposes in a wide range of plant species, including onion and allied crops. Large chromosomes, which have made onion a cytological model, a high degree of genetic diversity, and the practical importance of this genus favour further investigation into haploid induction in cultivated species of the genus *Allium*.

12. Acknowledgements

One of the authors (L. Korzun – enzymatic embryo sac isolation) has been supported by grant no. 435 WER 17/2/94 of the Deutsche Forschungsgemeinschaft (DFG).

13. References

Allaway, W.G. and G. Setterfield, 1972. Ultrastructural observations on guard cells of *Vicia faba* and *Allium porrum*. Can. J. Bot. 50: 1405–1413.

Blakeslee, A.F., J. Belling, M.E. Farnham and A.D. Bergner, 1922. A haploid mutant in the Jimson weed, *Datura stramonium*. Science 55: 646–647.

Bohanec, B., M. Jakše, A. Ihan and B. Javornik, 1995. Studies of gynogenesis in onion (*Allium cepa* L.): induction procedures and genetic analysis of regenerants. Plant Sci. 104: 215–224.

Borisenko, V.A., V.D. Turkov and S.Y. Kraevoi, 1971. The use of induced parthenogenesis and polyembryony in the production of haploid forms of vegetable plants (Russ). In: V.S. Moshaeva and S.Y. Kraevoi (Eds.), Prakticheskie Zadachi Genetiki v Sel'skom Xoziajstve, pp. 141–148. Nauka, Moscow.

Butterfaß, T.H., 1959. Ploidie und Chloroplastenzahlen. Ber. Deutsch. Bot. Ges. 72: 440–451.
Campion, B., 1994. Production of doubled haploid lines of onion (*Allium cepa* L.) through *in vitro* gynogenesis. In: Abstr. VIIIth Intl. Congr. Plant Tissue Cell Cult., Florence, June 12–17, 1994, p. 92.
Campion, B. and C. Alloni, 1990. Induction of haploid plants in onion (*Allium cepa* L.) by *in vitro* culture of unpollinated ovules. Plant Cell Tiss. Org. Cult. 20: 1–6.
Campion, B. and M.T. Azzimonti, 1988. Evolution of ploidy level in haploid plants of onion (*Allium cepa*) obtained through *in vitro* gynogenesis. In: 4th EUCARPIA *Allium* Symposium, Wellesbourne, UK, September 6–9, 1988, pp. 85–89.
Campion, B., M.T. Azzimonti, E. Vicini, M. Schiavi and A. Falavigna, 1992. Advances in haploid plant induction in onion (*Allium cepa* L.) through *in vitro* gynogenesis. Plant Sci. 86: 97–104.
Campion, B., A. Falavigna, G.P. Soressi and M. Schiavi, 1984. Efforts for *in vitro* androgenesis in onion (*Allium cepa* L.). In: 3rd EUCARPIA *Allium* Symp. Wageningen, September 4–6, 1984, p. 110.
Cohat, J., 1994. Obtention chez l'échalote (*Allium cepa* L. var. *aggregatum*) de plantes gynogénétiques par culture *in vitro* de boutons floraux. Agronomie 14: 299–304.
Currah, L. and F.J. Proctor, 1990. Onions in tropical regions. Natural Resources Institute Bull, No. 35, Kent.
D'Amato, F., 1952. Endopolyploidy in differentiated plant tissues. Caryologia 4: 115–117.
Doré, C., 1986. Évaluation du niveau de ploidie des plantes d'une population de choux de Bruxelles (*Brassica oleracea* L. ssp. *gemmifera*) d'origine pollinique. Agronomie 6: 797–801.
Doré, C. and F. Marie, 1993. Production of gynogenic plants of onion (*Allium cepa* L.) after crossing with irradiated pollen. Plant Breed. 111: 142–147.
Dowker, B.D., 1990. Onion breeding. In: H.D. Rabinowitch and J.L. Brewster (Eds.), Onions and Allied Crops, Vol I, pp. 216–232. C.R.C. Press, Boca Raton.
Dunstan, D.I. and K.C. Short, 1977. Improved growth of tissue cultures of the onion, *Allium cepa*. Physiol. Plant. 41: 70–72.
Feulgen, R. and H. Rossenbeck, 1924. Mikroskopisch-chemischer Nachweis einer Nucleinsäure vom Typus der Thymonucleinsäure und die darauf beruhende elektive Färbung von Zellkernen in mikroskopischen Präparaten. Z. Physiol. Chemie 135: 203–248.
Flier, W., 1990. Onderzoek naar de ontwikkeling van in *vitro* haploidisatietechnieken in prei, *Allium porrum* L. Internat. Agrar. Highschool Larenstein. Report of a working stay in CPRO, Wageningen.
Gamborg, O.L., R.A. Miller and K. Ojima, 1968. Nutrient requirements of suspension cultured soybean root cells. Expt. Cell. Res. 50: 157–158.
Glushtchenko, G.I., 1957. Materialy po zito-embriologii vida *Allium cepa* L. Izv. Akad. Nauk SSSR 22: 220–233.
Guha, S. and B.M. Johri, 1966. *In vitro* development of ovary and ovule of *Allium cepa* L. Phytomorphology 16: 353–364.
Guha, S. and S.C. Maheshwari, 1964. *In vitro* production of embryos from anthers of *Datura*. Nature (Lond.) 204: 497.
Gvaladze, G.E., 1963. K isutcheniju poliembrionii v rode *Allium* L. Soobschtschenija Akad. Nauk Gruz. SSR 31: 393–398.
Gvaladze, G.E., 1970. Formy apomiksisa v rode *Allium*. In: Apomixis i Selekcija, pp. 155–160. Nauka, Moscow.
Haberlandt, G., 1923. Zur Kenntnis der Polyembryonie von *Allium odorum* L. Ber. Deutsch. Bot. Ges. 41: 174–179.
Hakansson, A., 1951. Parthenogenesis in *Allium*. Botaniska Notiser, Lund. 1951: 143–179.
Hakansson, A., 1953. Die Samenbildung bei *Nothoscordum fragrans*. Botaniska Notiser, Lund. 1953: 129–139.
Hanelt, P., 1990. Taxonomy, evolution, and history. In: H.D. Rabinowitch and J. L. Brewster (Eds.), Onions and Allied Crops, Vol. I, pp. 1–26. C.R.C. Press, Boca Raton.

Hegelmaier, F., 1897. Zur Kenntnis der Polyembryonie von *Allium odorum* L. Bot. Ztg. 55: 133–139.

Kanne, J., 1991. Onderzoek naar de ontwikkeling van in *vitro* haploidisatietechnieken in prei, *Allium porrum* L. Internat. Agrar. Highschool Larenstein. Report of a working stay in CPRO, Wageningen.

Keller, J., 1990a. Results of anther and ovule culture in some species and hybrids in the genus *Allium* L. Arch. Züchtungsforsch. 20: 189–197.

Keller, J., 1990b. Culture of unpollinated ovules, ovaries, and flower buds in some species of the genus *Allium* and haploid induction via gynogenesis in onion (*Allium cepa* L.). Euphytica 47: 241–247.

Keller, J., 1990c. Haploids from unpollinated ovaries of *Allium cepa* – single plant screening, haploid determination, and long term storage. In: H.J.J. Nijkamp, L.H.W. van der Plas and J. van Aartrijk (Eds.), Progress in Plant Cellular and Molecular Biology. Proc. VII. Intl. Congr. Plant Tissue and Cell Cult. Amsterdam, June 24–29, 1990, pp. 275–279. Kluwer Academic Publishers, Dordrecht.

Keller, J., 1992a. Einige Aspekte zum Einsatz von Saccharose in der pflanzlichen *In vitro*-Kultur aus der Sicht angewandter Forschung. In: I. Dahse (Ed.), Vom Organismus zum Molekül. Physiologische Prozesse, pp. 243–252. Friedrich-Schiller-Universität, Jena.

Keller, J., 1992b. *In vitro* cultivation of *Allium* species – a method for application in plant breeding and germplasm conservation. In: P. Hanelt, K. Hammer and H. Knüpffer (Eds.), The Genus *Allium* – Taxonomic Problems and Genetic Resources. Proc. Intl. Symp. Gatersleben, June 11–13, 1991, pp. 137–152. Inst. Genet. Crop Plant Res., Gatersleben.

Keller, E.R.J., D.-E. Lesemann, H.I. Maaß, A. Meister, H. Lux and I. Schubert, 1995. Maintenance of an *in vitro* collection of *Allium* in the Gatersleben genebank – problems and use. In: M. Terzi, R. Cella and A. Falavigna (Eds.), Current Issues in Plant Molecular and Cellular Biology. Proc. VIIIth Intl. Congr. Plant Tissue Cell Cult., Florence, June 12–17, 1994, pp. 347–352. Kluwer Academic Publishers, Dordrecht.

Kohlenbach, H.W. and T. Geier, 1972. Embryonen aus *in vitro* kultivierten Antheren von *Datura meteloides* Dun., *Datura wrightii* Regel und *Solanum tuberosum* L. Z. Pflanzenphysiol. 67: 161–165.

Kojima, A. and T. Kawaguchi, 1988. Dihaploids derived from unpollinated ovule culture of Chinese chives. In: 4th EUCARPIA *Allium* Symp. September 6–9, 1988, Wellesbourne, UK, pp. 212–217.

Kojima, A. and Y. Nagato, 1992. Pseudogamous embryogenesis and the degree of parthenogenesis in *Allium tuberosum*. Sex. Plant Reprod. 5: 79–85.

Kojima, A., Y. Nagato and K. Hinata, 1991. Degree of apomixis in Chinese chive (*Allium tuberosum*) estimated by esterase isozyme analysis. Jpn. J. Breed. 41: 73–83.

Maggioni, L. and G. Marchesi, 1984. Plantule di aglio (*Allium sativum* L.) ottenute *in vitro*. Sementi Elette 30: 29–31.

Martinez, L., C. Agüero and C. Galmarini, 1994. Obtention of haploid plants by ovaries and ovules culture in onion (*Allium cepa* L.). In: Proc. 1st Int. Symp. Edible Alliaceae, Mendoza, Argentina (in press).

Maryakhina, I.Y., L.I. Moskovkin, V.P. Tugolukov, E.I. Tugolukova and I.V. Polumordvinova, 1985. *In vitro* garlic polyploidization as method to obtain the initial material for selection (Russ.). Sel'skokhozjaistvennaja Biologija 10: 49–52.

Maryakhina, I.Y., I.V. Polumordvinova and N.M. Kozlova, 1983. Clonal propagation of onion plants and formation of polyploid forms (Russ.). Sel'skokhozjaistvennaja Biologija 6: 16–21.

Maryakhina, I.Y., V.P. Tugolukov, E.I. Tugolukova, I.V. Polumordvinova, M.D. Kalizhenkova and V.V. Merzlyakova, 1990. Morphogenetic features of the vegetative progeny of polyploid garlic obtained by means of colchicine treatment *in vitro* (Russ.). Sel'skokhozjaistvennaja Biologija 25: 144–151.

Mochizuki, A. and N. Sueoka, N. 1955. Genetic studies on the number of plastid in stomata. I. Effects of autopolyploidy in sugar beets. Cytologia 20: 358–366.

Modilewski, J., 1930. Neue Beiträge zur Polyembryonie von *Allium odorum*. Ber. Deutsch. Bot. Ges. 48: 285–294.

Murashige, T. and F. Skoog, 1962. A revised medium for rapid growth and bioassays with tobacco tissue cultures. Physiol. Plant. 15: 473–497.

Muren, R.C., 1989. Haploid plant induction from unpollinated ovaries in onion. HortScience 24: 833–834.

Novák, F.J., 1974. The changes of karyotype in callus cultures of *Allium sativum* L. Caryologia 27: 45–54.

Novák, F.J., 1983. Production of garlic (*Allium sativum* L.) tetraploids in shoot-tip *in vitro* culture. Z. Pflanzenzüchtg. 91: 329–333.

Novák, F.J. and P. Havránek, 1974. A cytological study on tissue cultures of *Allium sativum* L. *in vitro*. Acta Fac. Rer. Nat. Univ. Comenianae - Genetica 5: 143–147.

Phillips, C. and K.J. Luteyn, 1983. Effects of picloram and other auxins on onion tissue cultures. J. Amer. Soc. Hort. Sci. 108: 948–953.

Porter, T.R., 1936. Development of the megagametophyte and embryo of *Allium mutabile*. Bot. Gaz. 98: 317–327.

Rangan, T.S., 1982. Ovary, ovule, and nucellus culture. In: B.M. Johri (Ed.), Experimental Embryology of Vascular Plants, pp. 105–129. Springer-Verlag, Berlin/Heidelberg/New York.

Roy, S.C., 1980. Chromosomal variations in the callus tissues of *Allium tuberosum* and *A. cepa*. Protoplasma 102: 171–176.

Roy, S. 1989. Differentiation and anther culture of *Allium cepa*. In: Intl. Symp. Hortic. Germplasm, Cultivated and Wild. Part II. Vegetables, 1988, pp. 267–272. Acad. Publ., Bejing.

San Noeum, L.H., 1976. Haploïdes d'*Hordeum vulgare* L. par culture *in vitro*. Ann. Amélior. Plant. 26: 751–754.

Schum A., L. Mattiesch, E.-M. Timmann and K. Hofmann, 1993. Regeneration of dihaploids via gynogenesis in *Allium porrum* L. Gartenbauwissenschaft 58: 227–232.

Seo, B.-B. and H.-H Kim, 1988. Regeneration of amphidiploid plants from tissue cultures of *Allium wakegi*. Plant Cell Rep. 7: 297–300.

Smith, B.M., R.M. Godwin, E. Harvey and C.P. Werner, 1991. Gynogenesis from whole flower buds in bulb onions (*Allium cepa* L.) and leeks (*Allium porrum* L.). J. Genet. Breed. 45: 353–358.

Sopory, S.K. and P.G. Rogan, 1976. Induction of pollen divisions and embryoid formation in anther cultures of some dihaploid clones of *Solanum tuberosum*. Z. Pflanzenphysiol. 80: 77–80.

Stöldt, A., 1994. Untersuchungen zur *In-vitro*-Polyploidisierung von *Allium* sp. Thesis M.Sc., Univ. Hamburg, Hamburg.

Suh, S.K. and H.G. Park, 1986. Studies on the anther culture of garlic (*Allium sativum* L.). I. Callus formation and plant regeneration. J. Korean Soc. Hort. Sci. 27: 89–95.

Symko, S., 1969. Haploid barley from crosses of *Hordeum bulbosum* (2*x*) × *H. vulgare* (2*x*). Can. J. Genet. Cytol. 11: 602–609.

Tagaki, K., 1990. Garlic *Allium sativum* L. In: H.D. Rabinowitch and J.L. Brewster (Eds.), Onions and Allied Crops, Vol. III, pp. 109–146. C.R.C. Press, Boca Raton.

Tanaka, M. and K. Nakata, 1969. Tobacco plants obtained by anther culture and the experiment to get diploid seeds from haploids. Jpn. J. Genet. 44: 47–54.

Theunis, C.H., E.S. Pierson and M. Cresti, 1991. Isolation of male and female gametes in higher plants. Sex. Plant Reprod. 4: 145–154.

Tian, H.Q. and H.Y. Yang, 1989. Haploid embryogeny and plant regeneration in unpollinated ovary culture of *Allium tuberosum*. Acta Biol. Exp. Sinica 22: 139–147.

Tretjakow, S., 1895. Die Beteiligung der Antipoden in Fällen der Polyembryonie bei *Allium odorum*. Ber. Deutsch. Bot. Ges. 13: 13–17.

Turkov, V.D., V.A. Nushikyan and N.S. Drozdova, 1973. Spontaneous mutations in the caryotypes of vegetable crops. Doklady Vsesoyuznoy Akademii Sel'skokhozyaistvennych Nauk Imeni Lenina 12: 15–16.

Varshanina, T.P., 1973. Polyembryony in solanaceous vegetables and in onion. Byulleten' Vsesoyuznogo Ordena Lenina Instituta Rastenievodstva Imeni N.I. Vavilova 33: 90–98.

Varshanina, T.P., 1977. The phenomenon of polyembryony in solanaceous vegetable crops and in onion. Trudy po Prikladnoi Botanike, Genetike i Selektsii 60: 128–133.

Voorrips, R.E., 1991. Unpollinated ovule and ovary culture in leek (*Allium porrum* L.). Allium Improvement Newsl. 1: 75–76.

Walkey, D.G.A., 1990. Virus diseases. In: H.D. Rabinowitch and J.L. Brewster (Eds.), Onions and Allied Crops, Vol. II, pp. 191–212. C.R.C. Press, Boca Raton.

Yamane, Y., 1975. Chromosomal variation in calluses induced in *Vicia faba* and *Allium cepa*. Jpn. J. Genet. 50: 353–355.

Yamane, Y., 1979. Regenerated plants derived from tissue culture of *Allium fistulosum*. Jpn. J. Genet. 54: 475.

Yamane, Y. 1983. Induced differentiation of chives from calluses *in vitro*. Jpn. J. Genet. 58: 698.

Yang, H.Y., C. Zhou, D.T. Cai, H. Yan, Y. Wu and X.M. Chen, 1986. *In vitro* culture of unfertilized ovules in *Helianthus annuus*. In: H. Hu and H.Y. Yang (Eds.), Haploids in Higher Plants *In Vitro*, pp. 182–191. China Acad. Publ., Bejing/Springer-Verlag, Berlin/Heidelberg/New York/Tokyo.

Zhou, C., H.Y. Yang, H.Q. Tian, Z.L. Liu and H. Yan, 1986. *In vitro* culture of unpollinated ovaries in *Oryza sativa*. In: H. Hu and H.Y. Yang (Eds.), Haploids in Higher Plants *In Vitro*, pp. 165–181. China Acad. Publ., Bejing/Springer-Verlag, Berlin/Heidelberg/New York/Tokyo.

5. Anther and microspore culture in *Capsicum*

F. REGNER

Contents

1. Introduction

Pepper is an economically important vegetable and spice crop. Its high nutritional value, especially in vitamin A and C content, has increased fresh market demand for the sweet pepper (*Capsicum annuum* L.) in recent years. *Capsicum frutescens* L. remains valuable for the spice industry, especially in Spain, Hungary, and the United States. *C. annuum* was domesticated in Mexico circa 7000 BC (Eshbough *et al.*, 1983). During the Colombian period, *Capsicum* was introduced in Europe and eventually cultivated as a crop for various purposes. Taxonomic classification of the genus *Capsicum*, family Solanaceae, has relied primarily on morphological data. Many forms of wild *Capsicum* complicate its classification (Morrison *et al.*, 1986). Three main groups have been distinguished: one purple- and two white-flowered classes. The economically important *C. annuum*, *C. frutescens*, *C. chinense* and *C. baccatum* (Pickersgill *et al.*, 1979) are all white-flowered species. Few species exist in both cultivated and wild forms, necessitating interspecific hybridization for introduction of desirable traits for breeding purposes. There is a major crossing barrier between purple- and white-flowered species (Zijlstra *et al.*, 1991). In most cases, the apparent cause for failure of the cross has been inhibition of pollen tube growth, effectively preventing fertilization.

Capsicum annuum ($2n = 2x = 24$) is an annual herbaceous plant that grows indeterminately with bushy main stems and pinnately veined leaves. The perfect flower is pentamerous, composed of a short, thick calyx, a corolla of 5–7 petals, and 5–7 stamens. The anthers from one flower bud contain more than 100,000 microspores. Cultivar development has traditionally been through pedigree or backcross breeding followed by self-pollination for several generations to develop true breeding lines (Greenleaf, 1986). Hybrid cultivars have been developed more recently and now comprise a large share of the market. The homozygous lines necessary for hybrid production can

S.M. Jain, S.K. Sopory & R.E. Veilleux (eds.), In Vitro *Haploid Production in Higher Plants, Vol. 3,* 77–89.

now be produced by anther culture to extract haploids ($n = 1x = 12$) followed by chromosome doubling to derive inbred lines that can be test-crossed for their ability to generate acceptable hybrids.

Another purpose of doubled haploid breeding for pepper improvement is the use of interspecific hybrids among white-flowered forms as anther donors for the introgression of desirable traits from related species into inbred lines following anther culture. This approach is attractive in pepper breeding using *C. chinense* as a source of resistance to TMV (Boukema, 1980), or *C. frutescens* as a source of tolerance to tobacco etch virus (Greenleaf, 1975) and resistance to nematodes *Meloidogyne* spp. (Vito and Saccardo, 1982). Dominant alleles have been identified for production of capsaicin, which is a major constituent of pungency in peppers, red mature fruit colour, and round fruit shape. Segregation of such traits among doubled haploid plants would enable the selection of homozygous lines with the most desirable horticultural characteristics but also expressing the greatest levels of resistance.

Plant regeneration from protoplasts and callus of pepper has not been very dependable. Regeneration procedures have been established from leaf discs using cotyledons and hypocotyls of young seedlings (Arroyo and Revilla, 1991; Ebida and Hu, 1993). Despite the development of a regeneration protocol, the genetic transformation of *C. annuum* is still hindered (Fari *et al.*, 1992). By comparison, plant regeneration from haploid or spontaneous diploid embryos following anther culture has been well-established allowing the utilization of microspores as a vehicle for gene transfer.

2. Haploid production by parthenogenesis

The first report on spontaneous haploid pepper from a twin embryo was published by Christensen and Bamford (1943). Morgan and Rappleye (1950) identified naturally occurring haploids as maternal types resulting from $2n + n$ pairs of embryos. The $2n$ partner indicated occurrence of hybridization whereas the haploid was derived from ovarian tissue. Pochard and Dumas de Vaulx (1971) have used haploid pepper from twin seedlings for breeding purposes. They demonstrated that haploidy occurs spontaneously at approximately O.1% in cultivated pepper but that cultivars and growing conditions may significantly affect this frequency (Pochard and Dumas de Vaulx, 1979). In an effort to increase the haploid frequency, Dumas de Vaulx and Pochard (1974) treated female flowers after pollination with nitrous oxide (N_2O) under pressure and generated haploids by disturbing the fertilization process.

Parthenogenetic haploids can be doubled using a brief colchicine treatment to derive autodiploids (Greenleaf, 1986). The performance of such autodiploids has been disappointing, however, in comparison with conventionally derived diploid lines. They have been characterized by sterility in 20–30% of the first selfed generation and instability for various morphological characters (Greenleaf, 1986). The cause of the instability is unknown – colchicine has been shown to be mutagenic so that the instability could have resulted from the doubling process. Otherwise, complete homozygosity may be undesirable, at least in the genetic material used to derive these particular haploids.

3. Haploid production by anther culture

The success of anther culture in other solanaceous species encouraged its application to *Capsicum*. Two research groups, Wang *et al.* (1973) in China and George and Narayanaswamy (1973) in India, first reported the production of haploid *Capsicum* plants by anther culture. Later, however, Novak (1974) obtained only haploid callus in anther culture of *Capsicum* without succeeding in regeneration of plants. Saccardo and Devreux (1974) regenerated the first haploids of European cultivars by androgenesis.

Haploids have been obtained from various cultivars of *C. annuum* (Dumas de Vaulx *et al.*, 1981) although genotypic differences in responsiveness among cultivars have been reported (Kristiansen and Andersen, 1993; Novak, 1974). Some genotypes, especially interspecific hybrids, have not responded to anther culture using the techniques developed for cultivars of *C. annuum* (Morrison *et al.*, 1986).

The conditions under which anther donor plants have been grown have influenced the success of pepper anther culture. Chambonnet (1988) warned against pesticide treatment before anther isolation. Kristiansen and Andersen (1993) found that the photoperiod under which anther donor plants were grown did not influence androgenic embryo production; however, they found a significant decline in anther culture response with increasing donor plant age. They also found that embryo formation was optimal when donor plants were grown at a minimum air temperature of 26°C. In our laboratory, haploid pepper was produced following the protocol of Chambonnet (1988). The mother-plants were cultivated by a standardized fertilization and irrigation procedure in the greenhouse or under field conditions. The frequency of anthers producing embryos exhibited a seasonal influence. This phenomenon was observed even when pepper donor plants were cultivated in a climate controlled greenhouse. In spring, when natural illumination and photoperiod were increasing, the frequency of responding anthers (cv. Wanas F_1) increased from 20% in March to 50% in May. During summer months, the rate of responding anthers remained more constant at 30%. Responsiveness varied widely among flower buds on a plant. Reliable data on treatment effects can only be obtained if anthers from a single flower bud are distributed randomly to treatments.

The size relationship between sepals and petals has been utilized as an indicator of microspore developmental stage; equal sepal and petal length has been taken as a morphological marker of the late uninucleate stage of microspore development most responsive to anther culture (Sibi *et al.*, 1979). However, this relationship may not hold for all genotypes or for the same genotype under different environmental conditions. Light blue pigmentation of the free end of the stamens has been suggested as a more precise marker (Chambonnet, 1988). Such anthers have been found to contain mostly uninucleate stage with somewhat more than 10% binucleate microspores. Increased embryo yield has been obtained by selection of anthers exhibiting this marker.

Initial success with pepper anther culture was through indirect androgenesis, i.e., first by inducing callus from cultured anthers followed by regeneration of plants from callus (Wang *et al.*, 1973) or proembryonic structures (George and Narayanaswamy, 1973). A qualitative improvement was made by producing direct androgenic embryos without an intermediate callus stage by Sibi *et al.* (1979). The protocol included a 48 h cold (4°C) pretreatment of the flower buds and incubation of anthers on two different media. The first medium (Cm) was supplemented with 2,4-D (9 μM), kinetin (9.3 μM), and B_{12} vitamins (0.03 mg l^{-1}) and used for the first 12 days of incubation. In the second medium (Rm), 2,4–D and B_{12} vitamins were omitted and kinetin was decreased to 0.1 mg l^{-1}. This media change is still used routinely in pepper anther culture.

Dumas de Vaulx *et al.* (1981) modified these methods by initiating sporophytic development of immature pollen at an elevated temperature. Embryo production was stimulated by incubating anthers for eight days on medium C at 35°C in the dark, then reducing the incubation temperature to 25°C and retaining the change to medium R at 12 days from the start of incubation. Heat treatment has also been applied successfully to *Brassica campestris* (Keller and Armstrong, 1979). The level of 2,4–D in the initiation medium could be minimized (0.02 μM) as a consequence of the heat treatment. By use of this modified protocol, the frequency of androgenic plant production increased from three to 40 plants per 100 cultured anthers for several pepper lines and hybrid cultivars.

A further improvement to the culture medium was made by addition of activated charcoal (Vagera and Havranek, 1983, 1985) for absorption of toxic substances. Phillips *et al.* (1984) obtained rooted pepper plants on a culture medium containing IAA and BA. Cold pretreatment stimulated sporophytic development to a lesser degree than heat shock treatment (Dumas de Vaulx *et al.*, 1981) or a combination of both (Morrison *et al.*, 1986). As a single treatment, heat shock (35°C) stimulated the highest yield of embryos and haploid plants. Nevertheless, cold pretreatment was required to induce sporophytic development of microspores (see microspore culture section). Morrison *et al.* (1986) also demonstrated that an increase in the number of embryos per anther was dependent on the length of cold (4°C) treatment and genotype. Munyon *et al.* (1989) succeeded in stimulating androgenesis under continuous light at a temperature of 29°C. Matsubara *et al.* (1992) obtained embryos on MS medium containing various concentrations of 2,4–D and kinetin. They confirmed that the 35°C heat treatment was essential for better androgenesis.

In their efforts at resistance breeding in pepper, Morrison *et al.* (1986) derived androgenic haploids from an interspecific *C. annuum* × *C. chinense* hybrid. In their protocol, they used all of the previously reported factors for stimulating pepper androgenesis, i.e., four day cold pretreatment of flower buds, heat shock (35°C) treatment for eight days, and addition of charcoal to the medium. Because they failed to obtain androgenic plants using these methods, they developed a bilayer system. The bilayer medium consisted of an agarose-solidified (1%) layer of R medium according to Sibi *et al.* (1979), except by modification of FeEDTA according to Murashige and Skoog (1962) and adjustment to 6% sucrose, onto which R medium without charcoal or agarose was poured. Embryos that developed initially on the same C medium used by Sibi *et al.* (1979) could only be regenerated on the two-phase R medium. They obtained only diploid plants derived from anther culture, suggesting that the culture method induced genome duplication.

In our experience with anther culture following cold pretreatment to the buds, most embryos did not convert into plantlets but instead developed callus at the base of the shoot and eventually deteriorated. Nevertheless, 10% of the embryos were partly green and developed rapidly into plantlets on MS basal medium without growth regulators (Fig. 1). By excising the callus from the base of embryos that did not convert, and placing the shoot tip on MS basal medium, 7% could additionally be converted into plantlets.

The rate of spontaneous diploidisation in embryos or plants is influenced by culture treatment. A combination of cold and heat treatments favors the formation of diploid embryos from microspores (Morrison *et al.*, 1986). Additionally, diploid embryos are more vigorous and resulting plants grow faster (Matsubara *et al.*, 1992). Triploidy resulting from anther culture was reported by Sibi *et al.* (1979).

4. Utilization of dihaploid pepper in breeding

The main application of DH pepper breeding is to derive horticulturally valuable inbred lines incorporating resistance to biotic factors from various sources. Pochard and Dumas de Vaulx (1971) first reported the use of haploids in pepper breeding. The haploids were obtained parthenogenetically from a twin seedling of a "Yolo Wonder" cross and were treated with colchicine to obtain DH lines. Selected progeny of this cross were compared to the DH lines. Several of the DH plants were similar in yield and vigour compared to hybrid plants although seed yield was partly reduced.

In a later study, Dumas de Vaulx and Pochard (1986) derived more than two hundred doubled haploid plants from a hybrid between TMV susceptible and TMV resistant lines. However, most of the DHs lacked resistance. Daubeze *et al.* (1990) attempted to derive androgenic pepper lines resistant

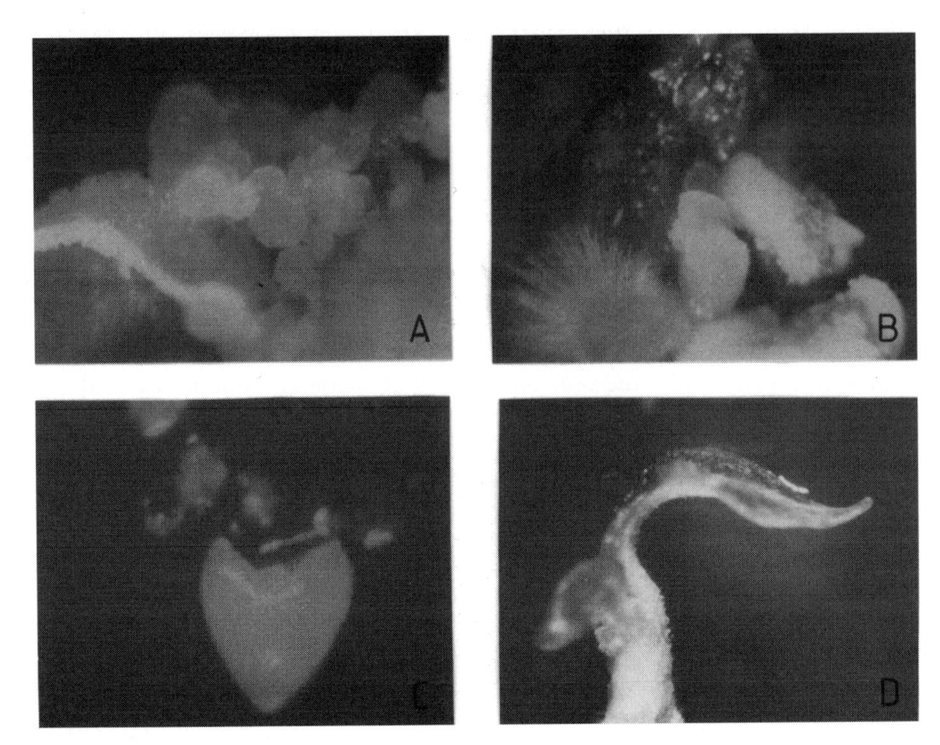

Figure 1. Anther culture. Burst anther with globular embryos on R1 medium (A, B), heart shaped embryo removed to MS O (C) developing plantlet derived from an embryo (D).

to *Phytophthora capsici* and TMV. The Mexican pepper line "CM 334" was crossed to "Yolo Wonder" and served as a source of resistance. They obtained heat-stable (32°C) resistance among DH lines in a 1:1 ratio. The resistance could be attributed to a single allele at the *L* locus.

Chen (1984) reported considerable variation among anther-derived DH lines. However, within a line, plants were uniform and stable. In a continued study of the combining ability of DH lines, Chen (1985) obtained early fruiting hybrids with high yield. Jiang and Li (1984) studied DH lines, which resulted from anther culture of a hybrid between sweet and hot pepper. They observed the main fruit characters of these DH lines and of the resulting five generations. Yield of DH plants was not higher than the hybrid. Morrison (1987) attributed both beneficial and deleterious gametoclonal variation among DH lines to callus formation during the regeneration process. Hendy *et al.* (1985) studied the inheritance of resistance to nematodes (*Meloidogyne* spp.) in two "Yolo Wonder" inbred lines. These lines were used to establish androgenic haploids for characterizing the resistance genes.

In addition to the potential of inbred lines of pepper derived from haploids for production of commercially useful F_1 hybrids, they are also important for genetic studies. Homozygosity improves genetic analyses by eliminating the confounding effects of allelic segregation from heterozygous parents in test crosses. Lefebvre *et al.* (1992) analysed segregation of markers in DH lines and found skewed segregation ratios ranging from 15 to 45% instead of the expected 50%. The amount of segregation distortion depended on the location of a particular locus within the genome. Androgenic DHs therefore are not a random sample of gametes. Moreover, the population derived by androgenesis of a hybrid does not necessarily express all genes that were in the hybrid. Use of molecular markers in conjunction with homozygous lines derived by anther culture will assist in increasing our genetic knowledge of pepper.

5. Microspore culture

Regeneration of pepper plants from isolated microspores has not been previously reported. Nevertheless, the applications of microspore culture are numerous. Microspores or microspore-derived embryos could function as a target for gene transfer by particle bombardment. This may be especially valuable because of the recalcitrance of pepper in attempts at regeneration from transformed leaf discs (Liu *et al.*, 1990). Microspores can also be used as a source of plant tissue for *in vitro* selection of mutants, for studies of embryo development, and for haploid breeding despite the well-established anther culture technology. An advantage of microspore culture over anther culture is that embryo formation from heterozygous somatic tissue is circumvented in microspore culture. Microspore culture of *Brassica* spp. has been extremely efficient with the production of thousands of haploid embryos from microspores from a single bud (Pechan and Keller, 1988; Kott *et al.*, 1988; Duijs *et al.*, 1992). We report now on attempts to develop a microspore culture system for pepper in our laboratory.

The donor plants require similar care as for anther culture. The best indication for a large fraction of late uninucleate microspores is again a light blue pigmentation at the apical end of the stamens. Isolation can be performed either from a single bud or microspores can be bulked from many buds from plants grown in a greenhouse or controlled environment. Material from the field has been unsuitable for microspore culture due to high contamination, variation in climatic conditions, and the effects of pesticides.

Flower buds were surface-sterilized with 2% NaOCl for 10 min, 96% ethanol for 1 min, followed by several washes with sterile water. Microspores were extracted by gently crushing the buds in a mortar and pestle. The microspores were suspended in culture medium, transferred to a filter (mesh size 40 μm) using a Pasteur pipette, and washed three times by centrifugation at 800 rpm for 3 min in the same culture medium used for plating. Alternatively, microspores from single buds could be obtained by squashing the anthers in a drop of medium on a microscope slide, again followed by filtering and washing. The microspores were counted in a haemocytometer, sedimented by centrifugation at 800 rpm for 5 min, and resuspended in culture medium. In liquid culture, microspores were cultured at a density of 4×10^4 per ml. On 2,5 ml solid Rm medium, 1×10^5 microspores were cultured per 3 cm Petri dish. Viability of the microspores was estimated under a Leitz epifluorescence microscope (excitation filter 450–490 nm, barrier filter 520 nm) after staining with FDA (Heslop-Harrison and Heslop-Harrison, 1970). The developmental stage was determined by observation of nuclei that had been stained with DAPI stain (Fig. 2) (Coleman and Goff, 1985). The viability of microspores ranged from 0–90%, even among flower buds from the same plant. Best response was obtained by culturing populations of microspores exhibiting 20–50% binucleate stage. Such microspores remained viable longer than those at earlier or later stages.

6. Liquid culture

Culture in liquid NLN medium (Polsoni *et al.*, 1987) as used for *Brassica* microspores or in liquid CP and R media (Chambonnet, 1988) failed to produce embryos or embryo-like structures. Adjusting the sucrose content (4, 6, 9, 13%) of the medium did not improve embryogenesis. Development up to four nuclei (two nuclear divisions) was observed when microspores were cultured on a medium containing 4% sucrose. However, the viability was reduced to 20% by the second day of culture and growth of the microspores stopped at a four- or eight-nucleate stage despite regular medium changes.

Microspores were given a 48 h temperature treatment at 4°C, 25°C or 35°C in the dark before incubation at 25°C in the dark. The frequency of microspores with more than two nuclei was recorded two and six days after this pretreatment (Table 1). Development of the microspores after the cold induction treatment was delayed. No apparent effect of the induction temperature treatments on sporophytic development of *Capsicum* could be observed.

7. Bilayer culture

Isolated microspores were also cultured on a bilayer system consisting of 2.5 ml agar-solidified (0,7%) medium (NLN or CP with 6% sucrose) and 1.5 ml liquid medium per 3 cm Petri plate. The liquid phase varied in different trials by supplementation of NLN or CP medium with 6, 9, or 13% sucrose. Fresh liquid medium (0.5 ml) was added weekly without removing the microspores. Even under this cultivation method, microspore development stopped at the four- or eight-nucleate stage. All six media used in the liquid layer gave the same result. None of the cultures contained developing

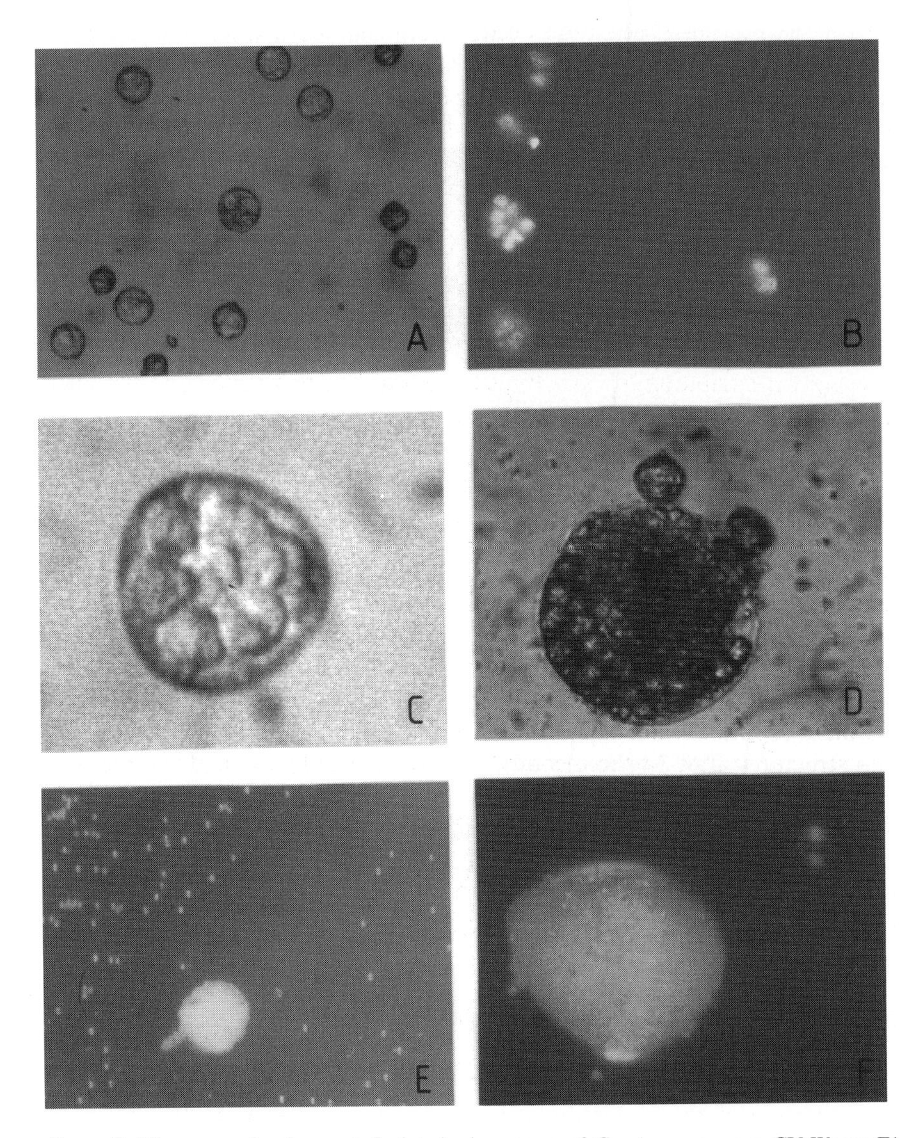

Figure 2. Microspore development. Isolated microspores of *Capsicum annuum*. CV Wanas F1 in liquid culture 2 days after plating (A), microspores 6 days after isolation, DAPI stained nuclei at the UV microscope (B), sporophytical microspore with eight nuclei (C), proembryo with suspensor (D, E), globular embryo on solid SPA agarose medium.

microspores with more than eight nuclei within a four week incubation period. During further incubation, no progress was observed and viability of the cultures decreased. Therefore, a new strategy was followed.

Preculturing of anthers before microspore isolation has been used success-

Table 1. Microspore development under different induction temperatures. The percentage of sporophytically divided microspores after a two day induction at three temperatures and two or six days of incubation at 25°C

Temperature	4°C	25°C	35°C
Day 4	21%	30%	42%
Day 8	62%	39%	45%

Table 2. Microspore development of precultured anthers. Number of embryo like structures as a function of incubation period and cultivation system

	Preculture treatment (days)		
	2	4	6
Bilayer culture system	–	7	2
Solid medium (agarose)	1	6	3

fully with cereals (Datta and Wenzel, 1987). The stimulatory influence of the preculture was attributed to the nurse-function of the anthers (Kohler and Wenzel, 1985). Following this approach, we cultured anthers in the same media as for anther culture. Preculturing was performed under the induction conditions at 35°C following the methods of Chambonnet (1988). Using precultured anthers for pepper microspore isolation, we observed embryo-like structures after 3–4 weeks of incubation in liquid CP medium with 6% sucrose (Table 2). Despite the use of a preculture treatment, full embryogenesis failed. In one experiment, cocultivation with ovaries was attempted as used successfully for wheat microspores (Mejza *et al.*, 1993). Ovaries produced a large amount of callus but did not improve microspore embryogenesis.

8. Solid culture medium

Cultivation of isolated microspores on solid medium was performed using two different methods. Firstly, sedimented microspores were suspended in 200 μl liquid culture medium (CP with 6% sucrose) and distributed on 2.5 ml solid CP in a 3 cm Petri plate covered with a nylon mesh (size < 10 μm). This thermostable mesh (SST Thal-AG, Nytal) permitted the transfer of microspores from induction to R1 incubation medium. Sporophytic development was observed but the structures were irregularly shaped, turned brown, and later deteriorated.

Alternatively, microspores were applied to a single solid medium for the full incubation time. The liquid medium for the purification and suspension of the microspores was again CP with 6% sucrose. The solid medium was R1 with a higher concentration of cytokinin (kinetin = 4.2 μM). Induction and incubation conditions were the same as for anther culture (Chambonnet, 1988). On agar medium, embryo-like structures appeared after two months of incubation. On a medium solidified with 0,7% sea plaque agarose, globular embryos were produced. On transfer to MS basal medium, no further development occurred. On transfer to fresh R1 medium, callus growth inhibited further embryogenesis.

9. Conclusion

Anther culture of *Capsicum* is now sufficiently reliable such that the technique can be applied routinely in a breeding program. Haploid peppers produced by this method in various laboratories have been used to establish homozygous lines and for studies concerning the inheritance of valuable traits. Our attempts at microspore culture resulted in embryos only after preculturing anthers for a few days; these microspore-derived embryos, however, could not be regenerated into plants under routine conditions. More research is required to develop an appropriate method for culturing and converting microspore-derived embryos. Such cultures offer the potential for using immature pollen as a target for gene transfer.

10. Acknowledgements

I am thankful to Mag. Schabner and A. Stadlhuber for their assistance while preparing this manuscript and to K. Zitta for reproducing photos from my slides.

11. References

Arroyo, R. and M.A. Revilla, 1991. *In vitro* plant regeneration from cotyledon and hypocotyl segments in two bell pepper cultivars. Plant Cell Rep. 10: 414–416.

Boukema, I.W., 1980. Allelism of genes controlling resistance to TMV in *Capsicum*. Euphytica 29: 433–439.

Chambonnet, D., 1988. Bulletin Interne de la Station d'Amélioration des Plantes Maraîchères d'Avignon Montfavet.

Chen, X.S., 1984. Genetic expression of major characters in sweet pepper lines derived by anther culture. Acta Hort. Sin. 11: 113–118.

Chen, X.S., 1985. Determination of combining ability and analysis of heterosis in pollen lines of *Capsicum annuum* var. *grossum* Sendt. Acta Hort. Sin. 12: 267–272.

Christensen, H.M. and R. Bamford, 1943. Haploids in twin seedlings of pepper *Capsicum annuum* L. J. Hered. 34: 99–104.

Coleman, A.W. and L.J. Goff, 1985. Applications of fluorochromes to pollen biology. I. Mithramycin and 4′6–diamidino-2–phenylindole (DAPI) as vital stains and for quantitation of nuclear DNA. Stain Technol. 60: 145–154.

Datta, S. and G. Wenzel, 1987. Isolated microspore derived plant formation via embryogenesis in *Triticum aestivum* L. Plant Sci. 48: 49–54.

Daubeze, A.M., A. Palloix and E. Pochard, 1990. Resistance of androgenetic autodiploid lines of pepper to *Phytophthora capsici* and tobacco mosaic virus under high temperature. Capsicum Newsl. 8–9: 47–48.

Daubeze, A.M., 1988. Utilisation de lignése haploides doubles issues d'androgenése par l'étude de l'expression de la résistance aux maladies. Mémoire de DESU, Acad. Montpellier, USTL.

Duijs, J., R. Voorrips, D. Visser and J. Custers, 1992. Microspore culture is successful in most crop types of *Brassica oleracea*. Euphytica 60: 45–55.

Dumas de Vaulx, R. and E. Pochard, 1974. Essai d'induction de la parthénogénèse haplöide par action du protoxyde d'azote sur les fleurs de piment (*Capsicum annuum* L.). Ann. Amélior. Plantes 24: 283–306.

Dumas de Vaulx, R., D. Chambonnet and E. Pochard, 1981. Culture *in vitro* d'anthères de piment (*Capsicum annuum* L.): amélioration des taux d'obtention de plantes chez différents génotypes par des traitments à +35°C. Agronomie 1: 859–864.

Dumas de Vaulx, R. and E. Pochard, 1986. Parthogénèse et androgénèse chez le piment. Role actuel dans les programmes de selection. Le Selectionneur Français 36: 3–16.

Ebida, A.I.A. and C.Y. Hu, 1993. *In vitro* morphogenetic responses and plant regeneration from pepper (*Capsicum annuum* L. cv. Early California Wonder) seedling explants. Plant Cell Rep. 13: 107–110.

Eshbaugh, W.H., S.I. Gutmann and M.J. McLeod, 1983. The origin and evolution of domesticated *Capsicum* species. J. Ethnobiol. 3: 49–54.

Fari, M., A. Szasz, J. Mityko, I. Nagy, M. Csany and M. Andrasfalvy, 1992. Induced organogenesis via the seedling decapitation method (SDM) in three solanaceous vegetable species. In: VIII EUCARPIA meeting on Genetics and Breeding of *Capsicum* and Eggplant, Sept. 1992, Rome, pp. 243–248.

George, L. and S. Narayanaswamy, 1973. Haploid *Capsicum* through experimental androgenesis. Protoplasma 78: 467–470.

Greenleaf, W.H., 1975. The tabasco story. HortScience 10: 98.

Greenleaf, W.H., 1986. Pepper breeding. In: M.J. Bassett (Ed.), Breeding Vegetable Crops, pp. 67–134. Avi Publishing Co., Inc., Westport, CT.

Hendy, H., E. Pochard and A. Dalmasso, 1985. Transmission de la resistance aux nématodes *Meloidogyne chitwood* (Tylenchida) portée par 2 lignées de *Capsicum annuum* L. Étude des descendances homozygotes issues d'androgenése. Agronomie 5: 93–100.

Heslop-Harrison, J. and H. Heslop-Harrison, 1970. Evaluation of pollen viability by enzymatically induced fluorescence; intracellular hydrolysis of fluorescein diacetate. Stain Technol. 45: 115–120.

Jiang, Z.R. and C.L. Li, 1984. Observations and experiments on later generations of sweet × hot pepper derived by anther culture. Acta Hort. Sin. 11: 191–194.

Keller, W.A. and K.G. Armstrong, 1979. Stimulation of embryogenesis and haploid production in *Brassica campestris*. Theor. Appl. Genet. 55: 65–67.

Kohler, F. and G. Wenzel, 1985. Regeneration of isolated barley microspores in conditioned media and trials to characterize the responsible factor. Plant Physiol. 121: 181–191.

Kott, L., L. Polsoni, B. Ellis and W. Beversdorf, 1988. Autotoxicity in isolated microspore cultures of *Brassica napus*. Can. J. Bot. 66: 1665–1670.

Kristiansen, K. and S.B. Andersen, 1993. Effects of donor plant temperature, photoperiod, and age on anther culture response of *Capsicum annuum* L. Euphytica 67: 105–109.

Lefebvre, V., T. Prevost and A. Palloix, 1992. Segregation of molecular markers in doubled haploid progenies of pepper. In: VIII EUCARPIA Meeting on Genetics and Breeding of *Capsicum* and Eggplant, Sept. 1992, Rome, pp. 232–235.

Liu, W., W. Parrott, D. Hildebrand, G. Collins and E. Williams, 1990. *Agrobacterium* induced gall formation in bell pepper (*Capsicum annuum* L.) and formation of shoot like structures expressing introduced genes. Plant Cell Rep. 9: 360–364.

Matsubara, S., K. Hu and K. Murakami, 1992. Embryoid and callus formation from pollen grains of eggplant and pepper by anther culture. J. Jpn. Soc. Hort. Sci. 61: 69–77.
Mejza, S., V. Morgant, D. DiBona and J. Wong, 1993. Plant regeneration from isolated microspores of *Triticum aestivum*. Plant Cell Rep. 12: 149–153.
Morgan, D.T. and R.O. Rappleye, 1950. Twin and triplet pepper seedlings. A study of polyembryony in *Capsicum frutescens*. J. Hered. 41: 91–95.
Morrison, R., R. Koning and D. Evans, 1986. Pepper. In: D.A. Evans, W.R. Sharp and P.V. Ammirato (Eds.), Handbook of Plant Cell Culture, Vol. 4, pp. 552–573. Macmillan, New York.
Morrison, R.A., R.E. Koning and D.A. Evans, 1986. Anther culture of an interspecific hybrid of *Capsicum*. J. Plant Physiol. 126: 1–9.
Morrison, R.A., 1987. Gametoclonal variation in pepper. Diss. Abstracts, Intl. (Sciences and Engineering) 48: 1226.
Munyon, I.P., J.F. Hubstenberger and G.C. Phillips, 1989. Origin of plantlets and callus obtained from chile pepper anther cultures. In Vitro Cell Develop. Biol. 25: 293–296.
Murashige, T. and F. Skoog, 1962. A revised medium for rapid growth and bioassays with tobacco tissue cultures. Physiol. Plant. 15: 473–497.
Novak, F.J., 1974. Induction of a haploid callus in anther cultures of *Capsicum* sp. Z. Pflanzenzüchtg. 72: 46–54.
Pechan, P.M. and W.A. Keller, 1988. Identification of potentially embryogenic microspores in *Brassica napus*. Physiol. Plant. 74: 377–384.
Phillips, G., S. Tanksley, I. Munyon and J. Hubstenberger, 1984. Influence of incubation environment and genotype on anther culture of chile pepper (*Capsicum annuum*). In Vitro 20: 277.
Pickersgill, B., C.B. Heiser and J. McNeil, 1979. Numerical taxonomic studies on variation and domestication in some species of *Capsicum*. In: J.G. Hawkes, R.N. Lester and A.D. Skelding (Eds.), The Biology and Taxonomy of the Solanaceae, pp. 679–700. Academic Press, New York.
Pochard, E. and R. Dumas de Vaulx, 1971. La monoploïdie chez le piment (*Capsicum annuum* L.). Z. Pflanzenzüchtg. 65: 23–46.
Pochard, E. and R. Dumas de Vaulx, 1979. Haploid parthenogenesis in *Capsicum annuum* L. In: J.G. Hawkes, R.N. Lester and A.D. Skelding (Eds.), The Biology and Taxonomy of the Solanaceae, pp. 442–455. Academic Press, New York.
Polsoni, L., L. Kott and W.D. Beversdorf. 1987. Large scale microspore culture technique for mutation-selection studies in *Brassica napus*. Can. J. Bot. 66: 1681–1685.
Saccardo, F. and M. Devreux, 1974. *In vitro* production of plantlets from anther culture of *Capsicum annuum*. In: Genetics and Breeding of *Capsicum*. C.R. Eucarpia Meeting, July, 1974, Budapest, pp. 45–50.
Sibi, M., R. Dumas de Vaulx and D. Chambonnet, 1979. Obtention de plantes haploïdes par androgénèse *in vitro* chez le piment (*Capsicum annuum* L.). Ann. Amélior. Plantes 29: 583–606.
Vagera, J. and P. Havranek, 1983. Stimulating effect of activated charcoal in the induction of *in vitro* androgenesis in *Capsicum annuum* L. Capsicum Newsl. 2: 63–65.
Vagera, J. and P. Havranek, 1985. *In vitro* induction of androgenesis in *Capsicum annuum* L. and its genetic aspects. Biol. Plant. 27: 10–21.
Vito, M. and F. Saccardo, 1982. Resistance of *Capsicum* to root-knot nematodes (*Meloidogyne* spp.). Capsicum Newsl. 1: 70–71.
Wang, Y.-Y., C.-S. Sun, C.-C. Wang and N.-F. Chien, 1973. The induction of the pollen plantlets of triticale and *Capsicum annuum* from anther culture. Sci. Sin. 16: 147–151.
Zijlstra, S., C. Purimahua and P. Lindhout, 1991. Pollen tube growth in interspecific crosses between *Capsicum* species. HortScience 26: 585–586.

6. Haploidy in cucumber (*Cucumis sativus* L.)

J.A. PRZYBOROWSKI

Contents

1. Introduction

The most important factor preventing the use of haploids in cucumber (*Cucumis sativus* L.) breeding is the lack of an effective method for their production on a large scale. The development of an effective production system of doubled haploids and its further application in breeding programmes of cucurbitaceous crops could reduce the time required for cultivar development (Sauton, 1988a). Sauton described an experimental process of haplo-diploidization in melon plants through which seeds of parthenogenetically derived doubled haploids were obtained 36 weeks after planting the parent plants (Fig. 1).

Haploid extraction from cucurbitaceous species has only rarely been reported, exclusively by parthenogenesis (Table 1). The first reported cucumber haploids occurred spontaneously after *in vitro* culture of embryos excised from normal seeds (Aalders, 1958). In France, some 30 years later, Sauton and Dumas de Vaulx (1987) found that pollination of female flowers of melon (*Cucumis melo* L.) with gamma-irradiated pollen resulted in the growth of fruits, some of which contained haploid embryos. Embryological analyses conducted to determine the effect of irradiated pollen on the processes of pollination and parthenogenetic development of the egg cell, proved that, although pollen tubes reached the embryo sac, gametic fusion did not occur normally (Niemirowicz-Szczytt and Dumas de Vaulx, in press). Subsequently, the egg cell was observed to undergo parthenogenetic development. Examination of the embryo sac every three days until 18 days after pollination revealed mostly degenerate embryo-like structures after pollination with irradiated pollen in comparison with normal embryo development after non-irradiated control pollinations. Such a phenomenon was usually accompanied by disturbances that limited or prevented endosperm development (Niemirowicz-Szczytt and Dumas de Vaulx, in press). In addition, endosperm was never noticed in the presence of haploid embryos (Sauton and Dumas de Vaulx, 1987; Sauton, 1989). Embryological research conducted on *Cucumis sativus* L. after pollination with irradiated pollen revealed simultaneous presence of endosperm and embryos at early stages; however, the endosperm was not typical and appeared underdeveloped (Przyborowski, unpublished).

S.M. Jain, S.K. Sopory & R.E. Veilleux (eds.), In Vitro *Haploid Production in Higher Plants, Vol. 3,* 91–98.

Table 1. Haploid plants in the Cucurbitaceae

Species	Process/method	Literature cited
Cucurbita maxima	• parthenogenesis induced by pollination with *Cucurbita moschata* pollen	Hayase 1954
Cucumis melo	• parthenogenesis induced by pollination with *Cucumis ficifolius* pollen	Dumas de Vaulx 1979
	• ovule culture	Dryanovska & Ilieva, 1983
	• parthenogenesis induced by pollination with irradiated pollen	Sauton & Dumas de Vaulx 1987; Sauton 1988a
Cucumis sativus	• spontaneous parthenogenesis	Aalders 1958
	• gynogenesis in vitro after pollination with irradiated pollen	Truong-Andre 1988
	• haploid parthenogenesis induced by pollination with irradiated pollen	Niemirowicz-Szczytt & Dumas de Vaulx 1989; Sauton 1989; Przyborowski & Niemirowicz-Szczytt 1992, 1994
Citrullus lanatus	• haploid parthenogenesis induced by pollination with irradiated pollen	Sari et al., 1994

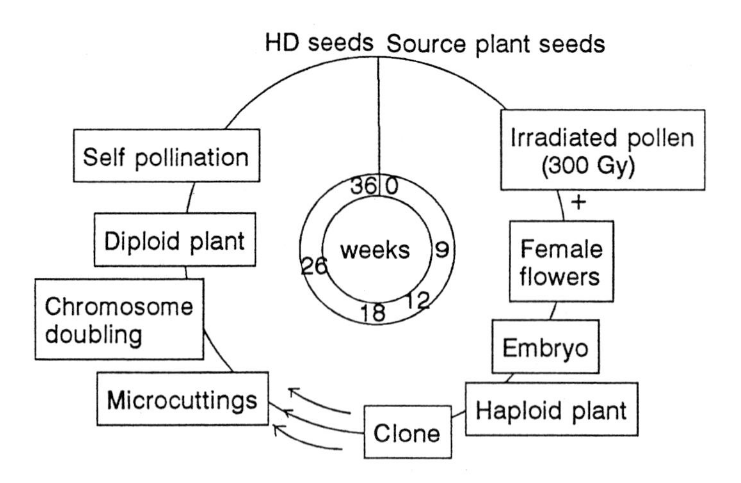

Figure 1. Experimental process for haplo-diploidization in melon (Sauton, 1988a).

Embryos were unable to develop completely, possibly due to insufficient nutrition from underdeveloped endosperms. For further growth of the embryo, embryo rescue on a nutrient medium must be practiced. E 20 A proved to be the best medium for haploid embryo cultures of muskmelon, cucumber and watermelon (Sauton and Dumas de Vaulx, 1987; Przyborowski and Niemirowicz-Szczytt, 1992, 1994; Sari *et al.*, 1994).

Application of irradiated pollen to induce haploids has been successful in cucumber although the frequency of haploids has been lower than for melon (Truong-Andre, 1988; Niemirowicz-Szczytt and Dumas de Vaulx, 1989; Sauton, 1989; Przyborowski and Niemirowicz-Szczytt, 1992, 1994). The first detailed descriptions of cucumber haploids have only recently been published (Przyborowski and Niemirowicz-Szczytt, 1992, 1994). The highest yield of haploids (1.3–1.75 per 100 cucumber seeds) has been obtained in cv. Polan F_1. By comparison, Sauton (1989) obtained 0.3 haploid embryos per 100 cucumber seeds whereas Sauton and Dumas de Vaulx (1987) and Sauton (1988a) reported nearly 3 embryos per 100 melon seeds, the species for which the method was first applied in the Cucurbitaceae. In comparison with other methods to obtain cucumber haploid plants, i.e., from normal seeds (Aalders, 1958) or by culture of ovules pollinated with irradiated pollen (Truong-Andre, 1988), induction of parthenogenesis by means of pollination with irradiated pollen seems to be the most effective and promising method.

The number of haploid cucumber plants obtained by this method depends both upon the cultivar (Przyborowski and Niemirowicz-Szczytt, 1992) and the environment at the time of pollination with irradiated pollen (Przyborowski and Niemirowicz-Szczytt 1992, 1994). Among four cultivars examined (Table 2), "Polan F_1" responded best as determined by the number of haploid embryos per 100 seeds (1.75), 45% of which developed into plants. The "Sweet Salad F_1" cultivar produced the fewest embryos per 100 seeds (0.37). The remaining two cultivars were intermediate as far as embryo yield per 100 seeds; however, "Fremont F_1" produced the greatest number of embryos which developed into plants (nearly 70%).

In a study of *Cucumis melo* L., no statistically significant difference was found for the frequency of haploid embryos among the cultivars after pollination with irradiated pollen (Sauton, 1988b). The summer season was more favourable for inducing haploid parthenogenesis in cucumber than spring (Table 3), despite a relatively lower efficiency of converting haploid embryos into plants. The tendency of cucumber towards parthenocarpy, especially in the spring, may likewise influence parthenogenetic embryo development after pollination with irradiated pollen. Sauton (1988b) also showed that the most advantageous months for haploid melon induction using similar methods were June, July and August.

Cytological analyses of cucumber plants obtained after pollination with irradiated pollen and embryo rescue revealed that most were haploid although some aneuhaploids were found (Table 4). The frequency of aneuhaploids increased at later stages of *in vitro* culture, reaching as much as

Table 2. Cucumber embryo development after pollination with irradiated pollen (Przyborowski and Niemirowicz-Szczytt, 1992)

Average for one experiment	Cultivar			
	Sweet Salad F_1	Polan F_1	Feno F_1	Fremont F_1
No. of plants	7.5	10.0	10.0	9.0
No. of pollinated flowers	12.0	37.8	33.7	27.0
No. of normal fruits	7.5	16.5	15.3	18.0
Pollination effect (%)	62.5	43.7	45.4	66.7
No. of seeds per 1 normal fruit	411.5	238.0	301.0	127.0
Total embryos obtained	11.0	54.3	36.3	21.0
Number developing into plants	6.5	26.8	15.3	14.5
Embryos per 100 seeds	0.37	1.75	0.75	0.96
Plants per 100 seeds	0.22	0.79	0.32	0.65

Table 3. Seasonal influence on cucumber embryo development after pollination with irradiated pollen – average from 3 years (Przyborowski and Niemirowicz-Szczytt, 1994)

Number of	Season		
	Spring	Summer	LSD (P=0.05)
No. of plants	9.0	9.4	
No. of pollinated flowers	21.3	28.4	
No. of normal fruits	14.5	15.0	10.67
Pollination effect (%)	68.1	52.8	47.0
No. of seeds per one normal fruit	290.5	199.3	
Total embryos obtained	10.5	47.4	22.08
No. developing into plants	8.0	22.7	11.76
Embryos per 100 seeds	0.31	1.66	0.99
Plants per 100 seeds	0.22	0.81	0.44

Table 4. Number of cucumber plants derived from embryo culture after pollination with irradiated pollen, classified according to ploidy level (Przyborowski and Niemirowicz-Szczytt, 1994)

Phase	F_1 cultivars or lines	Number of plants				
		Total	$2n<7$	$2n=7$	$7<2n<14$	$2n=14$
Phase 1	Sweet Salad	11	-	11	-	-
	Fremont	25	1	24	-	-
	Polan	93	2	90	1	-
	WPR 187	42	4	38	-	-
	Gy-3	12	3	6	3	-
	140	37	3	33	1	-
	Total	220	13	202	5	-
	(%)		5.9	91.8	2.3	-
Phase 2	Sweet Salad	1	1	-	-	-
	Fremont	2	-	1	1	-
	Polan	16	3	9	3	1
	WPR 187	11	1	9	1	-
	Gy-3	3	1	1	1	-
	140	3	1	1	1	-
	Total	36	7	21	7	1
	(%)		19.4	58.3	19.4	2.9
Phase 3	WPR 187	11	-	11	-	-
	Diploid (control)	3	-	-	-	3

Phase 1 - directly after rooting.
Phase 2 - at the time of micropropagation.
Phase 3 - after transplantation to a peat substrate.

50% in some instances. Chromosome instability as a consequence of duration of *in vitro* culture may have affected the frequency of aneuhaploids. In other studies of plants obtained from embryos resulting from pollination with irradiated pollen, all were reported to be haploid both in cucumber and melon (Sauton and Dumas de Vaulx, 1987; Sauton, 1988a, 1989). However, aneuhaploids may have been overlooked. Truong-Andre (1988) and Niemirowicz-Szczytt and Dumas de Vaulx (1989) mentioned that there were aneuhaploids among the cucumber plants so obtained. They also referred to plants with a diploid chromosome number which probably appeared as a result of a spontaneous chromosome doubling. Spontaneous doubling of chromosome number has also been observed in the course of micropropagation of haploid melon plants (Sauton, 1988a). Przyborowski and Niemirowicz-Szczytt (1994) reported only a single instance of spontaneous chromosome doubling among cucumber haploids.

The size of cells and tissues or even organs has often been correlated with the ploidy level of plants (Burns, 1972). In the case of haploids, the size of cells and organs is often reduced (Chase, 1964, 1969). Aalders (1958) stated that the guard cell length of his cucumber haploid plants was reduced by

Table 5. Main stem length and daily growth rate of haploid cucumber plants (Przyborowski and Niemirowicz-Szczytt, 1994)

Plant no.	Main stem length (cm)	Growth rate (cm/day) Phase 1	Phase 2	Phase 3	Average
Haploids					
H 1	205	2.23	1.96	0.27	**1.49**
H 2	217	0.19	1.90	0.20	**0.76**
H 3	184	0.66	1.36	0.50	**0.84**
H 4	125	0.37	0.82	0.89	**0.69**
H 5	136	0.49	0.65	0.57	**0.57**
H 6	90	0.14	0.74	0.36	**0.41**
H 7	255	0.74	1.25	1.61	**1.20**
H 8	236	0.57	1.31	1.36	**1.08**
H 9	236	0.19	0.93	2.46	**1.19**
H 10	213	0.39	1.17	2.16	**1.24**
Average	**190**	**0.60**	**1.21**	**1.04**	**0.95**
Diploids					
D1	285	3.95	2.12	0.71	**2.26**
D2	270	1.76	2.48	2.57	**2.27**
D3	332	2.17	1.57	2.50	**2.08**
Average	**296**	**2.63**	**2.06**	**1.93**	**2.21**

Phase 1 - < 70 days after transplantation to a peat substrate.
Phase 2 - 70 - 140 days after transplantation to a peat substrate.
Phase 3 - > 140 days to the end of the vegetative period.

25% compared to the diploid standard. Moreover, cucumber haploids had smaller leaves, a finding corroborated by Przyborowski and Niemirowicz-Szczytt (1992, 1994). Sauton (1989) claimed that her cucumber haploids grew quickly and normally, whereas Przyborowski and Niemirowicz-Szczytt (1992, 1994) noticed significant differences between haploid and diploid plants with regard to growth rate. The average daily increase in stem length of diploids was 2.5× greater than that of haploids, and the plants were about 100 cm shorter at the end of the vegetative period (Table 5). It should be noted that the vegetative period of haploid plants was considerably longer. This may have been due to several factors, the most important of which were differences in growth rate and continual trend to seed production. The latter factor was due to male and female sterility in haploids; the process of cucumber microsporogenesis in haploids was disturbed and led to production of sterile pollen (Niemirowicz-Szczytt *et al.*, 1993). The disturbances, caused by an odd chromosome number comprising only a single set of chromosomes, were initiated by the formation of an abnormal karyokinetic spindle. The abnormal spindle structure as well as partial division of univalents into chromatids and chromonemas lead to formation of dyads and polyads, characterized by nuclei and micronuclei of different sizes. As a result, sterile pollen comprising grains of various sizes was produced. Cucumber haploid plants developed male and female flowers which generally tended to be smaller than those of diploids (Aalders, 1958; Sauton, 1989; Przyborowski and Niemirowicz-Szczytt, 1992, 1994). Flower buds often turned yellow and dropped before opening. Similar features have been observed in melon haploid plants: small, narrow leaves, diminutive flowers, small anthers with no or few grains of sterile pollen (Dumas de Vaulx, 1979; Sauton and Dumas de Vaulx, 1987).

There have been several attempts to define optimal conditions for induction of haploid parthenogenesis by means of pollination with irradiated pollen (Niemirowicz-Szczytt *et al.*, 1993). Lowering the irradiation dose from 0.3 kGy to 0.2 or 0.1 kGy increased the number of embryos; however, the number of haploid plants capable of further development did not increase. It has also been observed that the greatest number of haploid embryos and plants could be induced in hybrids, which is probably related to their high fertility.

The method of obtaining cucumber haploids by inducing haploid parthenogenesis, first developed in France, has been successfully repeated under Polish conditions. Nevertheless, there is still a great need for its further improvement to increase the frequency of primary haploids and to examine the stability of haploids in the course of *in vitro* handling. It is also indispensable to work out effective methods of diploidization and to examine the process of gametic segregation.

2. References

Aalders, L.E., 1958. Monoploidy in cucumber. J. Hered. 48: 41–44.

Burns, G.W., 1972. Chromosomal aberrations. In: G.W. Burns (Ed.), The Science of Genetics: An Introduction to Heredity, Macmillan Co., New York, Collier Macmillan Ltd., London, pp. 214–217.

Chase, S.S., 1964. Monoploids and diploids of maize, a comparison of genotypic equivalents. Amer. J. Bot. 51: 928–933.

Chase, S.S., 1969. Monoploids and monoploid-derivatives of maize (*Zea mays* L.). Bot. Rev. 35: 117–167.

Dryanovska, O.A. and I.N. Ilieva, 1983. *In vitro* anther and ovule cultures in muskmelon (*Cucumis melo* L.). C.R. Acad. Bulgare Sci. 36: 1107–1110.

Dumas de Vaulx, R., 1979. Obtention des plantes haplöides chez le melon (*Cucumis melo* L.) après pollinisation par *Cucumis ficifolius* A. Rich. C.R. Acad. Sci. Paris 289: 875–878.

Hayase, H., 1954. Cucurbita crosses. Occurrence of a haploid twin pair from F_1 progeny of *Cucurbita maxima* × *Cucurbita moschata*. Jpn. J. Breed. 4: 55.

Niemirowicz-Szczytt, K. and R. Dumas de Vaulx, 1989. Preliminary data on haploid cucumber (*Cucumis sativus* L.) induction. Cucurbit Genet. Coop. 12: 24–25.

Niemirowicz-Szczytt, K., V. Nikolova and D. Czeżyk, 1993. Optymalizacja metod otrzymywania, mikrorozmnażania i diploidyzacji haploidów ogórka. Mat. konf. I Ogólnopolskiej Konferencji Biotechnologii Roślin i Przetwórstwa Żywności, Warszawa 1993: 42–45.

Niemirowicz-Szczytt, K. and R. Dumas de Vaulx, 1994. The influence of pollen irradiation on fertilization and embryo development in *Cucumis melo* L. (in press).

Przyborowski, J. and K. Niemirowicz-Szczytt, 1992. Haploids induction in *Cucumis sativus L.* and their preliminary description. In: Proc. Eucarpia Meeting on *Cucurbitaceae* 1992, Warsaw-Skierniewice, pp. 87–90.

Przyborowski, J. and K. Niemirowicz-Szczytt, 1994. Main factors affecting cucumber (*Cucumis sativus* L.) haploid embryo development and haploid plant characteristics. Plant Breed. 112: 70–75.

Sari, N., K. Abak, M. Pitrat, J.C. Rode and R. Dumas de Vaulx, 1994. Induction of parthenogenetic haploid embryos after pollination by irradiated pollen in watermelon. HortScience 29: 1189–1190.

Sauton, A. and R. Dumas de Vaulx, 1987. Obtention de plantes haplöides chez le melon (*Cucumis melo* L.) par gynogénèse induite par du pollen irradié. Agronomie 7: 141–148.

Sauton, A., 1988a. Doubled haploid production in melon (*Cucumis melo* L.). In: Proc. Eucarpia Meeting on Cucurbitaceae 1988, Avignon-Montfavet, pp. 119–128.

Sauton, A., 1988b. Effect of season and genotype on gynogenetic haploid production in muskmelon (*Cucumis melo* L.). Sci. Hort. 35: 71–75.

Sauton, A., 1989. Haploid gynogenesis in *Cucumis sativus* L. induced by irradiated pollen. Cucurbit. Genet. Coop. 12: 22–23.

Truong-Andre, L., 1988. *In vitro* haploid plants derived from pollination by irradiated pollen on cucumber. In: Proc. Eucarpia Meeting on *Cucurbitaceae* 1988, Avignon-Montfavet, pp. 143–144.

7. Microspore culture in chicory (*Cichorium intybus* L.)

R. THEILER-HEDTRICH and C.S. HUNTER

Contents

1. Introduction

1.1. *Botany*

The genus *Cichorium* belongs to the Compositae family and covers 9 species (Hegi, 1929). *C. endivia* (endive) and *C. intybus* var. *foliosum* Hegi (chicory, radicchio, witlof and roodlof) are used as vegetable crops, whereas *C. intybus* cv. *sativum* DC is used for its roots; this species has potential as a new industrial crop because it synthesizes inulin, a fructosan, which may be suitable for many industrial purposes, e.g., the production of glucose and fructose (Frese *et al.*, 1991; Delesalle and Dhellemmes, 1992). *C. intybus* is also of biochemical interest for its possible use as a medicinal plant (Krivenko *et al.*, 1989; Kaur *et al.*, 1989), and for use of the bitter sesquiterpene lactones (Van Beek *et al.*, 1990).

According to Schoofs and De Langhe (1988) witlof chicory is a selection product of wild chicory, a plant originally growing around the Mediterranean Sea, but now distributed through Western, Central and Southern Europe, North Africa, the temperate regions of Asia and the United States (Corey *et al.*, 1990). Smith (1976) states that chicory is native to Europe and has been cultivated as a salad plant at least since Greek and Roman times, whereas endive was cultivated by the ancient Egyptians. In Western European countries chicory is cultivated for its salad leaves and also for its sweet roots which when roasted are used with, or instead of, coffee (Gibault, 1912).

The discovery of Breziers (ca. 1850) at the State Botanical Garden in

S.M. Jain, S.K. Sopory & R.E. Veilleux (eds.), In Vitro *Haploid Production in Higher Plants, Vol. 3,* 99–113.

Brussels, that chicory roots stored in the dark would produce long white leaves (Dutch: witlof), slightly yellow or creamy at their margin, was the beginning of the story of witlof chicory as a winter vegetable (Gibault, 1912; Kiss, 1963; Schoofs and De Langhe, 1988). This cultivation of chicory was initially restricted to the area of Brussels but later introduced to France and to the Netherlands (Anonymous, 1979).

The culture of witlof chicory starts in the first year by growing vegetative plants from seeds: at harvest (late autumn) the leaves are cut off and the roots harvested. These roots are stored in the dark at low temperatures (0 ± 1°C) and then forced in the dark to grow by raising the temperature during winter. This leads to the formation of the stunted inflorescence with its characteristic white leaves (chicons of witlof chicory) which become yellow and green when exposed to light (Schoofs and De Langhe, 1988).

All wild and cultivated species of chicory are diploid (2n = 18) and readily intercross.

1.2. *Plant breeding objectives*

These differ according to the purposes for cultivation. For the industrial root crop, yield and total sugar content are the most important objectives (Frese *et al.*, 1991; Delesalle and Dhellemmes, 1992). For the vegetable crops, agricultural traits such as productivity, uniformity of root growth, duration of culture period and time of harvest, chicon production, etc. are of greatest importance. Additional selection criteria are for tolerance against diseases – mildew (*Erysiphe cruciferarum*), leaf blight (*Pseudomonas marginalis*), wet-rot (*Erwinia carotovora*), *Phytophthora erythroseptica*, *Sclerotinia sclerotiorum* (Vissers, pers. comm.), and resistance against rust (*Puccinia cichorii* [D.C.] Bell) (Cirulli *et al.*, 1989).

A further objective is to breed new forms of witlof chicory, e.g. chicons with red leaves (so-called "roodlofs"; Kiss, 1963; De Coninck and Bannerot, 1989), of which one commercial cultivar is "Robin" (S&G Seeds, Enkhuisen, NL).

1.3. *Breeding techniques*

For witlof seed production, the first attempts at F_1-hybrid production were made in the 1960s (Bannerot and De Coninck, 1971): these were obtained by classical inbreeding and back-crossing. Although chicory is regarded as a self-incompatible plant, self-fertility appears sporadically (Schoofs and De Langhe, 1988). The compatibility-incompatiblility system within chicory constrains both inbreeding and crossbreeding in the production of F_1-hybrids. The introduction of cytoplasmic male sterility (cms) to chicory (*C. intybus* var. Magdebourg) by fusion of protoplasts from chicory leaves and hypocotyls of cms-containing sunflower plants (Rambaud *et al.*, 1990) represented a major development. *In vitro* cultures have been used for maintaining and cloning selected parent plants of chicory (Doré, 1984; Mix, 1985; Theiler-Hedtrich and Badoux, 1986; review by Schoofs and De Langhe, 1988). A further improvement in chicory breeding is expected from the introduction both of herbicide resistance *via* gene transfer and protoplast culture (Vermeulen *et al.*, 1992), and the transfer of engineered genes for fertility control (Reynaerts *et al.*, 1993): transgenic plants are already being evaluated (Dale *et al.*, 1993).

As with other crops, tetraploid chicory could be of breeding and agronomic value, either with respect to higher inulin content in the roots (Rambaud *et al.*, 1992), or for the production of triploid hybrid seeds (Reheul *et al.*, 1992). Tetraploid chicory has been obtained both by colchicine treatment (Gobbe *et al.*, 1987) and by protoplast fusion (Rambaud *et al.*, 1992).

1.4. *Haploids*

Within the Compositae reports on haploid plant production are limited to only a few species, e.g., *Helianthus* spp. (review by Guerel *et al.*, 1991), *Gerbera jamesonii* (Preil *et al.*, 1977; Meynet and Sibi, 1984; Cappadocia and Vieth, 1990) and *Chrysanthemum* (Watanabe *et al.*, 1972). For chicory, anthers were cultured *in vitro* in attempts to raise haploid plants (Theiler-Hedtrich and Badoux, 1986; Guedira *et al.*, 1989), but such plants were not obtained. Desprez (1993) reported that one dihaploid plant of androgenic origin was found (out of $> 10^5$ plants) in Magdebourg chicory, as well as after intergeneric hybridization of *Cichorium intybus* × *Cicerbita alpina*.

The research reported here on microspore culture started in 1991 and was repeated in 1992 using different selections from witlof, Treviso and Robin chicory. From this work dihaploid plants were raised in 1992 and 1993 (Theiler-Hedtrich and Hunter, 1995). In this chapter a summary of the procedure for chicory microspore culture and results on the ploidy levels of regenerated plants are presented.

2. Microspore cultures

2.1. *Microspore isolation and culture*

Microspores have been isolated and cultured from three genotypes: witlof (WI-BO) from a green-leaf cultivar, Robin (R-4) from a commercial red-leaf cultivar, and a genotype of Treviso (TR-1) with intense red leaves. These were kept *in vitro* (Theiler-Hedtrich and Badoux, 1986) and also grown in the greenhouse as flowering plants. Following the method prescribed previously (Theiler-Hedtrich and Hunter, 1995) capitula (3.0 ± 0.2 mm in length) were harvested throughout the flowering season (June until October), surface sterilized in NaOCl (0.75% available Cl^-) for 20 min and washed three times in sterile distilled water. Single florets with microspores at the unin-

Table 1. Composition of media used for microspore cultures *in vitro*: quantity in mg/l (except where otherwise indicated)

Compounds	R92.01	R92.25	D91.1	PM-109
$CaCl_2 \cdot 2H_2O$	440	440	220	
$Ca(NO_3)^2 \cdot 4H_2O$				600
KCl			750	
KH_2PO_4	170	170	85	135
KNO_3	1900	1900		900
$MgSO_4 \cdot 7H_2O$	370	370	185	180
NH_4NO_3	1650	1650	825	200
NaFeEDTA	20	20	20	40
$CoCl_2 \cdot 6H_2O$	0.025	0.025	0.025	0.025
$CuSO_4 \cdot 5H_2O$	0.025	0.025	0.025	0.025
H_3BO_3	3.0	3.0	3.0	6.2
KI	0.75	0.75	0.75	0.83
$MnSO_4 \cdot 4H_2O$	13.2	13.2	13.2	22.3
$Na_2MoO_4 \cdot 2H_2O$	0.25	0.25	0.25	0.25
$ZnSO_4 \cdot 7H_2O$	2.0	2.0	2.0	8.6
glutamine			250	
myoinositol	100	100	100	100
thyamine-HCl	1.0	1.0	1.0	1.0
pyridoxine-HCl	1.0	1.0	1.0	0.5
nicotinic acid	1.0	1.0	1.0	0.5
biotin	0.01	0.01	0.01	
Ca D-pantothenate	1.0	1.0	1.0	
sucrose (g/l)	20	20	20	30
2,4–D (μM)	0.5 (2.26)			
IAA (μM)	0.5 (2.85)	0.5 (2.85)	0.2 (1.14)	
IBA (μM)				0.2 (1.0)
zeatin (μM)	2.0 (9.1)			
kinetin (μM)			0.4 (1.86)	
BA (μM)		0.5 (2.2)		

The pH for all media was adjusted to 5.6 before sterilization by autoclaving. If not otherwise indicated, liquid media were used. For gelling, either 1.6 g/l Gelrite (Sigma & Co), or agar 6.2 g/l (Gibco, bacteriological grade) was used.

uclear stage (prechecked microscopically in one floret from the capitulum before isolating the microspores from other florets) were excised from the capitulum and the microspores were released in a drop (50–100 μl) of medium R92.01 (Table 1). Microspores from 4–6 florets were collected and sieved through a 45 μm nylon mesh and dispensed into 7.5 ml liquid medium in a Petri dish (50 mm diam) at a density of approximately 5×10^4 microspores/ml. One to two ml of medium R92.01 was added to the Petri dish at two monthly intervals. Calli grown from microspores (Plate 1) were transferred to a gelled regeneration medium (R92.25, Table 1) for 4 to 8 weeks, then nodular calli were subcultured to shoot induction medium (D91.1, Table 1) and small plantlets subcultured to rooting medium (PM-109, Table 1). Rooted plants, 2–3 cm tall, were transferred to pots containing a commercial

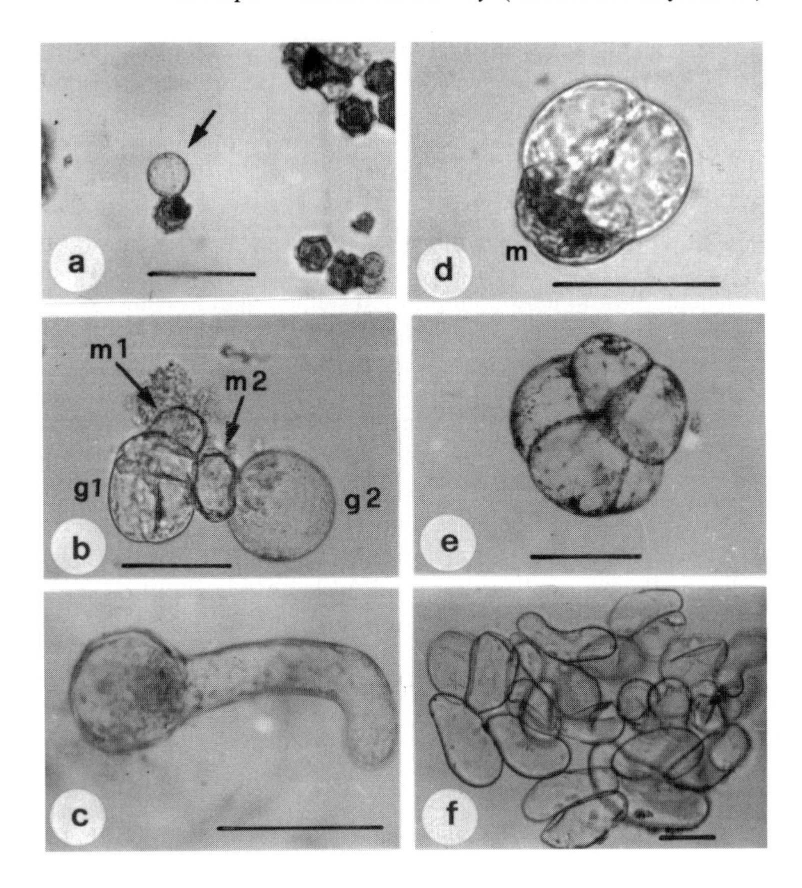

Plate 1. Culture and development stages of chicory microspore cultures *in vitro*. (a) Gametoplast emerging from pollen grain (arrow); (b) gametophytic cell development from gametoplast; gametoplast g1 emerged from microspore m1 and g2 from m2, where g1 has already divided; (c) gametic cell emerging from microspore, showing pollen tube-like development; (d) first mitotic division of gametophytic cell, m = remains of microspore; (e) microspore-derived cell colony; (f) microspore-derived calli. Size bar = 100 μm.

peat mixture (Brill No3, Floratorf) in the greenhouse and further cultivated in a nursery bed in the field.

2.2. *Culture conditions* in vitro *and* ex vitro

Initially, microspore cultures were kept either in darkness or dim light at 25 ± 1°C for 10–14 days. For the growth of callus, plant regeneration and rooting, cultures were incubated under a 16 h photoperiod at 20 $\mu mol\ m^{-2} \cdot s^{-1}$ (Durotest True lite) at 22 ± 1°C. Rooted plants were kept under high humidity (95 ± 5% RH) under a 16 h photoperiod (40 $\mu mol\ m^{-2} \cdot s^{-1}$) for 10–14 days prior to transfer to normal greenhouse conditions.

2.3. *Microscopic investigations, ploidy determination and RAPD analysis*

Pollen grains were stained with acetocarmine and their developmental stages observed microscopically. Ploidy of plants obtained either from seeds, clonal propagation or microspore-derived plants was detected by flow cytometry (Sharma *et al.*, 1983; Galbraith, 1984; Brown *et al.*, 1991). Leaf tissue (approximately 100 to 200 mg) was chopped with a razor blade in 1.5 ml citric acid solution (2.1 g citric acid plus 0.5 g Tween 20 dissolved in 100 ml distilled water), incubated for at least 15 min, then stained with 5–6 ml 4,6–diamidino-2–phenylindole (DAPI) solution (7.1 g $Na_2HPO_4 \cdot 2\ H_2O$, plus 175 mg DAPI dissolved in 100 ml distilled water), and filtered through a 15 μm mesh. The fluorescence of the stained nuclei was measured in a PAS-II-cytophotometer (Partec, Münster, Germany). The filter combination for DAPI analysis was UG1, TK420, TK560 and RG590. DNA histograms were automatically analyzed and plotted by the flow cytometer linked computer system. The coefficients of variation (CV) values for G1–cells were determined using the formula of Thornthwaite (Ulrich *et al.*, 1988).

Preliminary RAPD analysis using OPA10, OPA17 and OPA20 10-mer primers (Operon Technologies Inc., Alameda, CA, USA) followed the method of Hu and Quiros (1991).

2.4. *Leaf colour index of plants*

Regenerated plants growing vegetatively in the nursery bed were indexed according to their anthocyanin expression in the leaves: *viz*; green leaves, no anthocyanin (−); mainly green leaves with only a few cells with anthocyanin on leaf lamina or red coloured midribs (− / +); green-red speckled leaves with 25–75% of their surface area anthocyanin coloured (+ / −); and red leaves with $>75\%$ of the upper leaf lamina evenly coloured by anthocyanin (+).

2.5. *Microspore development and growth*

Microspores grown in the Petri dishes exhibited slow rates of cell division during periods up to many (6–12) months. Within a Petri dish, cell cultures originating from single microspores could not be separated from other colonies, consequently subcultures were made non-selectively with respect to the progeny from single microspores.

After isolation of the microspores and their culture in liquid medium, a gradual degradation of the exine was recorded, leading to high quantities of debris in the Petri dishes. Microspores which remained actively growing emerged from the pollen cell wall revealing a gametoplast (Plate 1a) whose volume slowly expanded up to six-fold (Plate 1b) before cell division occurred. Some of these gametophytic cells remained enlarged during periods of up to six months, or formed a pollen tube-like structure (Plate 1c). From

Table 2. Number of capitula per genotype for microspore culture, number of capitula with microspores that divided and number of plants raised and investigated

Year	Genotype	No. of capitula excised for microspore culture	No. of capitula with microspores that divided (capitulum no.)	No. of regenerated plants tested by flow cytometry
1991	witlof	13	1 (WI-BO/2)	80
	Treviso	11	2 (TR-1/16)	50
			(TR-1/22)	120
	Robin	10	1 (R-4/3)	205
1992	Robin	59	5 (R-4/11)	68
			(R-4/20)	40
			(R-4/33)	39
			(R-4/66)	3
			(R-4/69)	13

divided gametophytic cells (Plate 1d) small colonies (Plate 1e) and calli formed (Plate 1f).

Calli of approximately 0.5–1.0 mm diam were then transferred to gelled regeneration medium (R92.25), on which further callus growth was induced before meristematic nodules formed. Only if this stage could be reached were plants regenerated relatively easily on either medium R92.25 or D91.1. Medium R92.25 enhanced clonal propagation rates, whereas on medium D91.1 only a small proportion (10–20%) of the subcultured nodular calli formed shoots. Therefore a two-step subculture regime was chosen to raise at least 1–2 plants per subcultured callus.

From the three genotypes, witlof (WI-BO), Robin (R-4) and Treviso (TR-1), microspores from 93 capitula were isolated and cultured during 1991 and 1992, but only microspores from florets from 9 capitula formed calli that developed into plants (Table 2).

After transfer of calli originating from microspores to regeneration medium R92.25, high rates of cell division occurred, leading to nodular callus. Shoots arising from these calli must have been derived either from single microspores or from more than one microspore, in which case they could be chimeric, even though the phenotype would not necessarily show that feature. Calli were subcultured for three to four passages to raise as many plants per callus as possible, thus the regenerant plants may or may not be clonal.

During subculture on regeneration medium R92.25 and shoot induction on medium D91.1, up to 50% of the plants formed inflorescences *in vitro*: of these flowering plants (several hundreds), some were transferred to soil and grew on to flower in the greenhouse, but many of the plants transferred to soil died.

2.6. *Determination of ploidy and leaf colour index*

For the determination of ploidy (Fig. 1a–d) only the results from vegetative plants are given here (Table 3). In most cases the haploid plants contained diploid cells indicating that they were chimeric, originating from haploid cells of which some had diploidized spontaneously. The ploidy state of plants with haploid cells (Fig. 1b) was compared with a diploid standard either from witlof or Treviso plants (Fig. 1a) to confirm that haploid cells were present (Fig. 1c). In addition, there was a high number of tetraploid plants (Fig. 1d) especially in the progeny of Robin R-4/3.

Microspore-derived plants revealed some differences in leaf phenotype, either in lamina size or form. During the growing period of the plants, this was not constant for haploid and diploid plants and could not be related to ploidy level, but for tetraploid plants, their leaf area was often smaller in size compared to diploids and the leaf texture was crisp (Plate 2). Additionally, stomata on fully expanded tetraploid and haploid leaves were significantly ($P = 0.05$) larger ($4n$) or smaller ($1n$) than on diploids. The most visible differences were in leaf coloration (anthocyanin expression) which could be seen as a genetic marker. For progeny derived from microspores of the two fully-red donor plants Robin (capitulum No. R-4/3) and Treviso (capitulum No. TR-1/22), leaf-colour was indexed and grouped according to ploidy level (Table 4).

3. Conclusions and prospects

Ploidy determination of regenerated plants by flow cytometry (Pickering and Devaux, 1992) gave clear evidence that they had been raised from microspores (Fig. 1b). Most of these plants had polyploidised spontaneously *in vitro* leading to 1n + 2n, 2n, 2n + 4n and other mixoploid plants. This phenomenon is well-known from microspore cultures in which plant organogenesis occurs from callus (Mix *et al.*, 1978): this leads to difficulty in distinguishing among different genotypes originating from single microspores. During the callus phase and further plant propagation, clones of the same genotype could occur; therefore methods to identify different genotypes need to be applied, either by using isozyme systems (Baes and Van Cutsem, 1993) or random amplified polymorphic DNA (RAPD) markers (Williams *et al.*, 1990; Bellamy *et al.*, 1992). Preliminary genetic analysis by RAPD markers showed clear differences in DNA banding between parent plants and their microspore-derived progeny (Plate 3). From the DNA samples analysed, it appeared that one haploid and one tetraploid plant showed the same DNA pattern, indicating that they originated from the same microspore. Further investigations are necessary to verify these results. For the results presented here, anthocyanin expression was used as a genetic marker (Lespinasse and Chevreau, 1984) and here, too, clear differences were shown (Table 4).

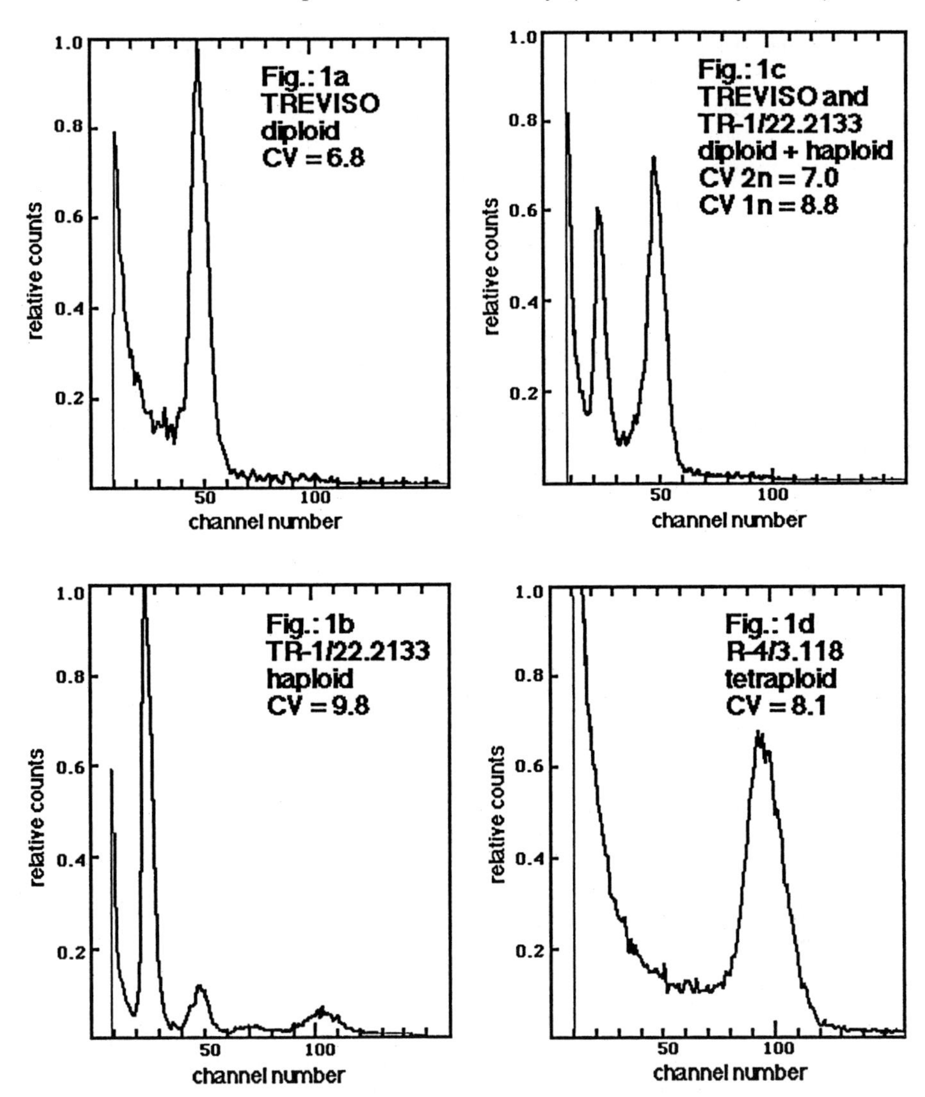

Figure 1. Flow cytometric histograms of chicory leaves from (A) Treviso (diploid standard), (B) haploid microspore-derived plant from capitulum No. TR-1/22, plant No. 133, (C) control of diploid and haploid tissue from Treviso and TR-1/22, plant No. 133, (D) tetraploid microspore-derived plant from Robin capitulum No. R-4/3, plant No. 118. Horizontal axis: channel number, showing relative intensity of nuclei fluorescence, for G1 cells, diploid standard at channel 50, G1 cells of haploid nuclei at channel 25 and for G1 cells of tetraploid nuclei at channel 100. Vertical axis: relative counts of impulses per channel. Counts below channel 20 are negligible background impulses from debris.

Table 3. Ploidy determination by flow cytometry of chicory progeny derived from microspore cultures. Number of plants at various ploidy states

Genotype	Capitulum number	Haploid ($1n$[a])	Diploid ($2n$)	Tetraploid ($4n$)	Mixoploid[b]	Sum
witlof	WI-BO/2	2	74	–	4	80
Treviso	TR-1/16	2	47	–	1	50
	TR-1/22	24	85	5	6	120
Robin	R-4/3	16	98	83	8	205
	R-4/11	3	44	18	3	68
	R-4/20	–	28	9	3	40
	R-4/33	1	32	4	2	39
	R-4/66	3	–	3	–	#M3
	R-4/69	1	10	1	1	13

[a] Haploid plants with a clear haploid peak (Fig. 1b) or as in the control (Fig. 1c), where the number of $1n$ and $2n$ nuclei were approximately equal.
[b] Mixoploid: leaf samples in which different ploidy levels could be detected e.g., $2n + 3n$, $2n + 4n$, $2n + 4n + 6n$.

Table 4. Leaf colour index of microspore-derived plants from Robin (capitulum No. R-4/3) and Treviso (capitulum No. TR-1/22): numbers of plants

Microspore-derived plants from capitulum number	Ploidy level	Leaf lamina mainly green with a few anthocyanin containing cells (− / +)	Green-red speckled leaves (25–75%) anthocyanin (+ / −)	Red leaves (+)
R-4/3 (+ / −)	haploid	3	11	2
	diploid	4	93	1
	tetraploid	28	55	0
TR-1/22 (+)	haploid	3	15	6
	diploid	4	48	33
	tetraploid	3	2	0

Even though this microspore culture protocol is still in its infancy, it has been established that dihaploid chicory plants can be raised. Further improvement of the method is essential, especially with respect to controlled growth conditions for the donor plants, culture conditions of the microspores *in vitro* and in reducing the culture period from microspore isolation to plant regeneration.

The advantages of the use of dihaploid plants in plant breeding are well-established and reviewed (Dunwell, 1986; Foroughi-Wehr and Wenzel, 1991; Pickering and Devaux, 1992) but for the Compositae reports of haploid plant induction are limited (see introduction). With the above protocol for dihaploid chicory plant regeneration, a considerable improvement in chicory breeding should be achievable either for increased inulin content (Rambaud *et al.*, 1992), or for the expression of recessive characters and the induction

Plate 2. Microspore-derived progeny from *C. intybus* var. *foliosum* Hegi, cv. Robin. (a) Haploid, but containing some diploid cells; leaf colour fully red. (b) Diploid, leaf colour fully red. (c) Predominantly tetraploid but containing some haploid cells; leaf colour red/green speckled. (d) Tetraploid: colour mostly green with only a few red cells. All plants are ca. 8 months *ex vitro*. Size bar: 5 cm.

of homozygosity for polygenic traits, coupled with recurrent selection alternating with haploid steps (Wenzel *et al.*, 1992). For chicory, especially when breeding for roodlof-type chicons, dihaploids should lead to improved F_1-hybrids. Conventionally inbred progeny from a fully red-leaf selection from Robin (R-4) always led to segregation of plants with green-red speckled leaves and red leaves; in backcrosses with witlof (green leaves) polygenic segregation was observed. From a total of 279 plants, chicons were produced, of which 3.6% showed green leaves, 70.6% green leaves with a few anthocyanin containing cells, 21.9% green-red speckled leaves and only 3.9% red

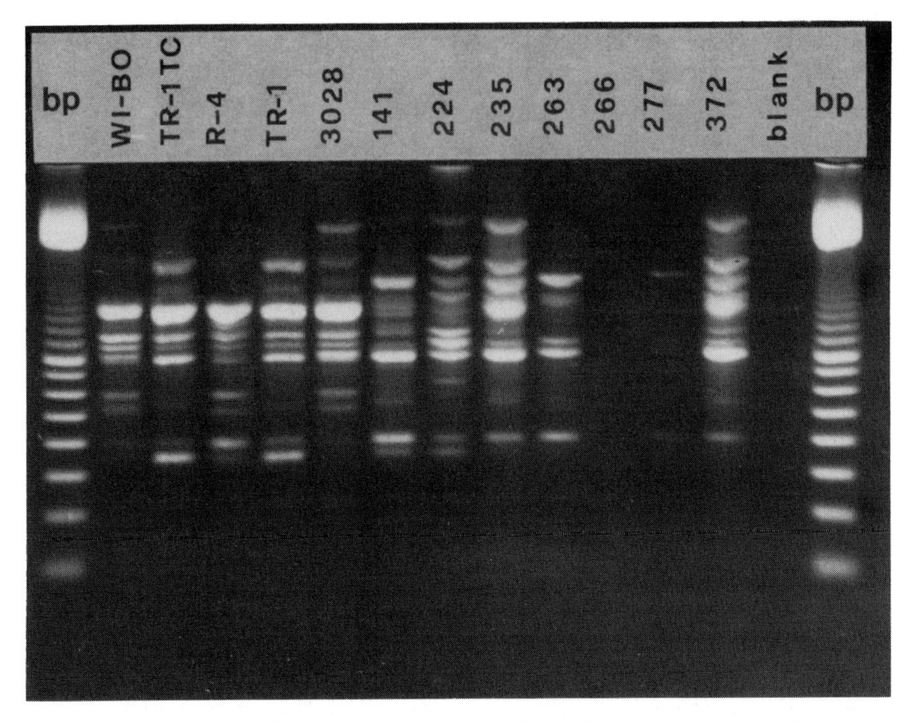

Plate 3. RAPD analysis of chicory genotypes with primer OP-A17. Key: bp – 100 base pair ladder (Pharmacia Biotech Ltd. Milton Keynes, UK); blank – PCR-reaction without genomic DNA; Parental genotypes WI-BO – witlof; R-4 – Robin; TR-1TC – Treviso (*in vitro* grown plant), and TR-1 – Treviso (field grown plant). Microspore-derived genotypes: plant number 3028 (diploid) from capitula WI-BO/2 and from capitula R-4/3 plant numbers 141 (tetraploid); 224 (haploid); 235 (haploid); 263 (diploid) and 372 (diploid + tetraploid [chimeric in ploidy levels]); Nos. 266 and 277 did not respond. Note that plants No. 235 (haploid) and No. 372 (diploid and tetraploid) show identical pattern of DNA fragments.

leaves (Theiler-Hedtrich, unpublished data). Further improvement might be achieved by raising dihaploid plants which are completely self-incompatible.

4. Acknowledgements

We thank Iris Finger and H.U. Bisang of the tissue culture laboratory of the Swiss Federal Research Station, Wädenswil, for their assistance with some aspects both of tissue culture work and the field plantation work, and Andrea Bzonkova for her skilful work with the flow cytometry analysis of leaf-samples. Also we thank P. Fisher of the University of the West of England, Bristol, for printing the microspore plates. For valuable discussions on chicory breeding we thank J. Vissers of S&G seeds, Enkhuizen, the Netherlands, and P. Devaux, of Florimond Desprez, France.

5. References

Anonymous, 1979. De witlofteelt. Minist. Landbouw, Dienst Inf. Brussels (cited in Schoofs and De Langhe, 1988).

Baes, P. and P. Van Cutsem, 1993. Electrophoretic analysis of eleven isozyme systems and their possible use as biochemical markers in breeding chicory (*Cichorium intybus* L.). Plant Breed. 110: 16–23.

Bannerot, H. and B. De Coninck, 1971. L'utilisation des hybrides F_1: nouvelle méthode d'amélioration de la chicorée de Bruxelles. In: Eucarpia. Symp Int. Chicorée de Bruxelles, Gembloux, 1970, pp. 99–118.

Bellamy, A., H. Bannerot, C. Mathieu and F. Vedel, 1992. Fingerprinting in *Cichorium intybus* L. based on nuclear and cytoplasmic RFLP and RAPD molecular markers. In: Proceedings of EUCARPIA Congress 1992, Angers, pp. 607–608.

Brown, S.C., P. Devaux, D. Marie, C. Bergounioux and P.X. Petit, 1991. Cytométrie en flux: application à l'analyse de la ploidie chez les végétaux. Biofutur 105: 2–16.

Cappadocia, M. and J. Vieth 1990. *Gerbera jamesonii* H. Bolus ex Hook: *In vitro* production of haploids. In: Y.P.S. Bajaj (Ed.), Biotechnology in Agriculture and Forestry, Vol. 12. Haploids in Crop Improvement I, pp. 417–427. Springer-Verlag, Berlin/Heidelberg.

Cirulli, M., F. Ciccarese and A. Siniscalco, 1989. Identificazione di fonti di resistenza alla ruggine della ciccoria. La difesa della piante 12: 241–244.

Corey, K.A., D.J. Marchant and L.F. Whitney, 1990. witlof chicory: a new vegetable crop in the United States. In: J. Janick and J.E. Simon (Eds.), Advances in New Crops; First National Symposium, New Crops: Research, Development Economics, October 23–26, 1988, Indianapolis, Indiana, pp. 414–418. Timber Press, Inc., Portland, Oregon.

Dale, P.J., J.A. Irwin and J.A. Scheffler, 1993. The experimental and commercial release of transgenic crop plants. Plant Breed. 111: 1–22.

De Coninck, B. and H. Bannerot, 1989. Création d'hybrides F_1 de chicorée de Bruxelles (*Cichorium intybus* L.) à limbe rouge adaptés au forsage hydroponique. Acta Hort. 242: 191–192.

Delesalle, L. and C. Dhellemmes, 1992. La chicorée: une source d'inuline et de fructose. Cultivar No. 310: 40–41.

Desprez, B., 1993. Recherche de méthodes d'obtention de plantes haploïdes chez la chicorée (*Cichorium intybus* L.). Thesis, Université de Paris-Sud Centre d'Orsay.

Doré, C., 1984. Multiplication végétative *in vitro* de la chicorée (*Cichorium intybus*) au service de la sélection. In: Colloque Eucarpia sur les légumes à feuilles, Versailles, pp. 123–125. INRA, Versailles.

Dunwell, J.M., 1986. Pollen, ovule and embryo culture as tools in plant breeding. In: L.A. Withers and P.G. Alderson (Eds.), Plant Tissue Culture and its Agricultural Applications, pp. 375–404. Butterworths, London.

Foroughi-Wehr, B. and G. Wenzel, 1991. Andro- and parthogenesis. In: M.D. Hayward, N.O. Bosmemark and I. Romagosa (Eds.), Plant Breeding – Principles and Prospects. Chapman and Hall, London.

Frese, L., M. Dambroth and A. Bramm, 1991. Breeding potential of root chicory (*Cichorium intybus* L. var. *sativum*). Plant Breed. 106: 107–113.

Galbraith, D.W., 1984. Flow cytometric analysis of cell cycle. In: I.K. Vasil (Ed.), Cell Culture and Somatic Cell Genetics in Plants. Vol. 1, pp. 765–777. Academic Press, New York.

Gibault, G., 1912. Chicorée endive. In: G. Gibaut (Ed.), Histoire de Légumes, pp. 107–120. Librairie Horticole, Paris.

Gobbe, J., B. Evrard and G. Coppens d'Eeckenbrugge, 1987. Polyploidisation. In: B. Longly and B.P. Louant (Eds.), Mécanismes de reproduction chez la Chicorée de Bruxelles: Fondements et Application à la Sélection, pp. 67–79. I.R.S.I.A., Bruxelles.

Guedira, M., T. Dubois-Tylski, J. Vasseur and J. Dubois, 1989. Embryogenèse somatique directe à partir de culture d'anthères de *Cichorium* (Asteraceae). Can. J. Bot. 67: 970–976.

Guerel, A., K. Nichterlein and W. Friedt, 1991. Shoot regeneration from anther culture of sunflower (*Helianthus annuus*) and some interspecific hybrids as affected by genotype and culture procedure. Plant Breed. 106: 68–76.

Hegi, G., 1929. Illustrierte Flora von Mitteleuropa. VI. Band, 2. Haelfte, pp. 992–1000. J.F. Lehmanns Verlag, Muenchen.

Hu, J. and C.F. Quiros, 1991. Identification of broccoli and cauliflower cultivars with RAPD markers. Plant Cell Rep. 10: 505–511.

Kaur, N., A.K. Saijpaul, S. Indu and P.P. Gupta, 1989. Triglyceride and cholesterol lowering effect of chicory roots in the liver of dexamethasone-injected rats. Med. Sci. Res. 17: 1009–1010.

Kiss, P.D., 1963. Genetische Untersuchungen zur Züchtung einer rotblättrigen Treibzichorie. Dissertationsschrift aus der Eidg. Versuchsanstalt für Obst-, Wein- und Gartenbau, Wädenswil.

Krivenko, V.V., G.P. Potebnya and V.V. Loiki, 1989. Experience in treating some digestive diseases with medicinal plants. Vrachebnoe Delo 0: 75–78.

Lespinasse, Y. and E. Chevreau, 1984. Utilisation d'un gène marqueur pour la recherche de plantes haploïdes de pommier: comparaison entre pollinisation manuelle et pollinisation par les abeilles. In: Compte Rendu de Ve Symposium sur la Pollinisation, 27–30 Septembre, 1983, Versailles, pp. 391–396.

Meynet, J. and M. Sibi, 1984. Haploid plants from *in vitro* culture of unfertilized ovules in *Gerbera jamesonii*. Z. Pflanzenzüchtg. 93: 78–85.

Mix, G., 1985. Regeneration von *in vitro* Pflanzen aus Blattrippensegmenten der Zichorie (*Cichorium intybus* L.). Landbauforsch. Völkenrode 35: 59–62.

Mix, G., H.M. Wilson and B. Foroughi-Wehr, 1978. The cytological status of plants of *Hordeum vulgare* L. regenerated from microspore callus. Z. Pflanzenzüchtg. 80: 89–99.

Pickering, R.A. and P. Devaux, 1992. Haploid production: approaches and use in plant breeding. In: P.R. Shewry (Ed.), Barley: Genetics, Molecular Biology and Biotechnology, pp. 511–539. CAP Int. Publ., Wallingford.

Preil, W., W. Huhnke, M. Engelhardt and M. Hoffmann, 1977. Haploide bei *Gerbera jamesonii* aus *in vitro*-Kulturen von Blütenköpfchen. Z. Pflanzenzüchtg. 79: 167–171.

Rambaud, C., J. Dubois and J. Vasseur, 1992. The induction of tetraploidy in chicory (*Cichorium intybus* L. var. Magdebourg) by protoplast fusion. Euphytica 62: 63–67.

Rambaud, C., J. Vasseur, J. Dubois, B. Lejeune and F. Quetier, 1990. Obtention de chicorées male-steriles (*Cichorium intybus* var. Magdebourg) par fusion de protoplastes. Cinquantenaire de la Culture *in Vitro*, 24–25 Oct., 1989, Versailles, pp. 315–316. INRA, Paris (les colloques de l'INRA, no. 51).

Reheul, D., J. Baert, E. Van Bockstaele, G. Rijckaert G., H. Vandepitte, A. Ghesquiere and M. Malengier, 1992. Rassen van het Rijksstation voor Plantenveredeling (Rv̇.P.): Vlaamse bescheidenheid met grote resultaten. Landbouwtijdschrift – Revue de l'Agriculture 45: 747–758.

Reynaerts A., H. Van de Wiele, G. De Sutter and J. Janssens, 1993. Engineered genes for fertility control and their application in hybrid seed production. Sci. Hort. 55: 125–139.

Schoofs, J. and E. de Langhe, 1988. Chicory (*Cichorium intybus* L.). In: Y.P.S. Bajaj (Ed.), Biotechnology in Agriculture and Forestry, Vol. 66. Crops II, pp. 294–321. Springer Verlag, Berlin/Heidelberg.

Sharma, D.P., E. Firoozabady, N.M. Ayres and D.W. Galbraith, 1983. Improvement of anther culture in *Nicotiana*: Media cultural conditions and flow cytometric determination of ploidy levels. Z. Pflanzenphysiol. 111: 441–451.

Smith, P.M., 1976. Chicory and endive *Cichorium* spp. (Compositae). In: N.W. Simmonds (Ed.), Evolution of Crop Plants, pp. 304–305. Longman, London.

Theiler-Hedtrich, R. and S. Badoux, 1986. Utilisation de la multiplication *in vitro* dans un programme de sélection de la chicorée amère rouge (*Cichorium intybus* L.). Revue Suisse Vitic. Arboric. Hortic. 18: 271–275.

Theiler-Hedtrich, R. and C.S. Hunter, 1995. Regeneration of dihaploid chicory (*Cichorium intybus* L. var. *foliosum* Hegi) via microspore culture. Plant Breed. 114: 18–23.

Ulrich, I., B. Fritz and W. Ulrich, 1988. Application of DNA fluorochromes for flow cytometric DNA analysis of plant protoplasts. Plant Sci. 55: 151–158.

Van Beek, T.A., P. Maas, B.M. King, E. Leclerq, A.G.J. Voragen and A. de Groot, 1990. Bitter sesquiterpene lactones from chicory roots. J. Agric. Food Chem. 38: 1035–1038.

Vermeulen, A., H. Vaucheret, V. Pautot and Y. Chupeau, 1992. *Agrobacterium* mediated transfer of a mutant *Arabidopsis* acetolactate synthase gene confers resistance to chlorsulfuron in chicory (*Cichorium intybus* L.). Plant Cell Rep. 11: 243–247.

Watanabe K., Y. Nishii and R. Tanaka, 1972. Anatomical observations on the high frequency callus formation from anther culture of *Chrysanthemum*. Jpn. J. Genetics 47: 249–255.

Wenzel, G., A. Graner, F. Fadel, J. Zitzlsperger and B. Foroughi-Wehr, 1992. Production and use of haploids in crop improvement. In: J.P. Moss (Ed.), Biotechnology and Crop Improvement in Asia, pp. 169–179. International Crops Research Institute for the Semi-Arid Tropics, Patancheru, Andhra Pradesh, India.

Williams, J.G.K., A.R. Kubelik, K.J. Livak, J.A. Rafalski and S.V. Tingey, 1990. DNA polymorphisms amplified by arbitrary primers are useful as genetic markers. Nucleic Acids Res. 18: 6531–6535.

8. Haploidy in eggplant

GIUSEPPE LEONARDO ROTINO

Contents

1. Introduction

Solanum melongena L. ($2n = 2x = 24$) is also known as eggplant, aubergine, brinjal or Guinea squash. This species is probably a native of Asia; the Indo-Burma Asiatic region, China and Japan are considered the centers of origin. It has been cultivated in Asia from the 5th century, and was introduced to Europe by the Arabs during the dark ages (Khan, 1979).

The annual worldwide production of eggplant has increased in the last ten years by about 1.1 million metric tons (Table 1). The largest producer is China with 40% of total world production followed by Turkey, India, Japan and Mediterranean countries such as Egypt, Italy, Spain, etc. Eggplant is an important and popular vegetable in the diets of these countries.

Although most of the produce is marketed fresh, the use of frozen pre-cooked eggplant, mainly in the developed countries, is spreading. The fruits and leaves lower the blood cholesterol level; some other medicinal uses of eggplant tissue and extract are treatment of diabetes, asthma, bronchitis and dysury (Porcelli, 1986).

Eggplant is a slow-growing perennial vegetable crop in the tropical coun-

S.M. Jain, S.K. Sopory & R.E. Veilleux (eds.), In Vitro *Haploid Production in Higher Plants, Vol. 3,* 115–141.

Table 1. Worldwide production of eggplant (based on FAO Production Yearbook, 1992)

	1981		1992	
	Area (ha)	Yield (t/ha)	Area (ha)	Yield (t/ha)
World	366,000	12.7	409,000	14.0
Asia	310,000	11.5	359,000	12.9
China	155,000	9.6	185,000	13.1
Turkey	39,000	16.9	44,000	16.8
Japan	22,000	29.9	16,000	31.2
Africa	28,000	15.4	28,000	19.2
Egypt	14,000	21.2	17,000	20.4
Europe	24,000	24.4	19,000	27.3
Italy	13,000	25.0	9,000	25.9
Spain	5,000	24.8	4,000	30.9
North America	3,000	22.5	4,000	18.5
South America	1,000	14.0	1,000	13.1

tries, while in the temperate zones it behaves as an annual. However, its growing season can be extended under controlled conditions in the greenhouse. The plant grows to a height of 50 to 150 cm and bears long, oval, pear-shaped or spherical fruits of very different size; skin may be purple, white, green or yellow. It is a day-neutral plant and flowering starts at the 6th–10th leaf stage and lasts for a long period. Although it is considered an autogamous species, the frequency of natural cross-pollination is estimated to vary from 0.2 to 48% (Quagliotti, 1979; Franceschetti and Lepori, 1985).

2. Breeding and crop improvement

Eggplant breeding is mainly focused on F_1 hybrid cultivars, which have nearly replaced the open-pollinated cultivars, particularly in the intensive growing areas. In fact, hybrids automatically safeguard the rights of breeders and seed companies even in the countries lacking plant variety protection laws. In eggplant, crosses are relatively easy and a satisfactory number of seeds is obtained by hand emasculation and pollination of the flowers; besides, some cross combinations may provide heterotic advantages (Quagliotti, 1992).

The major objectives of breeding are to develop high-quality and disease resistant varieties. The procedures usually followed are: pedigree, backcross, recurrent selection and single seed descent (Kalloo, 1993).

Fruit colour, proportion of seed to pulp, flesh consistency and browning, and quantity of compounds with solasonine-like structure are important characters for assessing the quality of eggplant fruits (Aubert *et al.*, 1989a,b).

In the countries where intensive and successive cropping is practised, the main breeding goal is to develop cultivars resistant to soil-born diseases (*Verticillium* and *Fusarium* wilt, bacterial wilt and nematode infections). Eggplant is severely damaged also by insects (*Leucinodes orbonalis*, *Leptinotarsa decemlineata* and *Trialeurodes vaporariorum*), mites and fruit rot.

Identification of resistance sources, knowledge of their genetics and incorporation into eggplant cultivars have been limited. Many attempts have been made to introgress resistance genes displayed by wild *Solanum* species by means of both sexual and somatic hybridization (Kalloo, 1993).

Other important goals are the development of cultivars suitable for protected cultivation and postharvest technologies.

Since eggplant is highly responsive to *in vitro* manipulation (Hinata, 1986) and genetic transformation (Rotino and Gleddie, 1990), biotechnological progress may help to solve several of its agronomic problems.

3. Induction of haploids through anther culture

Spontaneous parthenogenesis leading to haploid plant production has never been reported in eggplant. Success in plant regeneration through anther culture was described by Raina and Iyer (1973). The first haploids in eggplant were obtained by the Chinese Research Group of Haploid Breeding (1978) and by Isouard *et al.* (1979). Subsequently, Dumas de Vaulx and Chambonnet (1982) greatly increased the yield of *in vitro* anther-derived plantlets by using a protocol similar to that applied to pepper (Dumas de Vaulx *et al.*, 1981). This method is based on a high temperature (35°C) treatment during the first period of anther culture. The beneficial effects of heat treatment in determining haploid production were previously demonstrated in Brassicaceae (Hansson, 1978; Keller and Armstrong, 1979).

Studies on isolated microspore culture in eggplant have been carried out; plants were regenerated following either anther preculture (Gu, 1979) or direct culture of microspores (Miyoshi, 1994).

Minor modifications of the Dumas de Vaulx and Chambonnet (1982) method have resulted in a reliable protocol to produce androgenic plants and enabled a successful integration of doubled haploid lines in eggplant breeding programs.

According to this method, excised anthers are cultured in induction medium C (Table 2) supplemented with appropriate growth regulators, and placed in the darkness at 35°C for the first eight days of culture. In the following days, petri dishes are kept in the growth chamber at 25°C under 16 h illumination (50 μmol m^{-2} s^{-1}, fluorescent light). On the thirteenth day, anthers are transferred to differentiation medium R (Table 2). Embryos become visible from the anthers one month after the beginning of culture

Table 2. Macronutrients, micronutrients and vitamins of the three basal media utilized for eggplant anther culture (mg/l)

	Medium C	Medium R	Medium V3
Macroelements*			
KNO_3	2150	2150	1900
NH_4NO_3	1238	1238	1650
$MgSO_4$-$7H_2O$	412	412	370
$CaCl_2$-$2H_2O$	313	313	440
KH_2PO_4	142	142	170
$Ca(NO_3)_2$-$4H_2O$	50	50	-
$NaH_2PO_4.H_2O$	38	38	-
$(NH_4)_2SO_4$	34	34	-
KCl	7	7	-
Microelements*			
$MnSO_4$-H_2O	22.130	20.130	0.076
$ZnSO_4$-$7H_2O$	3.625	3.225	1.000
H_3BO_3	3.150	1.550	1.000
KI	0.695	0.330	0.010
Na_2MoO_4-$2H_2O$	0.188	0.138	-
$CuSO_4$-5 H_2O	0.016	0.011	0.030
$CoCl_2$-$6H_2O$	0.016	0.011	-
$AlCl_3$-$6H_2O$	-	-	0.050
$NiCl_2$-$6H_2O$	-	-	0.030
Vitamins and amino acids			
myo-inositol	100.00	100.00	100.00
pyridoxin HCl	5.500	5.500	5.500
nicotinic acid	0.700	0.700	0.700
thyamine HCl	0.600	0.600	0.600
Calcium panthotenate	0.500	0.500	0.500
vitamin B12	0.030	-	-
biotin	0.005	0.005	0.005
glycin	0.100	0.100	0.200
Chelated iron			
Na_2 EDTA	18.65	18.65	37.30
$FeSO_4$-$7H_2O$	13.90	13.90	27.80

* = from Chambonnet, 1985.

and the embryo production lasts for three-four months. Well formed embryos of 4–6 mm are cultured in growth regulator-free medium V3 (Table 2) for further development. Complete plantlets can easily be propagated *in vitro*, by cuttings, or transferred to soil.

3.1. *Stage of microspore culture*

All investigators who have reported success with eggplant anther culture agree that microspores are responsive at the uninucleate stage. This corresponds, roughly, to a flower bud with closed sepals still almost completely covering the petals.

We also have experienced large deviations related to genotype, age and growth condition of the donor plant. Thus, routinely applying anther culture to different genotypes, we have used anthers coming from a wide range of

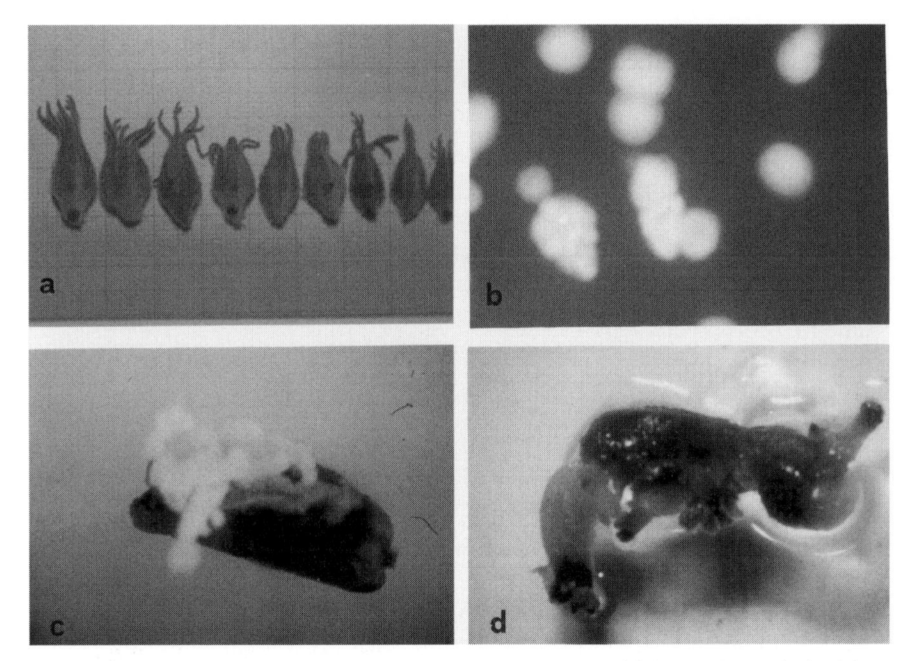

Figure 1. (a) Size of flower buds employed for anther culture. (b) Pro-embryos derived from microspore division (stained with DAPI). (c,d) Views of embryos emerging from an anther.

flower sizes (Fig. 1a). This ensures some anthers containing microspores at the right stage, the others being less, but still, responsive (early uni- or bi-nucleate microspores).

3.2. *Age and growing conditions of donor plant*

Vigor and physiological state of donor plant have been reported to influence anther culture response (Maheshwari *et al.*, 1980). Generally, healthy and vigourous donor eggplants provide anthers with the highest androgenic potential. It is important to prevent seed-setting and plant aging by removing open flowers and small fruits. In other species a significant reduction of androgenic potential has been associated with plant age (Sunderland, 1971; Rashid and Street, 1973). This phenomenon has been ascribed to a deterioration in the general condition of the plant.

Satisfactory results have been obtained by growing vigourous and healthy donor eggplants either in the open field or in the greenhouse. It is important to control insects and mites; however anthers should be collected several days after spraying.

Seasonal variation has been observed in the yield of eggplant androgenic embryos. Tuberosa *et al.* (1987b) reported that during the period July–

Table 3. Sugar and growth regulator concentration of 8 induction media (Rotino *et al.*, 1987)

Compounds		Induction media							
		1	2	3	4	5	6	7	8
Sucrose	(g/l)	120	30	60	60	60	60	60	24
Glucose	(g/l)	-	-	63	63	63	63	63	6
KIN	(μM)	13.9	0.05	23.3	23.3	23.3	-	-	0.05
2,4 D	(μM)	-	0.03	-	-	22.6	-	13.6	0.03
IAA	(μM)	5.7	-	28.5	-	-	-	-	-
NAA	(μM)	-	-	-	5	-	3	-	-
ZEA	(μM)	-	-	-	-	-	1	1	-

October the highest numbers of responding anthers were found in the middle of September and continued until the middle of October.

These results are in accordance with our observations that, in the Mediterranean climate, higher androgenic frequencies are obtained during cooler months, and the best periods are spring and autumn (unpublished). Most likely, the photoperiod as well as the day/night temperature, affect anther response; and precise information may be obtained by growing donor plants in a phytotron.

3.3. *Media*

For regeneration of haploid eggplants the mineral costituents of C (induction medium) and R (regeneration medium) (Table 2) have generally been employed. However, in the first reports, the MS (Murashige and Skoog, 1962) basal medium was used. Among the vitamins, cyanocobalamin (Vit. B12) seems to be necessary for a good response (Chambonnet, 1985).

Induction medium contains both auxin and cytokinin while differentiation medium is supplemented with cytokinin. Various combinations and concentrations of several growth regulators have been tested by different investigators (Chambonnet, 1985; Rotino *et al.*, 1987). Auxins 2,4-D and NAA as well as cytokinins KIN, ZEA and BA have been shown to support androgenesis. Tables 3 and 4 show that embryos were obtained on all the media supplemented with different kinds and ratios of auxin/cytokinin. (Rotino *et al.*, 1987). Eggplant is very responsive to growth regulators; callus induction and plant regeneration can be obtained from any somatic explant (Gleddie *et al.*, 1983, 1986). This is probably the cause of obtaining low frequency of androgenesis when different growth regulators are used. It is most likely that the initiation of the androgenic process has no link with the specificity of both nutrition and growth regulators in the culture medium. An increase in

Table 4. Effect of the sugar and growth regulator concentration reported in Table 2 on anther culture response of four eggplant genotypes (Rotino *et al.*, 1987)

Media	Genotypes									
	Lunga Violetta Barbentane		Dourga x Ronde de Valence		WIR 768 x Viserba		WIR 768 x Picentia		Total	
	anth. n°	embr. %	anth, n°	embr. %	anth. n°	embr. %	anth. n°	embr. %	anth. n°	embr. %
1	180	3.9	100	3.0	130	-	-	-	410	2.4
2	60	-	200	2.5	-	-	120	-	380	1.3
3	60	5.0	170	0.5	110	2.7	110	-	450	1.5
4	160	120.0	-	-	130	23.7	-	-	290	76.5
5	260	18.0	180	5.0	-	-	120	7.5	560	11.6
6	50	32.0	480	7.5	170	14.0	230	-	930	8.2
7	30	136.0	330	1.2	170	1.2	300	-	830	5.7
8	110	-	-	-	200	18.0	-	-	310	11.6

the shift of microspores from the gametophytic to the sporophytic pathway seems mainly due to the high temperature treatment, which probably allows metabolism of inhibiting factors. Further development of the cultured microspore is more influenced by nutritional and hormonal factors. This process appears to be influenced by the genotype; androgenic plants of particular genotypes have been regenerated only on a medium supplemented with specific growth regulators. For example, Ronde de Valence cultivar produced plants only in the presence of IAA (Chambonnet, 1985), the hybrid WIR 768 × Picentia regenerated plants only on one of the five media tested (Tables 3 and 4).

Furthermore, high concentration of sugar in the induction medium is recommended to enhance anther response. The differentiation medium R is supplemented only with the same cytokinin as in the induction medium.

In addition to the standard medium proposed by Chambonnet and Dumas de Vaulx (1983), we simultaneously tried alternative media (see Table 3) for anther culture of eggplant cultivars or segregating progenies.

3.4. *Genotype*

Wide variation in anther response of different genotypes has been reported (Chambonnet, 1985; Rotino *et al.*, 1987). Using a partial balanced diallelic crossing scheme with eight lines, it was demonstrated that the androgenic process is mainly under the control of a genetic system with prevalence of

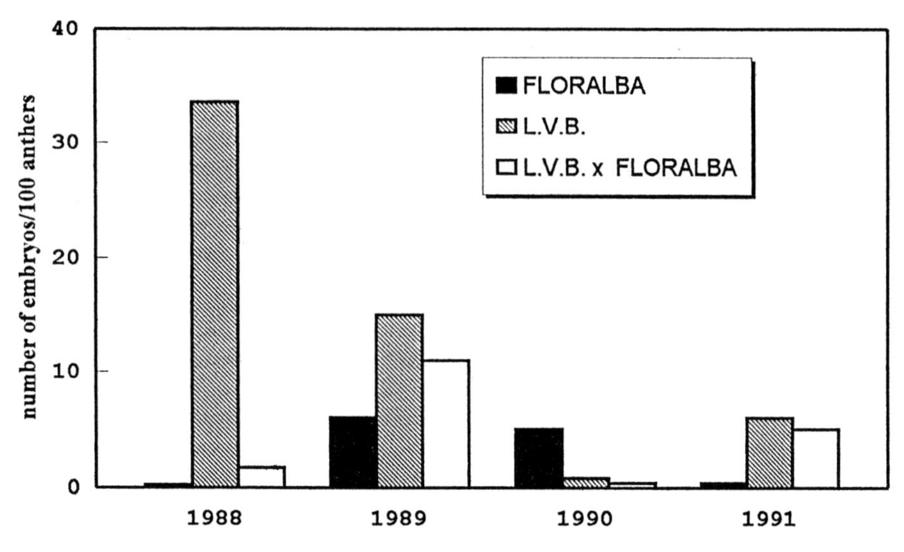

Figure 2. Average anther response of two inbred cultivars and their F_1 hybrid during four successive years of anther culture (a minimum of 100 anthers per year were cultured in the medium 1, 4 and 5 reported in Table 1; Rotino, unpublished).

non-additive effects. Thus the performance of hybrids cannot be predicted only on the basis of parental values (Tuberosa *et al.*, 1987a). On average, F_1 hybrids showed a higher level of anther response than breeding lines.

Studies aiming to improve anther response are complicated by the differential environmental effects exerted on genes controlling the various steps of androgenesis. An example is given in Fig. 2, which indicates a wide variability in the number of androgenic embryos developed from two lines and their F_1 hybrid during four successive years of anther culture (Rotino, unpublished).

3.5. *Regeneration of plants*

Aubergine anther-derived plants are mainly regenerated through direct embryogenesis. As demonstrated in other solanaceous species, actively dividing microspores (Fig. 1b) grow out of anthers as embryos (Fig. 1c,d). Additionally, microspore-derived compact structures resembling degenerated or malformed embryos are also found. The advantages of direct embryogenesis are: a) microspore proliferation can easily be identified; and b) few gross chromosomal changes.

Fundamental studies on the mechanism leading to the induction of eggplant microspore divisions were carried out jointly by French and Dutch groups. They demonstrated that the microtubules established polarity before the first division occurred (Traas and Chatelain, 1988).

The number of induced microspores per single responding anther has been much higher than the number of embryos recovered and it has been exceptional to obtain numerous well formed embryos from a single anther. Chambonnet (1985) regenerated 198 plantlets from a single anther of the cultivar Shinkuro; in our experiments the highest yielding anther (cv. Rimina) gave rise to 80 plantlets.

High frequency of early embryo abortion observed in several species has been ascribed to lethal genotypic factors and/or to inadequate culture medium composition (Heberle-Bors, 1985). An average percentage of egg-plant anther response has been between 2 to 25; each anther gives an average of four mature plantlets. The frequency of responding anthers and maturation of induced microspores are two of the main constraints to improvement of the yield of anther-derived plants.

Well-formed embryos generally develop into complete plants with normal roots and shoots upon transfer to growth regulator-free medium (Fig. 3a). The average rate of conversion of embryos into mature plants has been about seventy-five percent. Abortion and malformation are the most common embryo abnormalities (Fig. 3b). Embryo abortion either leads to callus formation or to death, but complete plants can be regenerated from malformed embryos by means of frequent subcultures on fresh culture medium. Slow growing embryos can normally be regenerated on modified medium R containing a high concentration of cytokinin.

3.6. *Transfer to field and genetic analysis*

Transfer to the autotrophic condition is a critical step in the production of eggplant haploid plants. During *in vitro* growth, plants can easily be multiplied by cuttings. Therefore, depending on time and space availability, plants can be maintained under *in vitro* conditions until plantlets are successfully transplanted into the greenhouse. Plantlets with four-five leaves and fast growing roots are the best material to transfer into the soil.

In vitro-grown plantlets are removed from vessels, and roots are gently rinsed of agar by washing with water. The plants are then transplanted into pots filled with a sterile mixture (3:1, v/v) of peat and agriperlite. They are kept for one week under 70–80% relative humidity, then ventilation is progressively increased and they can be transferred to the greenhouse within a few days.

Morphological abnormalities (Fig. 4) are rarely observed and, during further plant development, are often absent on the newly formed organs; furthermore, abnormal traits are not generally transmitted to the progeny.

3.7. *Cytological studies*

Regenerated plants have been mostly haploid or diploid; rarely different ploidies were recovered (Table 5). A few weeks after transfer to soil, haploid

Figure 3. (a) Plantlet derived from a well formed embryo. (b) Plantlet without a shoot derived from a malformed embryo.

plants can be morphologically distinguished from diploid ones (Fig. 5). Haploids are small with bushy habit and have narrow and long leaves compared with diploids. Unambiguous distinction can be made at flowering. Haploids produce small flowers with non-functional pollen grains and set small parthenogenic fruits (Fig. 6b). Occasionally, some of these fruits produce a few

Figure 4. A rare floral abnormality revealed by androgenic plant.

Table 5. Frequency of ploidy levels observed among anther-derived plants (in%)

Haploid	Diploid	Others	Note	Reference
84.6	15.4	0	2,4D as auxin	Dumas de Vaulx and Chambonnet (1982)
100.0	0	0	IAA as auxin	" " "
76.2	23.5	0.3	1 plant trisomic	Tuberosa *et al.* (1987)
74.3	25.6	0.01	1 aneuploid	Rotino (unpublished)

Figure 5. Morphological differences of haploid and diploid plants obtained from the same anther donor genotype.

seeds. Dumas de Vaulx (1986) obtained two trisomic plants ($2n + 1$) by pollinating haploid plants with pollen from a seed-derived plant.

Chromosome counting was done by staining root tips in order to verify the ploidy status of androgenic plants (Fig. 6a). Sometimes one single plant had roots with both haploid and diploid chromosome numbers. The diploidization occurred spontaneously either in the root apex or *in vitro* in the callus formed at the base of cuttings before rooting. Haploid plants have been found useful in the analysis of eggplant karyotypes (Siljak-Yakovlev and Isouard, 1983; Venora *et al.*, 1992).

Another reliable indirect approach for determining the ploidy level is by counting the number of chloroplasts per guard cell pair. Chloroplast number was highly correlated ($r = 0.92$) with the ploidy. Haploid plants revealed significantly fewer chloroplasts compared with diploids (Table 6). Chloroplast counts, performed on *in vitro*-grown plantlets, are a rapid and easy method for early identification of haploid and diploid plants. This technique is very useful because it requires little time, specialized equipment or technical skill.

In eggplant, the haploid condition is highly stable. Plants remain haploid over several years, even after repeated decapitations. In addition, most of the plants regenerated through somatic organogenesis from haploid leaf-disks have maintained the haploid state (Rotino, unpublished).

It has been demonstrated that the auxin employed in the induction

Figure 6. (a) Chromosome number in root tip of haploid plant. (b) Haploid plant bearing small parthenogenetic fruits.

medium influenced the ploidy of the regenerated plants; with IAA all plants were haploid (Dumas de Vaulx and Chambonnet, 1982; see also Table 5). On the contrary, the heat treatment and the time of appearance of the embryos did not affect the ploidy level.

By analysis of recessive morphological markers (e.g., fruit color) or pro-

Table 6. Mean chloroplast number per guard cell pair of six haploid and six diploid plants derived from the same anther donor genotype

Diploid plants		Haploid plants	
mean	±se	mean	±se
9.7 b	0.1	6.1 c	0.2
11.9 a	1.1	6.9 c	0.3
9.9 ab	0.2	6.2 c	0.6
10.5 ab	0.3	5.1 c	0.2
9.7 b	0.1	5.7 c	0.3
10.1 ab	0.3	5.7 c	0.1
10.3*	0.3	5.9*	0.2

Means bearing the same letter are not significantly different according to Tukey's test ($p = 0.05$). * = General mean.

geny tests, diploid anther-derived eggplants have been demonstrated to be mostly of gametophytic origin.

In order to obtain fertile doubled haploid homozygous plants, the chromosome number must be doubled. This is often a difficult step. Normally, dormant secondary auxillary buds are treated with colchicine (dissolved in lanolin paste). On average, 30–40% of the treated haploid plants can be expected to double their chromosome number. Cloning of haploids by *in vitro* cutting, increases the probability of their conversion to the diploid status. The best results have been obtained with young and fast growing plantlets.

4. Incorporation of doubled haploids in breeding

For breeding purposes a large number of homozygous plants are needed. Thus plants regenerated from microspores are ideal. Eggplant anther culture technique has been successfully incorporated in commercial breeding programs in France, Italy and other countries. Seed companies, mainly in France, are exploiting the potential of doubled-haploid homozygous lines for releasing hybrid cultivars. Prerequisites for success have been: availability of a broad range of germplasm; ease to convert haploids to fertile diploids; sufficient DH lines to allow breeders to select desirable traits.

4.1. *Doubled haploid line variability*

Genetic variation has been observed among doubled haploid (DH) lines derived from both inbred cultivars and heterozygous donors.

4.1.1. *Variability from inbred cultivars*

Genetic variation has been observed among DH lines arising from inbred cultivars (Sanguineti *et al.*, 1990; Rotino *et al.*, 1991a,b). Field evaluation of self-pollinated progenies derived from androgenic plants revealed high intra-line homogeneity. For example, Table 7 shows variation in DH lines derived from two different inbred cultivars. Taking into account six traits, the DH lines derived from Dourga showed significant modifications compared with the anther donor parental line, while the DH lines derived from Lunga Violetta di Napoli (L.V.N.) did not show any modification. Significant differences were also revealed in all traits of Dourga-derived DH lines and for two traits in L.V.N.-derived DH lines. Similar patterns of genetic variability have also been observed for other inbred cultivars (Rotino, unpublished). Therefore, the amount of genetic variation seems to be dependent on the genotype of anther donor parental line. Probably, these genetic changes took place during the haploid phase as no segregation occurred in the self-pollinated progenies.

This variation observed in eggplant DH lines, derived from inbred cultivars, could be due to partial heterozygosity of the anther donor (residual heterozygosity). Genetic change might also occur in the pollen DNA prior to, or during, the *in vitro* phase. In this case it can be termed gametoclonal variation (Evans *et al.*, 1984) in order to distinguish it from somaclonal variation (Larkin and Scowcroft, 1981). In other species gametoclonal variation was not due to residual heterozygosity, since it occurred again after subsequent cycles of androgenesis (Picard *et al.*, 1986; Evans, 1986).

Our studies on eggplant have revealed that morphological variation in androgenic lines was higher than that expected from residual heterozygosity. For instance, variation in fruit size shown in Fig. 7 and Table 7 may represent putative gametoclonal variation. More revealing information about factors involved in eggplant gametoclonal variation will be obtained from field tests and molecular analyses of DH lines derived from a second cycle of anther culture.

Whatever the cause, the occurrence of genetic variation from highly inbred genotypes may be exploited in practical breeding especially when either a single or a few traits need to be modified. Gametoclonal variation from homozygous genotypes could also be effectively employed in physiological and pathological studies.

4.1.2. *Variability from heterozygous genotypes*

When a heterozygous genotype is used as anther donor, a wide range of variability in both quantitative and qualitative traits can be observed among

Table 7. ANOVA significance, mean values of parental anther donors and their corresponding DH lines; minimum and maximum mean values of DH lines and number of DH lines significantly superior or inferior to parental lines for six traits

	DOURGA						LUNGA VIOLETTA DI NAPOLI					
Trait	Anova significance	Parental mean	DHs[a] mean	DHs[a] min.-max.	N° DH lines [c] superior	N° DH lines [c] inferior	Anova significance	Parental mean	DHs[b] mean	DHs[a] min.-max.	N° DH lines [c] superior	N° DH lines [c] inferior
Yield (kg/plant)	**	1.68	1.64	1.25-2.17	1	1	ns	1.91	1.90	2.11-1.57	0	0
Fruits/plant (n°x10)	**	1.04	1.00	1.21-0.82	0	2	ns	0.93	1.05	1.22-0.89	0	0
Fruit weight (hg)	**	1.60	1.65	2.25-1.50	2	1	*	2.03	1.82	2.08-1.68	0	0
Fruit shape[d]	**	2.96	2.87	3.23-1.87	0	1	*	3.72	3.76	3.96-3.53	0	0
Ped. length[e] (cm)	**	4.1	4.20	5.00-3.80	1	0	ns	6.20	6.20	6.50-5.80	0	0
Plant height (cm)	**	84	81	72-85	0	3	ns	92	94	101-87	0	0

a = from 9 DH lines (Rotino *et al.*, 1991); b = from 6 DH lines (Rotino, unpublished); c = according to two-tailed Dunnet's test (p=0.05); d = fruit polar/equatorial diameter ratio; e = fruit peduncle length; ** significant at p = 0.01, * significant at p = 0.05; ns = not significant.

Figure 7. Variability of fruit size and shape of DH lines derived from inbred cultivar Dourga. Note the different shape of the DH line in the middle of the lower row.

the DH lines. In addition to the variation discussed in the previous paragraph, androgenesis fixes genetic recombinations which occur during meiosis.

Compared with the classical breeding method of self-pollination, the main advantage of anther culture is that it saves time in obtaining pure lines. Two years after anther culture it is possible to include eggplant DH lines in comparative field trials. This is less than half the time required by sexual reproduction. Anther-derived DH lines may be released as self-pollinated cultivars or used as parents of F_1 hybrids.

For breeding purposes, DH lines should represent the same genetic variations that result from sexual recombination in the donor plant. Other factors to consider are: the heterozygosity of the anther donor (F_1, F_2 or advanced selected progenies) and the genetic inheritance of the desirable traits. By Correspondence Analysis, Borgel and Arnaud (1986) estimated that the total variability of 109 eggplant DH lines was wider than that of the corresponding 12 anther donor plants.

A wide range of variability has been recovered among DH lines derived from F_1 hybrids between genetically distant parents. This variability concerned vigour, plant structure and height, fruit color and shape (Fig. 8), intensity of anthocyanin, earliness, yield, peduncle length, presence of spines, etc. The DH line mean values for the quantitative traits considered were generally distributed around the mid-parent value (Figs. 9 and 10). For all

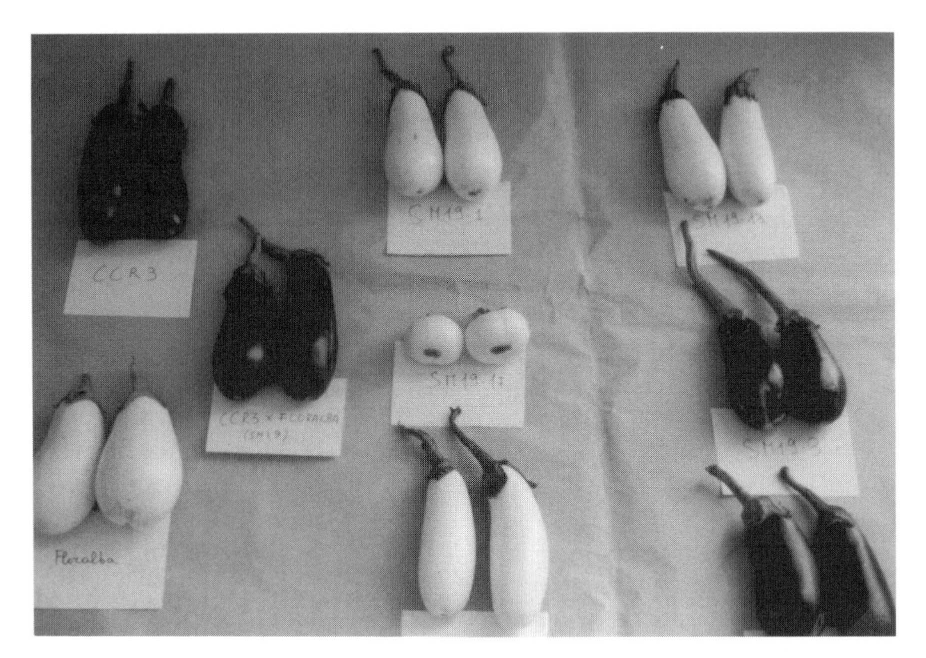

Figure 8. Wide variability of fruit characteristic of six DH lines obtained from anther culture of F_1 hybrid. On left: the two parents; beside them the F_1 hybrid. The six lines derived from the F_1 hybrid are on the right.

the traits except fruit shape, at least one DH line was significantly different from the values of the parents. With regard to yield, one line performed significantly better than the best parent and was not different from the F_1 hybrid. Moreover, the DH7 line stably fixed the desirable characteristics of Floralba (high yield and short plant) together with that of Lunga Violetta Barbentane (violet fruit) (Rotino *et al.*, 1994). High yielding DH lines derived from F_1 anthers have been reported previously (Sanguineti *et al.*, 1990; Rotino *et al.*, 1992). F_1-derived DH lines have also been used to determine the genetic basis of several traits (fruit and stem color, presence of spines, etc.) (Cabannes, 1983).

Although it is possible to obtain good recombinant DH lines from F_1 anthers, most of the DH lines recovered are useless for practical breeding. Therefore, it is advisable to exploit anther culture technique in a segregating plant population that has been previously selected. In such plants, there is a higher probability to find favourable gene combinations, since the parental chromosomes have already undergone at least two recombination cycles. Furthermore, individuals can be selected for disease resistance, earliness, cold tolerance, etc. before anther culture.

Production of DH lines can be effectively applied when a relatively small number of genes are involved with the breeding objective or when the

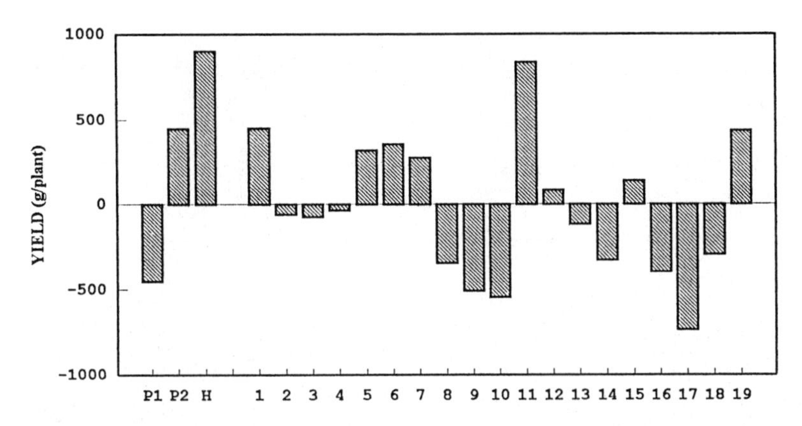

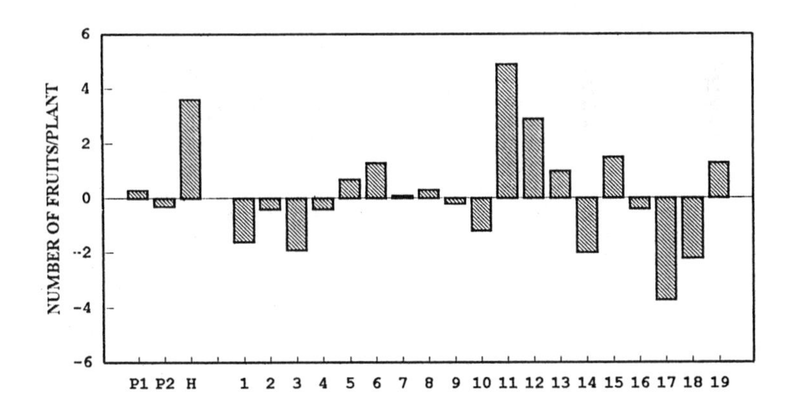

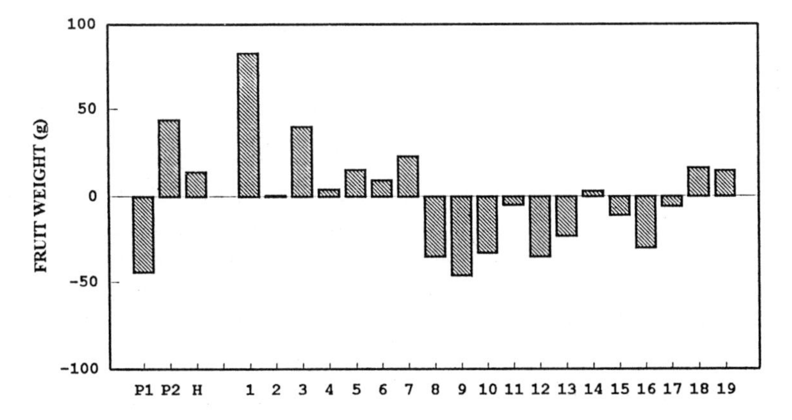

Figure 9. Variation of nineteen DH lines (1–19), the two parents (P1, P2) and the F_1 hybrid anther donor (H) with respect to the mid-parent value (x-coordinate) for yield, number of fruits per plant and fruit weight.

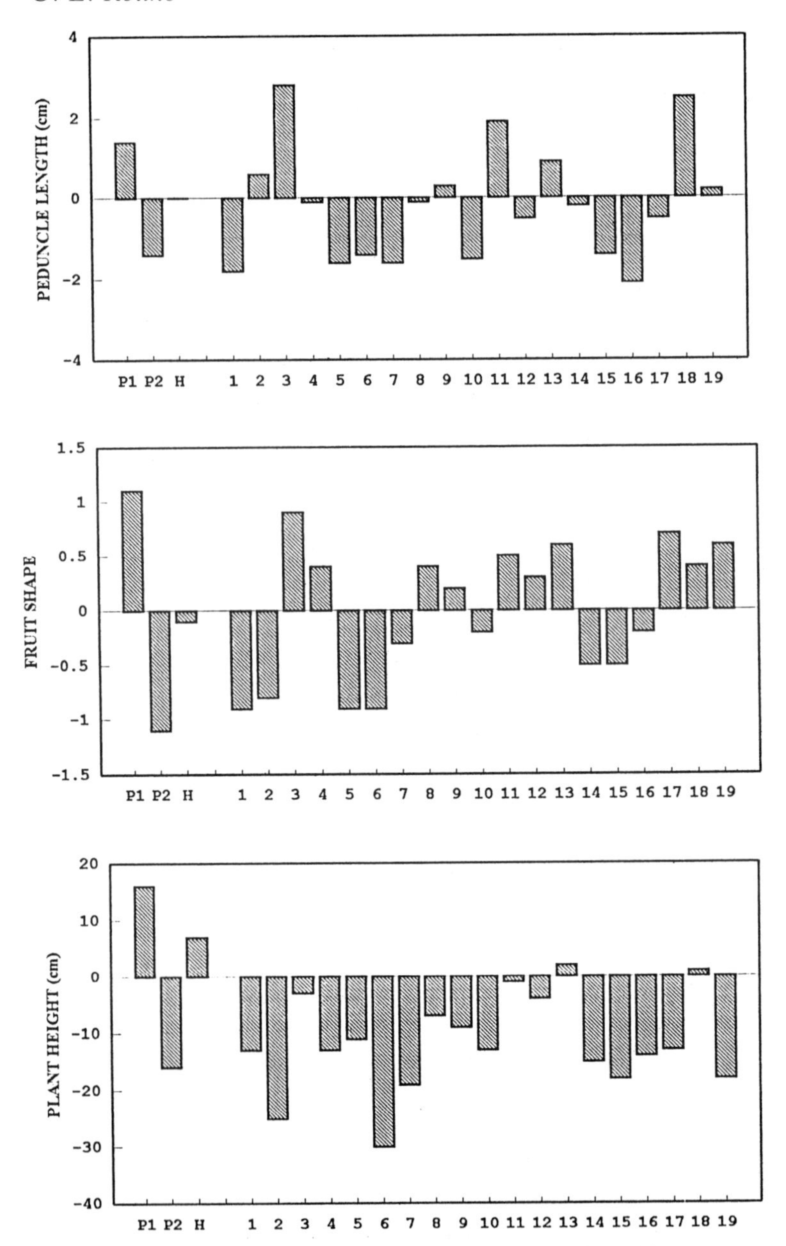

Figure 10. Variation of nineteen DH lines (1–19), the two parents (P1, P2) and the F_1 hybrid anther donor (H) with respect to the mid-parent value (x-coordinate) for fruit peduncle length, fruit shape and plant height.

desirable alleles are recessive and not closely linked. A useful application of anther culture technique is to isolate DH lines from advanced cycles of a recurrent selection breeding program.

4.2. *Field performance of DH line-derived hybrids*

In eggplant, although inbreeding depression does not represent a problem, F_1 hybrids can be advantageous for yield, disease resistance and quality. A breeding program based on the synthesis of hybrids was started in 1987 at Istituto Sperimentale per l'Orticoltura with the main objective of improving the tolerance to *Verticillium dahliae*, the most dangerous soil-borne disease in Italy. The parents of these hybrids are DH lines. *In vitro* androgenesis was performed using F_1 anthers from different crosses between cultivars and breeding lines (derived from recurrent selection) showing tolerance to *Verticillium* wilt (Restaino, 1986). The DH lines produced were evaluated in a naturally infected soil. F_1 hybrids were obtained by crossing the best DH lines. Artificial infection of some DH lines and F_1 hybrids showed a visible improvement of resistance to *Verticillium* (Cristinzio and Rotino, in preparation). Five F_1 hybrids between DH lines were recently tested at two locations: Montanaso Lombardo (Milano) and Monsampolo del Tronto (Ascoli Piceno), with the latter field naturally infected by *Verticillium*. In the non-infected soil (Montanaso L.) the early and total marketable yields of DH line-derived hybrids were comparable to those of commercial hybrids (Table 8). In the infected soil, three of the five DH line-derived hybrids yielded more than commercial and other experimental hybrids (Table 9), confirming the potential of homozygous DH lines (Nicklow, 1983) for releasing *Verticillium*-tolerant F_1 hybrids.

5. Concluding remarks

Androgenic eggplant can be obtained from a wide range of genotypes with frequencies between 8 to 40 plants per 100 cultured anthers. This allows incorporation of anther culture technique into practical breeding programs. Further improvement could be achieved by: (a) increasing the number of responding anthers; (b) having a high frequency of microspores which form well-developed embryos; and (c) increasing the efficiency of chromosome doubling and/or direct regeneration of diploid plants. Additional improvements could be obtained by isolated microspore culture. Cytological, physiological and biochemical studies would help to solve the problems described above by providing a better knowledge of the early events of *in vitro* androgenesis.

Another important topic concerns the molecular aspects of gametoclonal variation which could throw light on DNA changes during microsporogenesis and *in vitro* androgenesis.

Table 8. Agronomic comparison of five DH line-derived F_1 hybrids with five commercial F_1 hybrid cultivars (Rotino, unpubl.)

Hybrid	Marketable yield (g/plant) early	Marketable yield (g/plant) total	Total number of fruits/plant	Fruit weight
DH001	740 b	3012 a-c	12.8 a	236 de
DH002	690 b	3078 ab	10.6 a-c	289 b
DH003	515 b	2442 cd	9.0 c	270 bc
DH004	149 c	2545 b-d	10.3 bc	249 cd
DH005	643 b	2279 d	10.3 bc	223 e
Galine	694 b	3569 a	10.2 bc	351 a
Megal	503 b	2602 b-d	10.6 a-c	244 de
Ancha	250 c	2902 bc	8.7 c	340 a
Fabina	1069 a	3136 ab	12.3 ab	254 cd
Talina	600 b	2639 bd	11.5 ab	230 de

In the field: 4 replication, 2,5 plant/mq, 8 plants per plot. In each column means followed by at least one identical letter are not significantly different according to Duncan's multiple range test (p = 0.05).

In eggplant, regeneration from protoplasts derived from haploids plants provides an ideal single cell system for genetic manipulation, somatic hybridization and mutant selection both for applied and basic studies.

6. Protocol (based on modified procedure of Chambonnet and Dumas de Vaulx, 1983)

6.1. *Prerequisite*

Healthy and vigorous greenhouse- or field-grown donor plants.

6.2. *Stage of microspores for anther culture*

Preferably late-uninucleate microspores, but early uni- or bi-nucleate ones can also be used.

Cytological analysis: squash one anther on freshly prepared DAPI-TRIS buffer (TRIS buffer: 0.05 M TRIS-HCl, 0.5% Triton X-100, 5% sucrose, pH 7) and observe under UV light.

6.3. *Media* (*Table* 2)

- Induction medium: basal medium C supplemented with agar 0.9%, sucrose 120 g/l or sucrose 60 g/l + glucose 63 g/l, pH 5.9. As growth regulators

Table 9. Agronomic comparison of DH line-derived hybrids, experimental and commercial hybrids. Ranks in yield, fruit number and fruit weight (Acciari and Rotino, unpubl.)

Hybrid	Yield		Total fruit	Fruit
	early	total	number	weight
DH001	6	1	1	12
DH002	13	2	3	14
DH003	8	3	5	5
008-89	16	4	17	1
003-89	5	5	2	16
A57xA62	1	6	14	12
A63xA76	11	7	7	7
JERSEY KING	18	8	8	10
A63xA67	3	9	16	3
A67xA76	12	10	13	6
009-89	15	11	18	2
DH004	9	12	4	17
005-89	4	13	11	9
DH005	2	14	12	8
ARTICA	14	15	15	11
011-89	12	16	6	15
RIMINA	7	17	10	18
007-89	10	18	19	4
INDIRA	17	19	9	18

*In the field: 3 replications; mean range: yield (t/ha): early = 14.1-3.8, total = 80.2-49.7; fruit number/plant = 15.3-8.3; fruit weight (g) = 335-168.

use, (in order of priority) one of the following four combinations (μM): 23.3 KIN + 22.6 2,4-D; 4.6 ZEA + 16.1 NAA; 23.3 KIN + 26.8 NAA; 13.9 KIN + 5.7 IAA.

- Regeneration medium: basal medium R supplemented with 0.46 μM of the same cytokinin present in the induction medium.
- Medium for stimulating the germination of malformed embryos: basal medium R supplemented with 9.3 μM KIN.
- Medium for plantlet growth and multiplication: growth regulator-free V3 medium.

6.4. *Disinfection and culture*

Disinfect flower buds for 20 min in a solution of 30% commercial bleach (1% active chlorine), followed by three washings with sterile demineralized water. After flower dissection, plate 10–12 anthers in a 60 mm petri dish containing medium C, petri dishes are sealed using household plastic wrap.

6.5. *Condition of culture*

A– Keep plated anthers at 35°C in the dark for 8 days.
B– Transfer the anthers to 25°C, 16 h day (50 μE m^{-2} s^{-1}).
C– After 4 days transfer the anthers to R medium.
D– Every 5–6 weeks transfer anthers to fresh R medium.

6.6. *Cytological examinations*

- Assessment of microspore viability: dilute FDA (fluorescein diacetate) stock solution (5 mg/ml acetone) with equal volume of MS liquid basal medium 2% sucrose, squeeze anthers in the solution and observe microspore under fluorescent blue light.
- Assessment of microspore nuclear division: use DAPI stain as in Section 6.2.

6.7. *Assessment of ploidy level*

Chromosome counting in root-tip of acclimatized plants can be performed following Feulgen standard procedure (Venora *et al.*, 1992).

Chloroplast counting in guard cell pairs of leaves from *in vitro* grown plantlets: wet the lower epidermis with FDA stock solution diluted in water (1:50) using a small paint brush and quickly observe under fluorescent light.

6.8. *Chromosome doubling*

Dissolve colchicine in lanolin paste (0.5%). Decapitate apical bud, remove the axillary buds with a scalpel and treat the secondary buds with colchicine. Keep treated plantlets in fresh condition and darkness for 2 days.

7. Acknowledgements

The author is thankful to Dr. A. Falavigna for his useful suggestions and critical review of the manuscript. Appreciation is also expressed to the Italian Ministero per le Risorse Agricole Alimentari e Forestali for the untiring supports of researches in the framework of the Project "Sviluppo di Tecnologie Avanzate Applicate alle Piante".

8. References

Aubert, S., M.C. Daunay and E. Pochard, 1989a. Saponosides stèroidiques de l'aubergine (*Solanum melongena* L.). I. Intérêt alimentaire, méthodologie d'analyse, localisation dans le fruit. Agronomie 9: 641–651.

Aubert, S., M.C. Daunay and E. Pochard, 1989b. Saponosides stéroidiques de l'aubergine (*Solanum melongena* L.). II. Variations des teneurs liées aux conditions de récolte, aux génotypes et à la quantité de graines des fruits. Agronomie 9: 751–758.

Borgel, A. and H. Arnaud, 1986. Progress in eggplant breeding, use of haplomethod. Capsicum Newsl. 5: 65–66.

Cabannes, M., 1983. Utilisation des haploides issus de culture *in vitro* d'anthères chez l'aubergine (*Solanum melongena* L.). In: Mémorie de fin d'étude de l'École Nationale Supérieure Agronomique de Montpellier, pp. 1–41.

Chambonnet, D., 1985. Culture d'anthères *in vitro* chez trois Solanacées maraichères: Le piment (*Capsicum annuum* L.), l'aubergine (*Solanum melongena* L.), la tomate (*Lycopersicon esculentum* L.) et obtention de plantes haploïdes. Thèse de Docteur d'Université, Academie de Montpellier.

Chambonnet, D. and R. Dumas de Vaulx, 1983. A new anther culture medium performant on various eggplant (*Solanum melongena* L.) genotypes. In: Eucarpia Meeting on Capsicum and eggplant, 4–7 July 1983, Plovdiv, pp. 38–41.

Dumas de Vaulx, R., 1986. Rapport d'activité 1984–1985 INRA Station d'amélioration des plantes maraichères d'Avignon Montfavet, p. 94.

Dumas de Vaulx, R. and D. Chambonnet, 1982. Culture *in vitro* d'anthères d'aubergine (*Solanum melongena* L.): stimulation de la production de plantes au moyen de traitments à +35°C associés à de faibles teneurs en substances de croissance. Agronomie 2: 983–988.

Dumas de Vaulx, R., D. Chambonnet and E. Pochard, 1981. Culture *in vitro* d'anthères de piment (*Capsicum annuum* L.): amélioration des taux d'obtention de plantes chez différents génotypes par des traitments à +35°C. Agronomie 1: 859–864.

Evans, D.A., 1986. Somaclonal and gametoclonal variation. In: P.C. Augustine, H.D. Danforth and M.R. Bakst (Eds.), Biotechnology for Solving Agricultural Problems, pp. 63–96. Martinus Nijhoff Publishers, Dordrecht.

Evans, D.A., W.R. Sharp and H.P. Medina-Filho, 1984. Somaclonal and gametoclonal variation. Amer. J. Bot. 71: 759–774.

Franceschetti, U. and G. Lepori, 1985. Impollinazione naturale incrociata nella melanzana. Sementi Elette 6: 25–28.

Gleddie, S., W.A. Keller and G. Setterfield, 1983. Somatic embryogenesis and plant regeneration from leaf explants and cell suspension of *Solanum melongena* (eggplant). Can. J. Bot. 61: 656–666.

Gleddie, S., W.A. Keller and G. Setterfield, 1986. Somatic embryogenesis and plant regeneration from cell suspension derived protoplasts of *Solanum melongena* (eggplant). Can. J. Bot. 64: 335–361.

Gu, S.R., 1979. Plantlets from isolated pollen culture of eggplant (*Solanum melongena* L.). Acta Bot. Sin. 21: 30–36.

Hansson, B., 1978. Temperature shock: a method of increasing the frequency of embryoid formation in anther culture of Swede rape (*Brassica napus* L.). Sver. Utsaedsfoeren Tidskr. 3: 141–148.

Heberle-Bors, E., 1985. *In vitro* haploid formation from pollen – a critical review. Theor. Appl. Genet. 71: 361–374.

Hinata, K., 1986. Eggplant (*Solanum melongena* L.). In: Y.P.S. Bajaj (Ed.), Biotechnology in Agriculture and Forestry, Vol. 2, pp. 363–370. Springer-Verlag, Berlin.

Isouard, G., C. Raquin and Y. Demarly, 1979. Obtention de plantes haploïdes et diploïdes par culture *in vitro* d'anthères d'aubergine (*Solanum melongena* L.). C.R. Acad. Sci. Paris Série D, 288: 987–989.

Kalloo, G., 1993. Eggplant, *Solanum melongena* L. In: G. Kalloo and B.O. Bergh (Eds.), Genetic Improvement of Vegetable Crops, pp. 587–604. Pergamon Press, Oxford.

Khan, R., 1979. *Solanum melongena* and its ancestral form. In: J.G. Hawkes, R.N. Lester and A.D. Skelding (Eds.), The Biology and Taxonomy of Solanaceae, pp. 629–636. Academic Press, London.

Keller, W.A. and K.C. Amstrong, 1979. Stimulation of embryogenesis and haploid production in *Brassica campestris* anther culture by elevated temperature treatments. Theor. Appl. Genet. 55: 65–67.

Larkin, P.J. and W.R. Scowcroft, 1981. Somaclonal variation – a novel source of variability from cell culture for plant improvement. Theor. Appl. Genet. 60: 197–214.

Maheshwari, S.C., A.K. Tyagi, K. Malhotra and S.K. Sopory, 1980. Induction of haploidy from pollen grains in Angiosperms – the current status. Theor. Appl. Genet. 58: 193–206.

Miyoshi, K., 1994. Efficient callus induction and subsequent plantlet formation from isolated microspore of eggplant (*Solanum melongena* L.). In: VIII Intern. Congr. of Plant Tissue and Cell Culture, June 12–17, Firenze, Abstract S3–25, p. 87.

Murashige, T. and F. Skoog, 1962. A revised medium for rapid growth and bioassays with tobacco tissue cultures. Physiol. Plant. 15: 473–497.

Nicklow, C.W., 1983. The use of recurrent selection in efforts to achieve *Verticillium* resistance in eggplant. HortScience 18: 600 (Abstr.).

Picard, E., A. Rode, A. Benslimane and L. Parisi, 1986. Gametoclonal variations in doubled haploids of wheat: biometrical and molecular aspects. In: J. Semal (Ed.), Somaclonal Variations and Crop Improvement, pp. 136–147. Martinus Nijhoff Publishers, Dordrecht.

Porcelli, S., 1986. Problemi e prospettive della melanzana. Cenni introduttivi. Agricoltura Ricerca 60: 1–2.

Quagliotti, L., 1979. Floral biology of *Capsicum* and *Solanum melongena*. In: J.C. Hawkes, R.N. Lester and A.D. Skelding (Eds.), The Biology and Taxomony of Solanaceae, pp. 399–419. Academic Press Inc., New York.

Quagliotti, L. 1992. Pepper and eggplant breeding in Italy in the last twenty years. In: EUCARPIA Proceedings of the 7th Meeting on Genetics and Breeding of Capsicum and Eggplant, 7–10 September, Rome, pp. 18–31.

Raina, S.K. and R.D. Iyer, 1973. Differentiation of diploid plants from pollen callus in anther cultures of *Solanum melongena* L. Z. Pflanzenzüchtg. 70: 275–280.

Rashid, A. and H.E. Street, 1973. The development of haploid embryoids from anther culture of *Atropa belladonna* L. Planta 113: 263–270.

Research Group of Haploid Breeding, 1978. Induction of haploid plants of *Solanum melongena* L. In: Proc. Symp. Plant Tissue Cult., pp. 227–232. Sci. Press, Peking.

Restaino, F., 1986. Contributo al miglioramento genetico della melanzana per caratteri agronomici, tecnologici e per bioresistenze. Agricoltura Ricerca 60: 3–14.

Rotino, G.L. and S. Gleddie, 1990. Transformation of eggplant (*Solanum melongena*) using a binary *Agrobacterium tumefaciens* vector. Plant Cell Rep. 9: 26–29.

Rotino, G.L., A. Falavigna and F. Restaino, 1987. Production of anther-derived plantlets of eggplant. Capsicum Newsl. 6: 89–90.

Rotino, G.L., E. Pedrazzini and D. Perrone, 1994. Valutazione di linee diploaploidi di melanzana (*Solanum melongena* L.). In: Workshop "Biotecnologie avanzate applicate alle piante ortofrutticole", Proceeding of 2nd Giornate Scientifiche Società Orticola Italiana, 23 June, 1994, S. Benedetto del Tronto (Ascoli Piceno), pp. 25–30.

Rotino, G.L., M. Schiavi, E. Vicini and A. Falavigna, 1991a. Variation among androgenetic and embryogenetic lines of eggplant (*Solanum melongena* L.). J. Genet. Breed. 45: 141–146.

Rotino, G.L., F. Restaino, M. Gjomarkay, A. Falavigna, M. Schiavi and E. Vicini, 1991b. Evaluation of genetic variability in embryogenetic and androgenetic lines of eggplant. Acta Hort. 300: 357–361.

Rotino, G.L., M. Schiavi, A. Falavigna, E. Pedrazzini, D. Perrone, M. Salzano, A. Correale and F. Restaino, 1992. Comparison of eggplant doubled-haploid lines with their inbred and hybrid parental genotypes. In: EUCARPIA Proceeding of 7th Meeting on Genetics and Breeding of Capsicum and Eggplant, 7–10 September, 1992, Rome, pp. 283–288.

Sanguineti, M.C., R. Tuberosa and S. Conti, 1990. Field evaluation of androgenetic lines of eggplant. Acta Hort. 280: 177–181.

Siljak-Yakovlev, S. and G. Isouard, 1983. Contribution à l'étude caryologique de l'aubergine (*Solanum melongena* L.). Agronomie 3: 81–86.

Sunderland, N., 1971. Anther culture: a progress report. Sci. Prog. 59: 527–549.

Traas, J. and A.F. Chatelain, 1988. In: Annual Report of Institute for Horticultural Plant Breeding, Wageningen, pp. 48–49.
Tuberosa, R., M.C. Sanguineti and S. Conti, 1987a. Anther culture of egg-plant (*Solanum melongena* L.) lines and hybrids. Genet. Agr. 41: 267–274.
Tuberosa, R., M.C. Sanguineti, B. Toni and F. Cioni, 1987b. Ottenimento di apolidi in melanzana (*Solanum melongena* L.) mediante coltura di antere. Sementi Elette 3: 9–14.
Venora, G., C. Russo and A. Errico, 1992. Karyotype analysis in *Solanum melongena* L. In: Eucarpia Proceedings of the 7th Meeting on Genetics and breeding of Capsicum and Eggplant, 7–10 September, 1992, Rome, pp. 266–271.

9. Experimental haploidy in *Brassica* species

CONSTANTINE E. PALMER, WILFRED A. KELLER and PAUL G. ARNISON

Contents

1. Introduction

The genus *Brassica* comprises many familiar and commercially important food and oil-seed plants. Leaves, stems and roots have been cultivated as vegetables and animal fodder; both edible and industrial oils as well as condiments (mustards) have been prepared from the crushed seed.

Brassica species are recognized by abundant, typically bright yellow, cross-shaped flowers (hence crucifer). Brilliant yellow fields of oil-seed rape have become a common sight in western Canada, Europe and other parts of the world.

Brassica species have been selected and manipulated by breeders into remarkable arrays of morphological forms. Many people unfamiliar with *Brassicas* are impressed to find that broccoli, cauliflower, kales, cabbages, kohlrabi and Brussels sprouts are all varieties of the same species (*B. oleracea*). Additionally some species (e.g., *B. rapa*, *B. napus*) have both cultivated vegetable (turnip, rutabaga) and oil-seed forms.

The interested reader is referred to Brassica Crops and Wild Allies (1980, S. Tsunoda, K. Hinata and C. Gómez-campo, Eds., Japan Scientific Press, Tokyo) or texts on economic botany for further information.

S.M. Jain, S.K. Sopory & R.E. Veilleux (eds.), In Vitro *Haploid Production in Higher Plants, Vol. 2*, 143–172.

Anther culture and isolated microspore culture techniques have been developed and widely used for many *Brassica* species and varieties (Keller *et al.*, 1982; Maheshwari *et al.*, 1980). Factors that are known to influence the occurrence of haploids will be explored in this chapter whereas the use of haploids for experimental purposes and breeding will be the subject of the following chapter.

Haploid *Brassica* plants occur naturally from seed in low frequency and have been used for breeding; however, the capacity to truly exploit *Brassica* haploidy originated with the discovery of methods to induce the formation of embryos from anthers and isolated microspores *in vitro*. Haploid plants are currently produced in large numbers by culture of anthers or isolated microspores. These methods are now considered routine and have been adopted widely by *Brassica* breeders. Although there is considerable variation between species, varieties, cultivars and even plants of the same cultivar, anther and microspore culture techniques have been successful for most commercially important *Brassica* species and varieties, related species such as yellow mustard (*Sinapis*) and more distant weedy relatives (e.g., *Arabidopsis*). An overview of the range of *Brassica* and related materials that have been shown to respond to anther or microspore culture is shown in Table 1.

In addition to their use with *Brassica* species, anther and microspore culture techniques have been successful for closely related species such as *Arabidopsis thaliana* (Scholl and Amos, 1980; Gresshoff and Doy, 1972), *Sinapis alba* (Leelavathi *et al.*, 1984) and *Raphanus sativus* (Lichter, 1989).

Although anther culture techniques have been successful, not all plant material responds in the same way or to the same degree. Significant variation can be seen from plant to plant or even within the same plant. Analysis of the parameters that underlie these differences is needed to obtain a uniform and predictable response. Factors that are known to influence the success of anther culture include: growth and pre-culture conditions of donor plants, genetics of donor plants, anther culture temperature induction treatments, density of cultured anthers, composition of the culture medium and presence of plant growth regulators (Keller *et al.*, 1982).

2. Historical overview

The occurrence of haploid plants in angiosperms was first reported in *Datura stramonium* (Blakeslee *et al.*, 1922). Since then the spontaneous occurrence of haploids has been noted in several species (for reviews see Kimber and Riley, 1963; Melchers, 1972; Nitsch, 1972; Nitzsche and Wenzel 1977). In *Brassicas*, spontaneous haploidy occurs in a number of species including *B. napus*, *B. carinata*, *B. campestris*, and *B. oleracea* (Komatsu 1936; Kuriyama and Watanabe, 1955; Morinaga and Fukushima, 1933; Olsson and Hagberg, 1955; Ramanujam, 1941; Stringam and Downey, 1973; Thompson 1956,

Table 1. Anther or microspore culture of *Brassica* species

Species	Culture	Reference
B. carinata	Anther microspores	Renu-Arora *et al.*, 1988; Chuong and Beversdorf, 1985
B. juncea	Anther	Yadav *et al.*, 1988; Sharma and Bhojwani, 1985; George and Rao, 1983
B. napus		
Winter	Anther	Dunwell and Cornish, 1985; Thurling and Chay, 1984
	Microspores	Lichter, 1982
Spring	Anther	Lichter, 1981; Keller and Armstrong, 1978
	Microspores	Gland *et al.*, 1988; Swanson *et al.*, 1987
B. nigra	Anther	Govil *et al.*, 1986; Leelavathi *et al.*, 1987
	Microspores	Lichter, 1989; Hetz and Shieder, 1991; Margale and Chèvre, 1991
B. oleracea		
Broccoli	Anther	Keller and Armstrong, 1982; Arnison *et al.*, 1990a,b
	Microspores	Takahata and Keller, 1991
Brussels sprouts	Anther	Ockendon, 1984; Lelu and Ballon, 1985
Cabbage	Anther	Lillo and Hansen, 1987; Chiang *et al.*, 1985; Roulund *et al.*, 1990
Cauliflower	Anther	Bagga *et al.*, 1982; Phippen and Ockendon, 1990
Kohlrabi	Anther	Arnison and Keller, 1990
Kale	Anther	Keller and Armstrong, 1981; Arnison and Keller, 1990
B. rapa		
Spring canola	Anther	Keller and Armstrong, 1979
	Microspores	Burnett *et al.*, 1992; Baillie *et al.*, 1992
Chinese cabbage	Anther	Sato *et al.*, 1989a
	Microspores	Sato *et al.*, 1989b
Rapid cycling	Anther	Aslam *et al.*, 1990a,b

1969). These haploids are easily recognized by male sterile flowers and morphological characteristics, such as prolonged flowering, slender branching habit and narrow leaves (Olsson and Hagberg, 1955). They occur at low frequencies ranging from 0.05 to 19.4 per 1000 plants (Banga and Labana, 1986; Thompson 1969), and they undoubtedly arise through abnormal pollination and fertilization events or by mutations. It has been suggested that naturally occurring haploid plants can be found in all varieties of spring and winter rape (Thompson, 1974), although Banga and Labana (1986) reported that spontaneous haploids were found in only ten of 76 lines examined. A new cultivar and breeding lines have been developed using spontaneous haploids (e.g., Thompson, 1974); however there has been much less emphasis on isolation of spontaneous haploids recently. Attempts have been made to develop an efficient selection system for the isolation of spontaneous androgenic haploids at the seedling stage (Primard *et al.*, 1991).

In at least one instance parthenogenetic haploids have been produced in *Brassica* through interspecific crosses (Prakash, 1973, 1974); however, as for spontaneously occurring haploids, the frequency of parthenogenesis was low.

Following the initial success in the production of pollen plants from cultured anthers of *Datura innoxia* (Guha and Maheshwari, 1964, 1966), there were attempts to culture anthers from *Brassica* species. The initial report was by Kameya and Hinata (1970) who described callus and haploid plants from cultured anthers of *Brassica oleracea*. Subsequently microspore embryogenesis was reported in anther cultures of *Brassica napus* and *B. campestris* (Keller *et al.*, 1975; Thomas and Wenzel, 1975). In later reports haploids were produced from a number of *Brassica* species through anther culture (Jain *et al.*, 1980; Jain *et al.*, 1989; Keller and Armstrong, 1978, 1979; Leelavathi *et al.*, 1984; Lichter 1981; Ockendon, 1984; Sharma and Bhojwani, 1985). Procedures were developed for recovery of haploids from many *Brassica* species through anther culture and their value in rapeseed breeding has been evaluated (Charne and Beversdorf, 1991; Hoffmann *et al.*, 1982). These studies recognized the importance of genotype, donor plant growth conditions, culture medium and culture temperature to successful recovery of haploid embryos.

3. Microspore and anther culture

The microspore origin of haploid embryos was defined and it was recognized that, compared to the large number of microspores per anther, the frequency of embryogenesis was very low (Lichter, 1982). It has been suggested that the anther walls imposed a restriction on embryo development, allowing preferential embryogenesis of certain microspores (Hoffmann *et al.*, 1982). As a consequence methods were developed for the culture of isolated microspores and the first successful culture was reported by Lichter (1982). Using this technique, high frequency embryogenesis can now be obtained in a number of *Brassica* species. The nuclear stage of the microspore, donor plant growth conditions, genotype and culture conditions are critical considerations for efficient recovery of embryos from microspores.

The use of thermal shock to induce high rates of embryogenesis is of fundamental importance in both anther and microspore culture (Keller and Armstrong, 1978, 1979). The high temperature induction is now routinely employed in microspore and anther cultures of *Brassicas* followed by culture at lower temperatures. It was apparent, from early studies on *Hyacinthus orientalis*, that thermal shock diverts the developmental pathway of the microspore from a gametophytic to sporophytic pathway (Pandey, 1973).

Another key factor in microspore-derived embryogenesis is the carbohydrate level in culture media. In early studies of anther cultures in other species a relatively simple medium with 2–3 percent sucrose was employed (Nitsch, 1972). Also in the original experiments with *B. oleracea* (Kameya and Hinata, 1970) the culture medium contained low levels of sucrose. Subsequently, the beneficial effect of high levels of sucrose in the culture medium was recognized (Dunwell and Thurling, 1985; Keller *et al.*, 1975; Lichter, 1981; Matsubayashi and Kuranuki, 1975; Thomas and Wenzel, 1975). Dunwell and Thurling (1985) reported a comparison of the osmotic potential of the anther homogenate with that of the sucrose level in the culture medium. An optimal level for embryo induction and another for embryo development were established. In addition to its functions as a source of carbohydrates for general metabolism, sucrose at high levels may provide an osmotic stress necessary for embryo induction.

As a result of early experiments by Lichter (1981, 1982), the culture medium is now sufficiently well-defined to allow a high frequency of embryogenesis in many species and genotypes of *Brassica* (Gland *et al.*, 1988; Huang and Keller, 1989; Swanson *et al.*, 1987; Swanson 1990).

4. Factors influencing microspore embryogenesis

4.1. *Donor plant genotype*

One of the most important factors affecting successful microspore and anther culture in *Brassica* is the genotype of the cultured material. This phenomenon has been documented in all species of *Brassica* where microspore and anther culture have been attempted (Arnison *et al.*, 1990a,b; Arnison and Keller, 1990; Baillie *et al.*, 1992; Burnett *et al.*, 1992; Cardy, 1986; Chuong *et al.*, 1988; Dunwell *et al.*, 1985; Ockendon, 1985; Thurling and Chay, 1984). Even within genotypes there may be significant plant-to-plant variation (Ockendon, 1985; Phippen and Ockendon, 1990; Seguin-Swartz *et al.*, 1983). The situation is further complicated by genotype – environmental interactions (Dunwell *et al.*, 1985; Roulund *et al.*, 1990; Thurling and Chay, 1984). In *Brassica oleracea* convar. *capitata* (L.) (Alef.) field grown plants were generally more responsive than those grown under greenhouse conditions (Roulund *et al.*, 1990). This indicates that some genotypes may have specific environmental requirements in order to yield embryogenic microspores. Genotypic influence is manifested in the frequency of embryogenesis, the quality of embryos and the mode of plant regeneration (Arnison and Keller, 1990; Chuong *et al.*, 1988; Duijs *et al.*, 1992; Kieffer *et al.*, 1993).

Evidence that androgenesis is strongly influenced by the genotype of the donor material has been suggested by studies between highly responsive genotypes and non-responding ones (Cardy, 1986; Dunwell *et al.*, 1985; Wenzel, 1980). Ockendon and Sutherland (1987) reported that about half the variation they observed in Brussels sprouts was genetic and half was due to non-genetic factors. The impact of genetic predisposition to respond to culture has also been observed in a comparison of anther culture response of doubled-haploid plants derived from a common parent (Arnison and Keller, unpublished).

In order to reduce variation in response to culture in broccoli (Arnison *et al.*, 1990a,b) experimental materials were taken from secondary branches of individual plants, known to be either highly responsive or null, as well as from cell culture-derived vegetative clones of previously tested plants. This practice greatly increased the predictability of response. Groups of individual doubled-haploid plants characterized as highly responsive were crossed with doubled-haploids from the same parent stock with a null or zero response. As expected the crosses between highly responsive plants were all highly responsive, crosses between high and low plants showed variable but intermediate values and the crosses between the null plants resulted in only a null response (Arnison and Keller, unpublished).

The clear distinction in response between the groups and the intermediate values of the hybrids suggest the involvement of a small number of genes. Interpretation of these results is complicated, however, by the observation that the results for reciprocal crosses between plants were often significantly different; some plants apparently contributed a low response phenotype only when used as a female parent. In addition it was subsequently observed (Arnison *et al.*, 1990b) that some null or poorly responding materials showed a greatly increased response when low concentrations of cytokinins were added to the culture medium. The involvement of plant growth regulatory substances in the genotypic response to culture is now well documented (Biddington *et al.*, 1988, 1992; Dunwell and Thurling, 1985; Gland *et al.*, 1988).

Attempts to investigate the molecular events occurring during the embryo induction process that could be correlated with genetic differences were conducted using two-dimensional gel electrophoresis of proteins from anthers of doubled-haploid plants previously characterized as responsive and unresponsive to culture (Fabijanski *et al.*, 1992). These experiments demonstrated that the initial phase of culture involves a typical heat shock response that was similar to the one previously documented in other broccoli tissues (Fabijanski *et al.*, 1987). Differences in extent of heat shock protein expression and presence or absence of some individual heat shock proteins were detected between individual plants; however no obvious correlation with response to culture was evident (Fabijanski *et al.*, 1992).

The genetics of responsiveness in anther and microspore culture protocols is an area where little hard information is available and is certainly worthy of further study.

4.2. *Donor plant physiology*

The conditions under which donor plants are grown is an important consideration for successful microspore culture in *Brassica* as well as in other species. There is general agreement in the literature that plants grown at low temperatures yield the most responsive microspores (Dunwell *et al.*, 1985; Keller and Stringam, 1978; Thurling and Chay, 1984). The extent of such low temperature requirement may be genotype or species dependent since highly responsive microspores were obtained from *Brassica napus* plants grown at 10°C/5°C day/night temperature (Keller *et al.*, 1987) while for some genotypes of *B. juncea*, a temperature regime of 20°C/15°C day/night temperature was required (Bevis and Keller, unpublished). In an examination of donor plant physiology Keller *et al.* (1987) reported maximum microspore embryogenic frequency from *B. napus* plants grown at 10°C/5°C compared to 20°C/15°C and 15°C/10°C. Microspores from winter type rape may be more responsive to low growth temperatures compared to summer type rape (Dunwell *et al.*, 1985).

Some genotypes of *B. oleracea*, especially those that are typically grown under cool conditions, respond favourably to cool growth temperatures; favourable response was obtained from plants grown at 5 to 15°C (Duijs *et al.*, 1992) or from plants that had over-wintered under minimal greenhouse conditions (Arnison and Keller, 1990). Under environmentally controlled conditions the response of broccoli plants is greatly diminished by sudden changes in temperature regime; however the anther culture response is typically restored after 12 days at the new temperature (Arnison and Keller, 1990). Other reports confirm that donor plant growth temperatures of 10 to 20°C are suitable for high frequency microspore-derived embryo production (Roulund *et al.*, 1990; Yang *et al.*, 1991). In some instances there may be seasonal variations in microspore culture response and, with *B. campestris*, the best responses were obtained in spring and autumn (Sorvari, 1985). Even though low donor plant growth temperatures are generally required, responsive microspores can also be obtained from greenhouse and field-grown plants (Jain *et al.*, 1989; Sorvari, 1985).

Although plant growth temperature is often a critical factor, the general physiological state of the donor plant is important. This is reflected in the influence of donor plant age on microspore embryogenesis. In *B. napus* higher frequencies of embryogenesis were obtained with older inflorescences compared with younger ones (Takahata *et al.*, 1991) in the case of young plants, while with older plants new (young) inflorescences yielded more embryogenic microspores. In other studies, as inflorescence age increased, the frequency of embryogenesis declined (Chuong *et al.*, 1988). The relevance of plant and bud age to successful microspore culture deserves further study as this may be influenced by other factors such as the nutritional status of the plant. Factors such as light intensity, mineral nutrition and freedom from disease and pests can all influence the responsiveness of microspores to culture.

An important question to be answered is how treatment of donor plants at low temperature influences the responsiveness of microspores to culture. It is assumed that such temperatures increase the frequency of embryogenic microspores by a general slowing of metabolism. Evidence to support this contention is lacking. However, in *Brassica oleracea* donor plant growth conditions were shown to affect the distribution of microspores of a particular nuclear stage (Duijs *et al.*, 1992). High light intensity increased the number of binucleate microspores in the general microspore/pollen population (Duijs *et al.*, 1992).

Studies with *B. napus* indicated that donor plant growth temperature altered the cytological composition of microspores and as a consequence embryogenic microspores may be physiologically different from non-embryogenic ones (Lo and Pauls, 1992). The developmental characteristics of the donor plants were altered by varying the growth conditions. This indicates a change in the physiology of the plants as the highest embryo yields were correlated with growth conditions which delayed flowering and increased plant height (Lo and Pauls, 1992). Also with low growth temperatures embryos could be produced from a wider range of bud sizes compared to higher growth temperatures.

Low temperature requirement may relate to endogenous levels of growth regulators and nutrient status of the anthers and, consequently, the microspores (Lo and Pauls, 1992). Such correlations have not yet been measured and reported in the literature.

It is well-established that environmental stress alters the level of growth regulators, amino acids, polyamines, carbohydrates and lipids in plants (Aspinall and Paleg, 1981; Flores and Galston, 1982; Guerrero and Mullet, 1986; Guy *et al.*, 1992; Hanson and Hitz, 1982; Pearcy, 1978; Rikin *et al.*, 1993). Since some donor plant growth temperatures are stressful, these metabolic changes may be relevant to embryogenic potential. In this regard L-proline increases in embryogenic tissues and its analogue, thioproline is used to select embryogenic cells (Shetty and McKersie, 1993). It is likely that embryogenic microspores are rich in L-proline. It would therefore be useful to analyze the relationship between the metabolic status of the microspores and environmental growth conditions of the donor plants.

4.3. *Stage of microspore development*

A review of the literature on microspore or anther culture in *Brassica* reveals that embryogenesis is related to specific stages of microspore and pollen development (Fan *et al.*, 1988a; Kott *et al.*, 1988a; Telmer *et al.*, 1993). This varies from early uni-nucleate to the early binucleate stage. In a study of 2 genotypes of *B. napus*, the highest frequency of embryogenesis occurred with microspores at the late uni-nucleate stage (Kott *et al.*, 1988a). Similarly with the highly embryogenic cultivar of *B. napus*, Topas, late uni-nucleate to early binucleate stages were the most responsive (Fan *et al.*, 1988a; Hansen and Svinnset, 1993; Pechan and Keller, 1988; Telmer *et al.*, 1992). In other *Brassica* species this stage was also optimal (Baillie *et al.*, 1992; Duijs *et al.*, 1992; Hamaoka *et al.*, 1991; Kieffer *et al.*, 1993; Yang *et al.*, 1992); however, at least two reports for *B. oleracea* (Cao *et al.*, 1990; Kameya and Hinata, 1970) claimed that embryogenesis occurred from advanced tri-nucleate pollen grains.

Depending on growth conditions, flower buds containing microspores at the appropriate stage for culture can be visually selected and a fairly good correlation has been established between flower bud size and the frequency of microspore embryogenesis (Kott *et al.*, 1988a; Polsoni *et al.*, 1988; Telmer *et al.*, 1992). This allows ready selection of buds for large scale microspore isolation. Therefore, close monitoring of inductive growth conditions is required for successful use of parameters such as bud size and anther length.

It is not yet resolved whether microspores outside the late uni-nucleate to early binucleate stages have not acquired, or have lost competence for embryogenesis and why this phenomenon is restricted to this phase. The late uni-nucleate to binucleate stage is when the developing microspores undergo a mitotic cell division. The culture procedure involves temperature shocks that could physically alter the mitotic microtubule apparatus (Simmonds *et al.*, 1991). Zaki and Dickinson (1991) reported that embryogenesis was stimulated by colchicine which is known to affect microtubule organization. Other temperature-sensitive components of the cell, such as factors affecting gene regulation, may have an effect similar to that in insects known as phenocopy (Fabijanski *et al.*, 1992). The normal developmental process of the microspore is disturbed and, instead of a single division, the cell division process continues and an embryo develops. This change may be equivalent to a jump from the genetic program to develop into a gamete to the next genetic program that would ensue after fertilization to form an embryo. The number of individual microspore cells that develop into embryos represent a small fraction of the total. As this is true for doubled-haploid plants it is unlikely that the lack of response is because genetic potential is absent, but is more likely a requirement to be at a very specific stage of development.

4.4. *Pretreatments*

Pretreatment of isolated buds prior to microspore isolation and culture has been less frequently used in the *Brassicas* compared to other species, such as cereals. In early studies by Lichter (1982) the frequency of microspore-derived embryos was enhanced by subjecting flower buds to 4°C before microspore isolation. Also in *B. oleracea* a 4°C pretreatment of isolated buds for 40 h enhanced embryogenesis in a genotype-dependent manner (Osolnik *et al.*, 1993). Pretreatment of anthers at reduced atmospheric pressure appeared to enhance microspore embryogenesis in *Brassica hirta* (Klimaszewska and Keller, 1983); however, it would appear that if donor plants are grown under optimal environmental conditions, there is no significant advantage of flower bud pretreatment.

Pretreatment of isolated microspores with low levels of irradiation, ethanol, modified atmosphere and anti-microtubule agents such as colchicine have been shown to enhance embryogenesis in *Brassicas* (Klimaszewska and Keller, 1983; MacDonald *et al.*, 1988a,b; Pechan and Keller, 1989; Zaki and Dickinson, 1991). Such treatments are, however, not routinely used in microspore culture in these species and there is no evidence that these replace the requirement for high temperatures.

4.5. *Culture conditions*

4.5.1. *Medium composition*

The most commonly used medium for *Brassica* microspore and anther culture has been the formulation developed by Lichter (1982) with various modifications (Gland *et al.*, 1988; Huang and Keller, 1989; Lichter, 1989). These formulations supply the mineral salts, vitamins and carbon required for embryogenic development. Although there have been variations in the mineral component, designed to enhance embryogenesis (Gland *et al.*, 1988) the most critical component appears to be the carbohydrate quality and quantity.

4.5.2. *Carbohydrates*

Sucrose is the most commonly used carbohydrate for both anther and microspore culture of *Brassicas*. In earlier studies employing a number of *Brassica* species, Matsubayashi and Kuranuki (1975) established a level of 10% while in other studies optimal concentrations range from 10 to 14% (Duijs *et al.*, 1992; Gland *et al.*, 1988; Keller and Armstrong, 1978; Roulund *et al.*, 1990). In a study by Dunwell and Thurling (1985) using 4 cultivars of *B. napus* L., 20% sucrose gave the greatest embryo production from cultured anthers. However, lower levels were required for embryo development. These authors indicated that the osmotic potential of anther homogenates was equal to that of 17% sucrose and concentrations of exogenous sucrose in this range were probably favourable to embryo induction. There is an indication that culturing anthers/microspores at high sucrose levels can remove some response differences between cultivars and even those due to poor plant growth conditions (Dunwell and Thurling 1985). This is the most detailed study of sucrose requirements for embryo induction and development. In other *Brassica* species, *B. campestris*, *B. nigra* and *B. oleracea*, sucrose has been established as the most suitable carbohydrate while maltose, glucose and glucose + fructose were less effective (Baillie *et al.*, 1992; Gland *et al.*, 1988; Lichter, 1989; Roulund *et al.*, 1990; Yang *et al.*, 1992). Indications are that sucrose provides both a carbon source and an osmoticum during the induction process. There is no explanation for the apparent specificity for sucrose. It is known that sucrose accumulates in pollen (Hoekstra *et al.*, 1991), and is a protective agent during desiccation (Hoekstra *et al.*, 1989). However these findings do not explain its importance in haploid embryogenesis. There is also no information on the level of sucrose in *Brassica* pollen and its variation with growth conditions. Such information would be of value in establishing its specificity in pollen embryogenesis.

Sucrose is a common transport carbohydrate in higher plants and accumulates during low temperature stress (Guy *et al.*, 1992). It probably protects cells during desiccation by interacting with membrane phospholipids (Hoekstra *et al.*, 1991). It may have a similar function in microspores cultured at high temperature as Hamaoka *et al.* (1991) demonstrated a specific requirement for sucrose during microspore cell division in anthers cultured at high temperature. However, there is no information on the extent of sucrose utilization during microspore culture and whether high levels are driving starch synthesis during embryogenesis. It would be useful to separate the osmotic and nutritional roles of sucrose during embryogenesis and to define its role in embryo induction.

4.5.3. *Nitrogen supply*

Organic nitrogen is a beneficial component of the culture medium and both L-glutamine and L-serine are generally included in most media (Gland *et al.*, 1988). However there are no reports of a specific requirement for these components. In anther cultures, deteriorating anther tissues may supply these amino acids.

4.5.4. *Plant growth regulators*

An absolute requirement for growth regulators such as auxins and cytokinins, has not been rigorously established. In some circumstances embryogenesis occurs without exogenous auxins or cytokinins (Keller *et al.*, 1987; Swanson *et al.*, 1987). In many instances however, these components have been included in the culture medium (Gland *et al.*, 1988; Lichter, 1981, 1982, 1989; Yang *et al.*, 1992). In studies with *B. napus* microspores, BA enhanced embryo yield while auxin was without effect (Charne and Beversdorf, 1988). However cytokinins suppressed embryogenesis in anther cultures of cauliflower (*B. oleracea* var. *botrytis*) (Yang *et al.*, 1992) whereas it acted in a genotype dependent manner in broccoli (*B. oleracea* var. *italica*) anther cultures (Arnison *et al.*, 1990b). Response varied from stimulation to inhibition or no effect.

At least in *B. oleracea* auxin response may be genotype dependent. Low levels enhanced embryogenesis in anther cultures of some genotypes (Arnison and Keller, unpublished; Ockendon and McClenaghan, 1993), whereas high auxin levels reduced embryogenesis and increased anther callusing (Yang *et al.*, 1992).

Ethylene is inhibitory to embryogenesis in *B. oleracea* anther cultures and inhibitors of its action or biosynthesis enhanced embryogenesis (Biddington and Robinson, 1990, 1991). It appears that *B. oleracea* genotypes that are unresponsive to culture produce large amounts of ethylene and conditions which enhance ethylene production are inhibitory to embryogenesis (Biddington and Robinson, 1990; Biddington *et al.*, 1988, 1992, 1993). Abscisic acid, known to affect certain phases of both zygotic and somatic embryogenesis, was shown to inhibit embryogenesis in anther cultures of *B. oleracea* (Biddington *et al.*, 1992). This was related to stimulation of ethylene production and the effect was negated with fluridone, an inhibitor of ABA biosynthesis. In other species of *Brassica*, where microspore embryogenesis is genotype dependent, the involvement of ethylene has not been established.

Endogenous ethylene production may be a limiting factor in *B. oleracea* anther culture. It is not clear whether isolated microspore culture of this species is also limited by endogenous ethylene production. However, both anther filament and wall tissue are rich sources of endogenous ethylene (Biddington, 1992). Because the microspore is a rich source of auxin (Stead, 1992), an exogenous source may not always be required for embryogenesis. The importance of auxins in the control of both zygotic and somatic embryogenesis has been well established (Liu *et al.*, 1993). Since pollen embryogenesis is developmentally similar to that of zygotic embryogenesis, it is difficult to envisage the absence of auxin participation. Therefore it can be assumed that the developing embryo is competent for endogenous auxin production. This capacity for endogenous hormone production may extend to ABA. Although this hormone inhibits embryogenesis in *B. oleracea* (Biddington *et al.*, 1992) the low osmotic potential of the culture medium coupled with high temperatures during the early stages of incubation are conducive to stress-induced ABA biosynthesis. This would explain the observed positive effect of inhibitors of ABA biosynthesis on embryogenesis (Biddington *et al.*, 1992). It has also been observed that addition of physiological concentrations of gibberellins, also known to antagonize the effect of ABA, can significantly enhance the embryogenic response of broccoli anthers (Arnison and Keller, unpublished). The effect was concentration and genotype dependent and showed a synergistic interaction with added auxin (IAA).

4.5.5. *Other factors*

The pH of the culture medium appears to be critical for both anther and microspore cultures. A value of 6.2 was optimal for microspore embryogenesis in cultures of *Brassica campestris* (Baillie *et al.*, 1992). With anther cultures of *B. oleracea* a pH range of 5 to 6.5 allowed embryogenesis in responding genotypes but the frequency declined at the extreme values (Arnison *et al.*, 1990a). In several genotypes of *B. napus* a pH of 5.8 to 6.6 was favourable to embryogenesis with microspore cultures but no embryos were produced at pH 5.3 (Gland *et al.*, 1988). Although both agar and agarose solidified media have been used, liquid medium is preferred in anther cultures (Yang *et al.*, 1992). Microspores are invariably cultured in liquid medium.

Addition of activated charcoal to the culture medium improved embryo development (Gland *et al.*, 1988). Its effect is probably related to absorption of toxic materials produced during culture. Frequent replenishment or exchange of the culture medium is beneficial to embryogenesis (Hansen and Svinnset, 1993; Lichter, 1989). With *B. napus* microspore culture periodic medium exchange improved embryogenesis 200 to 800 percent (Hansen and Svinnset, 1993). The positive effect of this procedure on embryogenesis is undoubtedly related to removal of toxic metabolites which may be produced by non-embryogenic microspores (Kott *et al.*, 1988b).

4.6. *Culture environment*

4.6.1. *Temperature*

Culture temperature appears to be an overriding factor in successful microspore or anther culture of *Brassicas*. An initial period at 30 to 35 °C for 24 h to 72 h is usually required to induce embryos which can then develop at lower temperatures (Arnison *et al.*, 1990a; Baillie *et al.*, 1992; Duijs *et al.*, 1992; Dunwell *et al.*, 1985; Keller and Armstrong 1977; Roulund *et al.*, 1990; Yang *et al.*, 1992). Within the range of 30 to 35 °C the optimal temperature seems to depend on genotype. For *B. oleracea* 32 to 35°C was optimal (Arnison and Keller, 1990a; Duijs *et al.*, 1992; Yang *et al.*, 1992). In *Brassica campestris* a temperature of 32 °C for 48 to 72 h gave the best results (Baillie *et al.*, 1992).

The time lapse between isolation of microspores and high temperature treatment can radically affect embryo induction. If microspores of *B. napus* were held at 25 °C for 24 h before the start of high temperature treatment, embryogenesis was completely inhibited (Pechan *et al.*, 1991). The duration of time that microspores are held at induction temperatures is also critical. A minimum of 6–8 h at 32 °C was required to elicit embryogenesis. Similarly with *B. oleracea* an exposure time to elevated temperatures of at least 12–18 h was required (Arnison *et al.*, 1990a; Duijs *et al.*, 1992; Ockendon, 1984).

The physiological and biochemical basis for this high temperature requirement has not been investigated in great detail; however, it has been determined (Fabijanski *et al.*, 1992) that a heat shock response occurred during the first 90 min of culture at 35 °C. If this heat treatment was disrupted by non-inductive high temperature shocks (40 °C) of short duration, embryo formation was blocked (Arnison *et al.*, 1990a). After the initial heat shock response passed (2 h), the embryo induction process was not significantly disrupted by brief high temperature (40 °C) shocks (Arnison *et al.*, 1990a).

There are other aspects of high temperature treatment that may be related to embryo induction. Evidence indicates that elevated temperature treatment increased frequency of symmetric cell divisions of the microspores (Hause *et al.*, 1993; Telmer *et al.*, 1993) and that this favours embryogenesis. Starch metabolism and specific gene expression may also be related to high temperature treatment (Pechan *et al.*, 1991; Telmer *et al.*, 1993). In Brussels sprout (*B. oleracea*) anther cultures, recalcitrant genotypes maintained a high level of ethylene production at high temperatures, compared to low levels by responsive genotypes (Biddington *et al.*, 1988; Biddington and Robinson,

1990). Inhibitors of ethylene biosynthesis and action, AVG (aminovinyl glycine) and $AgNO_3$, enhanced embryogenesis in non-responding genotypes; however there was still a requirement for high temperature (Biddington *et al.*, 1988). This suggests that other factors besides ethylene production are involved in the high temperature response. It has been shown for anther cultures of *Brassica campestris* that the effectiveness of high temperatures on embryo induction was dependent on the presence of sucrose (12%) in the culture medium (Hamaoka *et al.*, 1991). Glucose, fructose and maltose were effective but not sugar alcohols. It is not clear whether this apparent carbohydrate requirement is nutritional, osmotic or protective. In *Nicotiana tabacum* microspores/pollen, stress resulting from nutrient starvation induced embryogenesis (Zarsky *et al.*, 1992). It is uncertain in *Brassicas* if temperature stress serves a similar function.

4.6.2. *Aeration*

There is little definitive information on the importance of aeration to microspore embryogenesis in *Brassicas*. In liquid cultures the volume is usually sufficiently low to allow adequate aeration without agitation. In some instances agitation and medium exchange improved the frequency of embryogenesis and the quality of embryos (Hansen and Svinnset, 1993; Lichter, 1989; Mathias, 1988).

4.6.3. *Light*

Routinely anther and isolated microspore cultures have been initiated in darkness; however the importance of light and the potential impact of the quality of light have not been thoroughly investigated. In *B. oleracea* anther cultures, incubation in the dark was more favourable to embryogenesis compared to light of 10 w/m^2 for 16 h (Yang *et al.*, 1992). Once embryos have been formed, cultures are generally transferred to light for further embryo development, although it has not been clearly demonstrated that such a transfer is essential.

4.6.4. *Microspore density*

For both anther and microspore culture, density is important to embryogenesis and embryo quality. For anther cultures of broccoli, the optimal density was 3 to 6 anthers per ml of medium (Arnison *et al.*, 1990a). With microspore cultures of *Brassica napus* optimal density was 5×10^4 cells per ml (Fan *et al.*, 1988; Kott *et al.*, 1988b; Polsoni *et al.*, 1988). In highly responding genotypes of *B. napus*, a minimum of 3×10^3 cells per ml was established (Huang *et al.*, 1990). There are probably two important factors in the establishment of the optimal culture density. One is to establish a density sufficient to provide a putative conditioning factor and the other is to prevent overcrowding and build-up of toxic metabolites (Huang *et al.*, 1990; Kott *et al.*, 1988b). Culture manipulations, such as the use of nurse culture, feeder layers and conditioned medium, improved embryogenesis at sub-optimal culture densities (Huang *et al.*, 1990; Simmonds *et al.*, 1991). Nurse cultures of the highly embryogenic *B. napus* genotype, Topas, enhanced embryogenesis and normal embryo development in one recalcitrant *B. napus* genotype (Simmonds *et al.*, 1991). This suggests the production of embryogenic factors by embryogenic microspores. However, any conditioning effect may not be general as there was no effect of media conditioning in other experiments (Hansen and Svinnset, 1993).

5. Developmental aspects

In isolated microspores of *Brassica napus*, cytological differences between potentially embryogenic and non-embryogenic cells have been detected (Kott *et al.*, 1988a; Lo and Pauls, 1992; Telmer *et al.*, 1993; Zaki and Dickinson, 1990). These include increase in the abundance of starch grains and the volume of the vacuole which is often fragmented (Telmer *et al.*, 1993). It is well-established that microspores at the late uni-nucleate to early binucleate stage are capable of embryogenic induction (Fan *et al.*, 1988; Kott *et al.*, 1988; Telmer *et al.*, 1992). At this stage plastids are present but the extent of starch synthesis is uncertain (Telmer *et al.*, 1993; Zaki and Dickinson, 1990).

Conditions that favour a high proportion of late uni-nucleate to early binucleate microspores in the general microspore population will allow a high frequency of embryogenesis. Bud selection, gradient centrifugation and flow cytometry have been used to enrich the population of embryogenic microspores (Deslauriers *et al.*, 1991; Fan *et al.*, 1988; Pechan *et al.*, 1988). Since the microspore is originally destined to become a gametophyte, a switch in developmental pathway towards embryogenesis probably requires dedifferentiation and the development of embryogenic competence.

It is unclear whether all microspores at the appropriate stage have this developmental capacity or if this is the property of only some microspores. In tobacco, *Nicotiana tabacum* L., distinct embryogenic microspores occur in the general microspore population (Horner and Street, 1978). During culture the pattern of cell division is important to embryogenesis. Embryogenic microspores divide symmetrically instead of the asymmetric division characteristic of the non-embryogenic microspores (Fan *et al.*, 1988; Telmer *et al.*, 1993; Zaki and Dickinson, 1991). High temperature treatment during the early stages of culture increased the frequency of symmetric cell divisions (Hause *et al.*, 1993; Telmer *et al.*, 1993). This was evident with isolated microspores and those from intact plants exposed to high temperature. The arrangement of microtubules and the cytoskeleton appear to be important for symmetric cell division and embryogenesis (Hause *et al.*, 1993; Zaki and Dickinson, 1991). Short-term exposure of isolated microspores to the anti-microtubule agent colchicine increased the number of symmetric cell divisions and the frequency of embryogenesis in *B. napus* (Zaki and Dickinson, 1991). The optimal time for such treatment was the first 12 h after isolation. Although symmetric cell division appears to be important for embryogenesis it may not be the only factor (Telmer *et al.*, 1993). It is also generally recognized that the zygotic embryo develops through an initial asymmetric cell division. Although many cell divisions in cultured mesophyll protoplasts are asymmetric, regeneration can still occur at high frequency. Therefore other factors are involved in the induction of embryogenesis besides the plane of cell division.

In culture there are certain events associated with early microspore embryogenesis (Telmer *et al.*, 1993; Zaki and Dickinson, 1990). These include nuclear movement, starch synthesis by plastids, development of a thick wall and appearance of globular structures within the cytoplasm. Commitment to an embryogenic pathway of development requires the synthesis of specific proteins associated with embryogenesis. One of the early events would be the establishment of polarity. This is undoubtedly related to the phytohormone auxin which is responsible for polarity during early zygotic embryogenesis (Liu *et al.*, 1993). Therefore exogenous or endogenous auxin may be critical to embryogenesis of isolated microspores. In embryogenesis of microspores of *N. tabacum* L., it appears that the nuclear behaviour of the generative and vegetative cells regulates embryogenesis. Stress induced by nutrient starvation allowed DNA replication in the vegetative cell which was not the case under non-stressed conditions (Zarsky *et al.*, 1992). Once such cells were returned to a nutrient-rich medium, embryogenic development continued. Associated with embryo induction in *N. tabacum* was the transcription of specific genes but their translation into proteins was not detected (Garrido *et al.*, 1993). However, the appearance of phosphoproteins in embryogenic microspores has been reported (Kyo and Harada, 1990). With *Brassicas* starvation-induced development of embryogenic capacity has not been reported. High temperature exposure during the initial stages of culture elicited the appearance of a number of mRNAS and proteins in embryogenic microspores (Pechan *et al.*, 1991). These were absent or less abundant in microspores maintained under non-inductive conditions. It appears that inductive conditions are required for the expression of the genes for these proteins and this expression is lost or reduced the longer cells are maintained at non-inductive temperatures after isolation.

6. Plant regeneration from haploid embryos

In most instances plant regeneration from cultured anthers and microspores occurs through direct embryogenesis and all developmental stages characteristic of zygotic embryogenesis can be identified in these cultures; however development may proceed via secondary embryogenesis (Loh and Ingram, 1982; Loh *et al.*, 1983; Lott and Haube, 1983). A few examples have been reported of callusing before plant regeneration occurred (Govil *et al.*, 1986;

Sunderland, 1971). This response may be related to carbohydrate and auxin levels in the culture medium. Under optimal conditions embryos reach the cotyledonary stage by 21 days in culture. However, there may not be synchronous development and embryos at various stages of development are still detectable at the end of the culture period. Initial culture density and competition for nutrients may be factors in this non-synchronous pattern. The quality of embryos can be improved by low temperature, partial desiccation treatments, and culture agitation (Brown *et al.*, 1993; Coventry *et al.*, 1988; Gland *et al.*, 1988; Kott and Beversdorf, 1990; Lichter, 1989; Mathias, 1988; Polsoni *et al.*, 1988).

In terms of plant development from embryos, the stage at which embryos are transferred to solidified medium is critical. Embryos transferred at the cotyledonary stage resulted in the highest frequency of plant regeneration (Burnett *et al.*, 1992; Coventry *et al.*, 1988; Polsoni *et al.*, 1988; Zhang *et al.*, 1991). Plant development usually occurs on growth regulator-free medium with reduced sucrose. Regeneration frequencies vary widely and may be a function of the quality of embryos transferred to solid medium. With *B. napus* frequencies varied from 1 to 47 percent (Chuong *et al.*, 1988; Kott and Beversdorf, 1990) while with *B. campestris* frequencies ranged from 5 to 20 percent (Baillie *et al.*, 1992; Burnett *et al.*, 1992). With *B. campestris* var. *pekinensis*, only 5 to 10 percent of the embryos produced plants (Sato *et al.*, 1989b). In *B. oleracea* plant recovery from microspore culture was genotype dependent and varied from 35 to 70 percent (Duijs *et al.*, 1992).

In general, it appears that culture manipulations such as agitation, use of activated charcoal and reduction of the embryo population during the latter stages of liquid culture can improve embryo quality and conversion to plants. However, there can be a high frequency of embryo abnormality, i.e., embryos without a shoot or root apex, some with multiple or no cotyledons and other aberrations. This will limit the recovery of useful plants. Fully formed embryos with shoot and root axes and cotyledons can be expected to develop directly into plants. In some cases, however, development into plants is by secondary shoots (Duijs *et al.*, 1992; Siebel and Pauls, 1989a) which may also be genotype dependent.

7. Methods of dihaploid production

For haploid microspore-derived plants to be useful in a breeding program, chromosome doubling to produce homozygous lines is necessary. In some cases diploidization occurs spontaneously and a portion of the regenerated plants has been recovered as homozygotes (Charne *et al.*, 1988; Lichter *et al.*, 1988). In *Brassica napus* a large percentage of spontaneously diploidized plants were recovered following cryopreservation of isolated microspores (Chen and Beversdorf, 1992). The reason for this response is unexplained and, so far, it has not been used as a method of homozygote production in other *Brassicas*. The most widely used method of chromosome doubling has been the use of colchicine (Coventry *et al.*, 1988). This is usually done at the plantlet stage when a 0.1 to 0.2% solution of colchicine can be administered either by immersing roots in the solution for 5 to 6 h or by injecting it into secondary buds (Lichter *et al.*, 1988; Polsoni *et al.*, 1988; Swanson *et al.*, 1988). Microspore-derived plants can also be effectively diploidized by colchicine treatment *in vitro* before plants are transferred to soil (Mathias and Röbbelen, 1991). Alternatively, methods are being developed involving the use of anti-microtubule chemicals such as trifluralin, oryzalin, and pronamide, in addition to colchicine, to effect chromosome doubling during the early stages of microspore culture (Chen *et al.*, 1993; Eikenberry, 1993; Hannig, 1993; Zhao and Simmonds, 1992). Any method of diploidization must be efficient in the production of normal homozygous plants and should not reduce embryogenesis.

8. Evaluation of doubled-haploids in *Brassica*

Production of homozygous lines for plant breeding is usually the aim of haploidy. Provided there is no selection or preferential regeneration during embryogenesis, the resulting plants should represent a genetic array of recombinants produced during meiosis. Progeny from individual plants should be uniform; however there may be variations due to mutation. If this occurs before chromosome doubling then homozygous mutants are recovered. Additionally it may also be possible to recover heterozygous plants from unreduced gametes which will show a segregation of characters in subsequent generations.

In a few reports it appeared that anther-derived plants showed a segregation distortion pattern for enzyme markers. This has been reported for anther-derived doubled haploids of *B. oleracea* var. *italica* (Orton and Browers, 1985) as well as for 5 genotypes of *B. napus*, that showed inheritance pattern different from the expected ratios for five isozymes (Foisset *et al.*, 1993). As no such distortion was detected with F_2 plants from the same genotypes it was assumed that the distortion was culture related. This led to the assumption of a linkage between the distorting isozyme loci and genes controlling *in vitro* embryogenesis. In earlier studies with anther cultures of *B. napus*, Hoffmann *et al.* (1982) detected genetic selection in anther-derived plants. This was explained as competition between the microspores in the anther resulting in regeneration of only some genotypes. With the development of microspore culture techniques this bias should be removed.

In comparative studies of microspore-derived doubled haploid plants and parental lines of *Brassica napus*, no differences were evident for the characteristics analyzed (Charne and Beversdorf, 1991; Chen and Beversdorf, 1990, 1991; Gruber and Röbbelen, 1991; Iqbal and Mollers, 1991; Lichter *et al.*, 1988; Naleczynska and Cegielska, 1991; Scarth *et al.*, 1991; Siebel and Pauls,

1989; Stringam and Thiagarajah, 1991). In most studies there was no deleterious effect of homozygosity and a similar array of recombinants was observed as for single seed descent plants. This makes doubled haploids extremely useful for development of inbred lines. In an analysis of fatty acid content of microspore-derived doubled haploids and plants derived from single seed descent, similar coefficients of variation were found (Chen and Beversdorf, 1990). These authors concluded that microspore-derived populations were unbiased for fatty acid content and plants were derived from a random array of gametes. In terms of selection the genetic gain for unit time should be greater with microspore-derived plants compared to single seed descent because of the increased time to homozygosity in the latter.

It also appears that spontaneous doubled haploids behave the same as colchicine-doubled haploids indicating no deleterious effect of the latter treatment (Charne and Beversdorf, 1991; Lichter *et al.*, 1988; Siebel and Pauls, 1989b). In earlier studies it was suggested that culture conditions biased embryogenesis towards doubled haploids with higher glucosinolate content compared to parental lines (Hoffmann *et al.*, 1982). Subsequent reports did not detect such bias and reaffirmed the importance of doubled haploids for selection of a number of characteristics in *Brassicas* (Iqbal and Mollers, 1991; Lichter *et al.*, 1988; Siebel and Pauls, 1989a).

9. Comparison of anther and microspore culture

Although haploid plants can be derived from *in vitro* culture of whole flower buds and unfertilized ovules (Doré, 1989; Yang *et al.*, 1992) and can also arise spontaneously under field conditions, anther and microspore culture are the predominant routes to large scale production. During the early stages of development of this technology in *Brassicas*, anther culture was the method of choice and experiments were designed to increase the efficiency of haploids from anthers (Dunwell and Cornish, 1985; Keller *et al.*, 1975, 1982; Lichter, 1981; Thomas and Wenzel, 1975). This technique has now been largely superseded by culture of isolated microspores but it is still used for haploid production in some *Brassica* species (Arnison and Keller, 1990; Arnison *et al.*, 1990a,b; Biddington *et al.*, 1993; Kieffer *et al.*, 1993; Roulund *et al.*, 1990).

Embryo yields from anther cultures as high as 100 to 900 per anther have been reported (Dunwell and Cornish, 1985); however as the number of microspores per anther ranges from 10,000 to 17,000 or more (Chuong and Beversdorf, 1985; Pan *et al.*, 1991) the frequency of embryo formation is actually low. It was suggested that the anther walls are restrictive to microspore embryogenesis and in addition may impose selection pressure allowing only some microspores to undergo embryogenesis (Hoffmann *et al.*, 1982; Osolnik *et al.*, 1993). As a result procedures were developed for isolation and culture of microspores (Lichter, 1982) and, with refinements in genotype selection, donor plant growth conditions, microspore selection and culture manipulations, about 150,000 embryos per 100 anthers can be recovered (Swanson *et al.*, 1987; Swanson, 1990). In studies with *B. napus*, microspore culture is about 10 times as efficient as anther culture (Lichter *et al.*, 1988; Siebel and Pauls, 1989b). With anther cultures of *Brassicas* the number of primary regenerants was lower than with microspore culture (Siebel and Pauls, 1989b), even though more spontaneous diploids were reported from anther cultures (Lichter *et al.*, 1988). In field studies plant populations derived from anther and microspore culture were equivalent (Siebel and Pauls, 1989b); thus it can be concluded that the population of microspores giving rise to embryos is the same in both cases.

There are some inherent problems associated with anther culture in general (Pierek, 1987) and these are applicable to the *Brassicas*. The most obvious is that there may be regeneration from anther tissues and these regenerants may overgrow the haploid tissues. This may lead to difficulties in isolating the haploid embryos. The frequency of embryogenesis is usually lower with anther cultures. Even when sporophytic tissues do not undergo embryogenesis, extensive callusing can obscure the development of haploid embryos. There are indications that, depending on the location within the anther, selective microspore division and embryogenesis can occur (Hoffmann *et al.*, 1982; Osolnik *et al.*, 1993). Anther isolation is time consuming and both the anther walls and filament can produce growth regulators which interfere with embryogenesis (Biddington and Robinson, 1990).

In contrast there are distinct advantages to microspore culture that include: (1) increased frequency of embryogenesis; (2) less time consumed as whole buds can be easily homogenized and microspores recovered; (3) potential to enrich the embryogenic microspore population through gradient centrifugation and cell sorting (Deslauriers *et al.*, 1991; Fan *et al.*, 1988; Pechan and Keller, 1988); (4) easy adjustment of microspore culture densities compared to anther cultures where microspores are crowded in a confined space; (5) relative ease of monitoring the developmental of embryogenesis; (6) suitability for biochemical and physiological studies as the system constitutes a uniform population of single cells; and (7) convenience for mutagenesis and *in vitro* selection as all cells can be directly exposed to the selective agent or mutagen.

One potential limitation is the possible damage to the microspores during isolation; however, with the use of standard disruption methods, this should not be a significant problem.

10. Conclusions and future prospects

The major route to haploid production in the *Brassicas* is through anther and microspore culture. This technique has been well-established for most of the commercially important species, varieties and genotypes. There are a number of factors which have contributed to successful haploid production including: selection of genotype, donor plant growth conditions, manipulation of culture temperatures and the use of high sucrose in the culture medium. The stage of microspore development is critical for haploid embryogenesis with late uni-nucleate to early binucleate cells the most suitable. Although the above-mentioned conditions are generally applicable to a number of species, genotype is still perhaps the most important factor for high frequency embryogenesis. Growth of donor plants at low temperatures is often beneficial and undoubtedly these conditions alter the physiology of the plant in a manner which is reflected in the responsiveness of the microspores to culture. However, there is no information on the physiological and biochemical changes that occur in the flower buds at low temperatures. Some of the changes may be related to growth regulators such as IAA, cytokinin, GA and ABA; however, there have been no reports of the successful substitution of growth regulators for the high temperature induction requirement. Growth temperatures appear to alter the cellular composition of the microspores (Lo and Pauls, 1992).

For both anther and microspore culture, an initial period at high temperatures ranging from 30° to 35°C is required. This is may be related to disruption of the development of the microspore during a critical temperature-sensitive phase (Fabijanski *et al.*, 1992). Subsequently culture conditions may encourage the development of embryos through specific gene transcriptions (Hause *et al.*, 1993). There are indications that appropriate levels of ethylene are required for embryogenesis during anther culture, notably in *Brassica oleracea* (Biddington and Robinson, 1991). High temperatures during the early stages of culture reduced endogenous ethylene production (Biddington and Robinson, 1991); however chemical inhibition of ethylene production does not replace the need for high initial temperatures (Biddington *et al.*, 1993). There are also some indications that elevated temperatures influence the plane of cell division leading to embryogenesis; however it is not clear if this is a cause or an effect of embryo induction. It is clear, however, that events at high temperature trigger the sustained embryo development that proceeds subsequently at lower temperatures.

Apart from the obvious importance in rapid development of homozygous lines, the microspore culture system is useful in studies of basic biochemical and physiological aspects of embryogenesis. Microspores provide a uniform, synchronous and easily accessible population of cells for such studies.

Even though the haploid embryogenesis system is quite successful in many *Brassica* species, there is still strong genotypic control on the frequency of embryo production. There are some indications that this can be ameliorated through media manipulation (Gland *et al.*, 1988; Lichter *et al.*, 1988). Most of the medium formulations have been developed empirically and there has been no attempt to relate biochemical and physiological changes in the microspores, or flower buds, as a result of inductive growth conditions, to culture requirements.

It is not clear whether all microspores at the appropriate nuclear stage undergo embryogenesis and if there are limitations imposed on embryogenesis of some microspores by the early developing ones. A microculture system may provide the answer to this question. Also an effective system to enrich the embryogenic microspore population may improve the frequency of embryogenesis in less responsive genotypes.

Embryo quality and the frequency of conversion to plants are still problems. The ideal situation is to recover whole plants directly from normal embryos with well-developed shoot and root axes and cotyledons. In many lines the frequency of primary regenerants is low and most plants arise through secondary shoot formation. There is evidence that this may be genotype-dependent (Duijs *et al.*, 1992).

Even with these limitations, microspore and anther culture for haploid production in *Brassicas* are very successful. Doubled haploids arising spontaneously or by chemically-induced chromosome doubling are equal in agronomic and compositional characteristics to the parental lines. There is no significant influence of *in vitro* culture conditions on the doubled haploids and these represent an array of recombinants. The microspore culture technique is likely to remain prominent in *Brassica* breeding programs because of its simplicity and efficiency compared to other methods of haploid production.

11. References

Arnison, P.G., P. Donaldson, L.C.C. Ho and W.A. Keller, 1990a. The influence of various physical parameters on anther culture of broccoli (*Brassica oleracea* var. *italica*). Plant Cell Tiss. Org. Cult. 20: 147–155.

Arnison, P.G., P. Donaldson, A. Jackson, C. Semple and W.A. Keller, 1990b. Genotype-specific response of cultured broccoli (*Brassica oleracea* var. *italica*) anthers to cytokinins. Plant Cell Tiss. Org. Cult. 20: 217–222.

Arnison, P.G. and W.A. Keller, 1990. A survey of the anther culture response of *Brassica oleracea* L. cultivars grown under field conditions. Plant Breed. 104: 125–133.

Aslam, F.N., M.V. MacDonald, P.T. Loudon and D.S. Ingram, 1990a. Rapid-cycling *Brassica* species: Inbreeding and selection of *Brassica napus* for anther culture ability and an assessment of its potential for microspore culture. Ann. Bot. 66: 331–339.

Aslam, F.N., M.V. MacDonald and D.S. Ingram, 1990b. Rapid-cycling *Brassica* species: Anther culture potential of *B. campestris* L. and *B. napus* L. New Phytol. 115: 1–9.

Aspinall, D. and L.G. Paleg, 1981. Proline accumulation: Physiological aspects. In: L.G. Paleg and D. Aspinall (Eds.), Physiology and Biochemistry of Drought Resistance in Plants, pp. 205–241. Academic Press, Sydney.

Bagga, S., N. Bhalla-Sarin, S.K. Sopory and S. Guha-Mukherjee, 1982. Comparison of *in vitro* plant formation from somatic tissues and pollen grains in *Brassica oleracea* var. *botrytis*. Phytomorphology 32: 152–156.

Baillie, A.M.K., D.J. Epp, D. Hutcheson and W.A. Keller, 1992. *In vitro* culture of isolated microspores and regeneration of plants in *Brassica campestris*. Plant Cell Rep. 11: 234–237.

Banga, S.S and K.S. Labana, 1986. Spontaneous haploidy in *Brassica napus*. Cruciferae Newsl. No. 11: 54–55.

Biddington, N.L., 1992. The influence of ethylene in plant tissue culture. Plant Growth Regul. 11: 173–187.

Biddington, N.L. and H.T. Robinson, 1990. Variations in response to high temperature treatments in anther culture of Brussels sprouts. Plant Cell Tiss. Org. Cult. 22: 48–54.

Biddington, N.L. and H.T. Robinson, 1991. Ethylene production during anther culture of Brussels sprout (*Brassica oleracea* var. *gemmifera*) and its relationship with factors that affect embryo production. Plant Cell Tiss. Org. Cult. 25: 169–177.

Biddington, N.L., H.T. Robinson and J.R Lynn, 1993. ABA promotion of ethylene production in anther culture of Brussels sprouts (*Brassica oleracea* var. *gemmifera*) and its relevance to embryogenesis. Physiol. Plant 88: 577–582.

Biddington, N.L., R.A. Sutherland and H.T. Robinson, 1988. Silver nitrate increases embryo production in anther culture of Brussels sprouts. Ann. Bot. 62: 181–185.

Biddington, N.L., R.A. Sutherland and H.T. Robinson, 1992. The effects of gibberellic acid, fluridone, abscisic acid and paclobutrazol on anther culture of Brussels sprouts. Plant Growth Regul. 11: 81–84.

Blakeslee, A.F., F. Belling, M.E. Farnham and A.D. Bergner, 1922. A haploid mutant in the Jimson weed, *Datura stramonium*. Science 55: 646–647.

Brown, D.C.W., E.M. Watson and P.M. Pechan, 1993. Induction of desiccation tolerance in microspore-derived embryos of *Brassica napus*. In Vitro Cell Dev. Biol. 29: 113–118.

Burnett, L., S. Yarrow and B. Huang, 1992. Embryogenesis and plant regeneration from isolated microspores of *Brassica rapa* L. ssp. *oleifera*. Plant Cell Rep. 11: 215–218.

Cardy, B.J., 1986. Production of anther-derived doubled haploids for breeding oilseed rape (*Brassica napus* L.). Ph.D. Thesis, University of Guelph, Guelph, Ontario.

Cao, M.Q., F. Charlot and C. Doré, 1990. Embryogenèse et régénération de plantes de chou a choucroute (*Brassica oleracea* L. ssp. *capitata*) par culture *in vitro* de microspores isolées. C.R. Acad. Sci. Paris Série III, 310: 203–209.

Charne, D.G. and W.D. Beversdorf, 1988. Improving microspore culture as a rapeseed breeding tool: The use of auxins and cytokinins in an induction medium. Can. J. Bot. 66: 1671–1675.

Charne, D.G. and W.D. Beversdorf, 1991. Comparisons of agronomic and compositional traits in microspore-derived and conventional populations of spring *Brassica napus*. In: D.I. McGregor (Ed.), GCIRC, July 9–11, 1991, Saskatoon, Saskatchewan, pp. 64–69.

Charne, D.G., P. Pukacki, L.S. Kott and W.D. Beversdorf, 1988. Embryogenesis following cryopreservation in isolated microspores of rapeseed (*Brassica napus* L.). Plant Cell Rep. 7: 407–409.

Chen, J.L. and W.D. Beversdorf, 1990. A comparison of traditional and haploid-derived breeding populations of oilseed rape (*Brassica napus* L.) for fatty acid composition of the seed oil. Euphytica 51: 59–65.

Chen, J.L. and W.D. Beversdorf, 1991. Fatty acid inheritance in microspore-derived populations of spring rapeseed (*Brassica napus* L.). Theor. Appl. Genet. 80: 465–469.

Chen, J.L. and W.D. Beversdorf, 1992. Production of spontaneous diploid lines in isolated microspores following cryopreservation of spring rapeseed (*Brassica napus* L.). Plant Breed. 108: 324–327.

Chen, Z.Z., S. Snyder, Z.G. Fan and W.H. Loh, 1993. High efficiency chromosome doubling by direct colchicine treatment of isolated microspores of *Brassica napus*. In: Crucifer Genetics Workshop. Saskatoon, Saskatoon, July 21–24, 1993, p. 12.

Chiang, M.S., C. Frechette, G. Kuo, C. Chong and S.J. Delafield, 1985. Embryogenesis and haploid plant production from anther culture of cabbage (*Brassica oleracea* var. *capitata*). Can. J. Plant Sci. 65: 1033–1057.

Chuong, P.V. and W.D. Beversdorf, 1985. High frequency embryogenesis through isolated microspore culture of *B. napus* and *B. carinata* Braun. Plant Sci. 39: 219–226.

Chuong, P.V., C. Deslauriers, L.S. Ott and W.D. Beversdorf, 1988. Effects of donor genotype and bud sampling on microspore culture of *Brassica napus*. Can. J. Bot. 66: 1653–1657.

Coventry, J., L. Kott and W.D. Beversdorf, 1988. Manual for Microspore Culture Technique for *Brassica napus*. Dept. of Crop Sci. Techn. Bull. OAC Publication 0489, Univ. of Guelph, Guelph.

Deslauriers, C., A.D. Powell, K. Fuchs and K.P. Pauls, 1991. Flow cytometric characterization and sorting of cultured *Brassica napus* microspores. Biochim. Biophys. Acta 1091: 165–172.

Doré, C., 1989. Obtention de plantes haploïdes de chou cabus (*Brassica oleracea* L. ssp. *capitata*) après culture *in vitro* d'ovules pollinisés par du pollen irradié. C.R. Acad. Sci. Paris 309 Série III: 729–734.

Duijs, J.G., R.E. Voorrips, D.L. Visser and J.B.M. Custers, 1992. Microspore culture is successful in most crop types of *Brassica oleracea* L. Euphytica 60: 45–55.

Dunwell, J.M. and L.M. Cornish, 1985. Influence of preculture variables on microspore embryo production in *Brassica napus* ssp. *oleifera* cv. Duplo. Ann. Bot. 56: 281–289.

Dunwell, J.M., L.M. Cornish and A.G.L. Decourcel, 1985. Influence of genotype, plant growth, temperature and anther incubation temperature on microspore embryo production in *Brassica napus* ssp. *oleifera*. J. Exp. Bot. 36: 679–689.

Dunwell, J.M. and N. Thurling, 1985. Role of sucrose in microspore embryo production in *Brassica napus* ssp. *oleifera*. J. Exptl. Bot. 36: 1478–1491.

Eikenberry, E., 1993. Chromosome doubling of microspore-derived canola using trifluralin. In: Crucifer Genetics Workshop. Saskatoon, Saskatoon, July 21–24, 1993, Abstr. #54.

Fabijanski, S., I. Altosaar and P.G. Arnison, 1987. Heat shock response of *Brassica oleracea*, L. (broccoli). J. Plant Physiol. 128: 29–38.

Fabijanski, S., I. Altosaar and P.G. Arnison, 1992. Heat shock response during anther culture of broccoli (*B. oleracea* var. *Italica*). Plant Cell Tiss. Org. Cult. 26: 203–212.

Fan, Z., K.C. Armstrong and W.A. Keller, 1988a. Development of microspores *in vivo* and *in vitro* in *Brassica napus* L. Protoplasma 147: 191–199.

Fan, Z., L. Holbrook and W.A. Keller, 1988b. Isolation and enrichment of embryogenic microspores in *Brassica napus* L. by fractionation using percoll density gradient. In: 7th Intl. Rapeseed Congr. Poland, 1988, 92th Ed.

Flores, H.E. and A.W. Galston, 1982. Polyamines and plant stress: activation of putrescine biosynthesis by osmotic shock. Science 217: 1259–1260.

Foisset, N., R. Delourme, M.O. Lucas and M. Renard, 1993. Segregation analysis of isozyme markers on isolated microspore-derived embryos in *Brassica napus* L. Plant Breed. 110: 315–322.

Garrido, D., N. Eller, E. Heberle-Bors and O. Vicente, 1993. *De novo* transcription of specific mRNAs during the induction of tobacco pollen embryogenesis. Sex. Plant Reprod. 6: 40–45.

George, L. and P.S. Rao, 1983. *In vitro* induction of pollen embryos and plantlets in *Brassica juncea* through anther culture. Plant Sci. Lett. 26: 111–116.

Gland, A., R. Lichter and H.-G. Schweiger, 1988. Genetic and exogenous factors affecting embryogenesis in isolated microspore culture of *Brassica napus* L. J. Plant Physiol. 132: 613–617.

Govil, S., S.B. Babbar and S.C. Gupta, 1986. Plant regeneration from *in vitro* cultured anthers of black mustard (*Brassica nigra* Koch). Plant Breed. 97: 64–71.

Gresshoff, P.M. and C.H. Doy, 1972. Haploid *Arabidopsis thaliana* callus and plants from anther culture. Aust. J. Biol. Sci. 25: 259–264.

Gruber, S. and G. Röbbelen, 1991. Fatty acid synthesis in microspore-derived embryoids of rapeseed (*Brassica napus*). In: D.I. McGregor (Ed.), GCIRC, July 9–11, 1991, Saskatoon, Saskatchewan, pp. 1818–1820.

Guerrero, F. and J.E. Mullet, 1986. Increased abscisic acid biosynthesis during plant dehydration requires transcription. Plant Physiol. 80: 588–591.

Guha, S. and S.C. Maheshwari, 1964. *In vitro* production of embryos from anthers of *Datura*. Nature (London) 204: 297.

Guha, S, and S.C. Maheshwari, 1966. Cell division and differentiation of embryos in the pollen grains of *Datura in vitro*. Nature (London) 212: 97–98.

Guy, C.L., J.L.A. Huber and S. Huber, 1992. Sucrose phosphate synthase and sucrose accumulation at low temperature. Plant Physiol. 100: 502–508.

Hamaoka, Y., Y. Fujita and S. Iwai, 1991. Effects of temperature on the mode of pollen development in anther culture of *Brassica campestris*. Physiol. Plant 82: 67–72.
Hannig, A., 1993. Antimicrotubule agents for diploidizing haploid tissues in *Brassica napus*. In: Crucifer Genetics Workshop. Saskatoon, Saskatoon, July 21–24, 1993, Abstr. #57.
Hansen, M. and K. Svinnset, 1993. Microspore culture of Swede (*Brassica napus* ssp. *rapifera*) and the effects of fresh and conditioned medium. Plant Cell Rep. 12: 496–500.
Hanson, A.D. and W.D. Hitz, 1982. Metabolic response of mesophytes to plant water deficits. Ann. Rev. Plant Physiol. 33: 163–203.
Hause, B., G. Hause, P. Pechan and A.A.M. Van Lammeren, 1993. Cytoskeletal changes in induction of embryogenesis in microspore and pollen cultures of *Brassica napus* L. Cell Biol. Intl. 17: 153–168.
Hetz, E. and O. Shieder, 1991. Direct embryogenesis and plant regeneration through microspore culture of *Brassica nigra*. Crucifer Newsl. 14/15: 102–103.
Hoekstra, F.A., L.M. Crowe and J.H. Crowe, 1989. Differential desiccation sensitivity of corn and *Pennisetum* pollen linked to their sucrose content. Plant Cell Environ. 12: 83–91.
Hoekstra, F.A., J.H. Crowe and L.M. Crowe, 1991. Effect of sucrose on phase behaviour of membranes in intact pollen of *Typha latifolia* L. as measured with Fourier transform infrared spectroscopy. Plant Physiol. 97: 1073–1079.
Hoffmann, F., E. Thomas and G. Wenzel, 1982. Anther culture as a breeding tool in rape. II. Progeny analysis of androgenic lines and induced mutants from haploid cultures. Theor. Appl. Genet. 61: 225–232.
Horner, M. and E. Street, 1978. Pollen dimorphism-origin and significance in pollen plant formation by anther culture. Ann. Bot. 42: 763.
Huang, B., S. Bird, R. Kemble, D. Simmonds, W.A. Keller and B. Miki, 1990. Effects of culture density, conditioned medium and feeder cultures on microspore embryogenesis in *Brassica napus* L. cv. Topas. Plant Cell Rep. 8: 594–597.
Huang, B. and W.A. Keller, 1989. Microspore culture technology. J. Tiss. Cult. Methods 12: 171–178.
Iqbal, M.C.M. and C. Mollers, 1991. Selection for low glucosinolate content of rapeseed *Brassica napus* L. using haploid embryos from microspore culture. In: D.I. McGregor (Ed.), GCIRC, July 9–11, Saskatoon, Saskatchewan, pp. 187–190.
Jain, R.K., U. Brune and W. Friedt, 1989. Plant regeneration from *in vitro* cultures of cotyledon explants and anthers of *Sinapis alba* and its implications on breeding of crucifers. Euphytica 43: 153–163.
Jain, S.M., N. Bagga, N. Bhalla-Sarin, S. Guha-Mukherjee and S.K. Sopory, 1980. *In vitro* culture of anthers in *Petunia hybrida* and *Brassica oleracea*. In: P.S. Rao, M.R. Heble and M.S. Chadha (Eds.), Plant Tissue Culture, Genetic Manipulation and Somatic Hybridization of Plant Cells, pp. 85–92. Bhabha Atomic Research Centre, Bombay.
Kameya, T. and K. Hinata, 1970. Induction of haploid plants from pollen grains of *Brassica*. Jpn. J. Breed. 20: 82–87.
Keller, W.A. and K.C. Armstrong, 1977. Embryogenesis and plant regeneration in *Brassica* anther culture. Can. J. Bot. 55: 1383–1388.
Keller, W.A. and K.C. Armstrong, 1978. High frequency production of microspore-derived plants from *Brassica napus* anther cultures. Z. Pflanzenzüchtg. 80: 100–108.
Keller, W.A. and K.C. Armstrong, 1979. Stimulation and embryogenesis and haploid production in *Brassica campestris* anther cultures by elevated temperature treatments. Theor. Appl. Genet. 55: 65–67.
Keller, W.A. and K.C. Armstrong, 1981. Production of anther derived dihaploid plants in autotetraploid marrowstem kale (*Brassica oleracea* var *acephala*). Can. J. Genet. Cytol: 23:259–265.
Keller, W.A. and K.C. Armstrong, 1982. Production of haploids via anther culture in *Brassica oleracea* var. *italica*. Euphytica 32:151–159.
Keller, W.A., K.C. Armstrong and I.A. de la Roche, 1982. The production and utilization of

microspore-derived haploids in *Brassica* crops. In: K.L. Giles and S.K. Sen (Eds.), Plant Cell Culture in Crop Improvement, pp. 169–183. Plenum Pub. Corp., New York.

Keller, W.A., P.G. Arnison and B.K. Cardy, 1987. Haploids from gametophytic cells – recent development and future prospects. In: C.E. Green, D.A. Somers, W.P. Nackett and D.D. Biesboer (Eds.), Plant Tissue and Cell Culture, pp. 233–241. Allan R. Liss, New York.

Keller, W.A., T. Rajhathy and J. Lacapra, 1975. *In vitro* production of plants from pollen in *Brassica campestris*. Can. J. Genet. Cytol. 17: 655–666.

Keller, W.A. and G.R. Stringam, 1978. Production and utilization of microspore-derived plants. In: T.A. Thorpe (Ed.), Frontiers of Plant Tissue Culture, pp. 113–122. Intern. Assoc. Plant Tiss. Cult., Calgary.

Kieffer, M., M.P. Fuller, J.E. Chauvin and A. Schlesser, 1993. Anther culture of kale (*Brassica oleracea* L. con var. *acephala* D.C. Alef.). Plant Cell Tiss. Org. Cult. 33: 303–313.

Kimber, G. and R. Riley, 1963. Haploid angiosperms. Bot. Rev. 29: 480–531.

Klimaszewska, K. and W.A. Keller, 1983. The production of haploids from *Brassica hirta* Moench (*Sinapis alba*) anther cultures. Z. Pflanzenphysiol. 109: 235–241.

Komatsu, Y., 1936. Investigations on the progeny of haploid plants and embryo sac formation by diploid and haploid plants of *Brassica napus*. Proc. Crop Sci. Soc. Japan 8: 364–372.

Kott, L.S. and W.D. Beversdorf, 1990. Enhanced plant regeneration from microspore-derived embryos of *Brassica napus* by chilling, partial desiccation and age selection. Plant Cell Tiss. Org. Cult. 23: 187–192.

Kott, L.S., L. Polsoni and W.D. Beversdorf, 1988a. Cytological aspects of isolated microspore culture in *Brassica napus*. Can. J. Bot. 66: 1658–1664.

Kott, L.S., L. Polsoni, B. Ellis and W.D. Beversdorf, 1988b. Autotoxicity in isolated microspore cultures of *Brassica napus*. Can. J. Bot. 66: 1665–1670.

Kuriyama, H. and Y. Watanabe, 1955. Studies on the haploid plant of *Brassica carinata*. Ikushugaku Zasshi/Jpn. J. Breed. 5: 1–5.

Kyo, M. and H. Harada, 1990. Specific phosphoproteins in the initial period of tobacco pollen embryogenesis. Planta 182: 58–63.

Leelavathi, S., V.S. Reddy and S.K. Sen, 1984. Somatic cell genetics in *Brassica* species. I. High frequency production of haploid plants in *Brassica alba* (L.) H.F. + T. Plant Cell Rep. 3: 102–105.

Leelavathi, S., V.S. Reddy and S.K. Sen, 1987. Somatic cell genetic studies in *Brassica* species. II. Production of androgenetic haploid plants in *Brassica nigra* (L.) Koch. Euphytica 36: 215–219.

Lelu, M.-A. and H. Ballon, 1985. Obtention d'haploïdes par culture d'anthère de *Brassica oleracea* L. var. *capitata* et var. *gemmifera*. C.R. Acad. Sci. Paris 300, Série III, no. 2: 71–76.

Lichter, R., 1981. Anther culture of *Brassica napus* in a liquid culture medium. Z. Pflanzenphysiol. 103: 229–237.

Lichter, R., 1982. Induction of haploid plants from isolated pollen of *Brassica napus*. Z. Pflanzenphysiol. 105: 427–434.

Lichter, R., 1989. Efficient yield of embryoids by culture of isolated microspores of different Brassicaceae species. Plant Breed. 103: 119–123.

Lichter, R., E. DeGroot, D. Fiebig, R. Schweiger and A. Gland, 1988. Glucosinolates determined by HPLC in the seeds of microspore-derived homozygous lines of rapeseed (*Brassica napus* L.). Plant Breed. 100: 269–221.

Lillo, C. and M. Hansen, 1987. Anther culture of cabbage. Influence of growth temperature of donor plants and media composition on embryo yield and plant regeneration. Norwegian J. Agric. Sci. 1: 105–109.

Liu, C.-M., Z.H. Xu and N.-H. Chua, 1993. Auxin polar transport is essential for establishment of bilateral symmetry during early plant embryogenesis. Plant Cell 5: 621–630.

Lo, K.-H. and K.P. Pauls, 1992. Plant growth environment effects on rapeseed microspore development and culture. Plant Physiol. 99: 468–472.

Loh, C.S. and D.S. Ingram, 1982. Production of haploid plants from anther cultures and

secondary embryoids of winter oilseed rape, *Brassica napus* ssp. *oleifera*. New Phytol. 91: 507–516.
Loh, C.S., D.S. Ingram and D.E. Hanke, 1983. Cytokinins and the regeneration of plantlets from secondary embryoids of winter oilseed rape, *Brassica napus* ssp. *oleifera*. New Phytol. 95: 349–358.
Lott, C.S. and D.E. Haube, 1983. Cytokinins and the regeneration of plantlets from secondary embryoids of winter oilseed rape, *Brassica napus* ssp. *oleifera*. New Phytol. 95: 349–350.
MacDonald, M.V., M.A. Hadwiger, F.N. Aslam and D.S. Ingram, 1988a. The enhancement of anther culture efficiency in *Brassica napus* ssp. *oleifera* Metzg. (Sinsk) using low doses of gamma radiation. New Phytol. 110: 101–107.
MacDonald, M.V., D.M. Newsholme and D.S. Ingram, 1988b. The biological effects of gamma irradiation on secondary embryoids of *Brassica napus* ssp. *oleifera* Metzg. (Sinsk) winter oilseed rape. New Phytol. 10: 255–259.
Maheshwari, S.C., A.K. Tyagi, K. Malhotra and S.K. Sopory, 1980. Induction of haploidy from pollen grains in angiosperms – the current status. Theor. Appl. Genet. 58: 193–206.
Margale, E. and A.M. Chèvre, 1991. Factors affecting embryo production from microspore culture of *Brassica nigra* (Koch). Crucifer Newsl. 14/15: 100–101.
Mathias, R., 1988. An improved *in vitro* culture procedure for embryoids derived from isolated microspores of rape (*Brassica napus* L). Plant Breed. 100: 320–322.
Mathias, R. and G. Röbbelen, 1991. Effective diploidization of microspore-derived haploids of rape (*Brassica napus* L.) by *in vitro* colchicine treatment. Plant Breed. 106: 82–84.
Matsubayashi, M. and K. Kuranuki, 1975. Embryogenic response of the pollen to varied sucrose concentrations in anther culture. Sci. Rep. Fac. Agric. Kobe Univ. 11: 215–230.
Melchers, G., 1972. Haploid higher plants for plant breeding. Z. Pflanzenzüchtg. 67: 21–32.
Naleczynska, A. and T. Cegielska, 1991. Doubled haploid production and field experiments with homozygous lines of rapeseed. In: D.I. McGregor (Ed.), GCIRC, July 9–11, 1991, Saskatoon, Saskatchewan, pp. 1488–1491.
Morinaga, T. and E. Fukushima, 1933. Karyological studies on a spontaneous haploid mutant of *Brassica napella*. Cytologia 4: 457–460.
Nitsch, J.P., 1972. Haploid plants from pollen. Z. Pflanzenzüchtg. 67: 3–18.
Nitzsche, W. and G. Wenzel, 1977. Haploids in Plant Breeding. In: Fortschr. Pflanzenzüchtg. 8. Parey, Berlin.
Ockendon, D.J., 1984. Anther culture in Brussels sprouts (*Brassica oleracea* var. *gemmifera*). I. Embryo yields and plant regeneration. Ann. Appl. Biol. 105: 285–291.
Ockendon, D.J., 1985. Anther culture in Brussels sprouts (*Brassica oleracea* var. *gemmifera*). II. Effect of genotype on embryo yields. Ann. Appl. Biol. 107: 101–104.
Ockendon, D.J. and R. McClenaghan, 1993. Effect of silver nitrate and 2,4–D on anther culture of Brussels sprout (*Brassica oleracea* var. *gemmifera*). Plant Cell Tiss. Org. Cult. 32: 41–46.
Ockendon, D.J. and R.A. Sutherland, 1987. Genetic and non-genetic factors affecting anther culture of Brussels sprout (*Brassica oleracea* var. *gemmifera*). Theor. Appl. Genet. 74: 566–570.
Olsson, G. and A. Hagberg, 1955. Investigations on a haploid rape. Hereditas 41: 227–237.
Orton, T.J. and M.A. Browers, 1985. Segregation of genetic markers among plants regenerated from cultured anthers of broccoli (*Brassica oleracea* var. *italica*). Theor. Appl. Genet. 69: 637–643.
Osolnik, B., B. Bohancec and S. Jelaska, 1993. Stimulation of androgenesis in white cabbage (*Brassica oleracea* var. *capitata*) anthers by low temperature and anther dissection. Plant Cell Tiss. Org. Cult. 32: 241–246.
Pan, Q.Y., G. Seguin-Swartz, R.K. Downey and G.F.W. Rakow, 1991. Number of microspores in immature and mature flowerbuds in *Brassica* species. In: D.I. McGregor (Ed.), 8th Intl. Rapeseed Congress, Vol. 6, pp. 1836–1839.
Pandey, K.K., 1973. Theory and practice of induced androgenesis. New Phytol. 72: 1129–1140.
Pearcy, R.W., 1978. Effect of growth temperature on the fatty acid composition of leaf lipids in *Atriplex lentiformis* (Torr.) Wats. Plant Physiol. 61: 484–486.

Pechan, P.M., D. Bartels, D.C.W. Brown and J. Schell, 1991. Messenger RNA and protein changes associated with induction of *Brassica* microspore embryogenesis. Planta 184: 161–165.

Pechan, P.M. and W.A. Keller, 1988. Identification of potentially embryogenic microspores in *Brassica napus*. Physiol. Plant. 74: 377–384.

Pechan, P.M. and W.A. Keller, 1989. Induction of microspore embryogenesis in *Brassica napus* L. by gamma irradiation and ethanol stress. In Vitro Cell Devel. Biol. 25: 1073.

Pechan, P.M., W.A. Keller, F. Mandy and M. Bergeron, 1988. Selection of *Brassica napus* L. embryogenic microspores by flow sorting. Plant Cell Rep. 7: 596–398.

Phippen, C. and D.J. Ockendon, 1990. Genotype, plant, bud size and media factors affecting anther culture of cauliflower (*Brassica oleracea* var. *botrytis*). Theor. Appl. Genet. 79: 33–38.

Pierek, R.L.M., 1987. Test tube fertilization. In: R.L.M. Pierek (Ed.), *In Vitro* Culture of Higher Plants, pp. 234–257. Martinus Nijhoff Publishers, Dordrecht.

Polsoni, L., L.S. Kott and W.D. Beversdorf, 1988. Large-scale microspore culture technique for mutation-selection studies in *Brassica napus*. Can. J. Bot. 66: 1681–1685.

Prakash, S., 1973. Haploidy in *Brassica nigra* Koch. Euphytica 22: 613–614.

Prakash, S., 1974. Haploid meiosis and origin of *Brassica tournefortii* Gouan. Euphytica 23: 591–595.

Primard, C., C. Camilleri, M. Tepfer, C. Tourneur, M. Bonade-Bottino, A. Martin-Canadel, L. Jouanin, G. Pelletier and M. Renard, 1991. Towards *in situ* androgenesis in *Brassica napus*. In: D.I. McGregor (Ed.), GCIRC, July 9–11, 1991, Saskatoon, Saskatchewan, pp. 1128–1129.

Ramanujam, S., 1941. A haploid plant in Toria (*Brassica campestris*). Proc. Indian Acad. Sci. B 14: 25–34.

Rikin, A., J.W. Dillwith and D.K. Bergman, 1993. Correlation between the circadian rhythm of resistance to extreme temperatures and changes in fatty acid composition in cotton seedlings. Plant Physiol. 101: 31–36.

Renu-Arora, R., S.S. Bhojwani and R. Arora, 1988. Production of androgenic plants through pollen embryogenesis in anther cultures of *Brassica carinata* A. Draun. Biol. Plant. 30: 25–29.

Roulund, N., L. Hansted, S.B. Anderson and B. Farestveit, 1990. Effect of genotype, environment an carbohydrate on anther culture response in head cabbage (*Brassica oleracea* L. convar. *capitata* Alef.). Euphytica 49: 237–242.

Sato, T., T. Nishio and M. Hirai, 1989a. Varietal differences in embryogenic ability in anther culture of Chinese cabbage (*Brassica campestris* ssp. *pekinensis*). Jpn. J. Breed. 39: 149–157.

Sato, T., T. Nishio and M. Hirai, 1989b. Plant respiration from isolated microspore cultures of chinese cabbage (*Brassica campestris* ssp. *pekinensis*). Plant Cell Rep. 8: 486–488.

Scarth, R., G. Sequin-Swartz and G.F.W. Rakow, 1991. Application of doubled haploidy to *Brassica napus* breeding. In: D.I. McGregor (Ed.), GCIRC, July 9–11, 1991, Saskatoon, Saskatchewan, pp. 1449–1453.

Scholl, R.L. and J.A. Amos, 1980. Isolation of doubled-haploid plants through anther culture in *Arabidopsis thaliana* [analyzed genetically and cytologically]. Z. Pflanzenphysiol. 96: 407–414.

Seguin-Swartz, G., D.S. Hutcheson and R.K. Downey, 1983. Anther culture in *Brassica campestris*. In: Proc. 6th Intl. Rapeseed Congr., Paris, pp. 246–251.

Sharma, K.K. and S.S. Bhojwani, 1985. Microspore embryogenesis in anther cultures of two indian cultivars of *Brassica juncea* (L.) Czern. Plant Cell Tiss. Org. Cult. 4: 235–239.

Shetty, K. and B.D. McKersie, 1993. Proline, thioproline and potassium mediated stimulation of somatic embryogenesis in alfalfa (*Medicago sativa* L.). Plant Sci. 88: 185–193.

Siebel, J. and K.P. Pauls, 1989a. Inheritance of erucic acid content in populations of *Brassica napus* microspore-derived spontaneous diploids. Theor. Appl. Genet. 77: 489–494.

Siebel, J. and K.P. Pauls, 1989b. A comparison of anther and microspore culture as a breeding tool in *Brassica napus*. Theor. Appl. Genet. 78: 473–479.

Simmonds, D.H., C. Gervais and W.A. Keller, 1991. Embryogenesis from microspores of embryogenic and non-embryogenic lines of *Brassica napus*. In: D.I. McGregor (Ed.), 8th Intl. Rapeseed Congress Proc., July 9–13, Saskatoon, Saskatchewan, pp. 306–311.

Sorvari, S., 1985. Production of haploids from anther culture in agriculturally valuable *Brassica campestris* L. cultures. Ann. Agr. Fenn. 24: 149–160.

Stead, A.D., 1992. Pollination induced flower senescence. A. Review. Plant Growth Regul. 11: 13–20.

Stringam, G.R. and R.K. Downey, 1973. Haploid frequencies in *Brassica napus*. Can. J. Plant. Sci. 53: 229–231.

Stringam, G.R. and M.R. Thiagarajah, 1991. Effectiveness of selection for early flowering in F-2 populations of *Brassica napus* L. A comparison of doubled haploid and single seed descent methods. In: D.I. McGregor (Ed.), GCIRC, July 9–11, 1991, Saskatoon, Saskatchewan, pp. 70–75.

Sunderland, N., 1971. Anther culture: a progress report. Sci. Prog. Oxford 59: 527–549.

Swanson, E.B., 1990. Microspore culture in *Brassica*. In: J.W. Pollard and J.M. Walker (Eds.), Methods in Molecular Biology Vol. 6, Plant Cell and Tissue Culture, pp. 159–169. The Humana Press, Clifton, N.J., USA.

Swanson, E.B., M.P. Coumans, G.L. Brown, J.D. Patel and W.D. Beversdorf, 1988. The characterization of herbicide tolerant plants in *Brassica napus* L. after *in vitro* selection of microspores and protoplasts. Plant Cell Rep. 7: 83–87.

Swanson, E.B., M.P. Coumans, S.C. Wu, T. Barsby and W.D. Beversdorf, 1987. Efficient isolation of microspores and the production of microspore-derived embryos from *Brassica napus*. Plant Cell Rep. 6: 94–97.

Takahata, Y., D.C.W. Brown and W.A. Keller, 1991. Effect of donor plant age and inflorescence age on microspore culture of *Brassica napus* L. Euphytica 58: 51–55.

Takahata, Y. and W.A. Keller, 1991. High frequency embryogenesis and plant regeneration in isolated microspore culture of *Brassica oleracea* L. Plant Sci. 74: 235–242.

Telmer, C.A., W. Newcomb and D.H. Simmonds, 1993. Microspore development in *Brassica napus* and the effect of high temperature on division *in vivo* and *in vitro*. Protoplasma 172: 154–165.

Telmer, C.A., D.H. Simmonds and W. Newcomb, 1992. Determination of developmental stage to obtain high frequencies of embryogenic microspores in *Brassica napus*. Physiol. Plant 84: 417–424.

Thomas, E. and G. Wenzel, 1975. Embryogenesis from microspores of *Brassica napus*. Z. Pflanzenzüchtg. 74: 79–81.

Thompson, K.F., 1956. Production of haploid plants of marrow-stem kale. Nature (London) 178: 748.

Thompson, K.F., 1969. Frequencies of haploids in spring oilseed rape (*Brassica napus*). Hereditas 24: 318–319.

Thompson, K.F., 1974. Homozygous diploid lines from naturally occurring haploids. In: 4th Interul. Rapskongr Giessen, pp. 119–124.

Thurling, N. and P.M. Chay, 1984. The influence of donor plant genotype and environment on production of multicellular microspores in cultured anthers of *Brassica napus* ssp. *oleifera*. Ann. Bot. 54: 681–695.

Wenzel, G., 1980. Recent progress in microspore culture of crop plants. In: D.R. Davies and D.A. Opwood (Eds.), The Plant Genome, pp. 185–213. The John Innes Charity, Norwich.

Yadav, R.C., N.R. Yadav, P.R. Kumar and D.R. Sharma, 1988. Differential androgenic response in *Brassica juncea* (L.) Czern and Coss. Cruciferae Newsl. No. 13: 76.

Yang, Q., J.E. Chauvin and Y. Hervé, 1991. A study of factors affecting anther culture of cauliflower (*Brassica oleracea* var. *botrytis*). Plant Cell Tiss. Org. Cult. 28: 289–296.

Yang, Q., J.E. Chauvin and Y. Hervé, Y. 1992. Obtention d'embryos androgénétiques par culture *in vitro* de boutons floraux chez le broccoli (*Brassica oleracea* var. *italica*). C.R. Acad. Sci. Paris 314 Série III: 145–152.

Zaki, M. and H.G. Dickinson, 1990. Structural changes during the first divisions of embryos resulting from anther and microspore culture in *Brassica napus*. Protoplasma 156: 149–162.

Zaki, M.A.M. and H.G. Dickinson, 1991. Microspore-derived embryos in *Brassica*: the significance of division symmetry in pollen mitosis I to embryogenic development. Sex. Plant Reprod. 4: 48–55.

Zarsky, V., D. Garrido, L. Rihova, J. Tupy, O. Vicente and E. Heberle-Bors, 1992. Derepression of the cell cycle by starvation is involved in the induction of tobacco pollen embryogenesis. Sex. Plant Reprod. 5: 189–194.

Zhang, W.J., G.H. Fang, K.X. Tang, Z.Q. Zhang and M.J. Yu, 1991. Plant regeneration from embryoids through isolated microspore culture in *Brassica napus* L. In: D.I. McGregor (Ed.), GCIRC, July 9–11, Saskatoon, Saskatchewan, pp. 1071–1074.

Zhao, J.P. and D.H. Simmonds, 1992. *In vitro* manipulation of microspore embryogenesis to generate doubled haploid plants. In: 3rd Intern. Assoc. Plant Tiss. Cult. Workshop, June 17–20, 1992, Abstr. #6.

10. Utilization of *Brassica* haploids

CONSTANTINE E. PALMER, WILFRED A. KELLER and
PAUL G. ARNISON

Contents

1. Introduction

Haploid embryos can be produced at high frequency by anther and microspore culture of many *Brassica* species and commercial cultivars (Keller *et al.*, 1984, 1987; Lichter, 1989; Swanson, 1990; Arnison and Keller, 1990; Duijs *et al.*, 1992; Palmer *et al.*, 1994). The procedures for such production are routine although there are still significant genetic limitations to recovery of haploids from some genotypes. The development of procedures for efficient recovery of haploids was driven principally by their potential usefulness in the development of homozygous lines for breeding programs (Wenzel *et al.*, 1977). The high regenerative potential of haploid microspores, especially when cultured in isolation from the anther tissues, provides an ideal system for mutagenesis and the recovery of mutants. In addition, the recovery of doubled haploids through chromosome doubling techniques ensures that mutant genes are fixed in the homozygous state.

Isolated microspore culture and single cells derived from haploid tissues are analogous to single cell microbial systems with all the advantages. In haploid embryogenic systems of some *Brassicas*, the phenomenon of secondary embryogenesis is advantageous as it can be used for embryo cloning and rapid multiplication of novel genotypes or mutants (Loh and Ingram, 1983; Loh *et al.*, 1983; Ingram *et al.*, 1984; Shu and Loh, 1987; Prabhudesai and Bhaskaran, 1991; Loh and Lim, 1992).

Microspore-derived embryos, because of uniformity, accessibility, abundance and developmental similarity to zygotic embryos have been employed extensively in studies of biochemical and physiological aspects of embryo development. Several aspects of embryo maturation and storage product accumulation have been studied. Embryos have been used for gene transfer

S.M. Jain, S.K. Sopory & R.E. Veilleux (eds.), In Vitro *Haploid Production in Higher Plants, Vol. 3,* 173–192.

through recombinant DNA techniques as well as a source of protoplasts for fusion and recovery of somatic hybrids. The haploid system is an attractive alternative for germplasm storage as microspores and embryos can be maintained by cryopreservation without loss of embryogenic or morphogenic potential. It is therefore apparent that the haploid embryogenic system in *Brassica* is advantageous for both fundamental and applied research. The purpose of this chapter is to examine the current uses of haploid embryos and the haploid system in the *Brassicas*.

2. Breeding and genetics

2.1. *Overview*

Among the advantages of haploid embryo production is the ability to recover homozygous plants either spontaneously or by chemically induced chromosome doubling. These doubled haploids can be readily obtained for use in cultivar development. The time to develop such inbreds is reduced in comparison to conventional methods of inbreeding by recurrent selfing. In species such as *B. oleracea* and *B. campestris* (rapa) that express self-incompatibility, the production of doubled haploids is a particularly effective means of developing pure lines without time consuming bud pollination, salt or carbon dioxide treatments. The essentially instant production of pure lines allows the discovery of potentially recessive traits masked in heterozygous material (Choo *et al.*, 1985). Additionally the population size required for selection may be considerably smaller with doubled haploids compared with conventional F_2 populations (Siebel and Pauls, 1989a,b).

The use of doubled haploids for breeding and cultivar development has been embraced extensively by *Brassica* breeders around the world (e.g., Hoffmann *et al.*, 1982; Ockendon, 1983; Scarth *et al.*, 1991; Stringam and Thiagovajah, 1991) and cultivars derived from materials developed by experimental haploidy are currently on the market.

Comparison of doubled haploids with lines derived from single seed descent showed similarity in agronomic and compositional characteristics (Chen and Beversdorf, 1990; Charne and Beversdorf, 1991; Stringam and Thiagarajah, 1991) and some doubled haploid hybrids were potentially superior to conventional hybrids in some traits (Hoffmann *et al.*, 1982; Keller *et al.*, 1987).

2.2. *Representative field studies*

Some original observations (Arnison and Keller, unpublished) described below are representative of the results obtained by numerous institutional and commercial breeders. A doubled haploid population derived from heterozygous starting material provides a broad spectrum of gametic recombi-

Table 1. Comparison of doubled haploid broccoli lines with parent cultivars

Overall rating		Morphology		Maturity	
dh1–16	71.3	**cv-1**	**59.7**	dh1–1	+18
cv-1	**69.2**	dh1–16	59.3	dh7–29	+14
cv-7	**60.0**	dh7–1	59.2	dh1–16	+12
dh7–1	58.2	dh7–75	58.8	**cv-1**	**+ 6**
dh7–19	57.4	**cv-2**	**58.3**	dh1–18	+ 3
dh7–29	57.4	dh7–19	55.6	dh7–19	+ 2
dh7–75	53.3	dh1–18	50.2	dh7–27	+ 2
dh1–18	53.2	dh7–27	49.5	**cv-7**	**0**
dh1–1	51.9	dh7–34	47.9	dh7–34	0
dh7–27	51.5	dh7–29	43.4	dh7–1	− 6
dh7–34	47.9	dh7–18	40.8	dh7–74	−10
dh7–74	32.9	dh7–150	39.3	dh7–75	−10
dh7–150	31.3	dh7–74	38.4	dh7–150	−13
dh7–31	13.1	dh7–31	36.1	dh7–31	−23
dh7–18	8.2	dh1–1	33.9	dh7–18	−49

Selected data for two broccoli cultivars (cv-1 and cv-7) and doubled haploids (dhs) derived by anther culture. The overall rating combines values for seedling vigour, morphology, yield and seed production; the morphological rating is a combination of leaf quality, head form, bud size, bud quality, head density and colour. Maturity was determined in comparison with cv-7.

nants from which selections can be made. The embryos formed predominantly represent a random sample of the original gametes. There are limited in any *in vitro* selection pressures (Chen and Beversdorf, 1990), although some deviations from predicted ratios have been observed (Orton and Browers, 1985).

Partial results from field comparisons of broccoli cultivars (*B. oleracea*) and doubled-haploid lines developed from those cultivars are shown in Table 1. The values in the table are based on observations for 20 cloned plants of each line. Morphological features were assigned values according to an arbitrary scale but serve to demonstrate the enormous variability observed among doubled-haploid lines. The results showed generally that most doubled haploid lines were inferior to established cultivars but some lines were equivalent and possibly superior. Doubled haploids typically showed reduced yield, greater variation and exaggerated forms as would be expected from genetic considerations. The extent of variation observed correlated with the genetic heterozygosity of the parent materials. Doubled-haploid lines derived from inbred F_3 and F_4 lines were highly uniform by comparison but did segregate for characters of interest, such as disease resistance.

The potential value of doubled haploids for breeding was further demonstrated by examination of the performance of a random population of hybrids produced by crossing doubled haploids derived from differing genetic backgrounds.

Tables 2–4 show values for field performance of test cultivars, derived doubled haploids and hybrids formed from doubled haploid crosses. Table

Table 2. Comparison of morphology of selected broccoli cultivars, doubled haploids and doubled haploid hybrids

Cultivars and doubled haploids		Doubled haploid hybrids	
cv-8	83.1	hy 58–10	87.9
dh8–30	76.9	hy 17–2	86.8
cv-2	76.1	hy 38–26	86.2
dh5–13	75.1	hy 27–7	84.1
dh3–58	72.0	hy 23–12	82.7
dh5–8	71.8	hy 23–11	82.6
dh3–75	71.5	hy 37–7	82.6
dh8–33	70.0	hy 37–4	82.3
cv-1	68.7	hy 18–38	79.2
dh3–76	68.5	hy 37–20	78.8
dh7–2	67.7	hy 15–13	78.3
dh2–19	67.5	hy 78–31	77.5
dh3–12	67.5	hy 78–28	76.5
dh5–7	66.1	hy 38–3	76.3
dh2–7	65.5	hy 37–1	75.7
dh8–40	65.5	hy 17–12	75.5
cv-3	**62.5**	hy 17–4	74.7
dh1–18	61.8	hy 78–18	74.2
dh1–28	61.5	hy 58–5	74.0
dh7–19	59.6	hy 17–5	73.5
dh7–44	58.2	hy 17–8	72.8
cv-7	**57.8**	hy 37–10	72.6
dh7–150	57–8	hy 78–1	71.8
dh3–26	57.5	hy 37–2	71.1
dh8–17	57.3	hy 18–19	70.8
dh5–6	55.4	hy 57–16	70.4
dh3–47	53.2	hy 78–2	68.7
dh7–582	50.2	hy 23–4	68.0
dh2–9	49.0	hy 38–14	66.8
cv-5	**48.0**	hy 78–13	63.5
dh2–2	48.0	hy 57–11	59.0
dh7–160	45.9	hy 58–9	58.5
dh3–60	39.2	hy 78–20	57.6
dh7–1	27.8	hy 18–1	56.5

The six cultivars used as starting materials are designated as cv-1, etc., doubled haploid lines derived from each of the cultivars are designated as dh7–1, etc. (line 1 from cv-7). Doubled haploid hybrids are designated as hy 18–1, etc. (which represents line 1 from a cross of doubled haploids from cv-1 × cv-8). The rating is a combination of values for leaf quality, head form, bud quality, head density and colour.

2 shows morphological ratings based predominantly on facets of head appearance of interest to commercial growers.

The results again show that generally the morphology of doubled haploid lines is inferior to the parent cultivars but some are equivalent. The average score for the parent cultivars is 66.2 compared with 59.1 for the doubled haploids. The average for the doubled haploid hybrids was 71.9 from this

Table 3. Comparison of maturity of selected broccoli cultivars, doubled haploids and doubled haploid hybrids

Cultivars and doubled haploids		Doubled haploid hybrids	
dh2–2	+16.0	hy 58–5	+10.0
dh2–7	+16.0	hy 37–7	+10.0
dh2–9	+14.0	hy 17–2	+ 9.0
dh5–8	+11.9	hy 23–12	+ 8.0
dh2–19	+11.5	hy 23–11	+ 7.0
dh1–28	+11.0	hy 38–14	+ 7.0
dh3–58	+10.3	hy 18–38	+ 6.7
dh8–40	+ 9.6	hy 78–28	+ 6.6
cv-1	+ **7.4**	hy 58–10	+ 6.5
cv-5	+ **6.0**	hy 78–2	+ 4.6
dh5–6	+ 5.6	hy 27–7	+ 4.4
dh5–7	+ 5.3	hy 18–19	+ 4.3
dh7–582	+ 5.0	hy 37–10	+ 3.3
cv-3	+ **4.2**	hy 15–13	+ 2.0
dh1–18	+ 4.2	hy 78–1	+ 0.7
dh3–12	+ 4.0	hy 37–1	0
cv-2	+ **3.1**	hy 17–8	0
dh7–2	+ 3.0	hy 17.5	− 2.0
dh3–75	+ 3.0	hy 57–16	− 2.0
dh7–19	+ 0.5	hy 78–18	− 3.0
cv-8	**0**	hy 17–4	− 3.3
dh5–13	− 1.4	hy 78–31	− 4.0
dh3–60	− 2.3	hy 78–13	− 6.0
dh8–30	− 2.6	hy 37–2	− 6.0
dh3–47	− 4.7	hy 58–9	− 6.0
dh7–1	− 4.7	hy 18–1	− 6.0
cv-7	− **6.8**	hy 17–12	− 6.7
dh7–44	− 7.5	hy 38–3	− 6.7
dh8–33	− 9.7	hy 38–26	− 8.0
dh3–26	−13.0	hy 37–20	− 9.5
dh8–17	−13.0	hy 78–20	−11.5
dh7–160	−13.0	hy 37–4	−12.0
dh3–76	−13.0	hy 23–4	−12.0
dh7–150	−19.0	hy 57–11	−12.0

The six cultivars used as starting materials are designated as cv-1, etc., doubled haploid lines derived from each of the cultivars are designated as dh7–150, etc. (line 150 from cultivar cv-7). Doubled haploid hybrids are designated as hy 57–11, etc. (which represents line 11 from a cross of doubled haploids from cv-5 × cv-7). Maturity was determined in comparison with cv-8.

sample, indicating that they compare favourably with the starting materials. A number of promising lines were obvious.

Information collected relating to maturity is shown in Table 3. The maturity scores show the extended range of maturity values for doubled haploids compared with the parent cultivars and doubled haploid hybrids. Table 4 shows the results for gross yield (average weight of individual heads) of the same materials. The results for gross yield show the comparative vigour of

Table 4. Comparison of yield of selected broccoli cultivars, doubled haploid lines and doubled haploid hybrids

Cultivars and doubled haploids (gms)		Doubled haploid hybrids (gms)	
dh3–26	618	hy 18–1	1260
dh3–75	578	hy 58–9	661
dh8–17	544	hy 38–26	572
cv-5	**393**	hy 78–20	546
cv-8	**384**	hy 37–4	524
cv-3	**343**	hy 57–16	514
cv-2	**340**	hy 78–13	474
cv-7	**334**	hy 78–18	462
dh8–30	321	hy 17–5	447
dh7–160	293	hy 37–10	417
dh3–12	289	hy 38–3	413
dh8–33	246	hy 78–31	388
dh2–9	223	hy 18–19	388
cv-1	**217**	hy 37–2	385
dh2–2	217	hy 38–14	384
dh5–13	207	hy 17–4	348
dh3–58	204	hy 17–12	331
dh8–40	204	hy 23–4	328
dh7–19	184	hy 37–20	325
dh1–18	180	hy 23–11	294
dh5–7	154	hy 78–2	285
dh3–60	149	hy 58–5	284
dh7–1	124	hy 17–8	284
dh3–47	123	hy 27–7	281
dh3–76	123	hy 23–12	260
dh5–6	101	hy 17–2	235
dh7–2	101	hy 37–1	233
dh7–150	98	hy 78–1	230
dh1–28	75	hy 18–38	229
dh5–8	67	hy 57–11	222
dh7–582	64	hy 15–13	209
dh7–44	60	hy 58–10	195
dh2–19	53	hy 78–28	193
dh2–7	35	hy 37–7	144

The six cultivars used as starting materials are designated as cv-1, etc., doubled haploid lines derived from each of the cultivars are designated dh2–7, etc. (line 2 from cultivar cv-2). Doubled haploid hybrids are designated as hy 37–7, etc. (which represents line 7 from a cross of doubled haploids from cv-3 × cv-7).

doubled haploid hybrids in comparison with parent cultivars and doubled haploid lines. The most vigourous doubled haploid hybrids were frequently of poor form and it is noted that the parent cultivars were bred for uniformity, disease resistance and quality of head rather than yield of vegetative matter.

2.2.1. *General conclusions from typical field trials*

1. Doubled haploid plants showed the expected segregation of traits such that plants with potentially desirable agronomic traits can be selected.
2. Random hybrids of doubled haploid lines compared favourably with commercial cultivars in uniformity, maturity, and morphology.
3. Although not readily apparent from the information presented in the tables, both specific combining ability and general combining ability were observed for doubled haploid lines used as parents in hybrid crosses.
4. Anther and microspore culture are convenient tools for the rapid selection of germplasm with adaptation to local climates. However materials must be tested rigorously and widely to determine their commercial value.

2.3. *Breeding activities*

The practice of using doubled-haploid lines as part of *Brassica* breeding programs has become a widespread standard tool of the plant breeder.

Doubled haploid populations have been used to examine the inheritance patterns of fatty acids, glucosinolates and seed coat colour in *Brassica* (Siebel and Pauls, 1989a,b; Chen and Beversdorf, 1991; Henderson and Pauls, 1992). This approach is particularly useful as genetic ratios are much simpler in microspore-derived doubled haploids and there is a greater distinction between classes as there are no heterozygotes (Choo *et al.*, 1985). Additionally the use of doubled haploids also allowed production of yellow seeded canola plants from crosses involving high glucosinolate and low erucic acid rapeseed. Yellow seeded canola was recovered from a relatively small plant population (Henderson and Pauls, 1992). Thus microspore-derived haploids are valuable in the introgression of recessive characteristics. The inheritance patterns of fatty acids and glucosinolates were also analyzed using doubled-haploid populations and the genes governing erucic acid content were determined (Lichter *et al.*, 1988; Siebel and Pauls, 1989a; Chen and Beversdorf, 1990; Iqbal and Mollers, 1991).

Experimental haploidy may even be used to produce pure lines of some male sterile plants provided the death of microspores occurs after the stage that developing microspores can be induced to form embryos by culture treatments (Erickson *et al.*, 1986). Even where male sterility results in complete lack of microspore production, haploids may still be produced by gynogenesis (Doré, 1989).

3. Mutation and selection

Cell and tissue culture systems have been used to isolate mutants and variants having useful agronomic and biochemical characteristics (Larkin and Scowcroft, 1981; Widholm, 1983). Such variants can occur as a natural consequence of the culture process, i.e. somaclonal and gametoclonal variation, or by the use of chemical and physical mutagens. Haploid cell cultures provide an ideal system for mutant isolation since both recessive and dominant genes are expressed and traits can be readily fixed in the homozygous condition by chromosome doubling. In contrast, selection with diploid cells may allow recessive genes to remain undetected and to be carried unnoticed in the heterozygote.

In *Brassicas*, the efficient regeneration of plants from microspores provides a useful system for mutagenesis and selection. In addition, haploid suspension cell cultures of *Brassicas* and haploid embryos that are capable of sustained secondary embryogenesis (Hoffmann, 1978; Ingram *et al.*, 1984; Shu and Loh, 1987, MacDonald *et al.*, 1988; Prabhudesai and Bhaskaran, 1991; Loh and Lim, 1992) are equally useful for such studies.

In *B. napus* haploid cells and tissues have been used to select variants for disease resistance (Sacristan and Hoffman, 1979; Sacristan 1982, 1985; MacDonald and Ingram, 1986; Newsholme *et al.*, 1989; MacDonald *et al.*, 1989; Ahmad *et al.*, 1991b; Shivanna and Sawhney, 1993). Such studies involved the use of culture filtrates of the pathogen as the selective agent. Although resistance to *Alternaria brassicicola* and *Leptosphaeria maculans* have been reportedly selected from haploid cell and microspore cultures, progress towards development of resistant cultivars using this approach has been modest. This may be due to the non-specificity of the toxin produced by these pathogens. Identification of specific toxins from virulent disease strains and characterization of the relationship of toxins to pathogenicity may improve results (Soledade *et al.*, 1992; Pedras *et al.*, 1993). Chemical and physical mutagenesis have been employed to augment selection (Sacristan, 1982; MacDonald *et al.*, 1989; Ahmad *et al.*, 1991a).

Additionally, in vegetable *Brassica*, gametoclonal variants for disease resistance in *B. oleracea* have been reported (Voorrips and Visser, 1990).

The production and selection of chlorosulfuron and imidazoloinone herbicide tolerance in *B. napus* was achieved using mutagenized microspores and cells derived from haploid embryos (Polsoni *et al.*, 1988; Swanson *et al.*, 1988, 1989; Saxena *et al.*, 1990; Ahmad *et al.*, 1991). Stable mutants resistant to these herbicides were isolated and characterized (Swanson *et al.*, 1988, 1989). In these studies selection pressure provided by the herbicide, *in vitro*, allowed recovery of resistant embryos and ultimately plants. This provides a convenient method for screening a large number of mutagenized cells in a small space. The few resistant embryos recovered following mutagenesis can be multiplied through secondary embryo proliferation, in genotypes exhibiting this capacity. An alternative is the use of protoplast culture for plant recovery (Swanson *et al.*, 1988). Even without mutagens, protoplasts of haploid microspore-derived embryos can be screened for herbicide resistance and plants recovered (Swanson *et al.*, 1988). One of the limitations to the widespread use of protoplasts from haploid cells in mutagenesis and selection of herbicide resistance is absence of efficient plant regeneration systems for some species and genotypes.

Another area in which the haploid embryogenic system of *Brassica* is applicable is for isolation of mutants having altered fatty acid composition. The fatty acid component is an important quality determinant in vegetable oils, including oilseed *Brassicas*. As the pathways of fatty acid biosynthesis are now known (Stumpf and Pollard, 1983), emphasis has been placed on seed mutagenesis to recover mutants with altered fatty acid composition (James and Dooner, 1990; Lemieux *et al.*, 1990). For reasons outlined elsewhere, the haploid system is superior to seed mutagenesis for these investigations. Mutants having high levels of oleic acid were isolated from mutagenized microspores of *B. napus* (Turner and Facciotti, 1990; Haung *et al.*, 1991).

These are the only reported instances of successful fatty acid alteration in *Brassica* using haploid mutagenesis. Once techniques have been refined for analysis of lipids from minute quantities of sample, one can expect the isolation of other fatty acid mutants. The *in vitro* screening of mutagenized embryos for fatty acid profile is particularly useful as mutants can be readily identified before plant regeneration by analysis of the cotyledons. It therefore appears that the haploid embryogenic system is amenable to mutagenesis and the isolation of mutants that result from gametoclonal variation.

4. Gene transfer

In *Brassica*, as in other species, haploid cells of somatic tissues may provide an advantageous system for gene transfer by a variety of established techniques. Firstly, these cells usually have a high regenerative potential, making plant recovery from transformed cells efficient. Secondly, the transgenes can be readily fixed in the homozygous state through chromosome doubling techniques. Direct gene transfer to microspores is also attractive as they represent a large population of uniform cells capable of regeneration in contrast to complex tissues where only a few cells may be capable of morphogenesis.

There are a number of techniques employed in *Brassica* transformation, in general, and some have been adapted to haploid cells and tissues. *Agrobacterium tumefaciens*-mediated gene transfer is the most commonly used method (Swanson and Erickson, 1989; Oelck *et al.*, 1991; Haung, 1992). Transgenic plants of *B. napus*, resistant to antibiotics and herbicides, have been recovered following co-cultivation of haploid embryo segments with *Agrobacterium* (Swanson and Erickson, 1989; Oelck *et al.*, 1991). Regeneration occurred by secondary embryogenesis of individual cells which reduced the possibility of chimeras.

Comparatively little information is available on the use of direct DNA transfer to microspores. Microspores of *B. napus* were co-cultivated with *Agrobacterium tumefaciens* (Pechan, 1989). The recovery of transformed cells was claimed but has not been confirmed and there was no indication of successful plantlet recovery. Recently transient GUS expression was achieved in microspores of *B. napus* subjected to electroporation, (Jardinaud *et al.*, 1993). This is potentially a very attractive system for transformation as all cells are directly exposed to the vector and the culture system can be easily manipulated.

Direct microinjection of DNA into epidermal cells of microspore-derived embryos of *B. napus* reportedly resulted in the recovery of transgenic plants, with efficiencies ranging from 27 to 51 percent (Neuhaus *et al.*, 1987). Although this technique is advantageous in the precision and amount of DNA delivery, it is technically demanding and only a limited number of cells or embryos can be processed conveniently. Additionally, the occurrence of chimeric plants was a frequent problem.

Other gene transfer techniques such as particle bombardment, imbibition and PEG-mediated DNA uptake have been successful in other species (Topfer *et al.*, 1989; Kuhlmann *et al.*, 1991; Fennell and Hauptmann, 1992; D'Halluin *et al.*, 1992). Recently, Chen and Beversdorf (1994) reported an increase in the frequency of homozygous transgenic *B. napus* plants from direct DNA transformation of microspore-derived embryo hypocotyls using a combination of micro-projectile bombardment and desiccation/DNA imbibition.

5. Protoplast fusion

Culture systems have been developed for protoplast isolation and culture from haploid tissues of *Brassica* (Thomas *et al.*, 1976; Kohlenbach *et al.*, 1982; Chuong *et al.*, 1987; Li and Kohlenbach, 1982). Regeneration occurred by both organogenesis and embryogenesis (Chuong *et al.*, 1987; Li and Kohlenbach, 1982). One obvious advantage of haploid protoplast culture is the availability of a large number of cells for mutation and selection in a manner similar to isolated microspores. This is particularly useful if plant regeneration frequency is efficient as is the case with secondary embryogenic systems of *Brassica* (Ingram *et al.*, 1984).

Another advantage is the use of haploid protoplasts for cell fusion followed by the recovery of somatic hybrids. This technique was used to recover *B. napus* plants having cytoplasmic male sterility and herbicide tolerance (Chuong *et al.*, 1988a,b). The diploid fusion products that were recovered were completely fertile. This is in contrast to diploid fusions where incomplete inactivation of the nucleus of one partner can result in aneuploidy, chromosomal aberrations and low fertility. Also, fusions involving diploid protoplasts often result in chromosome elimination and aneuploidy.

Attempts to avoid these problems by cytoplast – diploid protoplast fusion are unnecessary (Chuong *et al.*, 1988a).

In these studies, protoplasts were derived from haploid vegetative tissues. However, protoplasts have been isolated from microspores (Bajaj, 1983) and, although this has not been reported for *Brassica*, the potential exists for the isolation, culture and fusion of microspore-derived protoplasts. The haploid fusion system should also allow the development of triploids by diploid-haploid fusions. The only limitations are the isolation of viable haploid protoplasts from microspores and the availability of an efficient regeneration system.

6. Biochemical and physiological studies

It appears that microspore-derived embryos follow a developmental pathway similar to that of the zygote and the biochemical pathways leading to storage product accumulation are similar (Crouch, 1982). In *B. napus*, fatty acid composition of both types of embryos was similar and, during the early stages of development total fatty acid per unit weight was also comparable (Pomeroy *et al.*, 1991; Taylor *et al.*, 1990). Therefore microspore-derived embryos are useful for studies of basic biochemistry and physiology in relation to embryogenesis and the accumulation of storage products (Holbrook *et al.*, 1990). These embryos can be obtained in large numbers in responding genotypes. The *in vitro* culture system allows convenient isolation of specific developmental stages. The influence of growth regulators, media and environmental factors on embryogenesis and metabolism can be readily monitored. In contrast, the zygotic embryo is relatively inaccessible during development and, although it can be explanted and cultured, the microspore-derived embryo system is far more convenient.

Microspore-derived embryos of *B. napus* and *B. rapa* have been used extensively in studies of lipid biosynthesis and accumulation (Holbrook *et al.*, 1991; Weselake *et al.*, 1991; Taylor *et al.*, 1991, 1993). These embryos are a rich source of enzymes involved in lipid biosynthesis (Taylor *et al.*, 1990). One area of considerable interest in lipid biochemistry is the biosynthesis and regulation of long chain fatty acids, such as erucic acid. Erucic acid is a seed-specific fatty acid and accumulates in both zygotic and microspore-derived embryos (De la Roche and Keller, 1977). Microspore-derived embryos of *B. napus* have been used to define the enzyme systems involved in the biosynthesis of these fatty acids (Taylor *et al.*, 1990, 1992; Holbrook *et al.*, 1991).

Microspore-derived embryos of *B. napus* were also used for rapid screening for oil quality *in vitro* (Wiberg *et al.*, 1991). Under specific culture conditions, cotyledons of these embryos accumulated triacylglycerols at rates greater than those of zygotic embryos (Wiberg *et al.*, 1991). Since the lipid composition was similar in both systems, these authors proposed the use of microspore-derived embryos for convenient screening for oil quality. Among the advantages outlined were reduction in labour requirement for screening, ability to evaluate recessive characteristics, and that cotyledons could be analyzed and the rest of the embryo retained for plant regeneration. Also genetic manipulations of haploid embryos for oil modification allow screening at the embryo level in a shorter time compared to other systems (Wiberg *et al.*, 1991).

Studies on storage protein biosynthesis and accumulation in microspore-derived embryos have been less frequent but the relationship of growth regulators and osmotic potential in these processes has been examined (Taylor *et al.*, 1990; Wilen *et al.*, 1990; Van Rooijen *et al.*, 1992). Comparative studies on the role of pyruvate kinase isozymes in metabolite partitioning in developing embryos and germinating seeds also utilized microspore-derived embryos (Sangwan *et al.*, 1992).

There are a number of other uses of microspore-derived embryos and haploid cells of *Brassicas*. They have been used in the evaluation of chilling and freezing tolerance (Johnson-Flanagan *et al.*, 1986; Orr *et al.*, 1986, 1990; Johnson-Flanagan and Singh, 1987). Freezing tolerant plants of *B. napus* were recovered from embryogenic microspores exposed to low temperatures (Cloutier, 1990; Pauls, 1990). This is another example of gametoclonal variation with low temperature as the selection agent.

In oilseed *Brassicas*, high chlorophyll content of the mature seed drastically reduces the oil quality. Microspore-derived embryos were used in studies of chlorophyll metabolism and physiological aspects of degreening (Johnson-Flanagan *et al.*, 1992; Johnson-Flanagan and Singh, 1991, 1993). Another aspect of embryo metabolism amenable to investigation utilizing haploid embryos is glucosinolate quality and quantity. Oilseeds with low levels of glucosinolates are an objective of breeding programs and haploid embryos can be effectively used in studies of the regulation of metabolism and accumulation of these components (McClellan *et al.*, 1993).

7. Germplasm storage and artificial seed technology

The induction and recovery of haploid embryos at high frequencies from cultured *Brassica* microspores resulted in more embryos than can be conveniently handled at one time. Since some of these embryos may represent novel recombinants of use in crop improvement, a convenient method of storage followed by plant recovery is required. There are two methods generally employed, cryopreservation and desiccation storage. Microspore-derived embryos of *B. napus* have been cryopreserved and plants successfully recovered (Bajaj, 1983, 1987). Also isolated microspores of *B. napus* and *B. oleracea* showed no loss of embryogenic potential following cryopreservation (Charne *et al.*, 1988; Chen, 1991; Chen and Beversdorf, 1992; Skladal *et al.*, 1992). An advantage of low temperature storage of *B. napus* microspores was the recovery of a large number of spontaneously diploid plants following embryogenesis (Chen and Beversdorf, 1992). This removed the necessity for artificial chromosome doubling which may result in abnormalities. It is not clear whether such spontaneous diploids are widespread among other *Brassicas* following low temperature microspore storage.

Factors affecting microspore-derived embryo maturation and storage product accumulation have been thoroughly examined in some *Brassica* species (Gruber and Röbbelen, 1991; Taylor *et al.*, 1990; Eikenberry *et al.*, 1991). Such mature embryos can be desiccated, stored and subsequently germinated (Senaratna *et al.*, 1991; Anandarajah *et al.*, 1991; Takahata *et al.*, 1990, 1993; Brown *et al.*, 1993). An important aspect of this desiccation tolerance was exposure of cotyledonary stage embryos to abscisic acid. Once desiccated these embryos can be stored at normal temperatures which eliminates the need for expensive cryopreservation equipment.

Another attribute of large scale haploid embryo production in *Brassicas* is their possible use in artificial seeds. This use requires the high frequency generation of spontaneous diploids or artificial chromosome doubling prior to embryo development. Embryos are encased in a protective coating which allows use in a manner similar to true seeds. This technology has been applied to somatic embryos of some dicots (Janick *et al.*, 1989; Redenbaugh *et al.*, 1986; Redenbaugh, 1993) and to microspore-derived embryos of barley (Datta and Potrykus, 1989).

Recent studies indicate that microspore-derived embryos of broccoli (*B. oleracea* var. *italica*) can be treated as artificial seeds (Takahata *et al.*, 1993). Such embryos could be desiccated to approximately 10 percent water content, stored at room temperature and subsequently germinated with a high percentage of plant conversion. Critical factors for desiccation tolerance were stage of embryo development and the use of abscisic acid during maturation. With improvements in procedures for embryo maturation and recovery of normal embryos at high frequencies, it should be possible to extend artificial seed technology to other *Brassica* species.

8. Conclusions and future prospects

In the *Brassicas*, the most efficient means of doubled haploid production is by microspore culture and the availability of doubled haploids is of enormous value in the improvement of these crop species. Homozygous lines have been produced in many genotypes of several species and evaluated for agronomic and compositional characteristics. It is evident that the microspore-derived embryos represent a random array of gametes and offer a potential source of novel variants for use in crop improvement programs. In comparison to its application in other crop species *in vitro* haploid culture of *Brassica* has been among the most successful and cultivars have been developed using this technology (e.g., Thompson, 1975; Agriculture Canada Variety Description,

1993). Effective utilization is enhanced by the ability to generate rapidly large numbers of embryos and select for specific attributes *in vitro*. There is still, however, a need to improve normal plant recovery from the embryos generated.

The use of microspores and derived embryos in mutagenesis and selection is advancing. Mutagenesis is being utilized for the production of such traits as the alteration of fatty acids, glucosinolate levels, and disease resistance. There are a number of reports on *in vitro* screening of microspores and microspore-derived embryos for resistance to pathogens, particularly black leg, *Leptosphaeria maculans*. However, success is more likely if specific fungal toxins are used in such selections. The major advantages of haploid systems for such studies have been discussed above. Even if mutants are produced from somatic diploid cells, microspore culture can be utilized to fix the trait in the homozygous state. One potentially useful area of *in vitro* selection with haploids in *Brassica* is the utilization of amino acid analogues to select for modified amino acid accumulation. This has implications for protein quality improvement. Mutation experiments aimed at fatty acid and seed oil modification can effectively utilize haploid embryos as the cotyledons can be analyzed for storage lipids and the rest of the embryo retained for plant recovery (Wiberg *et al.*, 1991; Haung, 1992). Therefore mutagenesis and full embryo development can allow for rapid screening *in vitro* with all the detailed advantages (Wiberg *et al.*, 1991).

The potential of haploid cells of *Brassica* for somatic hybridization and development of triploids has not been fully exploited. The protoplast system would be a valuable tool for recovery of a large number of plants from mutagenized embryos.

The use of haploids in gene transfer experiments has been modest to date even though the advantages are evident. Microspore populations enriched for embryogenic cells can be readily subjected to a variety of gene transfer techniques, such as particle bombardment, electroporation and *Agrobacterium* cocultivation. So far most of the transformation experiments have been with microspore-derived embryo segments with transformant recovery through secondary embryogenesis. While this has been relatively successful, gene transfer to individual microspores prior to embryo development is potentially more advantageous.

The microspore-derived embryo system of *Brassica* is a useful model for studies of the regulatory control of embryogenesis as well as storage product biosynthesis and accumulation. Most of the current studies involve lipid and fatty acid metabolism. However the system is amenable to studies of carbohydrate and protein synthesis and accumulation. One attractive feature of this system is that mutations affecting storage product physiology and biochemistry can be readily evaluated *in vitro*.

Compared to other haploid systems, notably monocots, there are relatively few instances of extensive albinism among *Brassica* haploid embryos. Although there may be some abnormalities in cotyledon and embryo axis development, the embryos are fairly uniform and can be handled as artificial seeds. This technology has advanced for some species but is still largely unexploited in *Brassica*. However, increased knowledge of embryo maturation, desiccation tolerance, chromosome doubling and storage product accumulation patterns will allow wide application of this technique.

The microspore population is a rich source of genetic variation as a result of gametic recombination that serves admirably as a basis for cultivar improvement and development, and as a vehicle for basic research in plant developmental and biochemical processes.

9. References

Agriculture Canada Variety Description, 1993. Registration number 3421.

Ahmad, I., J.P. Day, M.V. MacDonald and D.S. Ingram, 1991a. Haploid culture and UV mutagenesis in rapid cycling *Brassica napus* for the generation of resistance to chlorsulfuron and *Alternaria brassicicola*. Ann. Bot. 67: 521–525.

Ahmad, I., M.V. MacDonald and D.S. Ingram, 1991b. *In vitro* selection of primary embryos derived from UV-treated microspores of rapid cycling *Brassica napus* for herbicide tolerance. Crucifer Newsl. 14/15: 86–87.

Anandorajah, K., L. Kott, W.D. Beversdorf and B.D. McKersie, 1991. Induction of desiccation tolerance in microspore-derived embryos of *Brassica napus* L. by thermal stress. Plant Sci. 77: 119–123.

Arnison, P.G. and W.A. Keller, 1990. A survey of the anther culture response of *Brassica oleracea* L. cultivars grown under field conditions. Plant Breed. 104: 125–133.

Bajaj, Y.P.S., 1983. Haploid protoplasts. In: K.L. Giles (Ed.), Intl. Rev. Cytol. Suppl. 16, pp. 113–141. Academic Press, New York.

Bajaj, Y.P.S., 1987. Cryopreservation of pollen and pollen-derived embryos, and the establishment of pollen banks. Intl. Rev. Cytol. 107: 397–420.

Brown, D.C.W., E.M. Watson and P.M. Pechan, 1993. Induction of desiccation tolerance in microspore-derived embryos of *Brassica napus*. In Vitro Cell. Devel. Biol. 29: 113–118.

Charne, D.G. and W.D. Beversdorf, 1991. Comparisons of agronomic and compositional traits in microspore-derived and conventional populations of spring *Brassica napus*. In: D.I. McGregor (Ed.), Proc. 8th Intl. Rapeseed Congr. Vol. 6, pp. 64–69.

Charne, D.G., P. Pubacki, L.S. Kott and W.D. Beversdorf, 1988. Embryogenesis following cryopreservation in isolated microspores of rapeseed (*Brassica napus* L.). Plant Cell Rep. 7: 407– 409.

Chen, J.L., 1991. Evaluation of microspore culture in germplasm preservation, lipid biosynthesis and DNA uptake in rapeseed *Brassica napus* L. Ph.D. Thesis, Dept. of Crop Sci., Univ. of Guelph, Guelph, Ontario.

Chen, J.L. and W.D. Beversdorf, 1990. A comparison of traditional and haploid-derived breeding populations of oilseed rape (*Brassica napus* L.) for fatty acid composition of the seed oil. Euphytica 51: 59–65.

Chen, J.L. and W.D. Beversdorf, 1991. Fatty acid inheritance in microspore-derived populations of spring rapeseed (*Brassica napus* L.). Theor. Appl. Genet. 80: 465–469.

Chen, J.L. and W.D. Beversdorf, 1992. Cryopreservation of isolated microspores of spring rapeseed (*Brassica napus* L.) for *in vitro* embryo production. Plant Cell Tiss. Org. Cult. 31: 141–149.

Chen, J.L. and W.D. Beversdorf, 1994. A combined use of microprojectile bombardment and DNA imbibition enhances transformation frequency of canola (*Brassica napus* L.) Theor. Appl. Genet. (in press).

Choo, T.M., E. Reinbergs and K.J. Kasha, 1985. Use of haploids in breeding barley. Plant Breed. Rev. 3: 219–252.

Chuong, P.V., W.D. Beversdorf and K.P. Pauls, 1987. Plant regeneration from haploid stem peel protoplasts of *Brassica napus* L. J. Plant Physiol. 130: 57–65.

Chuong, P.V., W.D. Beversdorf, A.D. Powell and K.P. Pauls, 1988a. The use of haploid protoplast fusion to combine cytoplasmic atrazine resistance and cytoplasmic male sterility in *Brassica napus*. Plant Cell Tiss. Org. Cult. 12: 181–184.

Chuong, P.V., W.D. Beversdorf, A.D. Powell and K.P. Pauls, 1988b. Somatic transfer of cytoplasmic traits in *Brassica napus* L. by haploid protoplast fusion. Mol. Gen. Genet. 211: 197–201.

Cloutier, S., 1990. *In vitro* selection for freezing tolerance using *Brassica napus* microspore culture. M.Sc Thesis, Dept. of Crop Science, Univ. of Guelph, Guelph, Ontario.

Crouch, M.L., 1982. Nonzygotic embryos of *Brassica napus* L. contain embryo-specific storage proteins. Planta 156: 520–524.

Datta, S.K. and I. Potrykus, 1989. Artificial seeds in barley encapsulation of microspore-derived embryos. Theor. Appl. Genet. 77: 820–824.

De la Roche, A.I. and W.A. Keller, 1977. The morphogenetic control of erucic acid synthesis in *Brassica campestris*. Z. Pflanzenzüchtg. 78: 319–326.

D'Halluin, K., E. Bonne, M. Bossut, M. DeBuckeleer and J. Leemans, 1992. Transgenic maize by tissue electroporation. Plant Cell 4: 1495–1505.

Doré, C., 1989. Obtention, de plantes haploïdes de chou cabus (*Brassica oleracea* L. ssp. *capitata*) après culture *in vitro* d'ovules pollinisés par du pollen irradié. C.R. Acad. Sci. Paris, 309 Série III: 729–734.

Duijs, J.G., R.E. Voorrips, D.L. Visser and J.B.M. Custers, 1992. Microspore culture is successful in most crop types of *Brassica oleracea* L. Euphytica 60: 45–55.

Eikenberry, E.J., P.V. Chuong, J. Esser, J. Romero and X. Ram, 1991. Maturation, desiccation, germination and storage lipid accumulation in microspore embryos of *Brassica napus* L. In: D.I. McGregor (Ed.), Proc. 8th Intl. Rapeseed Congr. Vol. 6, pp. 1809–1814.

Erickson, L., I. Grant and W.D. Beversdorf, 1986. Cytoplasmic male sterility in rapeseed (*Brassica napus* L.) 2. The role of a mitochondrial plasmid. Theor. Appl. Genet. 72: 151–157.

Fennell, A. and R. Hauftmann, 1992. Electroporation and PEG delivery of DNA into maize microspores. Plant Cell Rep. 11: 567–570.

Gruber, S. and G. Röbbelen, 1991. Fatty acid synthesis in microspore-derived embryoids of rapeseed (*Brassica napus*). In: D.J. McGregor (Ed.), Proc. 8th Intl. Rapeseed Congr. Vol. 6, pp. 1818–1820.

Haung, B., 1992. Genetic manipulation of microspores and microspore-derived embryos. In Vitro Cell. Develop. Biol. 28: 53–58.

Haung, B., E.B. Swanson, C.L. Baszczynski, W.D. Macrae, E. Bardour, V. Armavil, L. Wohe., M. Arnaldo, S. Rozakis, M. Westecott, R.F. Keats and R. Kemble, 1991. Application of microspore culture to canola improvement. In: D.I. McGregor (Ed.), Proc. 8th Intl Rapeseed Congr. Vol. 6, p. 298.

Henderson, C.A.P. and K.P. Pauls, 1992. The use of haploidy to develop plants that express several recessive traits using light-seeded canola (*Brassica napus*) as an example. Theor. Appl. Genet. 83: 476–479.

Hoffmann, F., 1978. Mutation and selection of haploid cell culture systems of rape and rye. In: A.W. Alfermann and E. Reinhord (Eds.), Production of Natural Compounds by Cell Culture Methods, pp. 319–329. Ges. und Strahlen und Umweltforsch., Munich.

Hoffmann, F., E. Thomas and G. Wenzel, 1982. Anther culture as a breeding tool in rape. II. Progeny analyses of androgenetic lines and induced mutants from haploid cultures. Theor. Appl. Genet. 61: 225–232.

Holbrook, L.A., W.R. Scowcroft, D.C. Taylor, M.K. Pomeroy, R.W. Wilen and M.M. Moloney, 1990. Microspore-derived embryos: A tool for studies in regulation of gene expression in zygotic embryos. In: N.J.J. Nijkamp, L.H.W. Van Der Plas and J. Van Aartrijk (Eds.),

Progress in Plant Cellular and Molecular Biology, pp. 402–406. Kluwer Academic Pubishers, Dordrecht.
Holbrook, L.A., G.J.H. Van Rooijen, R.W. Wilen and M.M. Moloney, 1991. Oil-body proteins in microspore-derived embryos of *Brassica napus*, Hormonal, osmotic and developmental regulation for synthesis. Plant Physiol. 97: 1051–1058.
Ingram, D.S., C.S. Loh, M.V. MacDonald and D.M. Newsholme, 1984. Secondary embryogenesis in *Brassica*: A tool for research and crop improvement. Annual Proc. Phytochem. Soc. Europe. 23: 219–242.
Iqbal, M.C.M. and C. Mollers, 1991. Selection for low glucosinolate content of rapeseed, *Brassica napus* L., using haploid embryos from microspore culture. In: D.I. McGregor (Ed.), Proc. 8th Intl. Rapeseed Congr. Vol. 1, pp. 187–190.
James, D.W. Jr. and H.K. Dooner, 1990. Isolation of EMS-induced mutants in *Arabidopsis* altered in seed fatty acid composition. Theor. Appl. Genet. 80: 241–245.
Janick, J., S.L. Kitto and V.A. Dim, 1989. Synthetic seed production by desiccation and encapsulation. In Vitro Cell Devel. Biol. 25: 1167–1172.
Jardinaud, M.-F., A. Souvré and G. Alibert, 1993. Transient GUS gene expression in *Brassica napus* electroporated microspores. Plant Sci. 93: 177–184.
Johnson-Flanagan, A.M., L.I. Barran and J. Singh, 1986. L-Methionine transport during the induction of freezing hardiness by abscisic acid in *Brassica napus* cell suspension cultures. J. Plant Physiol. 124: 309–319.
Johnson-Flanagan, A.M., Z. Huiwen, X.-M. Geng, D.C.W. Brown, C.L. Nykifarutk and J. Singh, 1992. Frost, abscisic acid and desiccation hasten embryo development in *Brassica napus* L. Plant Physiol. 94: 700–706.
Johnson-Flanagan, A.M. and J. Singh, 1987. Alteration of gene expression during the induction of freezing tolerance in *Brassica napus* suspension cultures. Plant Physiol. 85: 699–705.
Johnson-Flanagan, A.M. and J. Singh, 1991. Degreening and its inhibition by stress in haploid embryos of *Brassica napus* cv. Topas and Jet Neuf. In: D.I. McGregor (Ed.), Proc. 8th Intl. Rapeseed Congr., p. 743.
Johnson-Flanagan, A.M. and J. Singh, 1993. A method to study seed degreening using haploid embryos of *Brassica napus* cv. Topas. J. Plant Physiol. 141:487–493.
Keller, W.A., K.C. Armstrong and A.I. de la Roche, 1984. The production and utilization of microspore-derived haploids in *Brassica* crops. In: K.L. Giles S.K. Sen (Eds.), Plant Cell Culture in Crop Improvement, pp. 169–183. Plenum Pub. Corp., New York.
Keller, W.A., P.G. Arnison and B.J. Cardy, 1987. Haploids from gametophytic cells – Recent developments and future prospects. In: C.E. Green, D.A. Somers, W.P. Hackett and D.D. Biesboer (Eds.), Plant Tissue and Cell Culture, pp. 223–241. Alan R. Liss Inc., New York.
Kohlenbach, H.W., G. Wenzel and F. Hoffmann, 1982. Regeneration of *Brassica napus* plantlets in cultures from isolated protoplasts of haploid stem embryos as compared with leaf protoplasts. Z. Pflanzenphysiol. 105: 131–142.
Kuhlmann, U., B. Foroughi-Wehr, A. Garner and G. Wenzel, 1991. Improved culture system for microspores of barley to become a target for DNA uptake. Plant Breed. 107: 165–169.
Larkin, P.J. and W.R. Scowcroft, 1981. Somaclonal variation – a novel source of variability from cell cultures for plant improvement. Theor. Appl. Genet. 60: 197–214.
Lemieux, B., M. Miguel, C. Somerville and J. Browse, 1990. Mutants of *Arabidopsis* with alterations in seed lipid fatty acid composition. Theor. Appl. Genet. 80: 234–240.
Li, L.-C. and H.W. Kohlenbach, 1982. Somatic embryogenesis in quite a direct way in cultures of mesophyll protoplasts of *Brassica napus* L. Plant Cell Rep. 1: 209–211.
Lichter, R., 1989. Efficient yield of embryoids by culture of isolated microspores of different *Brassicaceae* species. Plant Breed. 103: 119–123.
Lichter, R., E. DeGroat, D. Fiebeg, R. Schweiger and A. Gland, 1988. Glucosinolates determined by HPLC in the seeds of microspore derived homozygous lines of rapeseed (*Brassica napus* L). Plant Breed. 100: 209–221.
Loh, C.S. and D.S. Ingram, 1983. The response of secondary embryoids and secondary em-

bryogenic tissues of winter oilseed rape to treatment with colchicine. New Phytol. 95: 359–366.

Loh, C.S., D.S. Ingram and D.E. Hanke, 1983. Cytokinins and the regeneration of plantlets from secondary embryoids of winter rape, *Brassica napus* ssp. *oleifera*. New Phytol. 95: 349–358.

Loh, C.S. and G.K. Lim, 1992. The influence of medium components on secondary embryogenesis of winter oilseed rape, *Brassica napus* L. ssp. *oleifera* (Metzg.) Sinsk. New Phytol. 121: 425–430.

MacDonald, M.V., I. Ahmad and D.S. Ingram, 1989. Mutagenesis and haploid culture for disease resistance in *Brassica napus*. In: Science for Plant Breeding XII. Eucarpia Congress, Feb. 27 - March 4, 1989, Göttingen, pp. 22–27.

MacDonald, M.V. and D.S. Ingram, 1986. Towards the selection *in vitro* for resistance to *Alternaria brassicicola* (Schus.) Wilts. in *Brassica napus* ssp. *oleifera* (Metzg.) Sinsk. Winter Oilseed Rape. New Phytol. 104: 621–629.

MacDonald, M.V., D.M. Newsholme and D.S. Ingram, 1988. The biological effects of gamma irradiation on secondary embryoids of *Brassica napus* ssp. *oleifera* Metzg. (Sinsk.) winter oilseed rape. New Phytol. 110: 255–259.

McClellan, D., L. Kott, W. Beversdorf and B.E. Ellis, 1993. Glucosinolate metabolism in zygotic and microspore-derived embryos of *Brassica napus* L. J. Plant Physiol. 141: 153–159.

Neuhaus, G., G. Spangenberg, O. Mittelstein-Sheid and G. Schweiger, 1987. Transgenic rapeseed plants obtained by the microinfection of DNA into microspore-derived embryoids. Theor. Appl. Genet. 75: 30–36.

Newsholme, D.M., M.V. MacDonald and D.S. Ingram, 1989. Studies of selection *in vitro* for novel resistance to phytotoxic products of *Leptosphaeria maculans* (Desm.) Ces. and De Not. in secondary embryogenic lines of *Brassica napus* ssp. *oleifera* (Metzg.). Sinsk, Winter Oilseed Rape. New Phytol. 113: 117–126.

Ockendon, D.J., 1983. Use of anther culture in Brussels sprout breeding. Crucifer Newsl. 8: 58–60.

Oelck, M.M., C.V. Phan, P. Eckes, G. Donn, G. Rakow and W.A. Keller, 1991. Field resistance of canola transformants (*Brassica napus* L.) to Ignite (Phosphinothricin). In: D.I. McGregor (Ed.), Proc. 8th Intl. Rapeseed Congr., p. 293.

Orr, W., W.A. Keller and J. Singh, 1986. Induction of freezing tolerance in an embryonic cell suspension culture of *Brassica napus* by abscisic acid at room temperature. J. Plant Physiol. 126: 23–32.

Orr, W., A.M. Johnson-Flanagan, W.A. Keller and J. Singh, 1990. Induction of freezing tolerance in microspore-derived embryos of winter *Brassica napus*. Plant Cell Rep. 8: 579–581.

Orton, T.J. and M.A. Browers, 1985. Segregation of genetic markers among plants regenerated from cultured anthers of broccoli (*Brassica oleracea*, var. *italica*). Theor. Appl. Genet. 69: 637–643.

Palmer, C.E., W.A. Keller and P.G. Arnison, 1994. Experimental haploidy in *Brassica*. In: S.M. Jain, S.K. Sopory and R.E. Veilleux (Eds.), *In Vitro* Haploid Production in Higher Plants Vol 3, pp. 143–172. Kluwer Academic Publishers, Dordrecht.

Pauls, K.P., 1990. Cell culture techniques and canola improvement. In: J. Rattray (Ed.), Biotechnology of Plant Fats and Oils, pp. 36–51. Amer. Oil Chemists' Society, Champaign, Ill.

Pedras, M., C. Soledade, J.L. Taylor and T.T. Nakashima, 1993. A novel chemical signal from the blackleg fungus: Beyond phytotoxins and phytoalexins. J. Org. Chem. 58: 4778–4780.

Pechan, P.M., 1989. Successful cocultivation of *Brassica napus* microspores and proembryos with *Agrobacterium*. Plant Cell Rep. 8: 387–390.

Polsoni, L., L.S. Kott and W.D. Beversdorf, 1988. Large-scale microspore culture technique for mutation-selection studies in *Brassica napus*. Can. J. Bot. 66: 1681–1685.

Pomeroy, M.K., J.K.G. Kramer, D.J. Hunt and W.A. Keller, 1991. Fatty acid changes during development of zygotic and microspore-derived embryos of *Brassica napus*. Physiol. Plant. 81: 447–454.

Prabhudesai, V. and S. Bhaskaran, 1991. A continuous culture system of direct somatic embryogenesis in microspore-derived embryos of *Brassica juncea*. Plant Cell Rep. 12: 289–292.

Redenbaugh, K., 1993. Syn Seeds. Applications of Synthetic Seeds to Crop Improvement. CRC Boca Raton, Ann Arbor/London/Tokyo.

Redenbaugh, K., B.D. Paasch, J.W. Nichol, M.E. Kossler, P.R. Viss and K.A. Walker, 1986. Somatic seeds. Encapsulation of asexual plant embryos. Biotechnology 4: 797–801.

Sacristan, M.D., 1982. Resistance response to *Phoma lingam* of plants regenerated from selected cells and embryogenic culture of haploid *Brassica napus*. Theor. Appl. Genet. 61: 193–200.

Sacristan, M.D., 1985. Selection for disease resistance in *Brassica* cultures. Hereditas Suppl. 3: 57–63.

Sacristan, M.D. and F. Hoffmann, 1979. Direct infection of embryogenic tissue cultures of haploid *Brassica napus* with resting spores of *Plasmodophora brassicae*. Theor. Appl. Genet. 54: 129–132.

Sangwan, R., D.A. Gauthier, D.H. Turpin, M.R. Pomeroy and W.C. Plaxton, 1992. Pyruvate-kinase isozymes from zygotic and microspore-derived embryos of *Brassica napus*. Planta 187: 198–202.

Saxena, P.K., D. Williams and J. King, 1990. The selection of chlorsulfuron-resistant cell lines of independent origin from an embryogenic cell suspension culture of *Brassica napus* L. Plant Sci. 69: 231–237.

Scarth, R., G. Seguin-Swartz and G.F.W. Rakow, 1991. Application of doubled haploidy to *Brassica napus* breeding. In: D.I. McGregor (Ed.), Proc. 8th Intl. Rapeseed Congr. Vol. 5., pp. 1449–1453.

Senaratna, T., L. Kott, W.D. Beversdorf and B.C. McKersie, 1991. Desiccation of microspore-derived embryos of oilseed rape (*Brassica napus* L.). Plant Cell Rep. 1: 342–344.

Shivanna, K.R. and V.K. Sawhney, 1993. Pollen selection for alternaria resistance in oilseed *Brassicas*: responses of pollen grains and leaves to a toxin of *A. brassicae*. Theor. Appl. Genet. 86: 339–344.

Shu, W. and C.S. Loh, 1987. Secondary embryogenesis in long term cultures of winter oilseed rape, *Brassica napus* ssp. *oleifera*. New Phytol. 107: 39–46.

Siebel, J. and K.P. Pauls, 1989a. Inheritance patterns of erucic acid content in populations of *Brassica napus* microspore-derived spontaneous diploids. Theor. Appl. Genet. 77: 489–494.

Siebel, J. and K.P. Pauls, 1989b. Alkenyl glucosinolate levels in androgenic populations of *Brassica napus*. Plant Breed. 103: 124–132.

Skladal, V., M. Vyvadilova and J. Lachova, 1992. Cryopreservation of microspore suspension of *Brassica napus* and *Brassica oleracea* L. Biol. Plant. 34 Suppl. 548.

Soledade, M., C. Petras and G. Seguin-Swartz, 1992. The blackleg fungus: phytotoxins and phytoalexins. Can. J. Plant Pathol. 14: 67–75.

Stringam, G.R. and M.R. Thiagarajah, 1991. Effectiveness of selection for early flowering in F_2 populations of *Brassica napus* L. A comparison of doubled haploid and single seed descent methods. In: D.I. McGregor (Ed.), Proc. 8th Intl. Rapeseed Congr., pp. 70–75.

Stumpf, P.K. and M.R. Pollard, 1983. Pathways of fatty acid biosynthesis in higher plants with particular reference to developing rapeseed. In: J.K.G. Kramer, F.D. Sauer and W.J. Pigden (Eds.), High and Low Erucic Acid Rapeseed Oil, pp. 131–141. Academic Press, New York.

Swanson, E.B., 1990. Microspore culture in *Brassica*. In: J.W. Pollard and J.M. Walker (Eds.), Methods in Molecular Biology Vol. 6, Plant Cell and Tissue Culture, pp. 159–169. The Humana Press, Clifton, N.J., USA.

Swanson, E.B. and L.R. Erickson, 1989. Haploid transformation in *Brassica napus* using an octopine-producing strain of *Agrobacterium tumefaciens*. Theor. Appl. Genet. 78: 831–835.

Swanson, E.B., M.P. Coumans and G.L. Brown, 1988. The characterization of herbicide tolerant plants in *Brassica napus* L. after *in vitro* selection of microspores and protoplasts. Plant Cell Rep. 7: 83–87.

Swanson, E.B., M.J. Herrgesell and M. Arnoldo, D. Sippell and R.S.C. Wong, 1989. Microspore mutagenesis and selection: canola plants with field tolerance to the imidazolinones. Theor. Appl. Genet. 78: 525–530.

Takahata, Y., D.C.W. Brown and W.A. Keller, 1990. Desiccation tolerance in broccoli (*Brassica oleracea* L.) microspore-derived embryos. Jpn. J. Breed. 40: 92–93.

Takahata, Y., D.C.W. Brown, W.A. Keller and N. Kaizuma, 1993. Dry artificial seeds and desiccation tolerance induction in microspore-derived embryos of broccoli. Plant Cell Tiss. Org. Cult. 35: 121–129.

Taylor, D.C., D.L. Barton, K.P. Roux, S.I. Mackenzie, D.W. Reed, E.W. Underhill, M.K. Pomeroy and N. Weber, 1992. Biosynthesis of acyllipids containing very long chain fatty acids in microspore-derived and zygotic embryos of *Brassica napus* L. cv. Reston. Plant Physiol. 99: 1609–1618.

Taylor, D.C., A.M.R. Ferrie, W.A. Keller, E.M. Giblin, E.U. Pass and S.L. Mackensie, 1993. Bioassembly of acyllipids in microspore-derived embryos of *Brassica campestris*. Plant Cell Rep. 12: 375–384.

Taylor, D.C., N. Weber, D. Barton, E.W. Underhill, L.R. Hoffe, R.J. Weselake and M.K. Pomeroy, 1991. Triacylglycerol bioassembly in microspore-derived embryos of *Brassica napus* cv. Reston. Plant Physiol. 97: 65–79.

Taylor, D.C., N. Weber, D. Barton, E. Underhill and K. Pomeroy, 1990. Biosynthesis of triacylglycerols containing erucic acid in microspore-derived embryos of *Brassica napus*. In: P.J. Quinn and J.L. Harwood (Eds.), Plant Lipid Biochemistry, Structure and Utilization, pp. 210–212. Portland Press, London.

Thomas, E., F. Hoffmann, I. Potrykus and G. Wenzel, 1976. Protoplast regeneration and stem embryogenesis of haploid androgenetic rape. Mol. Gen. Genet. 145: 245–247.

Thompson, K.F., 1975. Homozygous diploid lines from naturally occurring haploids. In: Proc. 4th Intl. Rapeseed Congr., Giessen, 1974, pp. 119–124.

Topfer, R., B. Gronenboru, J., Shell and H. Steinbiss, 1989. Uptake and transient expression of chimeric genes in seed-derived embryos. Plant Cell 1: 133–139.

Turner, J. and D. Facciotti, 1990. High oleic acid *Brassica napus* from mutagenized microspores. In: J.R. McFerson, S. Kresovich and S.G. Dwyer (Eds.), Proc. 6th Crucifer Genetics Workshop, Geneva, NY, p. 24.

Van Rooijen, G.J.H., R.W. Wilen, L.A. Holbrook and M.M. Moloney, 1992. Regulation of accumulation of mRNAs encoding a 20 kDa oil-body protein in microspore-derived embryos of *Brassica napus*. Can. J. Bot. 70: 503–508.

Voorrips, R.E. and D.L. Visser, 1990. Doubled haploid lines with clubroot resistance in *Brassica oleracea*. In: J.R. McFerson, S. Kresovich and S.G. Dwyer (Eds.), Proc. 6th Crucifer Genet. Workshop, Geneva, NY, p. 40.

Weselake, R.J., D.C. Taylor, M.K. Pomeroy, S.L. Lewson and E.W. Underhill, 1991. Properties of diacylglycerol acyltransferase from microspore-derived embryos of *Brassica napus* L. Phytochemistry 30: 3533–3538.

Wenzel, G., F. Hoffmann and E. Thomas, 1977. Anther culture as a breeding tool in rape. I. Ploidy level and phenotypes of androgenetic plants. Z. Pflanzenzüchtg. 78: 149–155.

Wiberg, E., L. Rahlen, M. Hellman, E. Tillberg, K. Glimelius and S. Stymne, 1991. The microspore-derived embryo of *Brassica napus* L. as a tool for studying embryo-specific lipid biosynthesis and regulation of oil quality. Theor. Appl. Genet. 82: 515–520.

Widholm, J.M., 1983. Isolation and characterization of mutant plant cell cultures. In: S.K. Sen and K.L. Giles (Eds.), Plant Cell Culture in Crop Improvement, pp. 71–87. Plenum Press, New York.

Wilen, R.W., R.M. Mandel, R.P. Pharis, L.A. Holbrook and M.M. Moloney, 1990. Effects of abscisic acid and high osmoticum on storage protein gene expression in microspore embryos of *Brassica napus*. Plant Physiol. 94: 875–881.

11. *In vitro* haploidization of fruit trees

S. J. OCHATT and Y.X. ZHANG

Contents

1. Introduction and historical background

Thirty years have passed since the pioneering work by Guha and Maheshwari (1964), who obtained haploid embryos from anther cultures of the solanaceous species *Datura innoxia*. Ten years later, Nitsch and Norreel (1973) described the first successful example of microspore culture with this same species. Since then, *in vitro* production of haploid plants has been extended to more than 250 species belonging to about 100 genera and over 40 families (Foroughi-Wehr and Wenzel, 1989). However, progress has been achieved only for a limited number of species, essentially confined to the Cruciferae, Gramineae, Ranunculaceae, and Solanaceae (Sangwan-Norreel *et al.*, 1986). For most other species, including the rosaceous fruit crops, androgenesis has been feasible, but the androgenic rates have most often remained very low. There are other approaches to produce haploids. These include the spontaneous occurrence of haploids in angiosperms *in vivo* (Kimber and Riley, 1963), *in vitro* gynogenesis (San and Gelebart, 1986), and *in situ* parthenogenesis induced by irradiated pollen (Raquin, 1985). The induction of haploids via *in vitro* gynogenesis, first demonstrated for oats (San, 1976) and now extended to a score of species, permits the production of haploids for non-androgenic or male sterile genotypes (e.g., several *Compositae* such as gerberas, lettuce or sunflower) (San and Gelebart, 1986). The extention of *in vitro* gynogenesis to a larger range of species, however, is hampered by a

S.M. Jain, S.K. Sopory & R.E. Veilleux (eds.), In Vitro *Haploid Production in Higher Plants, Vol. 2,* 193–210.

lack of efficient strategies for the study of the embryo sac and its evolution (both *in vitro* and *in situ*). In addition, for many rosaceous fruit trees (*Malus*, *Pyrus*, *Prunus*), the small size of the embryo sac and the reduced number of female gametophytes constitute a further barrier to the successful exploitation of this approach for the production of haploid plants.

The induction of *in situ* parthenogenesis by pollination with irradiated pollen was first developed for *Petunia* (Raquin, 1985), and involves the *in vitro* culture of the pollinated ovaries or immature embryos resulting from such pollinations. For this approach, it is advisable that the pollen or female parent should carry one or several marker genes, in order to facilitate the subsequent screening of plants produced by parthenogenesis. This strategy is particularly helpful in offering the possibility of producing haploids for species where methods of androgenesis and gynogenesis have so far proven unsuccessful.

2. *In vitro* haploidization in fruit trees

Compared with research on herbaceous angiosperms, studies of *in vitro* haploidization with woody perennials are few (Bonga *et al.*, 1987; Chen, 1987) and have seldom been successful. Studies with fruit trees have involved mostly anther culture (Zhang and Lespinasse, 1992), although gynogenesis and *in situ* parthenogenesis induced by irradiated pollen have also been explored for a number of fruit crops. All three of these approaches will be discussed for temperate, tropical and subtropical fruits.

2.1. *Induction of androgenesis*

Androgenesis has been described as the deviation of development from a normal gametophytic to a sporophytic pathway, using *in vitro* techniques. This process leads to callusing or embryogenesis, whereby plants can be obtained indirectly by organogenesis from such androgenic calli or embryos, or directly by development of the androgenic embryo. Of the four techniques available to induce androgenesis *in vitro* (i.e., anther culture, culture of pollen isolated from a precultured anther, the float culture technique, and culture of isolated pollen), anther culture is by far the most widely used (Sangwan and Sangwan-Norreel, 1987), having led to the successful production of androgenic haploid plants from some 15 woody fruit genotypes to date.

In this context, genotype (including species, cultivar, clone or line) has played a paramount role in the induction of androgenesis, in line with most other biotechnological approaches (Ochatt *et al.*, 1990, 1992). For instance, in *Prunus* anther culture has generally resulted only in callus formation. Differences between species and hybrids in terms of androgenic competence have been commonplace in *Citrus*; Chen (1985) obtained embryo formation and plant recovery from two of nine genotypes involving eight different species and hybrids. In apple (*Malus Xdomestica*) two mutants of "Delicious" apple, "Topred" and "Starking", underwent androgenesis and produced embryos (Zhang *et al.*, 1987, 1990), but "Golden Delicious" and other genotypes related to it only produced callus. In grapevines (*Vitis vinifera*), only three of 27 clones tested by Gresshoff and Doy (1974) were capable of producing haploid callus.

Stage of microspore development plays a determinant role in the induction of androgenesis. Microspores or pollen at the uninucleate stage are the most responsive for nearly all the perennial fruit crop species studied to date. Detailed cytological studies conducted on poplars, rubber (Chen, 1986) and apple (Zhang, 1988; Zhang *et al.*, 1990) have shown that androgenic callus and embryos were mainly induced through a deviation of the first pollen mitosis to produce two undifferentiated nuclei. This is probably responsible for uninucleate microspores being more receptive to *in vitro* androgenesis in woody species.

The physiological state of the mother plants is also a factor. Chen (1985) demonstrated that anthers harvested from old and senescent *Citrus* trees had a lowered ability to undergo embryogenesis than those collected from young and vigorous trees. Anthers from spring flowers produced embryos while those from the second flowering season (summer) failed to do so. A cold pre-treatment of flower buds before *in vitro* culture of anthers improved both callus production and embryo formation. Although the mechanisms underlying such beneficial effects remain as yet unexplained, cytological studies for *Citrus* (Chen, 1985) and apple (Zhang, 1988) have shown an increase in the percentage of microspores having two identical nuclei following a cold pre-treatment of flower buds. Noteworthy, though, such pre-treatment can also lead to a delay of the first pollen mitosis, a modification of the microspore wall and to disorganization of the tapetum (Sangwan and Sangwan-Norreel, 1987), and even to the inhibition of spindle formation during mitosis (Sunderland and Dunwell, 1974). Whatever the case, a treatment of flower buds for at least two weeks at 3–4 °C prior to anther culture was necessary for androgenic embryo formation and has improved callus initiation in apple (Johansson *et al.*, 1987; Zhang, 1988; Ogata and Wang, 1989), and *Citrus* (Chen, 1985).

Another important factor influencing *in vitro* androgenesis is the composition of the medium. Most common tissue culture media have been tested for fruit crops (cf. Tables 1–4), but the medium used most frequently and for the largest number of species is that of Murashige and Skoog (1962). The N6 formulation (Chu *et al.*, 1975) has been used with success for some *Citrus* genotypes (Table 3). With respect to the carbon source, sucrose is the only sugar that has been employed in anther culture research with fruit trees, at a concentration of 20 g/l for cherimolla, 30 g/l for apple and longan, 50 g/l for lychee and *Poncirus*, 80 g/l for *Citrus* and 90 g/l for coconut.

The effect of growth regulators is also strongly genotype-dependent. Both auxins and cytokinins have typically been required for the successful induction of androgenic embryos and callus in woody fruit crop species. Thus, for apple, Xue and Niu (1984) obtained androgenic embryos for "Delicious" on a medium with 2,4-D, IAA and kinetin, but with IAA, BA and GA_3 for "American Summer Pearmain", "Jonathan", "Rainier" and "Ralls". Results from our group with "Topred" indicate that a large range of growth regulator combinations promote androgenesis. Low auxin concentrations (i.e., 0.5 mM IBA, NAA or 2,4-D) induce the formation of androgenic embryos. In *Citrus*, Chen (1985) concluded that the determining factor for embryo production was the quantity of auxin (2,4-D) in the medium rather than the auxin/cytokinin (2,4-D/BA) ratio.

Other medium additives may influence androgenesis. Activated charcoal (3–10 g/l) enhanced *in vitro* androgenesis for coconut, longan and apple. Bee royal jelly and glutamine stimulated embryo formation and plant recovery in lychee. Biotin and insulin promoted plant regeneration from the anther-derived callus tissues of *Malus prunifolia*.

2.1.1. *Temperate fruit crops*

2.1.1.1. *Rosaceae*. The most advanced anther culture research studies for temperate fruit crops in the Rosaceae have concerned various *Malus* genotypes (Table 1; Fig. 1). The first report on anther culture research for a perennial fruit crop was for the cultivated apple (*Malus Xdomestica* Borkh.). Nakayama *et al.* (1972) induced anther callus from which roots and bud-like structures were differentiated, but failed to sustain growth. Thereafter, several authors reported the production of callus or embryos from cultured anthers of apple (Table 1), but only a few obtained haploid plants (Fei and Xue, 1981; Wu, 1981; Xue and Niu, 1984; Zhang *et al.*, 1987, 1990). There are three articles published on pear (*Pyrus communis* L.); Jordan (1975) obtained multicellular structures after *in vitro* culture of anthers, whereas Braniste and Petrescu (1981) produced anther-derived callus for 5 scion cultivars. In a separate study, this same group identified and characterized two haploid plants that failed to survive *in vitro* (Braniste *et al.*, 1984).

Several articles have been published with stone fruit crops (*Prunus* spp.), (Table 2) but haploid plants are yet to be produced. The production of callus has been reported from anther cultures of apricot (Harn and Kim, 1972), sweet cherry (Jordan, 1975; Hoefer and Hanke, 1990), sour cherry (Seirlis *et al.*, 1979), almond (Michellon *et al.*, 1974), *P. Xpandora* (Jordan and Feucht, 1977), peach (Michellon *et al.*, 1974; Seirlis *et al.*, 1979; Hammerschlag, 1983), *P. cerasifera* (myrobalan) and plum (Seirlis *et al.*, 1979). The ploidy level of such callus has only been assessed in some cases. Michellon *et al.* (1974) obtained haploid, diploid, and triploid callus from anther cultures of almond and peach. Hammerschlag (1983) confirmed the haploidy of her peach anther-derived callus tissues by cytofluorometry. The only differentiation events reported for *Prunus* species were the regeneration of roots

Table 1. In vitro androgenesis in *Malus* genotypes

GENOTYPE	MEDIUM COMPOSITION[a]	RESULT[b]	REFERENCE
Malus Xdomestica	MS+IAA(0.6),NAA(1.6),Kin(0.5)	C, R	Nakayama et al,1972
	MS+IAA(5.7),Kin(4.7)	C, E	Kubicki et al, 1975
	MS+2,4-D(4.5),BA(0.4)	C	Braniste & Petrescu, 1981
	LS+CH(100) - LH(100)+BA(2.2)	C	James & Wakerell, 1981
	MS - MS +Kin(0.5) - MS+IAA (0.6) or NAA (0.5-5.4) plus Kin (0.5-4.7)	C, E	Hidano, 1982
	MS + 2,4-D(1.8),IAA(11.4-22.9), Kin (0.9), AdeS(108.7) MS+IAA(11.4-22.9).BA (0.9), AdeS(108.7),GA_3(1.2) => embryos - MS + IBA(0.5-2.5),BA(4.4),GA_3(0.3-5.8) => shoots	C, E, P	Fei & Xue, 1981; Xue & Niu, 1984
	MS+IBA(0.5) or NAA(0.5) or 2,4-D (0.5) or IAA(2.9), BA(2.2) => embryos	C, E, P	Zhang et al, 1987; 1990
	MS(double layer + 45 L-cysteine-HCl) - MS +2,4-D(2.3) - MS + Zeatin (4.6) => embryos	C, E	Johansson et al, 1987
	MS+2,4-D(0.9-1.8),Kin(0.9) +/- BA (0.9)	C, E	Hoefer & Hanke, 1990
M. pumila cl. M.9	N_6+NAA(0.3),BA(8.9)	C, E	Ogata & Wang,1989
M. prunifolia	MS+IAA(5.7-11.4) or NAA(5.4-10.8) or 2,4-D (4.5-9.1),Kin(9.3)=>callus - MS+NAA(0.5-2.7) BA(2.2-4.4),biotin(5000) - W + NAA(0.5-2.7), BA(4.4-8.9),insulin(8000), LH(500) => shoots - MS/2 + IAA(2.9-11.4) or IBA(0.5-2.5) => roots	C, P	Wu, 1981

[a] LS, Linsmaier and Skoog (1965); MS, Murashige and Skoog (1962); N_6, Chu *et al.* (1975), W, White (1963). Concentrations are given in mM for growth regulators and in mg/l for other compounds. AdeS = adenine sulphate,CH = casein hydrolysate, LH = lactoalbumin hydrolysate.
[b] C, callus; E, embryos; P, plants; R, roots.

from *P.* × *pandora* androgenic callus (Jordan and Feucht, 1977), and the recovery of one diploid plant from anther callus of *P. cerasus* cv. Ferracida (Seirlis *et al.*, 1979).

2.1.1.2. *Other temperate fruit crops*. There are four published reports of anther culture in grape. Gresshoff and Doy (1974) reported that three of the 27 genotypes studied gave callus. Hirabayashi *et al.* (1976) regenerated shoots from anther-derived callus of *V. thunbergii*, but the ploidy level of such shoots was not established. Rajasekaran and Mullins (1979) obtained callus tissues from anthers of *V. vinifera* × *V. rupestris* hybrid grapevines from

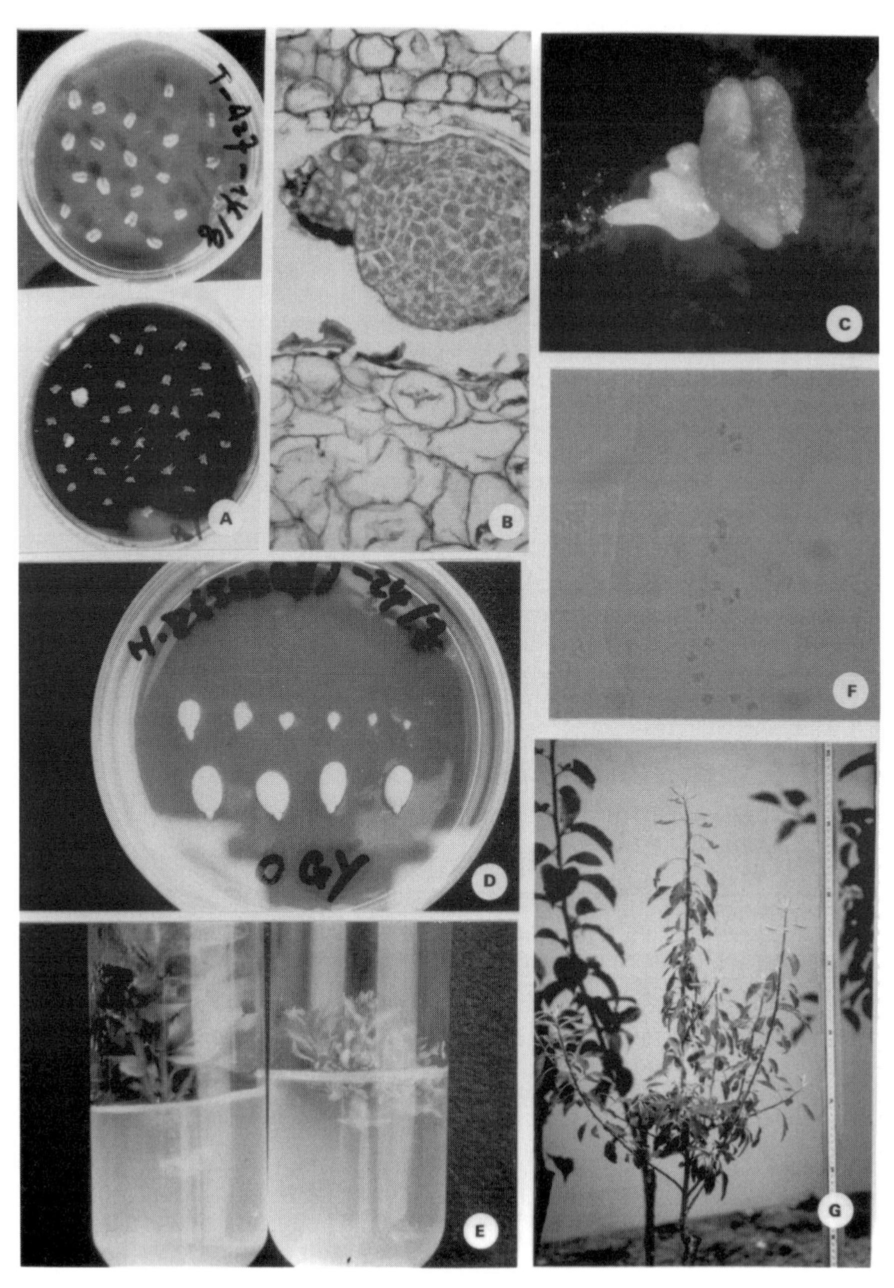

Figure 1. Haploid plant production in apple (*Malus Xdomestica* Borkh.). (A) Anther culture on a medium without (above) and with (below) activated charcoal. Notice enhancement of response from the use of charcoal. (B) Histological cut of a cultured anther showing the internal origin of an androgenic embryo and the absence of callus proliferation. (C) A cultured anther exhibiting embryo formation. (D) Rescued embryos at different developmental stages from a directed cross (non irradiated pollen) (E) Typical phenotvpe of haploid shoots and leaves (right) compared to a diploid shoot (left) of "Erovan" apple. (F) A plate showing the haploid chromosome complement ($n = 17$). (G) A field-grown "Golden Delicious" haploid apple tree (right) alongside a diploid tree of the same variety (left).

Table 2. In vitro androgenesis in stone fruit crops (*Prunus* spp.)

SPECIES	MEDIUM COMPOSITION[a]	RESULT[b]	REFERENCE
P. amygdalus	MS+IAA(0.6),2,4-D(5.0),Kin(0.9)	C	Michellon et al, 1974
P. armeniaca	MS+NAA(10.8) or 2,4-D(9.1),Kin (9.3)	C	Harn & Kim, 1972
P. avium	MS+NAA(5.4),BA(4.4)	MP	Jordan, 1975
	MS or NN + NAA(5.4) or 2,4-D (1.8-4.5),BA(2.2-4.4)	C	Hoefer & Hanke, 1990
P. X "Pandora"	N + NAA(5.4),BA(4.4)	MP	Jordan & Feucht, 1977
P. cerasifera	M + 2,4-D(4.5),BA(22.2)	C	Seirlis et al, 1979
P. cerasus	M+2,4-D(4.5-22.6),BA(22.2-44.4) => callus - M+BA(44.4) => shoots - M => roots	C, P	Seirlis et al, 1979
P. domestica	M + 2,4-D (4.5-45.2),BA(0-44.4)	C	Seirlis et al, 1979
P. persica	M+IAA(0.6),2,4-D (5.0),Kin(0.9)	C	Michellon et al, 1974
	M + 2,4-D(4.5),BA(4.4-22.2)	C	Seirlis et al, 1979
	Liquid medium	C	Hammerschlag, 1983

[a] M, Miller (1965); MS, Murashige and Skoog (1962); N, Nitsch (1972); NN, Nitsch and Nitsch (1969). Concentrations are given in mM for growth regulators and in mg/l for other compounds.
[b] C, callus, MP, multicellular pollen; P, plants.

which embryos and plants were subsequently regenerated. Callus from anthers was shown to be mixoploid, containing both haploid and diploid cells, while all the plants regenerated were diploid. Finally, Zou and Li (1981) produced haploid plants from anther callus of *V. vinifera* cv. Shenli (syn. "Victoria").

Callus tissues were induced from cultured anthers of *Ribes nigrum* (Jordan, 1975) and *Corylus avellana* (Salesses and Mouras, 1978) but, although embryos were produced for *Corylus*, haploid plants were not obtained for either species.

2.1.1. *Tropical and subtropical fruit crops*

2.1.2.1. *Rutaceae*. Tissue culture results for *Citrus* are summarized in Table 3. The first report for Rutaceous fruit crops concerned *Citrus limon* and *C. medica*, with callus being produced from cultured anthers of lemon only (Drira and Benbadis, 1975). The first androgenic plants in this family were obtained from anthers of *Poncirus trifoliata*, an important rootstock for many *Citrus* species, and among them, some were haploid ($2n = x = 9$) while others were aneuploid ($2n = x + 1$, $x + 2$) or diploid (Hidaka *et al*, 1979). Soon

Table 3. In vitro androgenesis in *Citrus* species

SPECIES	MEDIUM COMPOSITION[a]	RESULT[b]	REFERENCE
C. limon	MS + 2,4-D(4.5),BA(0.4),Glu(2.0)	C	Drira & Benbadis, 1975
C. madurensis	N_6 + BA(4.4),2,4-D (0.5) => embryos - MS + BA(1.1),IAA (0.6),LH(500) => shoots - MS + IBA (0.5-1.0) => roots	E, P	Chen, 1985
C. microcarpa	N_6 + Kin (9.3),2,4-D (2.3) or BA(8.9), 2,4-D(9.1) => embryo - MS + BA(8.9) or BA(1.1), IAA(0.6), LH (500) => shoots	E, P	Chen et al, 1980
C. reticulata X C. grandis (var. "Sunkan")	As for *C. madurensis*	E, P	Chen, 1985
Poncirus trifoliata	MS + IAA(1.1 or 11.4), Kin (0.9 or 9.3) => embryos and shoots - MS => roots	C, E, P	Hidaka et al, 1979

[a] MS, Murashige and Skoog (1962); N_6, Chu *et al.* (1975). Concentrations are given in mM for growth regulators and in mg/l for other compounds. Glu = glutamine, LH = lactoalbumin hydrolysate.
[b] C, callus, E, embryos; P, plants.

after this, Chen *et al.* (1980) produced haploid plants from anther cultures of *C. microcarpa*, a result that was later extended to eight other species and hybrids of which two commercially important *Citrus* genotypes, "Calamondin" (*C. madurensis*) and the hybrid cultivar "Sunkan" (*C. reticulata* × *C. grandis*), regenerated embryos first and then haploid plants from anther cultures (Chen, 1985). The last two published reports for this group of species are those of Hidaka *et al.* (1982), with *C. aurantium*, and Geraci and Starrantino (1990), with several different *Citrus* species. Hidaka *et al.* (1982) produced embryos and plants from anther cultures of sour orange; these plants were shown to be diploid and, hence, were not derived from pollen unless they doubled later. Geraci and Starrantino (1990) obtained only callus proliferation from somatic tissues of anthers.

2.1.2.2. *Other tropical fruits.* Anther culture research for this group of fruit crops has encompassed seven different species (Table 4). For coffee (*Coffea arabica*) cvs. Bourbon Amarelo and Mondo Novo, haploid ($x = 11$) and diploid ($2x = 22$) calluses were obtained, the latter exhibiting a highly embryogenic potential, but the proembryos thus produced failed to develop further (Sharp *et al.*, 1973). The first successful example of the production of anther-derived haploid plants in this group of fruit crops was for *Carica papaya* ($2n = 2x = 18$), with the chromosome complement of the regenerated plants ranging from eight to ten (Litz and Conover, 1978). A single strategy

Table 4. In vitro androgenesis of tropical fruit crops excluding *Citrus*

SPECIES	MEDIUM COMPOSITION[a]	RESULT[b]	REFERENCE
Annona squamosa	NN + NAA(28.6) => callus - NN + NAA (5.4) or IAA (0.6), BA (8.9) => shoots	C, P	Nair et al, 1983
Carica papaya	MS + NAA (5.4), BA (2.2), AC(10000)	P	Litz & Conover, 1978
Psidium guajava	MS + BA(4.4)	C	Babbar & Gupta, 1986
Cocos nucifera	B or K + NAA (10.8), AC (5000),CW (150000)	E	Thanh-Tuyen & De Guzman, 1983
	M+AC(3000),TIBA(4.0) Glu (1000)	E	Monfort, 1985
Coffea arabica	LS + 2,4-D (0.5) or NAA (10.8), Kin (9.3)	C	Sharp et al, 1973
Euphoria longan	MS+Kin(4.7),2,4-D (9.1) AC(5000) => callus - A + BA (2.2), NAA (0.5) => embryos - A + BA (1.3), NAA (0.6) => roots	C, E, P	Yang & Wei, 1984
Litchi chinensis	MS+Kin(9.3),2,4-D (9.1) NAA(2.7)=>callus - MS+Kin(2.3),NAA(0.5), bee royal jelly (400), LH (500) => embryos - MS + bee royal jelly (400), Glu (500) => shoots - B5/2 + Kin (0.5), IAA (0.06), LH (500),Glu(1600)=>roots	C, E, P	Fu & Tang, 1983

[a]A, Anderson (1978); B, Blaydes (1966); B5, Gamborg *et al.* (1968); K, Keller *et al.* (1975); LS, Linsmaier and Skoog (1965); M, Miller (1965); MS, Murashige and Skoog (1962); NN, Nitsch and Nitsch (1969). Concentrations are given in mM for growth regulators and in mg/l for other compounds. AC = activated charcoal, CW = coconut water, Glu = glutamine, LH = lactoalbumin hydrolsate.
[b] C, callus; E, embryos; P, plants.

was used to induce androgenesis, embryos and whole haploid plants ($2n = x = 15$) for two species of the Sapinaceae, *Litchi chinensis* cvs. Wilting and Old Purple (Fu and Tang, 1983) and *Euphoria longan* cv. Dongbi (Yang and Wei, 1984). In addition, androgenic embryos were produced from anther-derived callus of *Annona squamosa* (Nair *et al.*, 1983). In *Psidium guajava*, androgenic callus was obtained from the cultured anthers but whole plant regeneration from such tissues was not established (Babbar and Gupta, 1986).

2.2. *Induction of androgenic cultures in fruit crops*

2.2.1. *Malus*

When flower buds are collected for anther culture (i.e., upon emergence of petals from the king flower which, being the most advanced one is at a later stage of development, and hence is discarded), the majority of microspores in the peripheral buds are at the uninucleate stage. The extracted anthers are cold-treated (3–4 °C, 1–11 weeks), and then cultured on MS basal medium with 30 g/l sucrose and supplemented with phytohormones (at variable combinations and concentrations depending on the genotype and authors, cf. Table 1). Following a lag phase of one or two weeks, microspores enter division, and one to two months later multicellular structures and proembryos can be observed (Zhang, 1988). During this time, anthers increase in volume and, subsequently, green or white callus becomes evident, with green callus from the connective tissue or the anther walls always proliferating first (i.e., somatic origin), whereas the white callus is initiated inside the anthers (i.e., androgenic origin). Both the type (somatic or androgenic) and the rate of callus proliferation are dependent on the genotype and on the culture medium employed, especially with regard to the auxin. From 10 to 30 weeks pass before embryos appear inside the anther and develop directly from the microspores, i.e. without an intervening callus phase (Fig. 1A–C). The overall yield of embryos is typically low, ranging from 0.4% (seven embryos from 1590 anthers) according to Johansson *et al.* (1987), to a maximum of 6.6% (33 embryos from 500 anthers) for "Topred" (Zhang, 1988). In addition to this, plant regeneration from such androgenic embryos has been difficult and only three groups have reported successful results in this respect. At the Institute of Pomology of Xingcheng (China) Fei and Xue (1981) and Xue Niu (1984) induced germination of androgenic embryos on a MS-based medium with BA, IBA and GA_3 for "Delicious", "Golden Delicious", "Rainier" and "Ralls" while in our laboratory at INRA's Research Centre of Angers (France), plants were recovered from "Topred" on MS-based medium with thidiazuron, IAA and IBA. The only other example of haploid plant production in *Malus* concerned *M. prunifolia* and involved shoot regeneration via organogenesis from anther-derived callus (Wu, 1981).

2.2.2. *Citrus*

Anthers containing microspores at the mid to late uninucleate stage were cultured on N6 medium (Chu *et al.*, 1975) with 80 g/l sucrose and various combinations of growth regulators (Chen, 1985) The proliferation of somatic callus was concomitant with microspore development and, usually, inhibitory to it. Somatic callus was evident within about two weeks of culture, whereas microspores underwent the first or second division by days 20 to 30, with androgenic embryos appearing inside the anthers after 80–100 days (Chen, 1985). The frequency of androgenic embryo formation has generally been low, ranging from 0.14% at worst to 2.21% at best. Most androgenic embryos have been unable to develop into plantlets. A few embryos were, however, induced to partial development on MS medium plus BA, IAA, lactoalbumin, and sucrose. The shoots thus obtained could be rooted either on MS medium with IBA at a low concentration, or directly on a growth regulator-free MS medium following soaking of the shoots bases, for 2 h, in a concentrated

IBA solution. Monoembryonic genotypes tended to be androgenic while polyembryonic *Citrus* species were recalcitrant.

2.3. *Isolated microspore culture*

Because of the limited success of androgenesis from anther cultures of fruit trees, microspore culture has been envisaged as an alternative strategy to increase haploid plant production in our laboratory. This methodology was first developed for members of the Solanaceae by Sangwan and Norreel (1975) and Wernicke and Kohlenbach (1977). For several angiosperms, isolated microspore culture has permitted the production of abundant androgenic embryos capable of germination, e.g., *Brassica napus* (Siebel and Pauls, 1989) and *Hordeum vulgare* (Hoekstra *et al.*, 1992). Despite such encouraging results, isolated microspore cultures have not been fully established for most perennial species, much less fruit species. The methodologies employed for microspore culture and the conceptual framework for this approach closely resemble those employed for protoplast culture (Ochatt *et al.*, 1992).

A range of factors can modulate the responses of cultured isolated microspores. These include cold or mannitol pretreatment of anthers or flower buds, variation in the temperature during culture, and co-culturing of microspores with anthers or ovaries. In *Ginkgo biloba*, electroporation of isolated microspores stimulates their competence for proliferation (Laurain *et al.*, 1993). Other inductive factors include the renewing of culture medium, and the desiccation and agitation of cultures. However, the only parameters for which all authors seem to agree so far, are genotype and the cytological stage of microspore development. With the exception of some Solanaceae (Sangwan and Norreel, 1975) and Brassicaceae (Cao *et al.*, 1990), there is a consensus that the uninucleate stage is the most promising (Pechan and Keller, 1988; Kott *et al.*, 1987). The studies with woody species confirmed the selection of uninucleate microspores (Carneiro, 1993; Laurain *et al.*, 1993; Neuenschwander *et al.*, 1993; Pedroso and Pais, 1994).

Experiments on microspore culture in our group have concentrated on apple and have identified several factors with an important effect (Ravard, 1991; Bouvier, 1993; Oldani, 1993). The late uninucleate stage was shown to be best for microspore culture and was therefore used throughout, with microspores cold-treated prior to culture. In this respect, the duration of such cold pre-treatment varied with the genotype and its efficiency was dependent on the culture medium (Ravard, 1991; Bouvier, 1993). Gamma irradiation of flower buds exerted no effect on microspores other than a reduction of viability (Bouvier, 1993), whereas variation of temperature or light conditions during culture did not have any significant effect on microspore proliferation. The density of microspores played a significant role on the response, with an optimum at 5×10^4 microspores/ml medium. Plating microspores in liquid medium increased the efficiency of division compared to embedding in agarose, which led to cell plasmolysis after a few weeks of culture.

A major problem encountered during the culture of apple microspores was the presence of endogenous contaminants that systematically infected from 70 to 100% of the cultures. These were identified as three different bacteria (*Pseudomonas picketti*, *P. luteola* and *P. capacia*) and one yeast (*Rhodotorula rubra*). Interestingly, the presence of bacteria from the family Pseudomonaceae and of different species of *Rhodotorula* had previously been reported in shoot buds of apple (Andrews and Kenerley, 1979). After several studies, gentamycin and tetracycline (at 5 mg/ml) and the antifungal cycloheximide (at 10 mg/ml) were found to be effective in inhibiting the proliferation of the bacteria and the yeast, respectively, but were phytotoxic for apple microspores (Ravard, 1991). Bouvier (1993) tried the inclusion of these substances during microspore isolation, or during the first 24 h of culture but to no avail as, in the former case, contaminants were not controlled, while in the latter, contaminants were indeed mostly eliminated but the microspores were unable to divide. Oldani (1993) obtained non-contaminated and proliferating microspores of apple. Forced young twigs bearing flower buds at room temperature were placed in a ChrysalQ solution (15 g/l) supplemented with 5 mg/l gentamycin, 5 mg/l tetracycline and 10 mg/l cycloheximide. Then, following disinfection of flower buds (30 s mercryl lauryl + 30 s ethanol 70° + 15 min NaClO (0.8% active Cl_2) + three 5 min rinses in sterile distilled water), anthers were excised in a 2 ml liquid layer of growth regulator-free MS medium. Microspores were then extracted by strong agitation in a vortex of anthers suspended in 6 ml medium, for 45 s at 20500 rpm, followed by the addition of 2 ml of MS medium to the suspension and a further 25 s at 20500 rpm. The resulting suspension was then sieved (50 mm pore size) and centrifuged/resuspended twice, first at 200 × g (10 min, 5 °C) then at 120 × g (5 min, 5 °C). The final pellet was recovered, layered on top of a salt medium with 21% (w/v) sucrose (CPW21S medium, Power and Davey, 1990), covered with 2 ml of W5 medium (Negrutiu *et al.*, 1987), whereby microspores can be collected in the pellet while most contaminants float to the interphase between the CPW21S and W5 media. The "clean" microspores obtained were suspended in a solution of the antibiotics and the fungicide (at their respective appropriate concentrations as above), and kept for 1–3 h at room temperature, or overnight at 4 °C. Finally, the suspension was centrifuged once more at 120 × g (5 min, 20 °C) to wash away these substances, the microspore density determined, and then cultured. Although rather complex, this procedure (involving a chemical treatment and a physical separation of microspores from the contaminants) was the only strategy capable of furnishing cultures of non-contaminated microspores that, in addition, were unaffected in their proliferation efficiency. Such a methodology was successfully applied to "Golden Delicious" apple and also to "Delicious" strains ("Topred" and "Erovan") permitting the recovery, for the first time, of multicellular colonies and, more interestingly, of embryogenic structures for all three genotypes (Oldani, 1993). These appear to be the first successful attempts at isolation and culture of microspores for temperate fruit tree crops.

3. Other *in vitro* approaches for haploidization of fruit crops

3.1. *Induction of gynogenesis*

Only one published report could be found in this field for fruit trees, concerning apple. Thus, flower buds of "Erovan", "Lodi" and R1–49 were collected at the balloon stage (i.e., when the embryo sac contains 8 haploid nuclei), and ovaries, isolated ovules, or ovules excised (still attached to the placental tissue) were cultured on MS-based media supplemented with different combinations of growth regulators. Callus was successfully obtained from the micropylar end of the ovule for all three genotypes. In addition, two embryo-like structures were also obtained from the cultured ovules of "R1–49", but they failed to develop any further (Zhang and Lespinasse, 1988).

We later obtained haploid shoot cultures from callus of R1–49, and also extended this approach to other genotypes, whereby haploid shoot cultures of gynogenic origin are at present available for "Erovan" (where mixoploid shoots were also produced), "Golden Delicious" and "Florina".

3.2. *Induction of* in situ *parthenogenesis by irradiated pollen*

Haploid apple plants have been successfully induced in "Erovan" (Fig. 1, D–G), after pollination with pollen from a clone carrying a dominant marker gene (RR) coding for anthocyanin production in whole plant organs. Pollen was irradiated at 500 and 1000 Gray, followed by *in vitro* culture of immature embryos on a half-strength MS medium with GA_3 and BA (Zhang *et al.*, 1988). This strategy, unchanged, was successfully extended to several other genotypes (Zhang and Lespinasse, 1991).

Most recently, Bouvier *et al.* (1993) successfully recovered haploid plants of pear using the same strategy as employed for apple above, using clone TNR 12.40 (homozygous for the marker gene C, coding for anthocyanin production and whose phenotype is completely red) as the pollinator and "Williams" as the female parent. In addition, "Doyenne du Comice", "Harrow Sweet", "President Heron" and "Williams" were also used as female parents for directed crosses. Significant differences were observed dependent on the female parent, with "Harrow Sweet" yielding 10 of the 12 surviving haploid plants obtained, and crosses involving "Williams" and "Doyenne du Comice" as the female parent yielding two haploids.

For both apple and pear, an irradiation dose of 500 Gray was most efficient for both embryo and haploid plant production. Three months after pollination was the most appropriate stage for the rescue and *in vitro* culture of the immature embryos; about 70% of the embryos germinated.

The unique phenotype of haploid compared to diploid plantlets aids in recognition. Diploid hybrid plants from crosses involving an anthocyanin marker gene were red. Haploid maternals possess typical pointed leaves and short internodes and are always weak (Lespinasse and Godicheau, 1980). Obviously, the ploidy level of all putatively haploid plants, obtained by either this or other haploidization approaches discussed, must be confirmed through chromosome counts or by flow cytometry. The true maternal origin of haploid pear plants obtained after *in situ* parthenogenesis by irradiated pollen was confirmed through isoenzymatic study of leaf extracts (Bouvier *et al.*, 1993).

4. Conclusion

The production of haploid plants of perennial fruit crops offers new possibilities for genetic studies and for breeding in these species. Homozygous plants, obtained after doubling, may be particularly useful for studies on the inheritance of horticultural traits. Moreover, the F_1 and F_2 progenies from such homozygous genotypes may provide new insights on the genetics of desirable characters and, if they carry desirable genes, could also be employed as novel parental lines for conventional hybridization programmes. By transmitting such desirable genes to each gamete, such parents may increase the frequency of promising seedlings in their progenies. In addition, homozygous clones may be valuable in rootstock breeding, as uniform F_1 rootstock cultivars could be produced by seed propagation, ensuring freedom from virus. *In vitro* haploidization methods are also a source of morphological, biochemical and chromosomal variation, as observed among doubled haploids for both annual (San and Dattée, 1985) and perennial (Chen, 1986) crops. The mutagenic treatment of isolated microspores, as developed for *Brassica napus* (Polsoni *et al.*, 1988), followed by chromosome doubling, would enable the production of homozygous mutants that express all mutations, including the recessive ones, thereby contributing to genetic improvement of fruit crops. The same would apply to protoplasts isolated from haploid plants, particularly now that regeneration from such protoplasts is feasible, as recently shown with apple (Patat-Ochatt *et al.*, 1993). In this context, haploid protoplasts would also be most interesting in somatic hybridization technology, where they could be fused either with themselves (homofusion) as an alternative doubling strategy, or with protoplasts of different ploidy levels (heterofusions) for the creation of genetic novelty fruit trees. This approach is presently being studied in our group. Finally, comparing haploid and diploid genotypes could simplify the interpretation of DNA markers (RAPDs), through the direct identification of the allelic variations. Also, cloning genes by transposon-tagging would be eased with haploid genotypes, as each modification of a recessive or dominant allele would be expressed.

5. References

Anderson, W.C., 1978. Rooting of tissue culture propagation of *Rhododendron*. *In Vitro* 14: 334.

Andrews, J.H. and C.M. Kenerley, 1979. Microbial populations associated with buds and young leaves of apple. Can. J. Bot. 58: 847–855.

Babbar, S.B. and S.C. Gupta, 1986. Induction of androgenesis and callus formation in *in vitro* cultured anthers of a myrtaceous fruit tree (*Psidium guajava* L.). Bot. Mag. 99: 75–83.

Blaydes, D.F., 1966. Interaction of kinetin and various inhibitors in the growth of soybean tissues. Physiol. Plant 8: 748–753.

Bonga, J.M., P. von Aderkas and D. James, 1987. Potential application of haploid cultures of tree species. In: J.W. Hanover and D.E. Keathley (Eds.), Genetic Manipulation of Woody Plants, pp. 57–77. Plenum Press, New York.

Bouvier, L., 1993. Haploïdie chez le pommier (*Malus Xdomestica* Borkh.) et le poirier (*Pyrus communis* L.). Thèse de Doctorat de l'Université de Paris VI, Paris.

Bouvier, L., Y.X. Zhang and Y. Lespinasse, 1993. Two methods of haploidization in pear, *Pyrus commmunis* L.: greenhouse seedling selection and *in situ* parthenogenesis induced by irradiated pollen. Theor. Appl. Genet. 87: 229–232.

Braniste, N. and B.I. Petrescu, 1981. Preliminary data of obtaining pear and apple haploids by anther culture *in vitro*. Probleme de Genetica Teoritica si Applicata 13: 409–414.

Braniste, N., A. Popescu and T. Comant, 1984. Production and multiplication of *Pyrus communis* haploid plants. Acta Hort. 161: 147–150.

Cao, M.Q., F. Charlot and C. Doré, 1990. Embryogenèse et régénération de plantes de chou à choucroute (*Brassica oleracea* L. ssp. *capitata*) par culture *in vitro* de microspores isolées. C.R. Acad. Sci. Paris Série II, 310: 203–209.

Carneiro, M.F., 1993. Induction of double haploids of *Coffea arabica* cultivars via anther or isolated microspore culture. In 15ème Colloque Scientifique International sur le Café, Montpellier, p. B-16 (Abstr.).

Chen, Z.G., 1985. A study on induction of plants from *Citrus* pollen. Fruit Var. J. 39: 44–50.

Chen, Z.G., M.Q. Wang and H.H. Liao, 1980. The induction of *Citrus* pollen plants in artificial media. Acta Genet. Sin. 7: 189–191.

Chen, Z.H., 1986. Induction of androgenesis in woody plants. In: H. Hu and H.Y. Yang (Eds.), Haploids of Higher Plants *in Vitro*, pp. 42–66. China Academic Publishers, Beijing.

Chen, Z.H., 1987. Induction of androgenesis in hardwood trees. In: J.M. Bonga and D.J. Durzan (Eds.), Cell and Tissue Culture in Forestry, Vol. 2. Specific Principles and Methods: Growth and Developments, pp. 247–268. Martinus Nijhoff Publishers, Dordrecht.

Chu, C.C., C.C. Wang, C.S. Sun, C. Hsü, K.C. Yin, C.Y. Chu and F.Y. Bi, 1975. Establishment of an efficient medium for anther culture of rice through comparative experiments on the nitrogen sources. Sci. Sin. 18: 659–668

Drira, N. and A. Benbadis, 1975. Analyse, par culture d'anthères *in vitro*, des potentialités androgénétiques de deux espèces de *Citrus* (*Citrus medica* L. et *Citrus limon* L. Burm.). C.R. Acad. Sci. Paris, Série D, 281: 1321–1324.

Fei, K.W. and G.R. Xue, 1981. Induction of haploid plantlets by anther culture *in vitro* in apple cv. "Delicious". Chinese Agric. Sci. 4: 41–44.

Foroughi-Wehr, B. and G. Wenzel, 1989. Androgenetic haploid production. Newsl, Int. Assoc. Plant Tiss. Cult. 58: 11–18.

Fu, L.F. and D.Y. Tang, 1983. Induction of pollen plants of litchi trees (*Litchi chinensis* Sonn.). Acta Genet. Sin. 10: 369–374.

Gamborg, O.L., R.A. Miller and K. Ojima, 1968. Nutrient requirements of suspension cultures of soybean root cells. Exp. Cell Res. 50: 151–158.

Geraci, G. and A. Starrantino, 1990. Attempts to regenerate haploid plants from *in vitro* cultures of *Citrus* anthers. Acta Hort. 280: 315–320.

Gresshoff, P.M. and C.H. Doy, 1974. Derivation of a haploid cell line from *Vitis vinifera* and the importance of the stage of meiotic development of anthers for haploid culture of this or other genera. Z. Pflanzenphysiol. 73: 132–141.

Guha, S. and S.C. Maheshwari, 1964. *In vitro* production of embryos from anthers of *Datura*. Nature 204: 497.

Hammerschlag, F.A., 1983. Factors influencing the frequency of callus formation among cultivated peach anthers. Hort Sci. 18: 210–211.

Harn, C. and M.Z. Kim, 1972. Induction of callus from anthers of *Prunus armeniaca*. Korean J. Breed. 4: 49–53.

Hidaka, T., Y. Yamada and T. Shichijo, 1979. *In vitro* differentiation of haploid plants by anther culture in *Poncirus trifoliata* (L.) Raf. Jpn. J. Breed. 29: 248–254.

Hidaka, T., Y. Yamada and T. Shichijo, 1982. Plantlet formation by anther culture of *Citrus aurantium* L. Jpn. J. Breed. 32: 247–252.

Hidano, Y., 1982. Callus and embryoid induction by anther culture of apple. Bull. Fac. Educ. Hirosaki Univ. 48: 69–74.

Hirabayashi, T., I. Kozaki and T. Akihama, 1976. *In vitro* differentiation of shoots from anther callus in *Citrus*. HortScience 11: 511–512.

Hoefer, M. and V. Hanke, 1990. Induction of androgenesis *in vitro* in apple and sweet cherry. Acta Hort. 280: 333–336.

Hoekstra, S., M.H. van Zijderveld, J.D. Louwerse, F. Heidekamp and F. van der Mark, 1992. Anther and microspore culture of *Hordeum vulgare* L. cv Igri. Plant Sci. 86: 89–96.

James, D.J and I.J. Wakerell, 1981. Anther culture. In Rep. East Malling Res. Stn. pp. 163–164.

Johansson, L., A. Wallin, A Gedin, M. Nyman, E. Pettersson and M. Svensson, 1987. Anther and protoplast culture in apple and strawberry. Biennial Report 1986–1987, Division of Fruit Breeding-Balsgard, Swedish University of Agricultural Sciences, pp. 83–92.

Jordan, M, 1975. *In vitro*-Kultur von *Prunus*, *Pyrus* und *Ribes*-Antheren. Planta Medica Suppl. 59–65.

Jordan, M. and W. Feucht, 1977. Differenzierunsprozesse und Phenolstoffwechsel bei *in vitro* Kultivierten Antheren von *Prunus avium* und *Prunus X "Pandora"*. Angewandte Botanik 51: 69–76.

Keller, W.A., T. Rajhathy and J. Lacapra, 1975. *In vitro* production of plants from pollen in *Brassica campestris*. Can. J. Genet. Cytol. 17: 655–666.

Kimber, G. and R. Riley, 1963. Haploid angiosperms. Bot. Rev. 29: 480–531.

Kott, L.S., L. Polsoni, B. Ellis and W.D. Beversdorf, 1987. Autotoxicity in isolated microspore cultures of *Brassica napus*. Can. J. Bot. 66: 1665–1670.

Kubicki, B., J. Telezynska and E. Milewska-Pawliczuk, 1975. Induction of embryoid development from apple pollen grains. Acta Soc. Bot. Poloniae 44: 631–635.

Laurain, D., J. Trémouillaux-Guiller and J.C. Chénieux (1993) Embryogenesis from microspores of *Ginkgo biloba* L., a medicinal woody species. Plant Cell Rep. 12: 501–505.

Lespinasse, Y. and M. Godicheau, 1980. Création et description d'une plante haploïde de pommier (*Malus pumila* Mill.). Ann. Amélior. Plantes 30: 37–44.

Linsmaier, E.M. and F. Skoog, 1965. Organic growth factor requirements of tobacco tissue cultures. Physiol. Plant. 18: 100–127.

Litz, R.E. and R.A. Conover, 1978. Recent advances in papaya tissue culture. Proc. Florida State Hort. Soc. 91: 182–184.

Michellon, R., J. Hugard and R. Jonard, 1974. Sur l'isolement de colonies tissulaires de pêcher (*Prunus persica* Batsch., cultivars Dixired and Nectared IV) et d'amandier (*Prunus amygdalus* Stokes, cultivar Ai) à partir d'anthères cultivées *in vitro*. C.R. Acad. Sci. Paris, Série. D 278: 1719–1722.

Miller, C.O., 1965. Evidence for the natural occurrence of zeatin and derivative compounds from maize which promote cell division. Proc. Natl. Acad. Sci. USA 54: 1052–1058.
Monfort, S., 1985. Androgenesis of coconut: embryos from anther culture. Z. Pflanzenzüchtg. 94: 251–254.
Murashige, T. and F. Skoog, 1962. A revised medium for rapid growth and bioassays with tobacco tissue cultures. Physiol. Plant. 15: 473–497.
Nair, S., P.K. Gupta and A.S. Mascarenhas, 1983. Haploid plants from *in vitro* anther culture of *Annona squamosa* L. Plant Cell Rep. 2: 198–200.
Nakayama, R., K. Saito and R. Yamamoto, 1972. Studies on the hybridization in apple breeding. III. Callus formation and organ differentiation from anther culture. Bull. Fac. Agric. Hirosaki Univ. 19: 1–9.
Negrutiu, I., R. Shillito, I. Potrykus, G. Biasini and F. Sala, 1987. Hybrid genes in the analysis of transformation conditions. I. Setting up a simple method for direct gene transfer in plant protoplasts. Plant Mol. Biol. 8: 363–373.
Neuenschwander, B., M. Dufour and T.W. Baumann, 1993. Formation de colonies cellulaires haploïdes à partir de microspores de *Coffea arabica* isolées mécaniquement. In: 15ème Colloque Scientifique International sur le Café, Montpellier, p. BA9 (Abstr.).
Nitsch, J.P., 1972. Haploid plants from pollen. Z. Pflanzenzüchtg. 67: 3–18.
Nitsch, J.P. and C. Nitsch, 1969. Haploid plants from pollen grains. Science 163: 85–87.
Nitsch, C. and B. Norreel, 1973. Effet d'un choc thermique sur le pouvoir embryogène du pollen de *Datura innoxia* cultivé dans l'anthère et isolé de l'anthère. C.R. Acad. Sci. Paris, Série D. 276: 303–306.
Ochatt, S.J., M.R. Davey and J.B. Power, 1990. Tissue culture and top-fruit tree species. In: J.M. Walker and J. Pollard (Eds.), Methods in Molecular Biology. Vol. 6, Plant Cell and Tissue Culture, pp. 193–207. The Humana Press, Clifton, NJ.
Ochatt, S.J., E.M. Patat-Ochatt and J.B. Power, 1992. Protoplasts. In: F.A. Hammerschlag, R.E. Litz (Eds.), Biotechnology of Perennial Fruit Crops, pp. 77–103. C.A.B. International, Cambridge.
Ogata R. and L.C. Wang, 1989. *In vitro* anther culture of apple rootstock M.9. Moet-Hennessy Louis Vuitton Conference "Control of Morphogenesis", Aix-les-Bains, p. 36 (Abstr.).
Oldani, C., 1993. Culture de microspores isolées: un outil nouveau pour l'haploïdisation du pommier (*Malus Xdomestica* Borkh.). Memoire de D.E.A. de Biologie du Développement des Plantes, Université P. and M. Curie (Paris VI).
Patat-Ochatt, E.M., J. Boccon-Gibod, M. Duron and S.J. Ochatt, 1993. Organogenesis of stem and leaf protoplasts of a haploid Golden Delicious apple clone (*Malus Xdomestica* Borkh.). Plant Cell Rep. 12: 118–120.
Pechan, P.M. and W.A. Keller, 1988. Identification of potentially embryogenic microspores in *Brassica napus*. Physiol. Plant. 74: 377–384.
Pedroso, M.C. and M.S. Pais, 1994. Induction of microspore embryogenesis in *Camellia japonica* L. Plant Cell Tiss. Org. Cult. (in press).
Polsoni, L., L.S. Kott and W.D. Beversdorf, 1988. Large-scale microspore culture technique for mutation-selection studies in *Brassica napus*. Can. J. Bot. 66: 1681–1685.
Power, J.B. and M.R. Davey, 1990. Isolation, culture and fusion of protoplasts of higher and lower plants. In: J.W. Pollard and J.M. Walker (Eds.), Methods in Molecular Biology. Vol. 6: Plant Cell and Tissue Culture, pp. 237–259. The Humana Press, Clifton, NJ.
Rajasekaran, K. and M.G. Mullins, 1979. Embryos and plantlets from cultured anthers of hybrid grapevines. J. Exp. Bot. 30: 399–407.
Raquin, C., 1985. Induction of haploid plants by *in vitro* culture of *Petunia* ovaries pollinated with irradiated pollen. Z. Pflanzenzüchtg. 94: 166–169.
Ravard, T., 1991. Tentative d'androgenèse *in vitro* chez le pommier cultivé (*Malus Xdomestica* Borkh.) par culture de microspores isolées. Memoire de D.U.T. de Biologie Appliquée, Institut Universitaire de Technologie d'Angers.
Salesses, G. and A. Mouras, 1978. Tentatives de cultures d'anthères chez divers *Prunus* et

Corylus. In: Compte Rendu de Table Ronde "Multiplication *in vitro* d'Espèces Ligneuses", Centre de Recherches Agronomiques de Gembloux, Gembloux, pp. 267–268.

San, L.H., 1976. Haploïdes d'*Hordeum vulgare* L. par culture *in vitro* d'ovaires non fecondés. Ann. Amélior. Plantes 26: 751–754.

San, L.H. and Y. Dattee, 1985. Variabilité des haploïdes doublés par androgenèse et gynogenèse *in vitro*. Bull. Soc. Bot. France Actualités Botaniques 132: 23–33.

San, L.H. and P. Gelebart, 1986. Production of gynogenetic haploids. In: I.K. Vasil (Ed.), Cell Culture and Somatic Cell Genetics. Vol. 3: Plant Regeneration and Genetic Variability, pp. 305–322. Academic Press, New York.

Sangwan, R.S. and B.S. Norreel, 1975. Induction of plants from pollen grains of *Petunia* cultured *in vitro*. Nature 257: 222–224.

Sangwan, R.S. and B.S. Sangwan-Norreel, 1987. Biochemical cytology of pollen embryogenesis. Int. Rev. Cytol. 107: 221–272.

Sangwan-Norreel, B S., R.S. Sangwan and J. Paré, 1986. Haploïdie et embryogenèse provoquée *in vitro*. Bull. Soc Bot. France, Actualités Botaniques 133: 7–39.

Seirlis, G., A. Mouras and G. Salesses, 1979. Tentatives de culture *in vitro* d'anthères et de fragments d'organes chez les *Prunus*. Ann. Amélior. Plantes 29: 145–161.

Sharp, W.R., L.S. Caldas, O.J. Crocomo, L.C. Monaco and A. Carvalho, 1973. Production of *Coffea arabica* callus of three ploidy levels and subsequent morphogenesis. Phyton 31: 67–74.

Siebel, J. and K.P. Pauls, 1989. A comparison of anther and microspore culture as a breeding tool in *Brassica napus*. Theor. Appl. Genet. 78: 473–479.

Sunderland, N. and J.M. Dunwell, 1974. Pathways in pollen embryogenesis. In: H.E. Street (Ed.), Tissue Culture and Plant Science, pp. 141–167. Academic Press, London.

Thanh-Tuyen, N.T. and E.V. De Guzman, 1983. Formation of pollen embryos in cultured anthers of coconut (*Cocos nucifera* L.). Plant Sci. Lett. 29: 81–88.

Wernicke, W. and H.W. Kohlenbach, 1977. Versuche zur Kultur Isolierten Mikrosporen von *Nicotiana* und *Hyoscyamus*. Z. Pflanzenphysiol. 81: 330–340.

White, P.R., 1963. The Cultivation of Animal and Plant Cells. Ronald Press, New York.

Wu, J.Y., 1981. Induction of haploid plants from anther culture of crabapple. J. Northeast Agric. Coll. (China) 3: 105–108.

Xue, G.R. and J.Z. Niu, 1984. A study on the induction of apple pollen plants. Acta Hort. Sin. 11: 161–164.

Yang, Y.Q. and W.X. Wei, 1984. Induction of longan plantlets from pollen cultured in certain proper media. Acta Genet. Sin. 11: 288–293.

Zhang, Y.X., 1988. Recherche *in vitro* de plantes haploïdes chez le pommier cultivé (*Malus Xdomestica* Borkh.): androgenèse, gynogenèse, parthénogenèse *in situ* induite par du pollen irradié. Ph.D. Thesis, Université de Paris-Sud, Paris.

Zhang, Y.X. and Y. Lespinasse, 1988. Culture *in vitro* d'ovules non fécondés et d'embryons prélevés 8 jours après pollinisation chez le pommier cultivé (*Malus Xdomestica* Borkh.). Agronomie 8: 837–842.

Zhang, Y.X. and Y. Lespinasse, 1991. Pollination with gamma-irradiated pollen and development of fruits, seeds and parthenogenetic plants in apple. Euphytica 54: 101–109.

Zhang, Y.X. and Y. Lespinasse, 1992. Haploidy. In: F.A. Hammerschlag and R.E. Litz (Eds.), Biotechnology of Perennial Fruit Crops, pp. 57–75. C.A.B. International, Cambridge.

Zhang, Y.X., J. Boccon-Gibod and Y. Lespinasse, 1987. Obtention d'embryons de pommier (*Malus Xdomestica* Borkh.) après culture d'anthères. C.R. Acad. Sci. Paris, Série III, 305: 443–448.

Zhang, YX., Y. Lespinasse and E. Chevreau, 1988. Obtention de plantes haploïdes de pommier (*Malus Xdomestica* Borkh.) issues de parthénogenèse induite *in situ* par du pollen irradié et culture *in vitro* des pépins immatures. C.R. Acad. Sci. Paris, Série III, 307: 451–457.

Zhang, Y.X., Y. Lespinasse and E. Chevreau, 1990. Induction of haploidy in fruit trees. Acta Hort. 280: 293–305.

Zou, C.J. and P.F. Li, 1981. Induction of pollen plants of grape (*Vitis vinifera* L.). Acta Bot. Sin. 23: 79–81.

12. Haploidy in *Vitis*

ALFREDO CERSOSIMO

Contents

1. Introduction

Grapevine is one of the most widely cultivated plants in the world and man has been growing it since the beginning of his recorded history. Despite this, haploid plants have never been observed in *Vitis* (Olmo, 1976), which would lead to the conclusion that haploidy, the ability to generate haploid (n) plants (sporophytes) rather than diploids ($2n$), is not a natural property of this genus. To understand the reasons for this inability, it is necessary first to analyse the general phenomena leading to haploidy and compare them with the genetic features of the genus *Vitis*.

2. Haploidy

Haploidy may be considered as the consequence of aberrant forms of sexual reproduction, especially apomixy, defined as "the formation of a new individual from a gametic cell in the absence of gametic reproduction", that is to say by a purely vegetative process. In the narrow sense of the term, apomixy may take place in two ways, depending on whether the gametic cell from which it originates is diploid or haploid. The former case is known as apogamy, which is necessarily preceded by apospory, the formation of diploid tetrads in the archesporial cells, to suppress reductional division during meiosis. The latter case is known as parthenogenesis. This is considerably differ-

S.M. Jain, S.K. Sopory & R.E. Veilleux (eds.), In Vitro *Haploid Production in Higher Plants, Vol. 3*, 211–229.

ent from apogamy not only by virtue of the chromosome number of the sexual cells, but also because apogamy is connected to a process of apospory and is a recurrent phenomenon in the ontogenetic cycle on the same level as gamic reproduction, whereas parthenogenesis is a purely accidental phenomenon caused by exceptional factors.

The ovary of one species, for example, may receive pollen from another species which is genetically incompatible but able to allow pollen germination. The sperm nucleus is thus able to enter the oosphere cytoplasm, but because they are incompatible the two nuclei cannot fuse. When this happens one nucleus degenerates and dies, while the other beginning to divide, causes oosphere segmentation and forms a haploid embryo.

The two alternatives are:

1. if the sperm nucleus dies and the oosphere nucleus survives, there is "gynogenesis";
2. if the oosphere nucleus is replaced by the spermatic nucleus, there is "androgenesis".

3. Some genetic and physiological features of the *Vitis* genus

Cultivated grapevine belongs to the Vitaceae family, which includes more than one thousand species, grouped into 14 living and two fossil genera.

Vitaceae, also called Ampelideae or Ampelidaceae, are characterized by polygamous flowers that may become unisexual male, by the complete or partial abortion of the ovary, or unisexual female, as a result of the malformation of the stamens or the sterility of their pollen.

Pollen from "male" and hermaphrodite flowers has an elliptical form of varying elongation with a deep groove along the greater diameter. It has three germ pores and germinates easily (close to 100%) in sugar solution.

Pollen from "female" flowers (with outward-bent, short, weak, spiralled, etc., stamens) has a rounded, triangular shape with slightly curved sides, has no germ pores and never gives rise to germination, even in sugar solution. As it can never fertilize the ovule, it is completely sterile pollen.

Mature pollen is binucleate with 19 chromosomes in the haploid phase ($n = 19$). Only in the subgenus *Muscadinia* $n = 20$ chromosomes.

In hermaphrodite flowers pollen maturation generally precedes flower opening by about 6–7 days. Autogamous fertilisation would therefore be theoretically possible with the flower still closed. In practice, however, it has been observed that the numerous floral anomalies, physiological incompatibilies and critical environmental conditions favour cross-pollination and lead to a high level of heterozygosity. In such conditions it is not difficult to see that its genome has accumulated many recessive genes, masked and perpetuated over time by its heterozygosity and vegetative propagation, which is the traditional reproduction practice in viticulture. In a haploid or homozygous diploid state, in which they are able to express themselves, these genes are finally unmasked and, if possible, assessed in terms of their agronomic value. It is almost always the case that their expression gives rise to genetic conditions not conducive to cell development, as witnessed by the marked inbreeding-induced depression noted in the progeny of self-fertilized vines and by the low germination capacity of their seeds.

The inability of grapevine to produce natural haploids and the high expectations that this type of material may produce completely homozygous diploid lines have led many viticulture researchers to use experimental techniques of advanced biotechnology in an attempt to repeat the successes obtained with other species (Guha and Maheshwari, 1964, in *Datura innoxia*; Bourgin and Nitsch, 1967; Nitsch and Nitsch, 1968, in *Nicotiana tabacum*; Zenkteler, 1971, in *Atropa belladonna*; etc.). The *in vitro* culture of anthers is the most widespread method of haploid production but it is not, of course, the only way.

Haploids can also be obtained by means of the *in vitro* culture of isolated microspores, unfertilized ovules, protoplasts extracted from tetrads, etc. Here, however, the inductive efficiency is very low, or nil, even in species which give good results with anther culture.

No matter how they were obtained, haploids would be, particularly for grapevine, a powerful tool for new models of biological studies, for increasing knowledge about the species and for dealing with the difficult task of its genetic improvement. Literature reports only one case of haploid production in *Vitis* (Zou and Li, 1981) – from a local Chinese cultivar called "Shengli". This remains an entirely isolated case and still awaits confirmation. Haploid callus lines have been produced by a number of researchers (Gresshoff and Doy, 1974; Kim and Peak, 1981; Cersosimo, 1986), while others have obtained the first regenerations of whole plants (Rajasekaran and Mullins, 1979; Bouquet *et al.*, 1982; Hirabayashi and Akihama, 1982; Mauro *et al.*, 1986a,b; Cersosimo *et al.*, 1990). In all cases the regenerated plants turned out to be diploid. The production of haploid callus (n), diploid ($2n$) and tetraploid ($4n$) (Cersosimo, 1986) suggests that the cells (n and $2n$) doubled spontaneously, as in tobacco (Devreux *et al.*, 1975).

Microspore activity during *in vitro* culture was shown by a histological study (Altamura *et al.*, 1992) on callused anthers of *Vitis rupestris* du Lot (syn. St. George). After about a month in culture, few anthers showed any sign of change in their external morphology. The inside of the tapetal cells had degenerated and the surviving microspores had developed into mature pollen grains complete with exine (Fig. 1b). Other anthers had produced callus. As a result there was very little space inside the loculi; they contained pollen grains and vital and/or dead microspores. In addition to uninucleate microspores (Fig. 1c), bicellular structures (Fig. 1d,e) could occasionally be observed (about 10 per anther), produced by microspore division. Some of these structures showed signs of degeneration (vacuolisation, starch and tannin deposits; Fig. 1f).

In the microspore-derived bicellular structures, the position of the cell

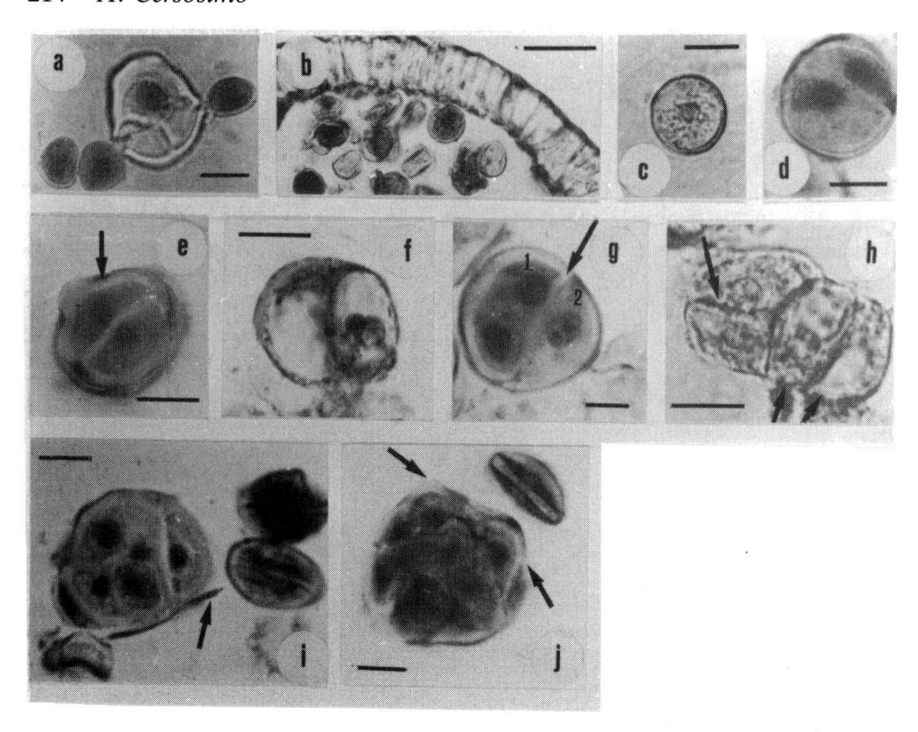

Figure 1. Cell division in microspores. Bars: (g) = 5 μm; (a, c-f, h-j) = 10 μm; (b) = 40 μm. (a) Uninucleate microspores emerging from the mother cell wall in an anther of *Vitis rupestris* du Lot crushed in iron-acetocarmin. (b) Anther loculi showing dead and viable pollen grains (CS, end of the first month of culture). (c) A uninucleate microspore within an anther loculus (end of 5th month). (d,e) First divisions in the microspores (the arrow shows the exine which differentiated during culture, 6th month). (f) Vacuolation and tannin accumulation in a two-celled microspore derivative (6th month). (g) Two-celled microspore derivative in which the nucleus of one cell (1) was already divided. The other cell (2) is presumed to be homologous to the basal cell of the zygotic proembryo. The first division plane is shown by the arrow (6th month). (h) Naked four-celled structure. The small arrows show the two cells derived from the basal cell of the two-celled stage. The large arrow shows the division plane between the two cells derived from the apical cell of the two-celled stage (LS, 6th month). (i) Early globular body derived from a microspore. The arrow shows the ruptured exine (6th month). (j) Globular body derived from a microspore (the arrow shows the remains of the exine; 6th month) (from: Altamura *et al.*, 1992).

division planes (Fig. 1g,h) was wholly similar to that of the tricellular zygotic embryo, which is of the "Asterad" type (Davis, 1966). Cells undergoing first vertical division (Fig. 1g) were considered homologous to the apical cells of the bicellular zygotic proembryo, while the other was the homologue of the basal cells (Fig. 1g). At the four-cell stage, two cells produced by the transverse division of the basal cells became vacuolated (Fig. 1h). Some of these tetracellular proembryonic structures were bare (Fig. 1h), while others were wrapped in exine, which had become evident at the bicellular stage (Fig.

1e). In some cases the exine remained until the globular stage (Fig. 1e,j), but was no longer present when the protoderm became visible. Similar results were obtained with *Atropa belladonna* by Misiura and Zenkteler (1973).

These results indicate that vine microspores can undergo embryogenic induction to the proembryonic spherical structure stage. It conflicts with difficulties in regeneration of haploid or diploid homozygote plants as a result of spontaneous duplication. The most likely explanation again lies in the heterozygous condition of the species and the high number of recessive genes accumulated in its genome. It must be assumed that most of these genes are deleterious and that if they are all able to manifest themselves concurrently (homozygosity) they bring about a genetic configuration incompatible with life.

4. Factors influencing haploid production

The embryogenic capacity of the gametophyte, like any other phenotypic expression, is constituted by a "genetic component" which is the determining factor for the presence of the character, and by an "environmental component", in the broadest sense, which can have a decisive influence on levels of expression of the former. These two components sum up the whole question of anther culture. An attempt will now be made to analyse this question in relation to grapevine.

4.1. *Genetic component*

Genotypic influence is of great importance in determining the inductive response of the gametophytes. As far as grapevine is concerned, interesting results were obtained by Rajasekeran and Mullins (1979, 1983):

a) There was a marked difference in the embryogenic capacity of anthers among the various species, hybrids and cultivars of the *Vitis* genus. This capacity was confined to certain species (*V. rupestris*, *V. longii*, etc.) rather than others (*V. labrusca*, *V. champinii*, etc.). Among the cultivars tested, only "Grenache noir" produced a reasonable number of embryos. Gresshoff and Doy (1974) found that out of 26 different species of vine only three genotypes produced calli. More recently, Cersosimo (1987a) observed that differentiation in anther cultures was dependent on growth regulators in the media. Furthermore, of 36 cultivars tested, the genotypes producing embryogenic calli manifested this capacity to a differing degree on all the culture media. Similarly, recalcitrant genotypes displayed their incapacity on all the culture media used.
b) The morphological expression of sex also proved to have a significant effect on embryogenic callus production. Grapevine anthers from male flowers produced more regenerative callus than those taken from hermaphrodite flowers. This occurred in both species and their hybrids. The

results obtained with the "Gloryvine" hybrid were even more significant. Its anthers were highly productive if they came from vines with male flowers but were non-responsive if they were from hermaphrodite flowers taken from the same male plants "feminized" with a cytokinin treatment.

c) The regenerative capacity of anthers proved to be a dominant hereditary factor. This was experimentally verified by the anther response of a number of male F_1 lines, obtained by crossing a non-regenerative wine grape female with a highly regenerative male hybrid. All the anthers produced embryogenic callus, although the frequency varied. The hereditary nature of the dominant character was evident with the embryogenic capacity of many hybrids having the highly regenerative *V. rupestris* as parent. This was also the case with complex hybrids in which the genetic contribution of the latter was rather limited.

The genetic component should certainly include the heterozygous nature of the species, and particularly the abundance of deleterious recessive genes in the genome. Although the literature does not include studies on the topic, it seems reasonable to hypothesize that a high frequency of these genes could severely jeopardise the success of the culture. Experiments are currently being conducted with anthers produced by plants subjected to a number of self-fertilization cycles in order to rid their genomes of a substantial part of these unwanted genes.

4.2. *Environmental component*

It is well-known that no environment can create characters that are not already built into the genetic code, and no genotype can manifest itself without appropriate environmental conditions. It is the environment that, because of its complexity, hampers the identification of the optimal levels of the parameters involved in haploid technology.

Also for grapevine the most important parameters of anther culture are:

1. physiological and health conditions of the mother plant;
2. stage of development of the microspores;
3. cold pre-treatments;
4. cultural medium and its various components:
   a) mineral nutrient salts,
   b) amino acids,
   c) vitamins,
   d) plant growth regulators,
   e) sugar,
   f) agar,
   g) activated charcoal;
5. physical incubation conditions:
   a) light,
   b) temperature,
   c) photoperiod.

4.2.1. *Physiological and health conditions of the mother plant*
Floral clusters are usually collected from field-grown plants or, less frequently, greenhouse-grown potted plants. These two cases imply profoundly different environmental conditions which produce different physiological responses and lead to differing technical procedures. In all cases, however, the mother plants must be perfectly healthy, especially free from viral or virus-like diseases.

On field vines, flowers should be cut in the morning, when air temperature is low. It is also advisable to collect flowers before any pesticide treatment or 10–15 days after it, especially if systemic products have been used.

Outdoor flowers are produced only once a year (late spring), whereas in the greenhouse floral clusters can be produced at any time and are generally less contaminated. A highly effective method in this regard is to use refrigerated cuttings (at 4–5°C minimum for 30 days) rooted under a controlled environment that stimulates root growth (basal heating at 25–26°C) but prevents the simultaneous sprouting of the distal buds (cooling of the distal buds at 4–5°C) (Mullins, 1966; Mullins and Rajasekaran, 1981; Cersosimo, 1987b) until the cuttings are rooted. Only under these conditions will the preformed cluster be supported by the endogenous cytokinins produced by the roots and be able to complete the physiological sporogenesis process. The anthers of these flowers generally give better results than those collected from the field because they are physiologically more uniform.

4.2.2. *Stage of microspore development*
It has been reported that the "mononucleate" stage of microspores, with the nucleus in *G2* phase, which means immediately after DNA synthesis with the nucleus ready to enter haploid mitosis, or the "early binucleate" stage immediately after mitosis, are the most responsive for embryogenic induction (Norreel, 1970).

It is known that microspores *in vivo*, as soon as they are released by the tetrads, have a dense cytoplasm with a haploid nucleus in a central position (Devreux *et al.*, 1975). The vacuole rapidly begins to form until it occupies most of the cellular volume, and pushes the nucleus towards the cell wall. In this peripheral position the nucleus undergoes its first haploid mitosis, generating two nuclei which immediately begin their natural differentiation. One, the "generative" nucleus, becomes very dense and spindle-shaped, wraps itself in its own membrane and becomes a cell, floating in the cytoplasm of the other. This nucleus, after a further division which usually takes place at the beginning of pollen germination (the second haploid mitosis), will form two sperm nuclei (the real male gametes) destined to carry out the double fertilization. The other, the "vegetative" nucleus, decreases in density and the cell containing it very rapidly begins to accumulate starch grains which make it difficult to observe under the microscope. Only a few days before anthesis does the action of amylase enzymes make it visible again with usual dyes. It is well-known that starch accumulation marks the beginning of

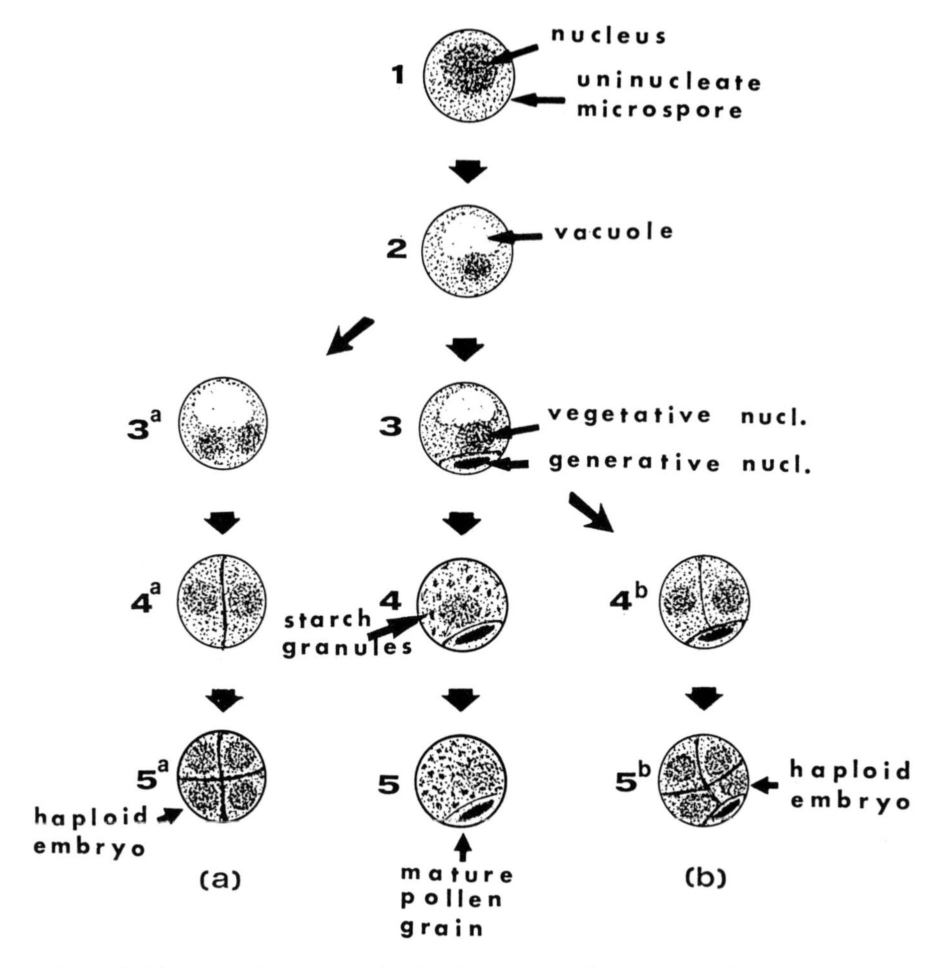

Figure 2. Microspore development *in vivo* (sequences 1–5) and *in vitro* (sequences 1–5a, 1–5b).

the irreversible stage of the differentiation process, in which the microspore becomes a mature pollen grain. Once this differentiation is completed, it is virtually impossible to induce cells to change their genetic programme (Fig. 2).

What happens during the *in vitro* culture of anthers? The process is more or less the same until the first haploid mitosis occurs. Most commonly the two nuclei differentiate (Fig. 2b) but there is no starch accumulation. In other cases, there is no differentiation and the two nuclei remain exactly the same (Fig. 2a). Their division is accompanied by the symmetrical segmentation of the cell. The same applies to many further division cycles until complete haploid embryos are formed and, after reaching the globular stage, the exine breaks and becomes free.

Where the two nuclei have differentiated, the vegetative one begins to undergo a series of divisions similar to those just described, leading to the formation of a haploid embryo. The generative one may remain inert or itself undergo some divisions. It is subsequently pushed against the cell wall and removed when the embryo frees itself of the exine. In both cases, therefore, the embryo is generated by the haploid somatic cell of the male gametophyte. To complete the picture we must consider at least two other developmental processes of the microspore *in vitro*:

- the first haploid mitosis and many subsequent divisions are sometimes not followed by cell segmentation, which leads to the formation of a multinucleate grain (up to 30–40 nuclei) with no separating walls. This results in abortion;
- on other occasions the generative nucleus, after undergoing several divisions, all with cell segmentations, produces a haploid embryo. This model was observed with *Hyoscyamus niger* (Raghaven, 1975; 1978).

The importance of the optimal microspore development stage for grapevine anther culture has been demonstrated by Gresshoff and Doy (1974). After sacrificing one loculus for every anther for microscopic observation, they placed the remaining loculi in culture, dividing them according to the following classes of microspore development: **pre-M* = Pre-meiosis, **M*-1 = Meiosis 1, **M*-2 = Meiosis 2, **T* = Tetrads, **u.m.* = uninucleate microspore, **m.p.g.* = mature pollen grains. As may be observed (Fig. 3), the highest incidence of callus production was obtained at u.m. and T stages. No activity was observed with m.p.g.

The microspore developmental stage is usually assessed by crushing anthers in a drop of iron-acetocarmine and observing microspores. For a better timing for taking cuttings (floral clusters), an attempt was made to find some correlation between the onset of the uninucleate or tetrad stage and the appearance of any morphological features of the flower or the whole cluster. Some studies (Gresshoff and Doy, 1974; Rajasekaran and Mullins, 1979) considered the metric characteristics of the floral bud (length, diameter, etc.). Mauro *et al.* (1986a) studied the anther length of three Cabernet Sauvignon clones, and found that for each clone the optimal stage of development for callus production was constant and corresponded to the two classes of intermediate length of 0.3–0.5 and 0.5–0.7 mm. The anthers of these two classes also produced callus with embryogenic capacity.

It seems clear, however, that the indications provided by these parameters cannot be generalized, since the variability among cultivars is wide and the environmental effect on these characters is great. A microscopic study of several cultivars (Cersosimo, 1987a) indicated that there is a correlation between the onset of the tetrad stage and the appearance along the petal suture lines of a thin layer of cells with a lighter colour than the dark green of the petals. This could also be seen with the naked eye or with a simple

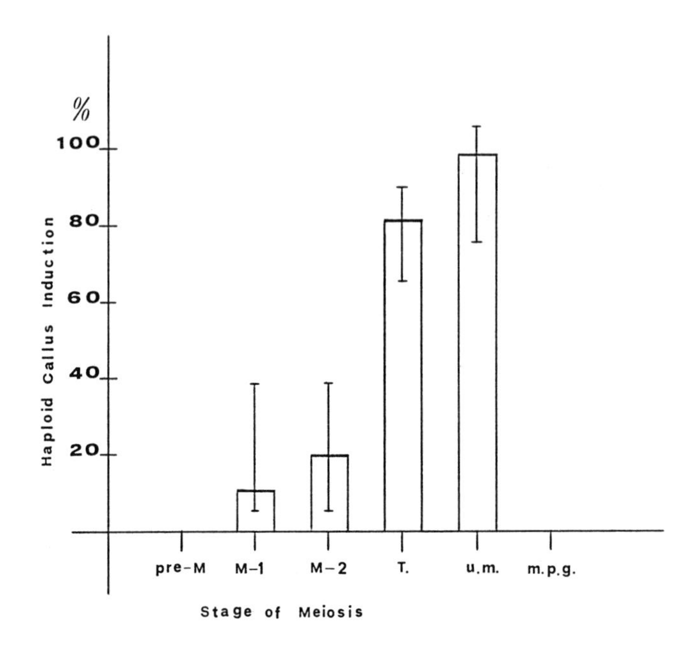

Figure 3. Haploid callus induction of grape vine (*Vitis vinifera* L.) to obtain microspore development (from Gresshoff and Doy, 1974 – modified).

magnifying glass. An open bud at this stage displayed translucent, almost transparent, anthers of a pale yellow-green colour. The microscopic observations indicated that they contained tetrads which were either still closed or just opened. When observed at a previous stage, the anthers appeared entirely transparent and colourless, containing mother cells in meiosis. At a subsequent stage, they were distinctly opaque as a result of the intense yellow colour caused by the presence of mature pollen rich in starch. At this stage, the external colour of the flower was virtually homogeneous as a result of the gradual enlargement of the light-coloured layer of cells and the simultaneous fading of the green colour over the petal surface.

4.2.3. *Cold pre-treatment*

Treatment of floral buds or anthers with physical agents can improve the inductive response of microspores. Cold treatment was most successful in improving this response.

Rajasekaran and Mullins (1979) observed that grape vines, like other species, benefited from cold pre-treatment to improve the inductive response of anthers. The best results were at 4°C for 72 h. Mauro *et al.* (1986b) working with Cabernet Sauvignon clones, reported the highest callogenesis rate (9%) with cold pre-treated anthers (flowers pretreated at 4°C for 3–8 days). The difference between treated and non-pretreated anthers, however,

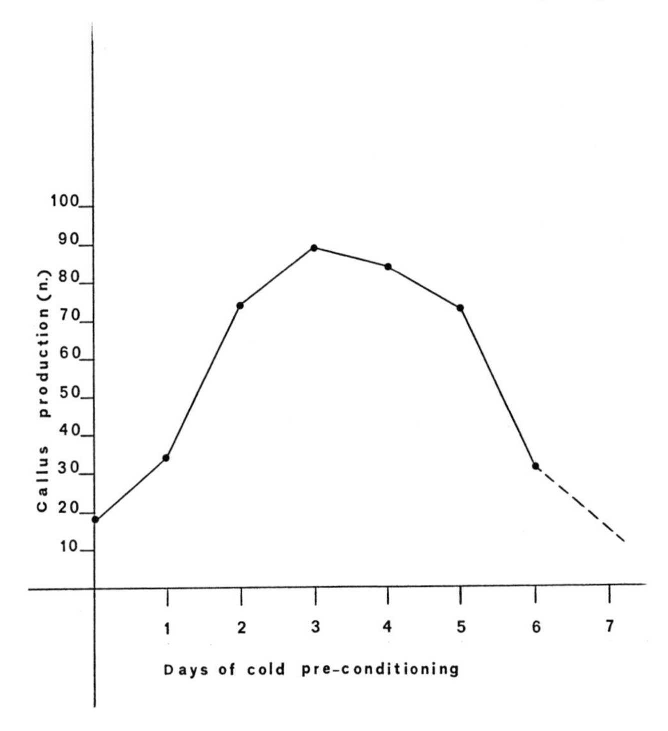

Figure 4. Effects of cold pretreatment on callus production.

was not significant. They also noted that pretreatment of cultured anthers on agar medium produced fewer results compared with flowers pre-treated before anthers were cultured. They also found that heat pretreatment (35°C for 3–8 days) inhibited callus formation. Results obtained by Cersosimo (1986) in anther culture of 36 different grape cultivars were consistent with those obtained by Rajasekaran and Mullins (1979). Floral clusters were cut, sterilized and refrigerated at 4–5°C. Every day for two weeks from the date of cutting, about 200 anthers for each genotype were cultured on agar medium. The positive effect of the cold was seen from the second to the fifth day of anther culture, with the optimum on days three and four (Fig. 4).

In Nitsch's view (1974), cold pretreatment increased the number of microspores with identical nuclei and contributed to maintaining their vitality.

4.2.4. *Culture substratum*

It has been shown (Nitsch, 1969; Nitsch and Nitsch, 1970; Raghaven, 1975) that a simple sugar solution may in some species trigger the first mitosis of microspores, provided that the latter are in the appropriate stage of development. But in general, the success of a culture (anthers, pollen or any

other) is always influenced by the composition of the substratum, that must provide:
- macro and micro nutrient salts;
- plant growth regulators (auxins, cytokinins);
- carbon energy source (sugars);
- adsorbent agent (activated charcoal);
- amino acids;
- vitamins;
- gelling agent (agar);
- others.

4.2.4.1. *Mineral salts*. Macro and micro salts are generally used in defined media formulations such as: *MS* (Murashige and Skoog, 1962); *NN-H* (Nitsch and Nitsch, 1968); *B5* (Gamborg *et al.*, 1968); *White* (White, 1963); *LS* (Linsmaier and Skoog, 1965) and a few others.

As far as grape vines are concerned, salts most frequently used in anther cultures are essentially *MS*, *NN-H* and *B5*, with a total nitrogen content of 60.0, 27.4 and 26.8 mM and an NO_3^-/NH_4^+ ratio of about 2:1, 2:1 and 12:1, respectively.

In a test carried out with three different culture media (modified MS, NN-H and B5), amended with 45.00, 27.40 and 13.35 mM of nitrogen and an NO_3^-/NH_4^+ ratio of 2:1, 2:1 and 12:1, respectively, the production of embryogenic callus from cultured anthers of grape was highest on NN-H medium (Cersosimo, 1986). Of the other two media, significantly better results were obtained on B5.

Although the nitrate fraction is considered more important, a certain amount of ammonia nitrogen is also needed. The presence of nitrate nitrogen alone could inhibit the activity of the enzyme nitrate and nitrite reductase, with a consequent accumulation of nitrites inside the cells, where they are notoriously toxic.

4.2.4.2. *Nitrogenous organic compounds*. There is a close relationship between availability of nitrogen and the use of organic compounds containing small quantities of nitrogen. In carrots, some amino acids played a positive role in the inductive processes of *in vitro* embryogenesis (Kamada and Harada, 1979). It was subsequently observed (Mauro *et al.*, 1986a) that in *Vitis* the addition of L-glutamine and DL-phenylalanine to the culture medium enhanced cell proliferation, and that adenine enhanced the efficiency of embryogenesis. Following this, the combined effect of these compounds was tested in Cabernet Sauvignon anther culture (Mauro *et al.*, 1986b). The results showed a significant increase in callus production, which was more marked in the medium containing all three amino acids. The poorest results were observed in the absence of glutamine, while the highest embryo percentage was obtained with the glutamine + adenine combination. Although no appreciable differences in callus production were noted as a result of the addition of casein hydrolysate, no callus was able to form embryos without it. Previous results showed that the effectiveness of casein hydrolysate in the embryogenic processes could be attributed to its high concentration of glutamine (Gamborg, 1970). It acted as a nitrogen donor and was able to support protein synthesis, a very intense activity during embryogenesis because of cell proliferation.

4.2.4.3. *Vitamin compounds*. Vitamins are essential for the biological activity of plant cells because the important catalytic function they perform in metabolic processes. Most media used in grape anther cultures contain vitamin compounds defined by Murashige and Skoog (1962), a mixture of 0.1 mg/l thiamine (B1), 0.5 mg/l nicotinic acid (niacin) and 0.5 mg/l pyridoxine (B6). Pantothenic acid, while making an important contribution to the growth of certain tissues and to the proliferation of callus (Gautheret, 1948), seems to have no effect on grape because it is endogenously synthesized. Folic acid seems to slow cell proliferation in the dark but enhance it under light because it is hydrolized by the light into p-aminobenzoic acid (Capozzi, 1953). Riboflavin, which appears in some vitamin compounds, seems to inhibit rooting, but it has not been found to have any specific effect on somatic and/or gametic embryogenesis. Myo-inositol is not a real vitamin but is almost always used in culture media. Although it is not considered an essential component of media (Murashige, 1974) it does seem to play a role in the stimulation of growth, in morphogenesis and in a number of biosynthetic processes leading to the formation of essential cell constituents such as ascorbic acid and pectins.

4.2.4.4. *Growth regulators*. Apart from the few familiar Solanaceae – *D. innoxia* (Sopory, 1972), *N. tabacum* (Nitsch, 1969) and *H. niger* (Corduan, 1975) – in which gametic embryos develop even when anthers or isolated microspores are cultivated on simple basal medium, all species studied so far need specific levels of growth regulators for induction. Those most widely used for grapes are auxins* and cytokinins.**

With combinations of the auxins 2,4-*D*, *NOA* and *NAA* and the cytokinins *BA* and 6-*BPA*, Rajasekaran and Mullins (1979) found that the optimal auxin/cytokinin ratio for grapes was 5:1. The highest production of embryogenic callus was obtained with the combination of 5 μM/l 2,4-*D* and 1 μM/l *BA*. Production was reduced virtually to zero when 2,4-D was replaced by an identical concentration of *NOA* or *NAA*.

---

*auxins

2,4-*D* – 2,4-Dichlorophenoxyacetic acid; *IAA* – 3-Indoleacetic acid; *IBA* – 3-Indolebutyric acid; *NAA* – 1-Naphthylacetic acid; *NOA* – 2-Naphthyloxyacetic acid.

**cytokinins

BA – Benzyladenine; *Kinetin* – 6-Furfurylaminopurine; *Zeatin* – 4-Hydroxy-3-methyl-trans-2-butenylaminopurine; 6-*PBA* – 6-Benzylamino-9-2(tetrahydropyranyl)-9H purine; 2-*iP* – N6-(2-Isopentyl)adenine).

A constant concentration of 2,4-*D* (5 μM/l) combined with 1 μM/l *BA*, kinetin or zeatin (Cersosimo, 1987a) confirmed that 2,4-D and BA provided the best auxin/cytokinin combination, callus produced in 54% of the cultured anthers, compared with 31% with kinetin and 15% with zeatin.

With regard to genotypes, most of the cultivars showed a high production of callus with BA, while "Granache Blanc" and the crosses for table grapes "Italia × Volta 199" and "Italia × Panse Precoce 129" gave the best results with kinetin. Only the hybrid "Seyve Villard 12.397" produced most calli with zeatin.

The general observation should be made that in processes of embryogenesis, of which androgenesis is a specific example, the induction stage is enhanced by high auxin contents, particularly 2,4–D. However, embryos are not formed until the concentration of auxin is reduced. This means that auxin induces cell embryogenic capacity but at the same time blocks their ability to divide. The subsequent reduction of auxin levels restores this lost ability (Sharp *et al.*, 1980).

4.2.4.5. *Sugars.* It is known that plant cells cultivated *in vitro* are not able to provide their own energy needs to support biosynthetic (metabolic and enzymatic) processes. For these reasons, culture substratum needs an alternative source of energy. The most suitable for this important role is sucrose, followed by glucose, maltose, fructose and galactose. The least effective seem to be mannose and lactose. Myo-inositol (a hexahedral cyclical alcohol) does not provide any form of energy (Meryl and Stone, 1973). Its effects are to be sought in biosynthesis, where it plays a biological role similar to that of vitamins.

In grape anther culture the sucrose concentration range is 2–3%. Higher concentrations (6–12%) are used in some Graminaceae (wheat and barley), in potato, in *Brassica*, etc. However, high sucrose concentration may act as an osmotic regulator, especially during the induction phase. Sucrose at 5–10% in the culture medium seems to have a positive effect on embryogenic induction of microspores (Sunderland and Dunwell, 1977). As soon as embryo development is manifested, high sugar levels are unnecessary and become inhibitory. The temporary presence of high sugar concentrations during induction may prevent the proliferation of somatic callus. This callus, if allowed to grow freely, would submerge the haploid embryos and make their development even more difficult. The transfer of anthers from a high-sugar-containing medium to one with a normal energy content is therefore appropriate for optimal embryo development.

4.2.4.6. *Agar.* Agar is commonly added as a gelling agent in solid or semi-solid media. There is no general agreement regarding the use of liquid rather than solid media in grape anther culture. Good results on solid media have been obtained by Gresshoff and Doy (1974), Gray and Mortensen (1986), Cersosimo (1987a), etc., while Rajasekaran and Mullins (1979) found these media to be unsatisfactory. Using a liquid medium with continuous agitation, they obtained high embryogenic callus production with the "Gloryvine" hybrid. The positive effect of the liquid medium may be explained (Devreux *et al.*, 1975) by the greater dilution of any inhibitors released by the degradation of the anther wall or present in agar. The rapid detachment of embryos from the remaining anther is also assisted by the liquid medium, and prevents any inhibitory influence of the anther tissue.

In a comparison of liquid and agar-solidified media (Cersosimo, 1987a), the solid medium produced better inductive results. The subsequent stages of embryo development (cotyledonary stage) and growth were better in the liquid medium. The subsequent transfer of the embryos to a solid medium, overlaid by a thin layer of the same liquid medium, favoured their complete development (Cersosimo *et al.*, 1990).

4.2.4.7. *Activated charcoal.* The addition of activated charcoal to the agar-solidified medium favors embryo development. Positive results in the induction response have been recorded with many species, although activated charcoal has not been used frequently in grape anther cultures. The positive results are explained by the adsorption of any inhibiting substances either produced by anther degradation or present in the agar.

Activated charcoal in the agar medium improves induction efficiency of microspores (Tyagi *et al.*, 1980). The increase is much greater than that observed in an agitated liquid medium. Induction is inhibited by adding activated charcoal in the liquid medium.

Although the mechanisms producing these positive effects are not yet known, it has been postulated (Weatherhead *et al.*, 1978) that activated charcoal either reduces or eliminates the negative effect of 5–hydroxymethylfurfural (HMF), an inhibitor of embryogenesis that forms as a result of sucrose degradation during sterilization in the autoclave. It should be remembered that a certain amount of growth regulators can be added to the culture medium adsorbed by the activated charcoal rendering them unavailable for the culture. This suggests that the growth regulator concentration in the medium should be increased when charcoal is present.

4.2.5. *Physical factors*

In the past, scant attention has been paid to the possible role of certain physical factors in embryogenesis. Only recently have researchers begun to assess their influence, and attention has been mostly concentrated on light and temperature.

4.2.5.1. *Light.* The effect of light on anther culture is essentially due to its regulatory capacity on embryo development in the post-inductive stages.

There are few, and discordant, references on the effect of light wavelength on anther or isolated microspore cultures. Nitsch (1977) found that red light was beneficial to *N. tabacum* pollen culture compared with blue or low-intensity white light or darkness. Contradictory results were obtained by Sopory and Maheshwari (1976), who obtained no embryo production at all in *D. innoxia* anther culture under red light.

Results obtained from light intensity and the photoperiod also varied according to the species tested. Sunderland (1971) with *N. tabacum*, Sopory and Maheshwari (1976) with *D. innoxia* and Corduan (1975) with *H. niger* found that the alternation of light and darkness had a positive effect on embryo formation. Continuous light was found to produce favourable results with *D. metel* (Narayanaswamy and Chandy, 1971) but inhibited *Anemone virginiana* embryo formation (Johansson and Eriksson, 1977).

With grapes the results have not been as clear, although some differentiation has been observed. Gresshoff and Doy (1974) found that, although callus induction could occur in darkness, there was a notable improvement in the induction rate when anthers were cultured in continuous light for the first 24 h and then in darkness. Rajasekaran and Mullins (1979) found that embryo production was better when anthers were initially cultured in the dark for several days compared to exposure to a 16 h photoperiod from the beginning. They also observed that callus formed 10 days earlier in the dark than in the light.

Grapes produce phenolic compounds in *in vitro* culture; it should be remembered that light oxidizes these compounds and they accumulate in the culture environment. The desirability of this phenomenon depends on the quantities accumulated. Low quantities would favour callogenesis and organogenesis, while higher concentrations could have a harmful effect on cell survival.

4.2.5.2. *Temperature*. In anther cultures of *Brassica napus* and *B. campestris* Keller and Armstrong (1978, 1979) found that initial temperatures of 30°C for the first two weeks or 35°C for the first 2–3 days, followed by a continuous temperature of 25°C, gave a notable increase in the production of haploids by direct androgenesis. Previous studies (Keller *et al.*, 1975) demonstrated no haploid production at 25°C at the beginning of anther culture. The optimal temperature for haploid induction in grape varies between 24 and 26°C. The lower the temperature used, the fewer haploids have been produced.

5. Conclusion

A general conclusion may assist in establishing a technique for the production of haploid plants in *Vitis*.

1. The genetic component plays a crucial role in determining inductive response; the choice of genotype is therefore of primary importance.
2. Once the genotype has been chosen, the success of the culture will depend entirely on microspore developmental stage. The optimal stage imme-

diately precedes or follows the first haploid mitosis and certainly occurs before the accumulation of starch in the cytoplasm.
3. Cold pre-treatment (4–5°C for 3–4 days) greatly improves the inductive efficiency of both embryos and callus formation.
4. The choice of culture medium does not appear to be particularly important. What is important is the choice of growth regulators and the optimal auxin-cytokinin ratio.

With regard to obtaining haploid plants in *Vitis*, the results obtained so far (Altamura *et al.*, 1992) show that grape can be induced into direct androgenic activity. Unfortunately this has not proceeded beyond the first stages of haploid embryo development. The greatest obstacle to the maturation of haploid embryos to haploid plant regeneration is the plant genotype, especially when its genome has many deleterious recessive genes leading to genetic disorders. The successful definition of this technique may, therefore, depend on the elimination of these genes by means of suitable self-fertilization programmes.

The production of haploid plants continues to be an important objective in bringing the genetic improvement of grapes into line with the present developments and requirements of modern viticulture.

6. References

Altamura, M.M., A. Cersosimo, C. Majoli and M. Crespan, 1992. Histological study of embryogenesis and organogenesis from anthers of *Vitis rupestris* du Lot cultured *in vitro*. Protoplasma 171: 134–141.

Bouquet, A., B. Piganeau and A.M. Lamaison, 1982. Influence du génotype sur la production de cals, d'embryons et de plantes entières par culture d'anthères *in vitro* dans le genre *Vitis*. C.R. Acad. Sci. Paris 259: 569–574.

Bourgin, J.P. and J.P. Nitsch, 1967. Obtention de *Nicotiana* haplöides à partir d'étamines cultivées *in vitro*. Ann. Physiol. Végét. 9: 377–382.

Capozzi, A., 1953. Acido folico e proliferazione cellulare in tessuti vegetali *in vitro*. Atti Istituto Botanico, Laboratorio Crittogam. Univ. Pavia 10: 211–216.

Cersosimo, A., 1986. Coltura *in vitro* di antere in *Vitis* sp. (primo contributo). Rivista di Viticoltura e di Enologia di Conegliano 39(12): 520–531.

Cersosimo, A., 1987a. Coltura *in vitro* di antere in *Vitis* sp. (secondo contributo). Agricoltura Ricerca 75–76: 55–60.

Cersosimo, A., 1987b. Prototipo di pre-radicatore di talee perla produzione di piante-test di Vite. Agricoltura Ricerca 75–76: 61–64.

Cersosimo, A., M.M. Altamura, M. Crespan and G. Paludetti, 1990. Embrogenesis, organogenesis and plant regeneration from anther culture in *Vitis*. Acta Hort. 280: 307–314.

Corduan, G., 1975. Regeneration of anther-derived plants from *Hyoscyamus niger* L. Planta 127: 27–36.

Davis, G.L., 1966. Systematic Embryology of the Angiosperms. Wiley, New York.

Devreux, M., U. Laneri and P. De Martinis, 1975. Réflexions sur nos cultures d'anthères de plantes cultivées. Giornale Botanico Italiano 109(6): 335–349.

Gamborg, O.L., R.A. Miller and K. Ojima, 1968. Nutrient requirements of suspension cultures of soybean root cells. Exp. Cell. Res. 50: 151–158.

Gamborg, O.L., 1970. The effects of amino acids and ammonium on the growth of plant cells in suspension culture. Plant Physiol. 45: 372–375.
Gautheret, R.J., 1948. Sur la culture indefinie des tissue de *Salix capraea*. Compt. Rend. Soc. Biol. 142: 807–808.
Gray, D.J. and J.A. Mortensen, 1986. Initiation and maintenance of long term somatic embryogenesis from anthers and ovaries of *Vitis longii* "Microsperma". Plant Cell. Tiss. Org. Cult. 9: 73–80.
Gresshoff, P.M. and C.H. Doy, 1974. Derivation of a haploid cell line from *Vitis vinifera* and the importance of the stage of meiotic development of anthers for haploid culture of this and other genera. Z. Pflanzenphysiol. 73: 132–141.
Guha, S. and S.C. Maheshwari, 1964. *In vitro* production of embryos from anthers of *Datura*. Nature 204: 497.
Hirabayashi, T. and T. Akihama, 1982. *In vitro* embryogenesis and plant regeneration from the anther-derived callus of *Vitis*. In: A. Fujiwara (Ed.), Plant Tissue Culture, pp. 547–548, Maruzen Co., Ltd., Tokyo.
Johansson, L. and T. Eriksson, 1977. Induced embryo formation in anther cultures of several *Anemone* species. Physiol. Plant. 40: 172–174.
Kamada, H. and H. Harada, 1979. Studies on the organogenesis in carrot tissue culture. 1. Effects of growth regulators on somatic embryogenesis and root formation. Z. Pflanzenphysiol. 91: 255–266.
Keller, W.A., T. Rajhathy and J. Lacapra, 1975. *In vitro* production of plants from pollen in *Brassica campestris*. Can. J. Genet. Cytol. 17: 655–666.
Keller, W.A. and K.C. Armstrong, 1978. High frequency production of microspore-derived plants from *Brassica napus* anther cultures. Z. Pflanzenzüchtg. 80: 100–108.
Keller, W.A. and K.C. Armstrong, 1979. Stimulation of embryogenesis and haploid production in *Brassica campestris*. Theor. Appl. Genet. 55: 65–67.
Kim, S.K. and K.Y. Peak, 1981. Studies on anther culture of grape. 1. Varietal differences in callus formation. J. Korean Soc. Hort. Sci. 22(2): 89–91.
Linsmaier, E.M. and F. Skoog, 1965. Organic growth factor requirements of tobacco tissue cultures. Physiol. Plant. 18: 100–127.
Mauro, M.C., G. Nef, C. Ambid and J. Fallot, 1986a. Utilisation de l'embryogénèse somatique en vue d'augmenter la variabilité chez *Vitis vinifera* var. Cabernet Sauvignon. In: Vignevini, 13, Suppl. 12, Atti del 4° Simposio Internazionale di Genetica della Vite, Verona, 1985, pp. 27–29.
Mauro, M.C., G. Nef and J. Fallot, 1986b. Stimulation of somatic embryogenesis and plant regeneration form anther culture of *Vitis vinifera* cv. Cabernet Sauvignon. Plant Cell Rep. 5: 377–380.
Meryl Smith, M. and B.A. Stone, 1973. Studies on *Lolium multiflorum* endosperm in tissue callium. 1. Nutrition. Aust. J. Biol. Sci. 26: 123–133.
Misiura, E. and M. Zenkteler, 1973. Studies on the androgenesis in cultured anthers of *Atropa belladonna*. L. Acta Sec. Bot. Polon. 42: 309–322.
Mullins, M.G., 1966. Test-plants for investigations of the physiology of fruiting in *Vitis vinifera* L. Nature (London) 209: 419–420.
Mullins, M.G. and K. Rajasekaran, 1981. Fruiting cuttings: revised method for producing test plant of grapevine cultivars. Amer. J. Enol. Viticult. 32: 35–40.
Murashige, T. and F. Skoog, 1962. A revised medium for rapid growth and bioassays with tobacco tissue cultures. Physiol. Plant. 15: 473–497.
Murashige, T., 1974. Plant propagation through tissue culture. Ann. Rev. Plant Physiol. 25: 135.
Narayanaswamy, S. and L.P. Chandy, 1971. *In vitro* induction of haploid, diploid and triploid androgenic embryoids and plantlets in *Datura metel* L. Ann. Bot. 35: 535–542.
Nitsch, J.P. and C. Nitsch, 1968. Haploid plants from pollen grains. Science 163: 85–87.
Nitsch J.P., 1969. Experimental androgenesis in *Nicotiana*. Phytomorphology, 19: 389–404.

Nitsch, J.P. and C. Nitsch, 1970. Obtention de plantes haploïdes à partir de pollen. Bull. Soc. Bot. France 117: 339–360.
Nitsch, C., 1974. Pollen culture, a new technique for mass production of haploids and homozygous plants. In: K.J. Kasha (Ed.), Haploids in Higher Plants: Advances and Potential, pp. 123–135. The University of Guelph, Guelph.
Nitsch, C., 1977. Culture of isolated microspores. In: J. Reinert and Y.P.S. Bajaj (Eds.), Plant Cell Tissue and Organ Culture, pp. 268–278. Springer-Verlag, Berlin/Heidelberg/New York.
Norreel, B., 1970. Étude cytologique de l'androgénèse expérimentale chez *Nicotiana tabacum* et *Datura innoxia*. Bull. Soc. Bot. France 117: 461–478.
Olmo, H.P., 1976. Grapes: *Vitis*, Muscadinia (Vitaceae). In: N.W. Simmond (Ed.), Evolution of crop-plants, pp. 294–298. Longman, London/New York.
Raghavan, V., 1975. Induction of haploid plants from anther cultures of henbane. Z. Pflanzenphysiol. 76: 89–92.
Raghavan, V., 1978. Origin and development of pollen embryoids and pollen calluses in cultured anther segments of *Hyoscyamus niger* (henbane). Amer. J. Bot. 65: 984–1002.
Rajasekaran, K. and M.G. Mullins, 1979. Embryos and plantlets from cultured anthers of hybrid grapevines. J. Exp. Bot. 30: 399–407.
Rajasekaran, K. and M.G. Mullins, 1983. Influence of genotype and sex-expression on formation of plantlets by cultured anthers of grapevines. Agronomie. 3: 233–238.
Sharp, W.R., M.R. Sondahl, L.S. Caldas, and S.B. Maraffa, 1980. The physiology of *in vitro* asexual embryogenesis. Hort. Rev. 2: 268–310.
Sopory, S.K., 1972. Physiology of development of pollen embyroids in *Datura innoxia*. Mill. Ph.D. Thesis, The University of Delhi, Delhi.
Sopory, S.K. and S.C. Maheshwari, 1976. Development of pollen embryoids in anther cultures of *Datura innoxia*. 1. General observations and effects of physical factors. J. Exp. Bot. 27: 49–57.
Sunderland, N., 1971. Anther culture: a progress report. Sci. Progr. 59: 527–549.
Sunderland, N. and J.M. Dunwell, 1977. Anther and pollen culture. In: H.E. Street (Ed.), Plant Tissue and Cell Culture, pp. 223–265. Blackwell Scientific, Oxford.
Tyagi, A.K., A. Rashid and S.C. Maheshwari, 1980. Enhancement of pollen embryo formation in *Datura innoxia* from isolated pollen grains by different culture conditions. Physiol. Plant 49: 296–298.
Weatherhead, M.A., J. Burdon and G.G. Henshaw, 1978. Some effects of activated charcoal as an additive to plant tissue culture media. Z. Pflanzenphysiol. 89: 141–147.
White, P.R., 1963. The Cultivation of Animal and Plant Cells. Ronald Press, New York.
Zenkteler, M., 1971. *In vitro* production of haploid plants from pollen grains of *Atropa belladonna* L. Experientia 27: 1087.
Zou, C.-J. and P.-F. Li, 1981. Induction of pollen plants of grape (*Vitis vinifera* L.). Acta Bot. Sin. 1: 79–81.

13. Polyhaploidy in strawberry

M.J. HENNERTY and A.J. SAYEGH

Contents

1. Introduction

As a result of octoploidy and inbreeding depression, little is known about the genetics of the strawberry, and breeding is based on conventional selection methods. Since many breeding procedures are based on segregating characters, selection at tetrahaploid level and resynthesis and testing at octoploid level would result in the fixing of desirable characters and increase the predictability in segregating populations.

Polyhaploids in strawberry have long attracted plant breeders because of their potential as tools for simplification and acceleration of breeding programmes. Polyhaploids can be used to speedily attain homozygosity, to expose and eliminate the genetic load, and to simplify the genetic system. Attempts to develop systems which increase the efficiency of polyhaploid production have been underway since shortly after Guha and Maheshwari (1964) produced their first haploids using anther culture.

Strawberry haploids have been produced by conventional breeding methods, but at such low frequencies that they have had little practical value. Plant tissue culture techniques have been used to attempt to produce haploids in different *Fragaria* species through anther culture, as well as through embryo rescue from conventional crosses, to increase the frequency of tetrahaploid production in the strawberry.

S.M. Jain, S.K. Sopory & R.E. Veilleux (eds.), In Vitro *Haploid Production in Higher Plants, Vol. 3*, 231–260.

2. The strawberry, *Fragaria* × *ananassa*

The commercial strawberry, *F.* × *ananassa* Duch., is a complex allopolyploid ($2n = 8x = 56$) derived from hybridization of the octoploid species *F. chiloensis* Duch. and *F. virginiana* Duch. It may have arisen in European gardens between 1714 and 1759, and thousands of cultivars have been developed from the originals, plus more recent infusions of genes from both species through hybridization and backcrosses (Staudt, 1962; Bradley, 1982). The cytogenetic evidence supports the conclusion that the modern cultivars are genetically very close to *F. virginiana*. Perhaps in the development of modern cultivars in the gardens of Europe, *F. virginiana* was extensively used as a pollen parent (Bradley, 1982).

Strawberry cultivars and selections are propagated vegetatively by runner plants rather than by seeds. This has influenced the breeding systems used for cultivar improvement since it is not necessary to have pure lines. Most cultivars and selections are highly heterozygous. Morrow and Darrow (1952) found that one generation of selfing produced selections that were weaker, less productive and smaller fruited than their parents and much weaker than seedlings obtained from outcrosses. Overall vigour in strawberry is strongly correlated with level of heterozygosity (Melville *et al.*, 1980). However, selections from the S1 to S5 generations have been used in breeding programmes and some have yielded desirable genotypes (Scott and Lawrence, 1975; Aalders and Graig, 1968; Spangelo *et al.*, 1971). Inbreeding is a valuable tool when used in genetic studies, or when there is a need to concentrate genes for a few particular characters (Aalders and Graig, 1968).

Both existing cultivars of strawberry and seedlings selected as parents in breeding programmes have a wide range of character variability in spite of a limited number of founding parents of pre 1960 released cultivars (Hancock and Luby, 1993). This, together with the variation found in the two wild parent species, makes blending precise characters into this highly heterozygous plant very difficult (Gooding, 1975, 1980). Rapidly changing consumer preferences and production technologies have compounded the breeding problems.

New cultivars are normally developed by crossing already highly heterozygous clones and picking superior phenotypes in the resulting progeny. Theoretically in an open-pollinated heterozygous species, recessive deleterious alleles will be carried as a part of the genetic load. Inbreeding should reveal undesirable combinations of genes, which, once eliminated, should leave plants with a high concentration of desirable characters. The great loss in vigour in the S1 generation, and its further decline over successive generations such that few lines can be carried to S5, is an indication of the deleterious load in strawberry. The most vigorous progeny are usually selected for selfing (Spangelo *et al.*, 1971).

Hancock *et al.* (1993) stated that the origin of the cultivated strawberry is arguably the most fascinating of all crop species histories. *F.* × *ananassa* was derived from the accidental hybridization of the New World species *F.*

chiloensis and *F. virginiana* in a European botanical garden approximately 250 years ago (Darrow, 1966). The first introductions of both *F. virginiana* and *F. chiloensis* were grown separately in Europe and were not very successful. The early collections of the Virginian strawberries had soft small fruits and were only marginally superior to the native European wood strawberry, *F. vesca*. The Chilean strawberry was a great disappointment because Francois Frezier had unknowingly brought back only female plants and no males to pollinate them. It was not until *F. virginiana* was identified as a successful pollinator that *F. chiloensis* began to show its true potential. A strong industry developed at Brest, France around 1750, where Frezier's female plants of *F. chiloensis* were interplanted with males of *F. virginiana*, and the first *F.* × *ananassa* hybrids arose.

The genus *Fragaria* is broadly adapted and substantial differentiation has occurred within and between species due to a wide range of growing environments (Hancock, 1990). Only by selection under specific environmental conditions have breeders attained the full expression of the characteristics sought in the strawberry. Strawberry yields and fruit quality are greatly influenced by a number of factors, including the effect of photoperiod and temperature on flower and stolon formation, tolerance to different soil conditions, winter hardiness, high temperature tolerance, and inherent vigour. Diseases and pests are also major factors limiting yield and fruit quality. One of the great challenges facing breeders is to incorporate several traits in a cultivar to make it an economic success. The increasing pressure from consumers to minimize fungicide and pesticide inputs into agricultural systems makes disease and pest resistance in commercially acceptable types a priority.

Deficiency in the knowledge of the genetics of *Fragaria*, and the fact that most traits of interest are encoded by several genes, may delay practical application of gene transfer (Wallin *et al.*, 1993). This is another reason for tetrahaploid production. Strawberry breeders, on the one hand, are fortunate that an array of wild germplasm from diverse environments and natural selection pressures are in existence. The octoploids are interspecifically compatible and give viable seeds; good horticultural types can usually be recovered within three generations of backcrossing to the cultivated parent (Bringhurst and Voth, 1984). In addition to the variability present in the octoploid, the variability available in other species can be exploited through modern biotechnology techniques (Wallin *et al.*, 1993). On the other hand, because of the complex genetics of many traits, strawberry breeding requires a generous allotment of time.

3. Strawberry breeding

Trends in strawberry breeding during the period 1970–1989 have been reviewed by Galletta and Maas (1990). Heritability estimates have been high for a number of yield components such as average berry weight, berries per flower stalk, yield per flower stalk and flower stalk number. It seems possible to improve total yield through selection of individual components as a linear function of the above-mentioned components, especially if positive genetic correlations exist between components of total yield (Spangelo *et al.*, 1971). Watkins and Spangelo (1971) studied strawberry selection index components. It was estimated that, for the average of the seven characters studied, and for the index, the non-additive variance constituted as much as 40% of the total genetic variance. The characters were: mildew resistance, leaf scorch, flower stalk number, berry appearance, berry number, marketable yield, and processing suitability They suggested, for economic reasons, that it is more efficient to exploit any major gene related to economic characters, if identified, and that it should be transferred by back-crossing.

Inbreeding has been used as a method for creating breeding lines in strawberry, and has been attempted to a limited extent for new cultivar production, though the cultivars produced did not survive (Scott and Lawrence, 1975). Lack of vigour is the main problem in working with inbred lines. Inbred selections are difficult to keep, as they may be affected quickly by viral infection due to their weakness. However, crosses between S1 selected lines yielded desirable selections. Inbreeding is the best way to concentrate genes for a few particular characters and is the most appropriate tool for genetic studies (Spangelo *et al.*, 1971; Aalders and Graig, 1968).

Largely because of all the limitations mentioned above, outcrossing techniques are the most widely used for the improvement of strawberry cultivars. The methods are always time-consuming and based on selection of the seedling which has the desired characters. Selection pressure is exerted simultaneously on each character. This necessitates testing of large numbers of seedlings for the target environment. Spangelo *et al.* (1971) proposed a two-step breeding procedure, which involves small scale testing of progenies, followed by production of larger numbers of progenies of the best ones, to make progress.

Breeding objectives have traditionally been centered on yield and disease resistance. Most of the plant and fruit characters of strawberry are inherited quantitatively, with several genes for each character. Some studies showed that the additive, dominance and epistatic components affected the expression of particular characters. The list of conventional breeding objectives includes yield components, suitability for mechanical harvesting, sex expression, variagation, runner production, hardiness, time of flowering and ripening, adaptability, rest requirement, pest and disease resistance, and quality parameters (size, colour, firmness, shelf life, use, vitamin C content, and *Botrytis* resistance) (Scott and Lawrence, 1975). The consumer pressure for less use of chemicals in pest control puts pressure on breeders to integrate inherent pest resistance in cultivars.

The objective of changing the plant characters dramatically has been tried through the use of *Potentilla* as the male parent. The success was limited due to the small scale of the research and to the sterility of the hybrids.

Bringhurst and Barrientos (1973) reported fertile plants of *F. ananassa* × *P. palistris* ($2n = 14x = 98$). This plant had a strawberry-like character.

4. Strawberry cytology

The widely accepted genomic designation for the octoploid strawberry is AAA′A′ BBBB (Senanayake and Bringhurst, 1967; Bradley, 1982), though Bringhurst (1990) has suggested a modified designation of AAA′A′ BBB′B′. Cytological studies suggest there is essentially normal meiotic behaviour of the diploid type in this crop, despite the octoploid chromosome number (Ibrahim *et al.*, 1981; Byrne and Jelenkovic, 1976). This could be explained by assuming genetic control over chromosome pairing in the octoploid strawberry which results in bivalent association (Byrne and Jelenkovic, 1976).

Genetic mapping of strawberry has been carried out on a diploid species, *F. vesca*, in order to circumvent the genetic complexity of the octoploid cultivars (Davis, 1993). Cytogenetic studies of chromosomal behaviour in cultivated strawberry at meiosis showed bivalent pairing (Ibrahim *et al.*, 1981; Byrne and Jelenkovic, 1976; Mok and Evans, 1971). The low frequency of some multivalent associations has been observed in all genotypes. The configurations of these associations were end-to-end, end-to-side and side-to-side, with no ring or chain configurations. Ibrahim *et al.* (1981) concluded that there was an absence of true homology in octoploid strawberry and attributed this to the effect of environmental factors on the chromosomal association. The presence of genetic control over chromosome pairing can explain the absence of multivalent configuration in *F. moschata*, which has three basic sets of homologous chromosomes (Bhanthamnian, 1965). Byrne and Jelenkovic (1976) suggested the presence of residual homology which is expressed either in advanced generations of selfing or in wide crossings. The absence of multivalents can be explained by preferential pairing which allows only homologous but not homeologous chromosomes to pair.

Islam (1960) and Senanayake and Bringhurst (1967) suggested unreduced gamete production as the possible evolutionary path for polyploid strawberries, including octoploids, in spite of the absence of tetraploid species in the New World. The table produced by Islam (1960) showed that unreduced gametes (pollen and ovule) and chromosome doubling can occur naturally. Ellis (1962) has provided some evidence that three of the four pairs of sets in the octoploid species may be cytologically distinct, but all the polyploid species appear to have at least one common pair of homologous sets.

Unreduced gametes are more common in interspecific and intergeneric hybrids if they have functional male or female flowers (Smith and Jones, 1985b; Jelenkovic *et al.*, 1984; Ellis, 1962; Senanayake and Bringhurst, 1967; Scott, 1951). The matroclinous seedlings which were recorded in interspecific and intergeneric hybridization showed considerable variation and, in general, reduced vigour and fertility relative to the maternal parents. Jelenkovic *et al.* (1984) suggested contamination as the source of octoploids. Doubling of the reduced chromosome number after fertilisation at the early stage of embryo development had been suggested by Asker (1971). The presence of one sterile seedling with 63 chromosomes from the crosses of cv. Tioga ($8x$) × *P. fruticosa* can be explained only on the basis of unreduced female gametes.

5. Haploidy and chromosome reduction in strawberry

Haploidy refers to the condition of any organism, tissue or cell having the chromosome constitution of the normal gametes of the species involved. The first haploid to be reported was found in *Datura stramonium* in 1921 (Blakeslee *et al.*, 1922). As early as 1924 Blakeslee and Belling wrote "Haploids . . . .. afford a new and rapid method of converting a heterogeneous (*sic*) stock into a pure line. Heretofore we have depended upon inbreeding for many generations" (quoted by Chase, 1974).

The origin of haploidy in higher plants might be either spontaneous, or induced. Kimber and Riley (1963) found that the list of reported haploids contained 71 species, representative of 39 genera in 16 families, of which haploids of 39 species occurred spontaneously. They stated that "it is perhaps the greatest obstacle to the wide use of haploids that no technique exists by which forms with gametic chromosome constitution can be freely produced when required". A promising technique became available soon afterwards when Guha and Maheshwari (1964) reported the occurrence of embryos in anther culture of *Datura innoxia*. Normally angiosperm pollen is a highly reduced gametophyte and distends to produce male gametes. Regeneration of pollen into an entire plant, a haploid sporophyte, is an induced function and is possible under experimental and controlled conditions. This culture technique, coupled to a lesser extent with isolated pollen and ovule culture, has been extended to over 200 plant species (Dunnwell, 1986).

Wide crosses have been used frequently to produce haploid plants. The mechanism by which this works and the inheritance of haploid production ability is poorly understood in most crops. The stimulation of parthenogenesis was suggested as a mechanism in potato (Rowe, 1974). Chromosome elimination in interspecific hybrids of *Hordeum vulgare* and *H. bulbosum* have resulted in haploid barley plants (Kasha and Kao, 1970). Chromosome elimination has been observed in interspecific hybrids in many genera, but it is not known whether the elimination was preferentially of the chromosomes of a particular genome, which would be required for the technique to have value (Davies, 1974). Only controlled elimination of one parental genome can lead to haploids which are useful in genetic studies and applications.

Haploids of higher plant species are sporophytes which have the gametophytic chromosome number. It was suggested that haploids derived from diploid species should be called monoploids, and those from polyploid species called polyhaploids. When considering differences between species in tolerance of haploidy, it is important to recognise that, except in certain aneuhaploids, every locus is in the hemizygous condition. In the case of outbreeding diploid species, recessive deleterious alleles, carried as part of the genetic load, will be exposed in haploids. A proportion of such haploids will, therefore, be non-viable or grossly abnormal. This will not occur in haploids derived from inbreeding, or in polyploids, because the parents are homozygous or polyploids respectively, in which there is informational redundancy. Thus the likelihood of obtaining viable haploids can be expected to relate to the breeding system and to the ploidy of the species, increasing with inbreeding and with increasing ploidy.

Strawberry plants have the elements to allow successful and easy polyploid production. It is generally believed that allopolyploid species originated by interspecific hybridization between two distantly related species, followed by doubling of the chromosome number, thereby restoring fertility to an otherwise sterile hybrid. Such an evolutionary history suggests that allopolyploid species may contain considerable duplication of all loci that were common in the parental species. The extent to which such duplication exists in modern cultivars depends upon crosses and diploidisation of duplicated material. The presence of duplicated genetic material may well be the reason for the ease with which haploids can be produced, and for the high survival rate of haploids in allopolyploid species. Various studies to estimate genetic variance in the allopolyploid species have shown most of the intra-allelic effects to be additive and most of the inter-allelic effects to be zero. In other words, without either type of interaction, duplicate genes at separate loci in a haploid would be just as efficient as two genes at the same locus in a diploid (Rowe, 1974). Though most of these elements are present in *Fragaria* × *ananassa*, there is no published record of spontaneous polyhaploid production in this species.

Haploids of *Fragaria* have been reported from interspecific hybridizations of *F. vesca* (Islam, 1960), from intergeneric hybridizations of *F.* × *ananassa* (Barrientos-Perez and Bringhurst, 1973; Hughes and Janick, 1974; Niemirowicz-Szczytt, 1984a,b, 1987) and *F. moschata* (Smith and Jones, 1985a), and from anther culture of *F. orientalis* (Anonymous, 1981) and *F.* × *ananassa* (Niemirowicz-Szczytt and Zakrzewska, 1980; Niemirowicz-Szczytt *et al.*, 1983; Popescu *et al.*, 1988). The numbers of confirmed haploid plants were small in all cases.

Strawberry plants with 49 chromosomes were produced from anthers, petioles and shoot tips cultured in medium containing para-fluorophenylalanine (Niizeki and Fukui, 1983). Smith and Jones (1985a) reported different aneuploid hybrids from *F. moschata* × *Potentilla fruticosa* crosses. In these studies the numbers of plants produced were very small.

It is thought that haploids, if produced as needed, could result in substantial change in strawberry breeding methods. It would be the most efficient technique for inbred line production (Jones, 1976). Strawberries are highly heterozygous and segregation in their progenies is not predictable. The production of haploids and their subsequent diploid lines would provide increased control and predictability over segregation, thus alleviating some of the vagaries of practical breeding endeavours with polyploid strawberries (Jelenkovic *et al.*, 1984).

6. Biotechnology in strawberry improvement

Tissue culture methods such as somatic hybridization, gene transfer, and somaclonal variation have been tried as methods of cultivar improvement in many crops, including strawberry (Wallin *et al.*, 1993; Jain and Pehu, 1992; Nyman and Wallin, 1992). Simon *et al.* (1987) and Sayegh (1989) reported somaclonal variation among strawberry plants regenerated from calli of anthers, and the variations were carried over by runners. Field evaluation of androgenic clones of selected cultivars has shown a wide range of variants from the original clone. These included variations in earliness, yield components, vigour, fruit firmness, shape and colour, and position of the fruit relative to the leaves (Foley, 1990; Foley and Hennerty, 1993). Nehra *et al.* (1992) found that the amount of somaclonal variation depended both on growth regulator type and concentration, and on callus age. A higher frequency of variants where there was no change in chromosome number was associated with 2,4–D compared with IAA (Sayegh, 1989; Foley, 1990). Temporary phenotypic variation which was not associated with chromosome changes have affected many characters in strawberry (Swartz *et al.*, 1981). Nehra *et al.* (1992) found that the induced variations were not of value but were mostly unstable chimeras; aneuploidy was common.

Strawberry plants propagated through tissue culture have not been used extensively in fruit production due to excessive vegetative growth, which could be carried through to daughter plants. In addition increased runnering and branching, longer petioles, and altered resistance to soil-borne diseases was noticed (Grout and Milliam, 1985; Scott *et al.*, 1985; Shoemaker *et al.*, 1985; Hennerty *et al.* 1983; Swartz *et al.*, 1981). The cause of this altered phenotype is unknown though it has been attributed to a change in the hormonal balance due to the use of growth regulators in the media. Marcotrigiano *et al.* (1984) suggested that, since field-grown strawberry plants are propagated from runners (axillary shoots), some cultivars may exist or arise as periclinal chimeras, originating from a spontaneous mutation in a single histogene of the shoot apical meristem. These cultivars remain stable if propagated by runners. In tissue culture propagation, homogeneous variants could have arisen as adventitious shoots.

The use of gibberellin reduced the number of runners in both *in vivo* and *in vitro* plants (Pankov, 1992; Mohamed *et al.*, 1991). Abscisic acid used exogenously on seedlings and in culture media reduced the runner number, while it had less clear effects on the runners themselves. The abscisic acid

Table 1. Media and growth regulators used successfully for callus induction and plant regeneration in strawberry anther culture

Initiation		Regeneration			References
Medium	Growth Regulators	Medium	Growth Regulators	Ploidy	
LS	NAA + BA	a	–	octo	Nishi *et al.*, 1974; Oosawa *et al.*, 1974, 1981.
LS	IAA + 2,4-D + kin.[b]	GD 1	NAA + kin.	octo	Rosati *et al.*, 1975.
GD 1	NAA + kin.	GD 1	NAA + kin.	octo	Rosati *et al.*, 1975.
GD 1	NAA + kin.	Kartha	BA	octo	Laneri & Damiano, 1980.
NN	NAA + 2,4-D + BA	MS	NAA + BA	octo	Mokra & Maliarcikova
LS	IAA + 2,4-D + BA	LS	BA	di,tetra, hexa	Niemirowicz-Sczcytt *et al.*, 1983; Niemirowicz-Sczcytt & Zakrewska, 1980.
LS[c]	NAA + BA	LS	BA	hepta[c]	Niizeki & Fukui, 1983.
LS[d]	No details	–	–	octo	Velchev & Milanov, 1984.
NN[d]	No details	–	–	octo	Velchev & Milanov, 1984.
NN	IAA + kin.	–	–	mixo	Douglas, 1985.
MS[d]	IAA + 2,4-D + kin.	–	–	tetra	Popescu *et al.*, 1988.
NN[d]	IAA + kin.	NN	IAA + kin.	octo	Sayegh, 1989
MS, LS	NAA + BA	–	–	octo	Kinugawa *et al.*, 1991a, 1991b.
GD	NAA + kin.	LS	BA, TDZ	hepta,hexa,octo, mixo	Svensson & Johansson, 1994

Legend: a: regeneration occurred in initiation medium; b: kin.=kinetin; c: the reduction in chromosome number was due to inclusion of para-fluorophenylalanine in the medium; d: modified medium.

increased the numbers of flowers in both tissue cultured plants and in runners to the same level. A gibberellin inhibitor (paclobutrazol) increased the flower number significantly in tissue cultured plants (Mohamed *et al.*, 1991).

7. Anther culture of strawberry

Strawberry anther culture has been tried as a technique for haploid production (Table 1), and polyhaploid production was reported by Niemirowicz-Sczcytt *et al.* (1983) and Popescu *et al.* (1988).

The factors affecting successful haploid production include growing environment and mother plant genotype, anther developmental stage, and anther position on the culture medium. Optimal culture medium composition and incubation environment are also critical for successful haploid production.

The first part of the strawberry flower to be affected by unfavourable conditions is the stamen (Darrow, 1966). The number of pollen grains found in sporogenic tissue was lower in plants with roots kept at 10°C than those at 32°C. The shape of pollen grains was affected by the nitrogen source (Ganmore-Neumann and Kafkafi, 1983). Strawberry plants treated with paclobutrazol showed decreased germination of pollen (McArthur and Eaton, 1987).

Scott (1951) showed that pollen grain diameter in strawberry is related to ploidy level, with greater variation in size among interspecific hybrids than among established species. Recent studies suggest that varietal differences in size could mask this relationship (Maas, 1977; Hebda *et al.*, 1988). Also, it was suggested that pollen grain diameter, coupled with pollen morphology, could be used in identification even at varietal level (Maas, 1977).

Douglas (1985) found that callus induction from anthers in the cultivar Aromel varied with the months of the year between October and April. The maximal callusing rate of anthers and the largest sized calli were obtained in April, which was nearest to the natural flowering time. It appeared that an additional variable was day-length.

Donor plant genotype is another factor affecting the response of anthers in culture. There is good evidence which suggests that there is genetic control of events during the culture phase (Brown and Atanassov, 1985). The earliest study on androgenesis of strawberry was by Fowler (1971) on one genotype only (cv. Surecrop), with no response except callusing. Similar procedures were subsequently applied to more than one genotype. Genotypic variation in callusing and plant regeneration, and environmental requirements for callus induction and plant regeneration were demonstrated (Rosati *et al.*, 1975; Nishi *et al.*, 1974; Velchev and Milanov, 1984). Douglas (1985) found a highly significant medium × cultivar interaction when she cultured two cultivars in three media. This reflects the need to optimize media for each cultivar. However, in strawberry, no genetic studies on inheritance of anther culture responses have been done, because the complexity of the octoploid system would make the usefulness of such studies doubtful.

Fowler (1971) cultured strawberry anthers at four stages of bud development which may correspond to the uninucleate to early binucleate stages of microspore development, and the maximal callusing occurred at the earlier stages. Rosati *et al.* (1975) selected anthers at the first haploid mitosis stage of the microgametophyte. Niemirowicz-Szczytt and Zakrzewska (1980) used anthers at different stages up to one day before flower opening. Laneri and Damiano (1980) used anthers from flowers with 5–6 mm long sepals, which corresponded generally to the uninucleate stage. Douglas (1985) used anthers with pollen at the uninucleate stage, testing one anther from each flower for stage confirmation. The result of these studies was that younger anthers gave better callusing and regeneration.

In strawberry anther culture, cold pretreatment of flower buds for 48 h (Rosati *et al.*, 1975) or 4 days (Nicmirowicz-Szczytt *et al.*, 1983) at 4°C have been suggested for higher rates of callus induction and plant regeneration. Douglas (1985) found that cold pretreatment reduced callus formation and growth rate, though this was not statistically significant. There was, however, a highly significant interaction between pretreatment and the subsequent culture environment. Cold pretreatment followed by culture in the dark gave the highest callusing.

Nitsch and Nitsch (1969) (N&N) medium has been used widely in anther culture. This medium was developed specifically for haploid production from pollen grains. The work in our laboratory suggested that there are two key factors affecting the growth of anthers in culture: these are (a) the iron form and concentration and (b) the growth regulators.

In standard Murashige and Skoog (1962) (MS) medium, an iron-phosphate complex will precipitate due to ineffective iron chelating. This can be prevented by changing the ratio of $FeSO_4 \cdot 7H_2O$ to Na_2 EDTA, or by using a more laborious technique for the sterilisation and preparation of chelated solution (Dalton *et al.*, 1983). Iron deficiency in culture medium is a limiting factor for cell division (Nitsch and Nitsch, 1969) and its increase causes necrosis of the anther walls in tobacco (Vagera and Havranek, 1983). The work in our laboratory used 10 μM $FeSO_4{\cdot}7H_2O$ and 100 μM Na_2 EDTA rather than an equimolar concentration as used in the MS formula. For plant regeneration the need for iron is greater, and 50 μM $FeSO_4{\cdot}7H_2O$ and 100 μM Na_2 EDTA may increase the regeneration rate. Equimolar concentrations of 100 μM each, as in the standard MS, will cause the death of anther walls in most cultures. The change of iron source to NaFe EDTA resulted in elimination of the blackening of the anther wall and improved callusing of the anthers. The callus diameter increased with concentrations of up to 200 μM NaFe EDTA. At this high concentration neither precipitation in the medium nor early anther wall necrosis was observed. The recommended concentration of iron is 100 μM in the form of NaFe EDTA.

The other factor affecting the results of anther culture is the growth regulator composition of the inoculation medium (Hennerty *et al.*, 1987). As regeneration always occurred through callus, callus induction was essential. Both auxin and cytokinin were necessary for callus induction, but pronounced varietal differences in the optimal concentrations of these growth regulators for callus initiation were observed. At concentrations of 0.4 mg l^{-1} 2,4–D and 0.5 mg l^{-1} BA, all genotypes used in our study callused in N&N medium supplemented with 100 μM NaFe EDTA. The use of IAA and kinetin, at concentrations of 1.6 mg l^{-1} and 0.4 mg l^{-1}, respectively, induced callus and plant regeneration in many cultivars. Increased concentrations of these growth regulators induced larger calli, but variability in the regenerants was noticeable. A ratio of IAA:kinetin of greater than 1.0 increased callusing and callus growth, while a ratio less than 1.0 increased regeneration.

Strawberry plants can be grown in a wide range of media and respond to various growth regulators at different concentrations (Boxus *et al.*, 1983). This also applies to the regeneration of plantlets from callus, though callus induction needs a higher concentration of auxin (Kinugawa *et al.*, 1991a,b). The presence of both auxin and cytokinin is a prerequisite for successful callus induction in a wide range of strawberry tissues with genotypic differences (Jones *et al.*, 1988; Rugini and Orlando, 1992).

Nehra *et al.* (1992) reported phenotypic variations in strawberry plants regenerated from callus initiated from leaf discs of cv. Redcoat at concentrations of 20 μM 2,4–D and BA, while no distinct phenotypic variants were observed at growth regulator concentrations lower than 10 μM.

In strawberry anther culture, various media have been used by different authors (Table 1). Linsmaier and Skoog (1965) medium (LS) has been used successfully most frequently (Rosati *et al.*, 1975; Nishi *et al.*, 1974; Oosawa *et al.*, 1974, 1981; Niemirowicz-Szczytt and Zakzewska, 1980; Niizeki and Fukui, 1983). Other media used were Gresshoff and Doy (1972) (GD1) by Rosati *et al.* (1975) and Laneri and Damiano (1980), and N&N medium by Fowler *et al.* (1971) and Douglas (1985). MS and LS media have high salt concentrations, while GD1 and N&N are intermediate. Minor elements are the same except for N&N, where KI and $CoCl_2$ are not present. The essential vitamin appears to be thiamine, which is the only one present in LS. The medium richest in vitamins is the N&N medium. The carbohydrate source is sucrose in all these media. Sucrose represents 92% of the phloem exudate from strawberry pedicels (Forney and Breen, 1985) and 15 Rosaceae species of 17 had better callus grown in media supplied with sucrose rather than sorbitol (Coffin *et al.*, 1976). Douglas (1985) found no significant differences when sucrose was supplied at different rates between 20 and 80 g l^{-1}.

Boron is an important element in pollen germination and pollen-tube growth. Fowler (1971) studied the effects of different concentrations of boron included in N&N medium on callus induction from strawberry anthers. He found that the optimal concentration was 1.008 mg l^{-1}.

Douglas (1985) tested LS, GD1 and N&N media with anthers from the cultivars Aromel and Cambridge Favourite. She found that the choice of induction medium had a significant effect on both the percentage of calli induced and the callus growth rate. The media tested had different hormonal compositions and N&N, her best medium, had 1/10 of the $Fe_2SO_4 \cdot 7H_2O$ contained in GD1 and LS media. The iron concentration may have been responsible for this significant effect. Iron was essential for growth and bud formation in *in vitro* propagation of strawberry. The absence of other microelements (B, Mn, Zn, Cu, Mo, Co, I) from culture media had no significant effect on strawberry micropropagation (Lee and De Fossard, 1977), although this diverges from the requirement for B reported by Fowler (1971) for anther callusing.

In all studies of strawberry anther culture, a combination of auxins and cytokinins was used, usually 2,4-D, NAA, or IAA as the auxin and BA, or kinetin as the cytokinin. The presence of both auxin and cytokinin was essential for callus induction in strawberry anther culture (Oosawa *et al.*, 1974; Hennerty *et al.*, 1987), in contrast to other strawberry explants which needed either auxin or cytokinin (Nishi and Oosawa, 1973; Foucault and Letouze, 1987; Jones *et al.*, 1988). Auxin had an inhibitory effect on shoot formation in these studies. Regeneration media had a higher concentration of cytokinin (Table 1).

Generally, callus induction and maximal callus proliferation required different auxin concentrations; 2,4-D has been the most effective auxin overall. NAA was as effective in callus induction as IAA, but was much more effective in stimulating callus proliferation (Fowler *et al.*, 1971; Oosawa *et *al.*, 1974). The treatments which maximized callusing did not necessarily maximize regeneration (Hennerty *et al.*, 1987).

The amino acid glycine is generally included in media, except for the N&N medium which has no amino acids. Studies in strawberry single cell cultures showed that inclusion of glutamine and adenine in the medium increased cell division (Sarwar, 1984). Vitamins, amino acids and inositol showed no significant effect on growth of meristems of strawberry (Lee and De Fossard, 1977).

In addition to optimisation of culture media, optimisation of the culture environment is important. The recommended culture temperature in strawberry micropropagation and callus regeneration is 21–25°C. In anther culture the temperature used for callus induction has been 24–25°C. Control of the three light parameters, duration, intensity, and spectral quality, has often been necessary to regulate morphogenetic processes in tissue culture.

For strawberry anther culture, researchers have used different light intensities and photoperiods. Fowler *et al.* (1971) found greater callus induction in darkness. Rosati *et al.* (1975) confirmed this and reported a light requirement for regeneration. Douglas (1985) found that incubation in the dark had given a significantly higher response only when accompanied by cold pretreatment. Other authors incubated their cultures in different regimes. An intensity of 300 lux and a photoperiod of 12 h was used by Oosawa *et al.* (1974, 1981), Oosawa and Takayanaji (1982) and Niizeki and Fukui (1983). A continuous low light intensity of 200 lux was used by Niemirowicz-Szczytt and Zakrzewska (1980), and darkness by Laneri and Damiano (1980). Hunter (1982) found that meristem cultures incubated in the light were more prolific, with maximal propagule production at 6000 lux and a 16 h photoperiod.

The work in our laboratory on polyhaploid production and androgenesis in strawberry was intended to repeat the results achieved by the authors Niemirowicz-Sczcystt *et al.* (1983) and Popescu *et al.* (1988), who reported polyhaploid production through androgenesis. We also studied the effects of factors generally affecting haploid production in other crop plants. The work began with 10 cultivars (Saladin, CL1, Pentagruella, Tyee, Tago, Bogota, Elsanta, Gorella, Elvira and Cambridge Favourite) and later concentrated on two cultivars selected for (a) easy callusing and regeneration (Saladin), and (b) difficult callusing and regeneration (breeding line CL1 from Clonroche Soft Fruit Research Station, Ireland). Throughout the work, hundreds of anthers from different treatments were either embedded and sectioned, or squashed, at different time intervals, and stained using different staining techniques. No signs of any type of active division of the microspores beyond the binucleate stage were observed, but the callus, if present, always developed from the somatic tissue of the anther. The most active callusing part was the remnant of the filament which remained attached to the anther, and this callus grew faster than callus generated from other parts of the anther. In the case of an anther divided into two parts – proximal (attached to the filament) and distal – both callused with no significant difference in callusing ability, though the callus size from the proximal part was significantly larger than that from the distal part. Svennson and Johansson (1994) observed up to 12% of cv. Red Gauntlet microspores dividing in culture, but obtained no haploid regenerant.

Foley and Hennerty (1993) evaluated 102 lines derived from anther callus of cv. Saladin and found 26 lines were significantly different for one or more characters. When daughter plants of these lines were studied in a subsequent trial 23 lines retained these significant differences. They concluded that such androgenic plants could contribute to strawberry breeding programmes.

Isolated pollen grains at different stages of growth which were cultured in liquid, solid, and conditioned media of a wide range of compositions, did not result in induction of any type of growth (Sayegh, 1989).

8. Technique for callus induction and plant regeneration

1. Mother plants were grown in soil-based medium.
2. Flower buds were collected at or before the stage where the white petals were just showing, preferably primary buds and during the natural flowering season. This stage of flower development corresponds to the uninucleate stage of pollen development and the anthers are white or have just begun to yellow. The deep yellow coloured anthers are mature and will not respond to culture. Pretreatment of the flower buds for three days at 4°C in moist filter paper in plastic bags can improve the response.
3. Sterilisation of the whole flower buds was carried out in 10% Domestos solution in distilled water (v/v) for 15 min followed by three rinses in sterile distilled water under sterile conditions.
4. Excision of anthers from flower buds was carried out in a laminar air flow cabinet under aseptic conditions. The calyx was cut at its base and an incision was made to remove the sepals. As a result of this cut, pieces of calyx with stamens attached could be easily removed. Single anthers were excised under a dissecting microscope by holding the calyx piece against the dissecting stage and gently pushing individual anthers upward using a fine pointed forceps. This process minimises the size of filament remnant attached to the anther, as well as minimising the damage to the anthers.
5. The anthers were inoculated on modified N&N medium. The modifications were changing the iron source to 100 μM NaFe EDTA and the addition of 20 mg l^{-1} of both L-proline and L-glutamine. Immersion of the anthers in the medium reduced the callus induction capacity of the anther tissue. Most cultivars callused when 0.5 mg l^{-1} 2,4–D and 1.0 mg l^{-1} BA were included in the medium. Using 1.6 mg l^{-1} of both IAA and kinetin induced both callusing and regeneration on the same medium.
6. The cultures were incubated in darkness for three weeks at 20–24°C. Incubation at a lower or a higher temperature in the range 16–32°C also induced callus.

7. This was followed by incubation at 20 ± 4°C at a light intensity of 40 μmol s^{-1} m^{-2} and a photoperiod of 16 h for 1–3 weeks.
8. If at this stage no regeneration had occurred, the callus was transferred to fresh N&N medium.
9. Regenerated plants were multiplied as recommended by Adams (1972).

9. Chromosome staining techniques

Strawberry chromosomes are small, and many techniques for their staining and separation have been described. These techniques were evaluated by Owen and Miller (1993) for use in root tip staining. Collection of 0.5 cm root tips in the morning (10–12 a.m.) was followed by pretreatment in saturated aqueous solution of α-bromonaphthalene, or 2 mM 8-hydroxyquinoline, for 5 h at 14°C followed by overnight fixation in Farmer's fluid (3 ethanol 95% : 1 acetic acid v/v), and hydrolysis for 15 min in 1 N HCl at 60°C. Staining with altered carbol fuchsin produced chromosome preparations superior to those from other treatments including alcoholic hydrochloric acid-carmine (2 days at 25°C). Stained root tips need tapping and squashing for 30–90 s.

In our laboratory the work was done with both field-grown and *in vitro*-rooted plants. Most of the procedures described by the different authors have been tested and a procedure, similar to one later recommended by Owen and Miller (1993), was developed. The best results were obtained with 2 h fixation at 4°C followed by digestion for 30 min in a 5% (w/v) solution of commercial wine pectinase. Extended hydrolysis for 30 min proved essential for good uptake of the stain, especially for root tips from *in vitro*-grown plants. The best stain was aged alcoholic hydrochloric acid-carmine (Snow, 1963).

Even with the best staining techniques, large numbers of cells at the right stage are needed to determine the exact chromosome number of somatic cells of strawberry. This is mainly due to two difficulties: (a) separation of chromosomes, and (b) getting all chromosomes at one focal level. These difficulties may explain the unexpected number of chromosomes reported by some authors.

More recently Nyman and Wallin (1992) used flow cytometry to determine the ploidy level of protoplast derived strawberries. This technique was also used by Svennson and Johansson (1994).

10. Interspecific and intergeneric hybridization

As mentioned previously, there is a wide range of possible hybridizations between different species of strawberry, which in many cases have given rise to viable seeds which developed to flowering plants. Unreduced gametes, which are common in *Fragaria* interspecific hybrids, can induce chromosome doubling, but with strong genetic control of sex expression. Colchicine treatment is a better route for production of fertile plants (Ahmedi and Bringhurst, 1991). Although interspecific hybridization has been used extensively in genetic studies and in genus evaluation, it has not been extensively used for breeding purposes in strawberry, except for work with the two parents *F. chiloensis* and *F. virginiana*. The opportunity to test a wide range of genetic characters at a simple genetic level (tetraploid) and then to utilise these by doubling the chromosome number (octoploid) could make the work of strawberry breeders easier. Evans (1977, 1982a,b) suggested four methods of producing synthetic octoploids from interspecific and trispecific hybridization of di-, tetra-, and hexaploid *Fragaria*. In order to achieve chromosome doubling, the resulting seeds needed colchicine treatment at germination. The rate of success was limited. Although he treated 6564 seeds, the number of established seedlings was 592, and only 6 of them proved to be octoploids. Some of the synthetic octoploids were used in crossings with octoploid cultivars. Evans (1977) stated that "the most striking characteristics of the first generation hybrids are upright flower stalks, drought resistance, and foliage intermediate between the parents in rugosity and colour. It was anticipated that the synthetic octoploids could be used as tester plants to determine the breeding potential of the cultivars, as the *F. moschata* × *F. nubicola* octoploids may be relatively homozygous for many characteristics, as complete genomes of the two parental species are duplicated in them. In this investigation the staminate nature of many of the progeny, and deformation of many fruits due to poor set, reduce the value of the synthetic octoploids as testers for fruit characters, but they may be useful testers for many other characters such as disease resistance, fruiting habit, hardiness and drought resistance".

The small number of octoploid plants in Evans' work might have been increased if *Fragaria* embryo culture and tissue culture techniques had been used. Embryo culture may help overcome many of the factors which result in weak embryo growth, while *in vitro* chromosome doubling should allow larger amounts of material to be available.

The other approach by which chromosomes from *Fragaria* species of lower ploidy have been transferred to cultivated strawberry is through decaploids – suggested by Scott (1951) to improve strawberry aroma – through interspecific hybridization. The best example of this is the cultivar Florika which is recommended by Bauer (1993) as meadow and ground cover. The decaploid approach for strawberry breeding eliminates the use of established strawberry cultivars as it gives 9*x* progeny. This type of progeny will not allow backcrossing which is prerequisite to any cultivar improvement. This also led to Bauer giving the decaploid its own name, *F. vescan*. Although there are many routes for creating them, all decaploids will include one or more of the chromosome complement of the homozygous status. *F. vescan* was produced by the following steps: tetraploids of *F. vesca*, produced by colchicine doubling, were crossed with octoploid *F. ananassa*; the hexaploid progeny, which was slightly fertile, was then backcrossed with octoploid strawberry, and this occasionally gave rise to decaploids due to unreduced gametes of the hexaploid (Bauer, 1993; Scott, 1951). This cross has double the chromosome complement of *F. vesca*.

Interspecific and intergeneric hybrids (with the genus *Fragaria* and the genus *Potentilla*) have been studied to elucidate phlogenetic and taxonomic relationships (Mangelsdorf and East, 1927; Jones, 1955; Harland, 1957; Ellis, 1962; Haskell, 1963; Sangin-Berezovskij, 1963; Senanyake and Bringhurst, 1967; Asker, 1970, 1971), and more recently for haploid production, or the integration of *Potentilla* characters within the cultivated strawberry (Asker and Denward, 1972; Barrientos-Perez and Bringhurst, 1973; Hughes and Janick, 1974; Niemirowicz-Szczytt, 1984a,b, 1987; Jelenkovic *et al.*, 1984; Smith and Jones, 1985a,b). Table 2 has a list of reported crosses between octoploid strawberry and different *Potentilla* species. Few generalisations can be drawn from these works. The seed set of octoploid strawberries diminished with each degree of increased ploidy in the *Potentilla* parent (Haskell, 1963; Asker, 1971). *P. fruticosa* was the most successful male parent and many mature hybrids were obtained. Sublethal hybrids with *P. anserina* were reported and suggested as a tool to screen hybrids (Sangin-Berezovskij, 1963; Barrientos-Perez and Bringhurst, 1973; Hughes and Janick, 1974; Jelenkovic *et al.*, 1984).

In intergeneric crosses of *Fragaria*, hybrids, as well as octoploid plants and a few haploids have been reported. The presence of octoploids has been explained as due to experimental shortcomings, apomixis, or both (Asker, 1971). Fowler and Janick (1972) found that improved isolation and control of wind movement would reduce the number of octoploids. Many mechanisms of apomixis have been proposed. The genus *Fragaria* belongs to the family *Rosaceae* which has a high frequency of apomixis. Work on the embryology of *Fragaria* has confirmed several phenomena associated with apomictic reproduction (Sukhareva, 1970). If these octoploids are matroclinous, a probable explanation is that they arose from aposporous and diplosporous sacs containing gametes with an unreduced chromosome number (Niemirowicz-Szczytt, 1984c). The presence of unreduced gametes was proposed as an explanation for increased ploidy levels in *Fragaria* (Islam, 1960). Naturally occurring hexaploids and nonoploids in *F. chiloensis* × *F. vesca* hybrids would support this theory (Senanayake and Bringhurst, 1967; Bringhurst and Gill, 1970). Asker and Denward (1972) suggested alternatively that restitution takes place at the first egg cell division. This was based on observed embryological behaviour in strawberry after pollination of tetraploids of cultivated strawberries with *P. fruticosa*. Unreduced gametes should lead to the production of true-to-type seeds, while restitution should lead to homozygous line production. None of the studies produced a genetic analysis of the offspring. The genetics of the octoploid strawberry are complicated, and well-expressed genetic markers at the early growth stages are difficult to find (Fowler *et al.*, 1971).

Aalders (1964) suggested utilisation of apomixis in strawberry to produce

Table 2. Octoploid strawberry × *Potentilla* species crosses

Strawberry species	*Potentilla* species	*Potentilla* ploidy	Seed set	Progeny type	Progeny survival	References
10x hybrid	*P. fruticosa*	2x	yes	hyb.	mat.	Harland, 1957
F. x *ananassa*	*P. fruticosa*	2x	yes	?	mat.	"
"	*P. anglica*	8x	yes	–	–	"
F. grandiflora	*P. fruticosa*	2x	yes	hyb.	mat.	Ellis, 1962
"	*P. erecta*	4x	yes	–	germ.	"
"	*P. reptans*	4x	no			"
"	*P. sterilis*	4x	no			"
"	*P. palustris*	6x	yes	hyb.	mat.	"
"	*P. anglica*	8x	yes	–	germ.	"
F. grandiflora	*P. anserina*	4x	yes	hyb.	germ.	Sangin-Berezovskij, 1963
F. x *ananassa*	*P. fruticosa*	2x	yes	hyb.	mat.	Haskell, 1963
F. x *ananassa*	*Potentilla*	2-14x	yes	–	–	"
F. virginiana	"	"	yes	–	–	"
F. chiloensis	"	"	yes	–	–	"
F. grandiflora	*P. glandulosa*	2x	yes	hyb.	mat.	Senanayake & Bringhurst, 1967
F. x *ananassa*	*P. fruticosa*	2x	yes	hyb.	mat.	Asker, 1971
"	*P. rupestris*	2x	yes	–	germ.	"
"	*P. erecta*	4x	yes	octp.	mat.	"
"	*P. anglica*	?	yes	octp.	mat.	"
"	*P. fruticosa*	4x	yes	octp.	mat.	"
"	*P. davurica*	?	yes	octp.	mat.	"
"	*P. palustris*	6x	yes	–	germ.	"
"	*P. anserina*	4x	yes	octp.	mat.	"
F. x *ananassa*	*P. anserina*	4x	yes	hap. + hyb.	mat. germ.	Barrientos-Perez & Bringhurst, 1973
"	*P. fruticosa*	2x	yes	hyb.	mat.	Barrientos-Perez, 1973
F. chiloensis	*P. glandulosa*	2x	yes	hyb.	mat.	"
"	*P. vrerna*	?	yes	–	–	"
F. x *ananassa*	*P. anserina*	4x	yes	hyb.	germ.	Hughes & Janick, 1974
				hap.	mat.	
				octp.	mat.	
F. x *ananassa*	*P. fruticosa*	2x	yes	hyb.	mat.	"
				hap.	mat.	
				octp.	mat.	
F. virginiana	*P. fruticosa*	2x	yes	–	mat.	Niemirowicz-Szczytt, 1984a, 1987
F. x *ananassa*	*P. geoides*	2x	yes	3,4,5, 6,8x	mat.	"
"	*P. rupestris*	2x	yes	2,3,4, 5,6,8x	mat.	"
"	*P. fruticosa*	2x	yes	4,5,6x	mat.	"
"	*P. glandulosa*	2x	yes	4,5x	mat.	"
"	*P. andicola*	6x	yes	–	germ.	"
"	*P. arguta*	2x	yes	–	mat.	"
"	*P. aurea*	?	yes	–	germ.	"
"	*P. purpur-eoites*	?	yes	4,5, 6,8x	mat.	"
"	*P. fragiformis*	6x	yes	–	mat.	"
"	*P. thur-ingiaca*	?	yes	–	germ.	"
"	*P. ambigens*	?	yes	–	germ.	"

Table 2. Continued

Strawberry species	*Potentilla* species	*Potentilla* ploidy	Seed set	Progeny type	Progeny survival	References
"	*P. atrosan-guinea*	?	yes	–	0	"
"	*P. nitidia*	?	yes	–	germ.	"
"	*P. nepalensis*	?	yes	–	germ.	"
"	*P. chry-santha*	?	yes	–	germ.	"
"	*P. argentea*	?	yes	–	germ.	"
"	*P. clusiana*	6x	yes	–	0	"
F. x *ananassa*	*P. fruticosa*	2x	yes	hyb.	mat. octp.	Jelenkovic *et al.*, 1984
"	*P. anserina*	4x	yes	octp.	mat.	"
F. virginiana	*P. fruticosa*	2x	yes	hyb.	mat. octp.	"
"	*P. anserina*	4x	yes	–	germ.	"
F. x *ananassa*	*P. fruticosa*	2x	yes	4,5x, aneu.	germ.	Sayegh & Hennerty, 1993

Legend: ?: no count for chromosome number was indicated; – : data not available because progeny died before maturity; hyb.: hybrid; octp.: octoploid; mat.: some progeny survived to maturity; germ.: some seeds germinated; 0: none of the seeds germinated; aneu.: aneuploid.

true-to-type seeds and eliminate non-seed-borne viral infections, and East (1930) suggested production of homozygotes through induced parthenogenesis. Lord and White (1962) attempted to promote apomixis in the strawberry by treatment with growth regulators. They obtained a number of seeds with embryos which gave octoploid plants and which differed from maternal plants in many characters. They concluded that these seeds resulted from accidental fertilisation. This could apply to other studies. Strawberry receptacles are capable of developing parthenogenetically in the presence of an exogenous auxin source (Nitsch, 1950; Tafazoli and Vince-Prue, 1979; Mudge *et al.*, 1981; Beech, 1983). Beech (1983) found that achenes removed from parthenocarpic fruits contained embryos which either withered or died, or did not develop. This has been confirmed by the work in our laboratory. The explanation for the appearance of haploids after pollination with pollen of different species may only be found in haploid pseudogamy (Sukhareva, 1970). Sangin-Berezovskij (1963) attributes the appearance of false hybrids to hybrid gynogenesis, that is, the phenomenon in which the pollen nuclei penetrate the embryo sac, but where the fusion of male and female nuclei does not take place.

Smith and Jones (1985a) worked with *F. moschata* (6*x*) × *P. fruticosa* (2*x*) crosses. The surviving progeny were expected to be hybrids, but considerable variation was found. Four plants had 28 chromosomes (hybrids), four were aneuploids with 23, 24, 25, and 27 chromosomes and one plant had 21 chromosomes (haploid). The presence of aneuploids could be explained by chromosome elimination. The progeny differed in vigour. All plants that flowered were male sterile, but pollen of both parents was able to germinate on their stigmas and the pollen-tubes were able to grow down the style to the region of the embryo sac, although they needed a much longer time than normal. Seeds developed, but none had a viable embryo. When the embryo culture technique was used to rescue some of the embryos at earlier stages of development, a number of progeny from the plant with 63 chromosomes were grown to maturity (Smith and Jones, 1985b). The chromosome number of these plants could be explained only on the basis of unreduced and doubled unreduced gametes.

11. Strawberry embryo culture

Seed set in most intergeneric hybridizations was reasonable, but seed germination and seedling viability were low. Smith and Jones (1985b) reported that they were able to get some plants from backcrosses of their *F. moschata* × *Potentilla* hybrids via embryo culture only. Wang *et al.* (1984) were successful in growing embryos from embryogenic callus on MS medium supplemented with casein hydrolysate and 1.0 mg l^{-1} GA_3. GA_3 also resulted in development of immature embryos. This medium was the most successful in embryo rescue used by Sayegh and Hennerty (1993).

Miller *et al.* (1992) found that cutting of surface-sterilised achenes of ripe fruits across the embryo axis, and then culturing the section of the achene containing the shoot apex on semi-solid MS medium without growth regulators, led to 97 to 100% germination of different crosses in five to fourteen days. This compared with seven days to seven weeks, and a maximum of 50% germination, in the case of whole achenes. The embryos collected from white or pink unripe fruits exhibited slower germination. The embryos from green fruits did not germinate. The inclusion of growth regulators (5 μM NAA, 5 μM BA, 5 μM BA + 5 μM NAA, or 1 μM GA_3) induced callus. Smith and Jones (1985b) reported the use of embryo cultures to obtain plants from backcrosses of *F. moschata* × *Potentilla* hybrids. Wang *et al.* (1984) successfully developed embryos to plantlets from embryogenic strawberry callus upon subculturing on GA_3 supplemented medium.

Strawberry achenes often exhibit variable and low germination rates. In experimental work this has resulted in comparisons of seedlings of different ages and the likely loss of valuable genotypes. Although many techniques have been suggested to improve germination, they are inefficient and genotype-dependent. Embryo culture offers the opportunity to increase germination and obtain a more uniform stand of seedlings. Embryo culture was done in our laboratory on a wide range of MS-based media. The inclusion of auxin and cytokinin enhanced growth and could induce callus, a phenomenon which is cultivar-dependent. GA_3 enhanced growth of the main shoot. None of the whole achenes germinated, possibly due to the phenolic exudate from the achene, which concentrates around the embryo in the medium. This exudate does not affect the germination in soil as it is probably absorbed by the growing medium.

In wide crosses, hybridization stimulates parthenogenesis and in most cases fertilisation of the polar nuclei occurs and endosperm forms (Rowe, 1974). Incompatibility or poor development of the endosperm and/or embryo requires additional nutrition to the embryo, which can be achieved through embryo culture.

Our work has shown that embryos resulting from self-pollination of strawberry plants can germinate *in vitro* with 100% success if excised from achenes pericarp and testa at the end of the third week after fertilisation (green fruit stage). It also confirmed the need for 1.0 mg l^{-1} GA_3 to enhance the growth of the main shoot and induce callusing in some genotypes.

12. Intergeneric hybridization for haploid production

Wide crossing is the conventional haploid production route which has been used extensively in strawberry breeding programmes for many objectives, e.g., evaluation of cytogenetics of strawberry, transferring of some required characteristics and sometimes for tetrahaploid production. The genus *Potentilla* is related to *Fragaria* and has the same basic number of chromosomes ($x = 7$) with different ploidy levels (Elkington, 1969; Harrison *et al.*, 1993). *P. fruticosa* is a small shrub having a polyploid complex of di-, tetra-, and hexaploid. Diploid *P. fruticosa* have been the most successful male parents in strawberry crosses and many mature hybrids have been produced, while *P. anserina* has only produced weak hybrids. The seed set of octoploid strawberries diminished with each degree of increased ploidy of *Potentilla* when it was used as the pollinator. The only reliably reported tetrahaploids were produced by crossing with either *P. fruticosa* ($2x$) or *P. anserina* ($4x$). The number of haploids compared to the number of total seedlings was very small and had no impact on breeding programmes (Hughes and Janick, 1974; Barrientos-Perez and Bringhurst, 1973).

The relatively large number of seeds produced has little impact on the number of viable tetrahaploids. The germination rate was very low with further losses due to weak seedlings. Hughes and Janick (1974) were able to raise two haploids out of 1,984 seeds from 112 crosses. The total number of seedlings in this study which survived the first 24 weeks after germination was 12 (2 tetrahaploids, 9 hybrids and an octoploid). Jelenkovic *et al.* (1984) using 11 lines of octoploid strawberries and 2 lines of *P. fruticosa*, were unable to produce any tetrahaploids, although some hybrids and octoploids were produced in addition to one nonoploid ($9x$) which displayed hybrid characters. Sayegh and Hennerty (1993) suggested that the loss in seed viability was due to either the genetic load of octoploid strawberry or poor development of the embryo. The former can be eliminated by using vigorous strawberry plants selfed for a few generations, and the latter by using embryo rescue techniques. By using this technique, we were able to recover both hybrids and tetrahaploids from a limited number of crosses. Embryo rescue is valuable in increasing the germination rate of both hybrid and tetrahaploid embryos. The appearance of different aneuploids, following intergeneric hybridization of octoploid strawberries, has recently been reported by Sayegh and Hennerty (1993). The presence of aneuploids suggests some kind of chromosomal elimination in the hybrids. The elimination of the *Potentilla* chromosome complement could result in haploid production. This chromosome elimination could take place at a very early stage of embryo development and could be under the genetic control of one parent (Davies, 1974). The strawberry is the likely female parent because its excessive heterozygosity and high ploidy gives a large spectrum of variability among the megaspores. Thus the presence of hybrids, aneuploids or haploids is to be expected due to the absence or presence of such controlling genes or gene complexes in strawberry. We suggested the feasibility of tetrahaploid production through *F. ananassa* × *Potentilla*. Embryo culture may eliminate some problems associated with germination. The identification of the plant type is not difficult.

13. Intergeneric hybridization technique

1. Mother plants of both *Fragaria* and *Potentilla* were forced by keeping plants under long days at a temperature of 24 ± 2°C. Forcing of strawberry plants in a protected environment eliminated the possibility of cross-pollination as long as unfertilised flowers were removed and emasculation was done prior to anther dehiscence.
2. Strawberry flowers which had not opened, but were just showing white petal tips, were at the right stage for emasculation. The anthers are uninucleate at this stage and are not as hard as they become after further development. Emasculation, preferably in the early morning, was done by using a fine pointed forceps to pull away the petals, then a scalpel was used to make an incision around the base of the sepals. The removal of the sepals resulted in the removal of most of the anthers with filaments attached. If some anthers remained they could be pulled upward and removed from the filaments. This procedure occasionally resulted in dehiscence of the anther. In such an event, the flower should be removed and discarded. The most successful pollination, in terms of seed setting, was in the primary flowers, followed by secondary flowers. Pollen of *Potentilla* was collected from fully open flowers. This pollen can be used immediately, or stored for up to four days at 4°C. Pollination was done by rubbing either whole *Potentilla* flowers over emasculated strawberry flowers, or by using a camel hair brush laden with *Potentilla* pollen. This process was repeated daily for three days. If the stigmas of the strawberry flower stayed yellow for one week, then the pollination was repeated on

the eight day. Controls of emasculated non-pollinated flowers were left, to monitor possible airborne cross pollination.

3. Fruit development is dependent on the number of successfully pollinated stigmas and on the number of developing achenes. Achene development is proportional to embryo development. Four weeks after the first pollination, whole strawberry fruits could be harvested and surface sterilised for embryo culture. Earlier than this it is difficult to excise embryos without causing damage to them. The longer the fruit remains on the mother plant, the greater the reduction in the number of live embryos. To separate the achenes containing live embryos from those which contain dead ones, simply remove the achenes from the fruit and place them in sterile distilled water. Those which sink contain live embryos (Morrow *et al.*, 1954). Floating achenes rarely contain live embryos.
4. Surface sterilisation of fruits was done using 10% (v/v) Domestos solution in sterile distilled water for 12 min, followed by three rinses in sterile distilled water. Embryos were excised from the achenes under sterile conditions. First the achenes were removed from the surrounding berry tissues with a fine-pointed forceps. Under a dissecting microscope an incision was made along the edge of the pericarp with a surgical scalpel and light pressure was applied to the achene to release the embryo. This step is the most crucial in avoiding damage to the embryo. The testa was carefully removed and the embryo placed on the medium.
5. The culture medium was standard MS medium supplemented with 200 mg l^{-1} casein hydrolysate and 1 mg l^{-1} GA_3. Addition of 0.2 mg l^{-1} IAA improved root development. The cultures were incubated at 24 ± 4°C under a 16 h photoperiod of 40 μmol s^{-1} m^{-2} for 4–5 weeks. Subsequently they were subcultured to the standard strawberry propagation medium of Adams (1972) at four week intervals.
6. Samples of root tips collected from cultures were analysed for chromosome number using the technique described earlier.

14. Tetrahaploid plant characters

Tetrahaploid plants produced from *F.* × *ananassa* resembled the maternal parents, but less vigorous, and comparable to weak plants produced by selfing strawberries (Barrientos-Perez and Bringhurst, 1973). The leaf length to width ratio was the main character by which hybrids could be distinguished from non-hybrids. This ratio was larger in the hybrids and near unity in juvenile strawberry leaves. The serration index which was suggested by Barrientos-Perez and Bringhurst (1973) was largest in the hybrids and smallest in the octoploids. This index is the ratio of the non-serrated edge length (basal portion of leaflet) to the total edge length of the leaflet. In the tetrahaploids, this index was intermediate between hybrids and octoploids, with a tendency to increase in aneuploids, as observed in our laboratory.

This can be used as a quick method for separating haploids from both hybrids and octoploids (Hughes and Janick, 1974).

15. Conclusions

Polyhaploid production in strawberry is feasible through intergeneric hybridization using octoploid *F.* × *ananassa* and *P. fruticosa* or *P. anserina*. The polyhaploid frequency has been low and the plants weak. Embryo culture can increase the frequency of survival of both hybrids and polyhaploids. Phenotypic characters based on leaf morphology make it possible to distinguish hybrids, polyhaploids, and octoploids from each other. A reduction in the number of serrations and serrations confined to the leaf apex indicate chromosome reduction. The leaf length to width ratio is near unity in both octoploids and polyhaploids while it is around 1.2 in hybrids.

To eliminate some of the genetic load and some of the deleterious genes from the strawberry mother plant, the use of selfed lines is required. The hybrids and the polyhybrids can be multiplied, and chromosome doubling achieved using colchicine in the medium. Tissue culture techniques can contribute to reduction of both time and effort needed in a breeding programme aimed at achieving high homozygosity in octoploid strawberries.

An approach to produce octoploids that have a relatively homozygous genetic structure has been the doubling of lower polyploids (Evans, 1977). This technique includes interspecific hybridization of different *Fragaria* spp., which can produce tetrahaploids either directly or through chromosome doubling. This should be followed by testing these tetrahaploids in the field before creating synthetic octoploids. In addition to the opportunity of testing the tetrahaploids at the lower level, this technique offers an easy means of octoploid production either by colchicine or by unreduced gametes. Unreduced gametes are widely reported in interspecific hybrids of *Fragaria*.

Another approach to obtain some degree of homozygosity in strawberry is breeding at decaploid level by pentaploid doubling. This approach eliminates the use of available and evaluated germplasm at the octoploid level, and increases the time needed to create the large gene pool of germplasm needed to produce the required variation for commercial cultivar production. Also, these decaploids, which can be fertile and are able to cross with octoploids, will produce nonoploids of little breeding value.

16. Acknowledgements

We wish to express our appreciation to David Llewellyn and Tom Moore, University College Dublin, for their help in preparing the manuscript.

17. References

Aalders, L., 1964. Production of maternal-type plants through crosses to apomictic species. Nature (London) 204: 101–102.

Aalders, L.A. and D.L. Graig, 1968. General and specific combining ability in seven inbred strawberry lines. Can. J. Genet. Cytol. 10: 1–6.

Adams, A.N., 1972. An improved medium for strawberry meristem culture. J. Hort. Sci. 47: 263–264.

Ahmedi, H. and R.S. Bringhurst, 1991. Genetics of sex expression for *Fragaria* species. Amer. J. Bot. 78: 504–514.

Anonymous, 1981. Haploid plants of strawberry derived from anther culture. China Institute of Pomology, Plant Journal 4, 10. In: Plant Breed. Abstr. 52, 5223.

Asker, S., 1970. An intergeneric *Fragaria* × *Potentilla* hybrid. Hereditas 67: 181–190.

Asker, S., 1971. Some viewpoints on *Fragaria* × *Potentilla* intergeneric hybridization. Hereditas 67: 181–190.

Asker, S. and T. Denward, 1972. Strawberry breeding and cytogenetics of *Fragaria*. Lantbrukshsskolan, Advelningen for Voxforadling av Frukt och Bor. Annual Report 1972, 21–24. Plant Breed. Abstr. 44,06173.

Barrientos-Perez, F., 1973. Development and exploitation of interspecific *Fragaria* and intergeneric *Fragaria-Potentilla* amphiploids in strawberry. Ph.D. Thesis, University of California, Davis.

Barrientos-Perez, F. and R.S. Bringhurst, 1973. A haploid of an octoploid strawberry cultivar. HortScience 8: 44.

Bauer, A., 1993. Progress in breeding decaploid *F. vescana*. Acta Hort. 348: 60–74.

Beech, M.G., 1983. The induction of parthenocarpy in the strawberry cultivar Red Gauntlet by growth regulators. J. Hort. Sci. 58: 541–545.

Bhanthamnian, K., 1965. A cytological study of polyploidy in the genus *Fragaria*. Ph.D. Thesis, University of Reading, Reading.

Blakeslee, A.F., J. Belling, M.E. Farnham and A.D. Bergner, 1922. A haploid mutant in Jimson weed, *Datura stramonium*. Science 55: 1433.

Boxus, P., C. Damiano and E. Brasseur, 1983. Strawberry. In: P.V. Ammirato, D.A. Evans, W.P. Sharp and Y. Yamada (Eds.), Handbook of Plant Cell Culture. Vol. 3, pp. 453–486. Macmillan Publishing Co., New York.

Bradley, G.L., 1982. Giemsa-band karyotyping and the cytogenetics and evolution of *Fragaria* species. Ph.D. Thesis, University of California, Davis.

Bringhurst, R.S., 1990. Cytogenetics and evolution in American *Fragaria*. HortScience 25: 879–881.

Bringhurst, R.S. and F. Barrientos, 1973. Fertile *F. chiloensis* × *P. glandulosa* amphiploids. Genetics 74: 530.

Bringhurst, R.S. and T. Gill, 1970. Origin of *Fragaria* polyploids. II. Unreduced and double-unreduced gametes. Amer. J. Bot. 57: 960–976.

Bringhurst, R.S. and V. Voth, 1984. Breeding octoploid strawberries. Iowa State J. Res. 58: 371–381.

Brown, D.C.W. and A. Atanassov, 1985. Role of genetic background in somatic embryogenesis in *Medicago*. Plant Cell Tiss. Org. Cult. 4: 111–122.

Byrne, D. and G. Jelenkovic, 1976. Cytological diploidization in the cultivated octoploid strawberry *Fragaria* × *ananassa*. Can. J. Genet. Cytol. 18: 653–659.

Chase, S.S., 1974. Production of homozygous diploids of maize from monohaploids. Agron. J. 44: 263–267.

Coffin, R., C.D. Tper and C. Chang, 1976. Sorbitol and sucrose as carbon source for callus culture of some species of Rosaceae. Can. J. Bot. 54: 547–552.

Dalton, C.C., K. Iqbal and D.A. Turner, 1983. Iron-phosphate precipitation in MS media. Physiol. Plant. 57: 472–476.

Darrow, G.M., 1966. The Strawberry. Holt, Rinehart, Winston, New York.
Davies, D.R., 1974. Chromosome elimination in interspecific hybrids. Heredity 32: 267–270.
Davis, T.M., 1993. Genetic mapping of the diploid strawberry using random amplified polymorphic DNA (RAPD) markers. Acta Hort. 348: 439–440.
Douglas, A.J., 1985. The production of strawberry plants by androgenesis. M. Agr. Sc. Thesis, University College, Dublin.
Dunwell, J.M., 1986. Pollen, ovule and embryo culture as tools in plant breeding. In: P.G. Alderson and L.A. Withers (Eds.), Plant Tissue Culture and its Agricultural Applications. Proceedings of the 41st University of Nottingham Easter School of Agriculture held in 1985, pp. 375–403. Butterworths, London.
East, E.M., 1930. Production of homozygotes through induced parthenogenesis. Science 72: 148–149.
Elkington, T.T., 1969. Cytotaxonomic variation in *Potentilla fruticosa*. New Phytol. 68: 151–160.
Ellis, J.R., 1962. *Fragaria-Potentilla* intergeneric hybridisation and evolution in *Fragaria*. Proc. Linn. Soc. (London) 173: 99–106.
Evans, W.D., 1977. The use of synthetic octoploids in strawberry breeding. Euphytica 26: 497–503.
Evans, W.D., 1982a. Guelph SO1 synthetic octoploid strawberry breeding clone. HortScience 17: 833–834.
Evans, W.D., 1982b. Guelph SO2 synthetic octoploid strawberry breeding clone. HortScience 17: 834.
Foley, I., 1990. Field performance of androgenic strawberry plants. M. Agr. Sc. Thesis, University College, Dublin.
Foley, I. and M.J. Hennerty, 1993. Field performance of androgenic strawberry plants. Acta Hort. 348: 188–195.
Forney, C.F. and P.J. Breen, 1985. Collection and characterisation of phloem exudate from strawberry pedicels. HortScience 20: 413–414.
Foucault, C. and R. Letouze, 1987. *In vitro* regeneration de plants de fraisier à partir de fragment de petioles et de bourgeons floraux. Biol. Plant. (Praha) 26: 409–414.
Fowler, C.W., 1971. Feasability studies for chromosome reduction in strawberry. Ph.D. Thesis, Purdue University, Purdue.
Fowler, C.W. and J. Janick, 1972. Wide crosses and pollen contamination in strawberry. HortScience 7: 409–414.
Fowler, C.W., H. Hughes and J. Janick, 1971. Callus formation from strawberry anthers. Hort. Res. 11: 116–117.
Galletta, G.J. and J.L. Maas, 1990. Strawberry genetics. HortScience 25: 871–879.
Ganmore-Neumann, R. and U. Kafkafi, 1983. The effect of root temperature and NO_3:NH_4 ratio on strawberry plants. Growth, flowering and root development. Agron. J. 75: 971–947.
Gooding, H.J., 1975. Soft fruit breeding. Sci. Hort. 27: 9–15.
Gooding, H.J., 1980. Strawberry cultivars – past, present and future. Scot. Hort. Res. Inst. Bull. 17: 1–7.
Gresshoff, P.M. and C.H. Doy, 1972. Development and differentiation of haploid *Lycopersicon esculentum* (tomato). Planta (Berl.) 107: 161–170.
Grout, B.W. and S. Milliam, 1985. Photosynthetic development of micropropagated strawberry plantlets following transfer planting. Ann. Bot. 55: 129–131.
Guha, S. and S.C. Maheshwari, 1964. *In vitro* production of embryoids from anthers of *Datura*. Nature 204: 249.
Hancock, J.F., 1990. Ecological genetics of natural strawberry species. HortScience 25: 869–871.
Hancock, J.F. and J.J. Luby, 1993. Genetic resources at our doorstep: the wild strawberries. BioScience 43: 141–149.
Hancock, J.F., A. Dale and J.J. Luby, 1993. Should we reconstitute the strawberry? Acta Hort. 348: 86–95.

Harland, S.C., 1957. Cytological investigation on soft fruits. Progress report of University of Manchester, Dept. of Botany for the period ending 31st May 1957, p. 6. In: Plant Breed. Abstr. 29, 1950.

Harrison, R.C., J.J. Luby and G. Furnier, 1993. Molecular investigation of the chloroplast genome in *Fragaria* spp. Acta Hort. 348: 395–402.

Haskell, G., 1963. Strawberry investigations. Tenth Annual Report, The Scot. Hort. Res. Inst.: 62.

Hebda, R.J., C.C. Chinnappa and B.M. Smith, 1988. Pollen morphology of the Rosaceae of western Canada. II. *Dryas*, *Fragaria*, *Holdiscus*. Can. J. Bot. 66: 595–612.

Hennerty, M.J., A.J. Douglas and A.J. Sayegh, 1987. Studies on androgenesis in strawberry. In: Proceedings of the Irish Botanists Meeting, 25–27 March, 1987, Cork, pp. 76–80.

Hennerty, M.J., S.A. Hunter and M.J. Foxe, 1983. Field performance of tissue cultured strawberry plants. In: P. Boxus and P. Larvor (Eds.), *In Vitro* Culture of Strawberry Plants, EUR 10871, pp. 41–46. CEC, Brussels.

Hughes, H.J. and J. Janick, 1974. Production of tetrahaploids in the cultivated strawberry. HortScience 9: 442–444.

Hunter, S.A., 1982. Micropropagation of the strawberry (*F. ananassa*) cv. Cambridge Favourite. Ph.D. Thesis, University College, Dublin.

Ibrahim, A.M.F., K. Sadanaga and E.L. Denisen, 1981. Chromosomal behaviour in octoploid strawberry progenies and their parental clones during meiosis. J. Amer. Soc. Hort. Sci. 104: 522–526.

Islam, A.S., 1960. Possible role of unreduced gametes in the origin of polyploid *Fragaria*. Biologia 6: 189–192.

Jain, S.M. and E. Pehu, 1992. The prospects of tissue culture and genetic engineering for strawberry improvement. Acta Agric. Scand., Sect. B 42: 133–139.

Jelencovic, G., M.L. Wilson and P.J. Harding, 1984. An evolution of intergeneric hybridization of *Fragaria* spp. × *Potentilla* spp. as a means of haploid production. Euphytica 33: 143–152.

Jones, J.K., 1955. Cytogenetics studies in the genera *Fragaria* and *Potentilla*. Ph.D. Thesis, Manchester University, Manchester.

Jones, J.K., 1976. Strawberry *Fragaria ananassa* (Rosaceae). In: N.W. Simmonds (Ed.), Evolution of Crop Plants, pp. 237–242. Longman, London.

Jones, O.P., B.J. Waller and M.G. Beech, 1988. The production of strawberry plants from callus culture. Plant Cell Tiss. Org. Cult. 12: 235–241.

Kasha, K.J. and K.N. Kao, 1970. High frequency haploid production in barley *Hordeum vulgare*. Nature (London) 225: 874–876.

Kimber, G. and R. Riley, 1963. Haploid angiosperms. Bot. Rev. 29: 480–531.

Kinugawa, K., E. Tanesaka and K. Shibao, 1991a. Production of young plants by anther culture of strawberry, variety Houkouwase. Mem. Faculty Agr. Kinki Univ. 24: 37–45.

Kinugawa, K., E. Tanesaka and M. Okamoto, 1991b. Capacity of different tissue strawberry plants to form callus. Mem. Faculty Agr. Kinki Univ. 24: 47–51.

Laneri, U. and C. Damiano, 1980. Strawberry anther culture. Annali dell "Instituto Sperimentale per la Fruticoltura" 11: 43–48.

Lee, E.C.M. and R.A. de Fossard, 1977. Some factors affecting multiple bud formation of strawbery *F.* × *ananassa in vitro*. Acta Hort. 78: 187–195.

Linsmaeier, E.M. and F. Skoog, 1965. Organic growth factor requirement of tobacco tissue culture. Physiol. Plant. 18: 100–127.

Lord, W.J. and D.J. White, 1962. The induction of parthenocarpy and an attempt to promote apomixis in the strawberry by treatments with growth regulators. Proc. Amer. Soc. Hort. Sci. 80: 350–362.

Maas, J.L., 1977. Pollen ultrastructure of strawberry and other small fruit crops. J. Amer. Soc. Hort. Sci. 102: 560–571.

Mangelsdorf, A.J. and E.M. East, 1927. Studies on the genetics of *Fragaria*. Genetics 12: 307–339.

Marcotrigiano, M., P. Morgan, H.J. Swartz, and J. Ruth, 1984. Histogenic instability in tissue culture proliferated strawberry plants. J. Amer. Soc. Hort. Sci. 112: 583–587.

McArthur, D.A.J. and G.W. Eaton, 1987. Effect of fertilizer, paclobutrazol, and chlormequat on strawberry. J. Amer. Soc. Hort. Sci. 112: 241–246.

Melville, A.H., G.J. Galleta, A.D. Draper and T.J. Ng, 1980. Seed germination and early seedling vigour in progenies of inbred strawberry selections. HortScience 15: 749–750.

Miller, A.R., J.C. Scheerens and P.S. Erb, 1992. Enhanced strawberry seed germination through *in vitro* culture of cut achenes. J. Amer. Soc. Hort. Sci. 117: 313–316.

Mohamed, H., H.J. Swartz and G. Buta, 1991. The role of abscisic acid and plant growth regulators in tissue culture induced rejuvenation of strawberry *ex vitro*. Plant Cell Tiss. Org. Cult. 25: 75–84.

Mok, D.W.S. and W.D. Evans, 1971. Chromosome associations at diakinesis in the cultivated strawberry. Can. J. Genet. Cytol. 13: 231–236.

Mokra, A. and V. Maliarcikova, 1981. Anther culture of strawberry I. Effect of medium composition on plant regeneration. Vedecke Prace vyskumneho Ustavo a Okrasny Drevin v Bojniciah 3: 91–98. In: Hort. Abstr. 53, 04932.

Morrow, E.B. and G.M. Darrow, 1952. Effects of limited inbreeding in strawberries. Proc. Amer. Soc. Hort. Sci. 39: 269–276.

Morrow, E.B., G.M. Darrow and D.H. Scott, 1954. A quick method of cleaning berry seed for breeders. Proc. Amer. Soc. Hort. Sci. 63: 265.

Mudge, K.W., K.R. Narayanan and B.W. Pooviah, 1981. Control of strawberry fruit set and development with auxins. J. Amer. Soc. Hort. Sci. 106: 80–84.

Murashige, T. and F. Skoog, 1962. A revised medium for rapid growth and bio-assays with tobacco tissue cultures. Physiol. Plant. 15: 473–497.

Nehra, N.S., C. Stushnoff, K.K. Kartha and K.L. Giles, 1992. The influence of plant growth regulator concentrations and callus age on somaclonal variation in callus culture regeneration of strawberry. Plant Cell Tiss. Org. Cult. 29: 257–268.

Niemirowicz-Szczytt, K., 1984a. The results of intergeneric pollination of *F. ananassa* Duch. and *F. virginiana* Duch. by *Potentilla* species. Acta Soc. Bot. Pol. 53: 443–454.

Niemirowicz-Szczytt, K., 1984b. Morphological and cytological characters of progeny obtained from pollination of *F. ananassa* Duch. with *Potentilla* species. Acta Soc. Bot. Pol. 53: 455–468.

Niemirowicz-Szczytt, K., 1984c. Embryological studies on *F. ananassa* Duch. after pollination by three spp. of *Potentilla*. Acta Soc. Bot. Pol. 53: 469–484.

Niemirowicz-Szczytt, K., 1987. Strawberry (*Fragaria* × *ananassa*) haploids and their generative progeny: induction and characteristics. Treatises and Monographs, Warsaw Agric. Univ., Poland No. 65.

Niemirowicz-Szczytt, K. and Z. Zakrzewska, 1980. *Fragaria* × *ananassa* anther culture. Bull. Acad. Pol. Sci. 5: 341–347.

Niemirowicz-Szczytt, K., Z. Zakrzewska, S. Malepszy and B. Kubick, 1983. Characters of plants obtained from *Fragaria* × *ananassa* anther culture. Acta Hort. 131: 231–337.

Niizeki, H. and K. Fukui, 1983. Elimination of chromosomes in strawberry plants by treatments with para-fluorophenylalanine. Jpn. J. Breed. 33: 55–61.

Nishi, S. and K. Oosawa, 1973. Mass production of virus-free plants through meristem callus. Jpn. Agric. Res. Quart. 7: 189–194.

Nishi, S., K. Oosawa and T. Toyoda, 1974. Studies on the anther culture of vegetable crops: I. Differentiation of young seedlings of some vegetable crops from calli formed by anther culture. Bull. Veg. Ornam. Crops Res. Stn., Jpn., Ser. A, 1: 1–40.

Nitsch, J.P., 1950. Growth and morphogenesis of strawberry as related to auxin. Amer. J. Bot. 37: 211–215.

Nitsch, J.P. and C. Nitsch, 1969. Haploid plants from pollen grains. Science 163: 85–87.

Nyman, M. and A. Wallin, 1992. Improved culture technique for strawberry protoplasts and the determination of DNA content in protoplast derived plants. Plant Cell Tiss. Org. Cult. 30: 127–132.

Oosawa, K. and K. Takayanaji, 1982. High yielding variants in strawberry derived from anther culture. In: Proc. 5th. Int. Cong. Plant Tissue and Cell Culture, pp. 453–462.

Oosawa, K., M. Toda, and S. Nishi, 1974. Studies on the anther culture of vegetable crops, II. Breeding of a great quantity of virus-free plants by means of strawberry anther culture method. Bull. Veg. Ornam. Crops Res. Stn., Jpn., Ser. A, 1: 41–57.

Oosawa, K., T. Kuriyama and Y. Sugahara, 1981. Clonal multiplication of vegetatively propagated crops through tissue culture. I. Effective balance of auxin and cytokinin in the medium and suitable explant part for mass reproduction in strawberry, garlic, scallion, Welsh onion, yam and tar. Bull. Veg. Ornam. Crops Res. Stn., Jpn., Ser. A, 9: 1–50.

Owen, H.R. and A.R. Miller, 1993. A comparison of staining techniques for somatic chromosomes of strawberry. HortScience 28: 155–156.

Pankov, V.V., 1992. Effect of growth regulators on plant production of strawberry. Sci. Hort. 52: 157–161.

Popescu, I., L. Nicolae, A. Popescu and M. Coman, 1988. A strawberry haploid obtained by androgenesis. In: Int. Strawberry Symp., Cesena, May 22–27, 1988. Programme and Abstracts, p. 57.

Rosati, P., M. Devreux and U. Laneri, 1975. Anther culture of strawberry. HortScience 10: 119–120.

Rowe, P.R., 1974. Parthenogenesis following interspecific hybridisation. In: K.J. Kasha (Ed.), Haploids in Higher Plants: Advances and Potential. Proc. First Int. Symp. on Haploids in Higher Plants, June 10–14, 1974. University of Guelph, Guelph, pp. 43–52.

Rugini, E. and R. Orlando, 1992. High efficiency shoot regeneration from callus of strawberry stipules of *in vitro* shoot cultures. J. Hort. Sci. 67: 577–582.

Sangin-Berezovskij, G.N., 1963. Distant hybridisation in strawberry. Trud. Inst. Genet. 30: 321–356. In: Plant Breed. Abstr. 34, 4747.

Sarwar, M., 1984. The effect of different media and culture techniques on plating efficiency of strawberry mesophyll cells in culture. Physiol. Plant. 60: 57–60.

Sayegh, A.J., 1989. Androgenesis and intergeneric hybridisation for strawberry haploid production. Ph.D. Thesis, University College, Dublin.

Sayegh, A.J. and M.J. Hennerty, 1993. Intergeneric hybrids of *Fragaria* and *Potentilla*. Acta Hort. 348: 151–154.

Scott, D.H., 1951. Cytological studies on polyploids derived from tetraploid *Fragaria vesca* and cultivated strawberries. Genetics 36: 311–331.

Scott, D.H., G.T. Galleta and H.J. Swartz, 1985. Tissue culture as an aid in the propagation of Tribute everbearing strawberry. Adv. Strawberry Prod. 4: 59–60.

Scott, D.H. and F.J. Lawrence, 1975. Strawberries. In: J. Janick and J.N. Moore (Eds.), Advances in Fruit Breeding, pp. 71–97. Purdue Univ. Press, W. Lafayette, Indiana.

Senanayake, Y.D.A. and R.S. Bringhurst, 1967. Origin of *Fragaria*. I. Cytological analysis. Amer. J. Bot. 54: 221–228.

Shoemaker, N.P., H.J. Swartz and G.J. Galletta, 1985. Cultivar dependent variation in the pathogen resistance due to tissue culture propagation of strawberries. HortScience 20: 253–254.

Simon, I., E. Racz and J. Zatyko, 1987. Preliminary reports on somaclonal variations of strawberry. Fruit Sci. Rep. 4: 151–154.

Smith, W.H.M. and J.K. Jones, 1985a. Intergeneric crosses with *Fragaria* and *Potentilla* I. Crosses between *Fragaria moschata* × *Potentilla fruticosa* and the original parents. Euphytica 34: 725–735.

Smith, W.H.M. and J.K. Jones, 1985b. Intergeneric crosses with *Fragaria* and *Potentilla* II. The progeny of *Fragaria moschata* × *Potentilla fruticosa* and the original parents. Euphytica 34: 737–744.

Snow, R., 1963. Alcoholic hydrochloric acid-carmine as a stain for chromosomes in squash preparation. Stain Technol. 38: 9–13.

Spangelo, L.P.S., C.S. Hsu, O.S. Fejer and R. Watkins, 1971. Inbred line X tester analysis and the potential of inbreeding in strawberry breeding. Can. J. Genet. Cytol. 13: 460–469.

Staudt, G., 1962. Taxonomic studies in genus *Fragaria*. Can. J. Bot. 40: 869–886.
Sukhareva, N.B., 1970. Elements of apomixis in strawberry. In: S.S. Khokhlov ((Ed.), Translator: B.R. Sharma), Apomixis in Higher Plants, pp. 120–144. Amerind Publishing Co. Pvt. Ltd., New Delhi.
Svennson, M. and L.B. Johansson, 1994. Anther culture of *Fragaria* × *ananassa*: Environmental factors and medium components affecting microspore divisions and callus production. J. Hort. Sci. 69: 417–426.
Swartz, H.J., G.T. Galletta and R.H. Zimmerman, 1981. Field performance and phenotypic stability of tissue culture propagated strawberry. J. Amer. Soc. Hort. Sci. 106: 667–673.
Tafazoli, E. and D. Vince-Prue, 1979. Fruit set and growth in strawberry, *F.* × *ananassa*. Ann. Bot. 43: 125–134.
Vagera, J. and P. Havranek, 1983. Regulation of androgenesis of *Nicotiana tabacum* L. var. White Burley and *Datura innoxia* Mill., Effects of bivalent and trivalent iron and chelating substances. Biol. Plant. (Praha) 25: 5–14.
Velchev, V. and E. Milanov, 1984. Regeneration of callus culture from anthers and pistils in strawberry. Gradinarska i lozarska Nauka 21: 29–35. In: Hort. Abstr. 54, 00669.
Wallin, A., H. Skjoldebrand and M. Nyman, 1993. Protoplasts as tools in *Fragaria* breeding. Acta Hort. 348: 414–421.
Wang, D., W.P. Wergin and R.Z. Zimmerman, 1984. Somatic embryogenesis and plant regeneration from immature embryos of strawberry. Hort. Sci. 19: 71–72.
Watkins, R. and L.P. Spangelo, 1971. Strawberry selection index components. Can. J. Genet. Cytol. 13: 42–50.

14. Haploidy in apple

MONIKA HÖFER and YVES LESPINASSE

Contents

1. Introduction

1.1. *Importance of haploids in apple*

Most temperate fruit trees are characterized by a long reproductive cycle with several years of a juvenile phase, a tendency to allogamy and a large tree size. They are generally highly heterozygous, outbreeding species, which are asexually propagated. For these reasons, their genetic improvement by conventional methods is time-consuming and limited by space for field experiments. The production of haploids offers new possibilities for genetic studies and breeding, especially in these perennial fruit species.

Homozygous plants of apple (*Malus domestica* Borkh.; $2n = 2x = 34$) obtained after doubling haploids would reveal both genotypic and phenotypic characters, allowing a direct selection even for quantitative traits. Selected homozygous clones derived from F_1 hybrids, which could be used as starting material for haploidization, may serve as parental lines for hybridization programs. Such clones would transmit desirable genes to each gamete and therefore increase the frequency of promising seedlings in backcrosses. This technique of recurrent selection alternating with haploid steps (Kuckuck *et*

S.M. Jain, S.K. Sopory & R.E. Veilleux (eds.), In Vitro *Haploid Production in Higher Plants, Vol. 3,* 261–276.

al., 1991) could be efficient in apple breeding programs. Furthermore, by producing F_1 and F_2 from crosses between homozygous genotypes, more information on the inheritance of desirable characters will be available.

Mutagenesis is a suitable method to improve a vegetatively propagated species (Van Harten and Broertjes, 1989) like apple. Mutagenic treatment of haploid plants followed by chromosome doubling will enable all mutations, including recessive ones, to be expressed and evaluated for possible contribution to genetic improvement. The culture of isolated microspores offers the possibility of combining mutation and single cell selection procedures with the advantages of a haploid system. Prerequisite to the application of this strategy is the development of an effective microspore culture technique.

Isolated haploid cells and protoplasts provide an ideal material for genetic transformation and somatic hybridization, respectively. The use of haploid protoplasts for fusion experiments has the advantage that somatic hybrids with the normal diploid level are possible. The first step, the regeneration of haploid apple protoplasts was reported by Patat-Ochatt *et al.* (1993).

Haploids represent an excellent experimental material for cytological and genetic studies (Chen, 1986). The basic chromosome number (x) of apple is 17. Several authors have tried to explain the origin of this genome. In haploids, studying the meiotic chromosome behaviour and pairing affinities, if any, will be an important step toward a better understanding of the genetic origin of the *Maloideae*.

For genetic mapping, haploids or homozygotes are excellent material to construct a linkage map comprising RFLP-markers as well as characters of importance to breeders.

1.2. *Spontaneous production of haploids*

Some spontaneous haploid plants have been obtained in pome fruit. In 1945, Einset reported two haploids among apple seedlings from open-pollination of a triploid cultivar but neither plant survived.

One haploid apple plant produced by *in vivo* gynogenesis was selected after pollination of *M. domestica* cv. Topred Delicious ($2n = 2x = 34$) by a hybrid derived from *M. pumila* var. *Niedzwetzkyana* which carries a homozygous dominant marker gene, *RR*, which determines anthocyanin pigmentation on leaves, stem and fruit (Lespinasse and Godicheau, 1980).

Therefore, green seedlings resulting from this cross are expected to lack the R allele and therefore be maternally derived. Alternatively, seedlings with anthocyanin pigmentation would be sexual hybrids. In similar experiments involving 14 apple cultivars, 22,000 seedlings were screened: 222 were green among which 8 maternal haploids were identified (Lespinasse and Chevreau, 1987). However, for practical purpose the low frequency of naturally produced haploids and the low induction frequency of *in vivo* gynogenesis are insufficient.

2. Techniques for the production of haploids in apple by *in vitro* methods

In vitro approaches to induce haploids in apple and success with such approaches have been rather limited until recently in comparison with other plant species. The methods used for haploid induction can be classified into three types: (1) *in vitro* androgenesis by anther culture, (2) *in vitro* gynogenesis by unfertilized ovule culture, and (3) *in situ* parthenogenesis by irradiated pollen followed by *in vitro* culture of immature embryos.

2.1. *Anther culture*

Anther culture in apple was pioneered by Nakayama *et al.* (1971, 1972), who induced calli capable of root formation. Subsequently Kubicki *et al.* (1975) and Milewska-Pawliczuk and Kubicki (1977) achieved androgenic induction directly from cultured anthers containing uninucleate microspores and obtained embryos from the globular to the torpedo stage.

Braniste and Popescu (1981) and Hidano (1982) cultured anthers of several apple cultivars, which also formed embryos. In 1981 and 1984 Chinese scientists (Fei and Xue, 1981; Xue and Niu, 1984) reported, for the first time, regeneration of androgenic plants of six apple cultivars. However, all the androgenic plants obtained were mixoploids (Zhang *et al.*, 1990).

In 1981 Wu regenerated haploid and aneuploid plants from anther-derived calli of crabapple (*M. prunifolia*). Zarsky *et al.* (1986), Johansson *et al.* (1987), Zhang *et al.* (1987), Höfer and Hanke (1990, 1994), and De Witte and Keulemans (1992) listed the details for induction of androgenic embryos of different apple cultivars, but only a few viable plants were produced from any of these embryos.

The main objectives in producing haploids via anther culture in apple are the increase of embryo induction frequency and the establishment of an embryo-to-tree-system. A summary of haploid research in anther culture in apple is given in Table 1. The only attempt to induce androgenesis via microspore culture with the production of two calli was reported by Bouvier (1993).

2.2. *Protocol for anther culture*

For anther culture of apple, flower buds are collected when microspores are at the uninucleate stage of development (Fig. 1). Morphologically this stage is characterized by the beginning of petal emergence on the king flower, which is the most advanced in development on an inflorescence.

To extend the anther culture period throughout the year, several methods are used. Flower buds are taken from forced bud wood and from trees grown in the orchard (Höfer and Hanke, 1990). Using this method, the flowers are available at almost any time from January to April, just until three weeks before blossom in Central Europe. Zhang (1988) and Bouvier (1993) col-

Table 1. Attempts at haploidization by *in vitro* androgenesis in apple

Species	Response	Reference
1. Anther culture		
Malus domestica	Callus and roots	Nakayama *et al.* (1971, 1972)
	Callus	Braniste and Popescu (1981)
	Callus and embryos	Kubicki *et al.* (1975)
	Callus and embryos	Hidano (1982)
	Callus and embryos	Zarsky *et al.* (1986)
	Callus and embryos	Johansson *et al.* (1987)
	Callus and embryos	Zhang *et al.* (1987)
	Callus and embryos	Höfer and Hanke (1990)
	Callus and embryos	De Witte-Keulemans (1992)
	Callus and embryos, haplo-, mixoploid plants	Fei and Xue (1981) Xue and Niu (1984)
	Callus and embryo, plantlets	Zhang *et al.* (1990)
	Callus and embryos, plants	Xue *et al.* (1990)
	Callus and embryos, plants	Höfer (1994)
Malus zumi	Callus and embryos	Höfer (1994)
Malus prunifolia	Callus and haplo- and aneuploid plants	Wu (1981)
2. Microspore culture		
Malus domestica	Callus	Bouvier (1993)

lected flower buds at an appropriate stage and placed them into glass petri dishes on wetted absorbent paper for cold storage at 4 °C. Branches collected at an appropriate stage and incubated at 2 to 6 °C and a relative humidity of 70–80% in a container were used as an alternative explant source. In a third method, potted trees 2 to 4 years of age were stored at the same conditions as cut branches mentioned above, just when flowers reached the appropriate stage. The last three methods enable the maintenance of anthers until two months past the ordinary flowering season.

In our experiments, following surface sterilization of buds with 0.1% mercuric hypochloride (3 s), anthers were excised from buds and incubated on MS basic medium supplemented with various growth regulator combinations at different light and temperature regimes. The cultured anthers were examined over a period of six months. The induced structures can be classified into calli and embryos and transferred to various solid media for plant regeneration.

2.3. *Process of androgenesis*

After one or two weeks, microspores undergo their first cell division. Based on cytological observations (Zhang, 1988), *in vitro* androgenesis of apple can

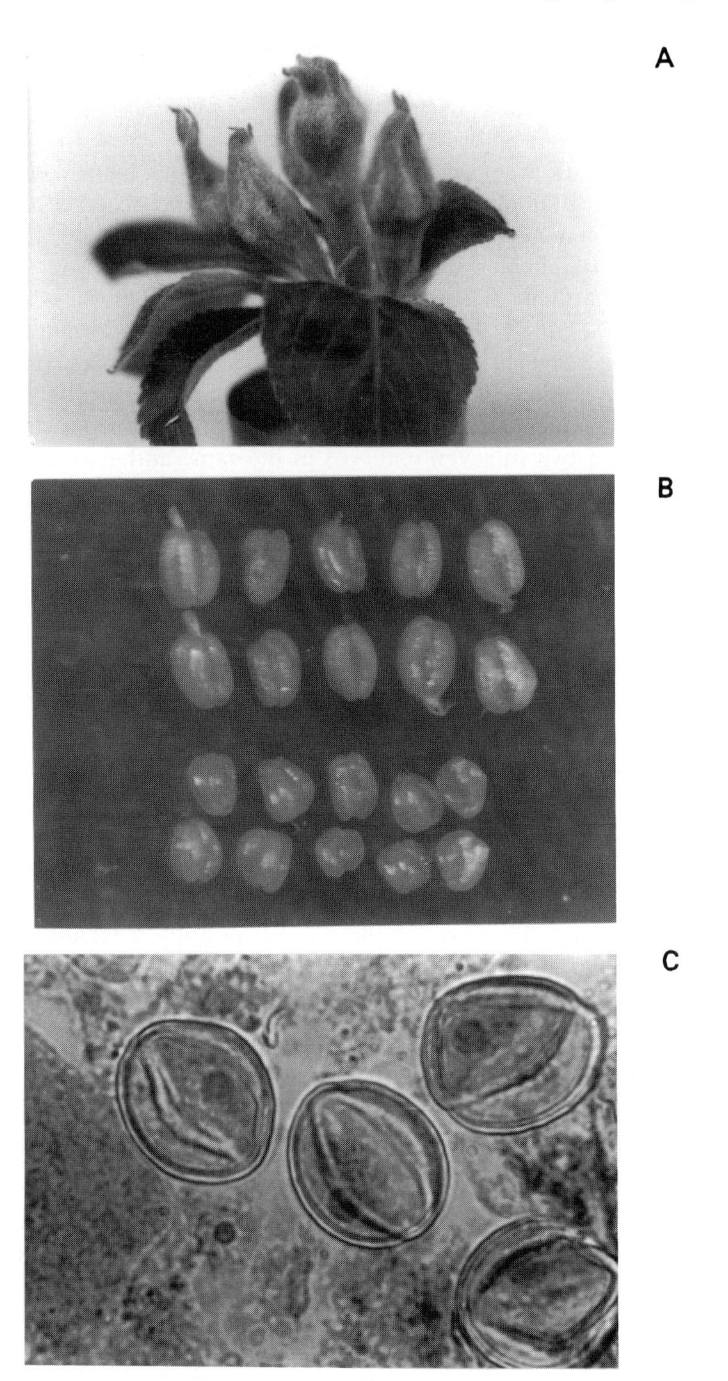

Figure 1. Characterization of buds and anthers at the incubation time of anther culture in apple; (A) Flower bud at the proper stage for *in vitro* androgenesis; (B) Two anther cycles with ten anthers from each cycle; (C) Microspore at the uninucleate stage of pollen development.

be induced in three different ways: (1) abnormal pollen mitosis leading to the formation of two identical nuclei, (2) division of the vegetative nucleus, resulting from a normal pollen mitosis, or (3) division of the generative nucleus, resulting from a normal pollen mitosis.

About four to six weeks after anther excision, an obvious increase in volume and the beginning of callus formation could be observed, in nearly all cases from the connective tissue. With progressive cultivation, the anther wall locally burst, due to callus growth inside the anther (Fig. 2A). The calli differed remarkably in colour and structure (Höfer, 1995). Histological investigations of Zhang (1988) confirmed that green callus originated from somatic cells, whereas white callus originated from microspores. In our own experiments with apple, after three to six months of anther incubation, white embryos emerged directly from inside the anther and not from calli. Some genotypes formed two to seven embryos in the same anther or even in the same theca. When first visible, most of the embryos were in the globular or torpedo stage of development (Fig. 2B). Morphologically aberrant embryos with a single cotyledon or fused cotyledons also occured (Höfer, 1995).

2.4. *Factors affecting androgenesis*

2.4.1. *Genotype*

In vitro androgenesis in apple, both callus initiation and embryo induction, is mainly determined by genotype. From our own experiences with apple, androgenic embryos have been induced in seven of 13 genotypes consisting of cultivars, clones and the wild type *Malus zumi* (Table 2). Zhang *et al.* (1990) achieved embryo formation in three of five cultivars tested, irrespective of the culture conditions. The overall efficiency of androgenic embryo formation has generally been low, often less than 1%. Our results ranged between 0.05% and 0.4% (ratio of the number of anthers yielding embryos compared to all those cultured) regarding the mean values of different genotypes (Table 2), reaching 6.4% with the best treatment for "Alkmene" (Höfer, 1995). Embryogenesis frequencies have been reported to be 0.4% (Johansson *et al.*, 1987), 3.4% (Xue and Niu, 1984) and 6.6% achieved with "Topred" (Zhang, 1988).

2.4.2. *Stage of pollen development and physiological stage of donor plants*

For nearly all perennial fruit crops studied, microspores at the uninucleate stage have been the most responsive for induction of androgenesis. In apple, embryo induction has been observed at the uninucleate stage of microspore development (Fei and Xue, 1981; Xue and Niu, 1984; Zhang *et al.*, 1987, 1990). The influence of the stage of pollen development on androgenic response has been studied in detail in our own experiments by independently incubating anthers from the tetrad to the late uninucleate stage of pollen development (Höfer and Hanke, 1994). While the highest rates of callus induction in apple were obtained at the early uninucleate stage, the induction

Figure 2. In vitro androgenesis in apple; (A) Callus induction from one anther of "Alkmene" after 2 months of culture; (B) Embryo of the cv. "Alkmene" at the torpedo stage; (C) Formation of secondary embryos; (D) Adventitious shoot formation from clusters of secondary embryos; (E) Rooted androgenic shoot.

of embryos in apple occurred nearly exclusively from microspores at the mid-uninucleate stage, before vacuole formation. The determination of the optimal stage of pollen development is an essential step to increase the efficiency of anther culture. However, routine identification of the stages of

Table 2. In vitro androgenesis in apple – effect of genotype on the induction of embryos

Genotype	Anthers with callus formation (%)	Number of embryos obtained	Embryo induction rate (%)
Alkmene	15.7	69	0.40
Remo	1.6	24	0.15
Malus zumi	0.5	10	0.24
Releika	8.9	4	0.05
Clivia	28.6	2	0.06
Reanda	9.4	2	0.10
Pinova	8.1	1	0.05
Average	10.4		0.21

pollen development is difficult because of the gradients within inflorescences. Thus, an indirect method associated with a morphological indicator was developed. On the basis of correlations between the stages of pollen development and easily determinable macroscopic characters of the flower bud, it was possible to define anther length, independent of genotype, as a marker of the optimal stage of pollen development in apple. This optimal length was 1.4 mm for anthers of buds taken from field-grown trees and 1.0 mm for anthers from forced bud wood (Höfer and Hanke, 1994).

In apple, Johansson *et al.* (1987) noted that anthers collected from branches cultivated under a 12 h light regime yielded a higher frequency of embryos than those from branches produced under continuous light or in darkness. Compared to results from culture of anthers taken from trees during the natural growth season, anther cultures from buds formed under artificial conditions resulted in a remarkably highler production of embryos (Johansson *et al.*, 1987; Höfer, unpublished). De Witte and Keulemans (1992) found no influence on embryo induction between flower buds from annual or bi-annual shoots.

2.4.3. *Culture medium*

The composition of the culture medium, is an important factor influencing *in vitro* androgenesis. For apple, Murashige and Skoog (MS) (1962) basic medium has usually been used, and positive results have been achieved (Xue and Niu, 1984; Johansson *et al.*, 1987; Zhang *et al.*, 1987; Höfer and Hanke, 1990). Other basal media, including B (Blaydes, 1966), N (Nitsch, 1972), N6 (Chu *et al.*, 1975), P (Potato-extract-medium, Anonymous, 1976) and W (White, 1963) have also been tested for culturing anthers, but embryo induction failed (Kubicki *et al.*, 1975; Fei and Xue, 1981; Höfer and Hanke, 1990).

The sugar concentration of the nutrient medium plays a significant role in anther culture. Milewska-Pawliczuk (1978) and Fei and Xue (1981) studied the effect of sucrose on androgenesis at concentrations between 5 and 120 g/l and determined an optimal sucrose concentration of 30 g/l. On the contrary, Höfer and Hanke (1990) observed only minor differences for callus induction

Table 3. Effect of cytokinins and auxins on androgenic embryo induction (%) of the genotypes "Alkmene" and "Remo" (MS basal medium; growth regulator content in mg/l)

	Embryo induction (%)					
Auxin	2,4-D		NAA		IBA	
	0.2	0.4	0.2	1.0	0.2	1.0
Cytokinin						
0.2 K	0.74	0.45	0.37	0.31	1.0	1.2
0.2 K + 0.2 BA	0.14	0	0.28	0	–	–
0.5 K	–	–	0.41	0	–	–
1.0 K	–	–	0.12	0	–	–
0.1 TDZ	0	0	0.24	0	–	–
0.2 TDZ	0.16	0	0.08	0	0	–
0.3 TDZ	0	–	0.3	–	–	–
0.2 BA	0.2	–	0.4	–	1.2	–
0.5 BA	0	–	0.47	–	0.8	–

– Not investigated.
Auxin: 2,4-D – 2,4-dichlorophenoxyacetic; NAA – *a*-naphthaleneacetic acid; IBA – indole-3-butyric acid.
Cytokinin: K – kenetin; BA – benzyladenine; TDZ – thidiazuron.
The androgenic embryo induction represents the average of 5 experimental years. In all, for each growth regulator combination, 1000–5000 anthers were incubated.

among different genotypes at 30 or 50 g/l sucrose, but embryos occurred only on MS medium supplemented with 50 g/l sucrose. Replacing sucrose with either 50 g/l maltose (Höfer, unpublished) or lactose or 30 g/l glucose (Zhang *et al.*, 1990) did not promote embryo formation.

Both auxins and cytokinins are usually required for induction of androgenic callus and embryos in apple (Zhang and Lespinasse, 1992). The optimal combination of growth regulators depends on different factors, especially on the genotype, such that a universal recommendation for apple anther culture may not be possible. In our own experiments the highly inductable genotypes "Alkmene" and "Remo" exhibited embryo induction in presence of all tested cytokinins and auxins (Table 3). The concentration of the single growth regulators was more important. Two important tendencies were derived. The addition of 0.2 mg/l kinetin and 0.2 mg/l benzyladenine (BA) resulted in embryo induction in all tested auxin modifications. Embryo development was favoured by low auxin concentrations of 0.2 mg/l 2,4-dichlorophenoxyacetic acid (2,4-D) or 0.2 mg/l *α*-naphthaleneacetic acid (NAA), independent of type and concentration of cytokinin (Höfer, 1995; Zhang, 1988).

Activated charcoal has a stimulating effect on *in vitro* androgenesis in many plant species (Johansson, 1983). The addition of 0.5% activated charcoal to double layer media (Johansson *et al.*, 1987) or solid media (Zhang *et al.*, 1990) was also beneficial in apple. Other components have been added to anther culture media of apple. Fei and Xue (1981) reported a promoting effect of adenine sulfate in the induction of embryos. The addition of biotin (5.0 mg/l) or insulin (8.0 mg/l) stimulated plant regeneration from anther-derived callus of *Malus prunifolia* (Wu, 1981).

2.4.4. *Pretreatment and physical conditions*

The application of various pretreatments either to flower buds before anther culture (e.g., cold treatment) or to anthers after excision (e.g., heat shock), has promoted microspore embryogenesis (Sangwan and Sangwan, 1987). For apple, different results have occurred in response to a cold treatment with regard to callus or embryo induction. According to Zhang (1988), a treatment of at least two weeks at 3 to 4 °C, applied to flower buds prior to culture of anthers, proved necessary for androgenic embryo formation. Similarly, two to six weeks of this kind of treatment also significantly improved callus initiation. Johansson *et al.* (1987) noted enhancement of both callus production and embryo formation in apple by treating flower buds at 7 °C for 6 to 12 days. De Witte and Keulemans (1992) reported that a short treatment (one week) seemed to be a good choice for embryo induction. Nevertheless, a cold treatment was unnecessary for haploid production in apple, since even untreated anthers gave rise to embryos. Corresponding conclusions could be drawn from our own results. Cold as well as heat pretreatment (34 °C for 24 h) can slightly increase embryo production, but the genotypic background plays a predominant role.

Zhang *et al.* (1992) examined the effect of low gamma dose irradiation of intact flower buds and extracted anthers of apple. They concluded that gamma irradiation to flower buds prior to anther culture can stimulate androgenic embryogenesis in recalcitrant apple genotypes. In responsive genotypes, anther culture efficiency could be slightly improved.

For embryo induction, different temperature and light/dark combinations have been used. The optimal temperature ranged from 23 °C (Zhang *et al.*, 1987, 1988) to 25–30 °C (Fei and Xue, 1981; Xue and Niu, 1984; Höfer and Hanke, 1990). With respect to the light-dark-regime, Kubicki *et al.* (1975) and Fei and Xue (1981) achieved embryo induction under continuous light (2000 lux), whereas Zhang *et al.* (1987) used a photoperiod of 16 h at 40 to 60 $\mu mol\ m^{-2}\ s^{-1}$ after four weeks in the dark. In contrast, our results showed a significantly better callus induction and improved embryoid induction in the dark (Höfer and Hanke, 1990). The abovementioned examples suggest that light is unnecessary for the induction process but it may affect regulatory processes during the post-induction.

2.5. *Morphogenic responses in* Malus *anthers*

Regeneration of plants from androgenic callus or from androgenic embryos in *Malus* spp. has been difficult to achieve. Many authors have reported on efforts towards the induction of organogenesis in calli but, with the exception of *M. prunifolia*, where haploid plants were regenerated from anther-derived calli (Wu, 1981), all experiments have been unsuccessful.

Androgenic plants via direct germination of anther-derived embryos were reported only by Fei and Xue (1981) and Xue and Niu (1984). On the other hand, the results of Zhang *et al.* (1990), Bouvier (1993) and our own experience (Höfer, 1994) demonstrated the occurence of adventitious shoot formation. After transfer to regeneration medium supplemented with 0.1 mg/l thidiazuron (TDZ), many embryos increased in size or changed colour from white to green and developed cotyledons; subsequently secondary embryogenesis has been observed (Fig. 2C). After subculture, the formation of adventitious shoots from secondary embryos has been obtained (Fig. 2D). The mean conversion frequency across genotypes for the induced androgenic embryos reached 54.2%. Therefore this is a reproducible protocol for plant regeneration from androgenic embryos in apple (Höfer, 1994). The adventitious shoots exhibited phenotypic variation with respect to shape and colour, suggesting different ploidy levels of the regenerants as reported also by Lespinasse and Noiton (1986) in *in situ* investigations. At present most of the shoots are in the phase of micropropagation and rooting (Fig. 2E); some have been already grafted.

A first determination of the ploidy level of the embryos and shoots showed 24% haploid or diploid, 28% triploid, and 48% tetraploid (Höfer, 1994). The shoots were analysed by isoelectric focusing and were found to be homozygous. A similar distribution of ploidy has been obtained with other plant species (D'Amato, 1985). Endo reduplication and nuclear fusion may possibly account for the variation in chromosome number as a result of anther culture (Sunderland *et al.*, 1974).

2.6. *Induction of gynogenesis*

In vitro culture of unpollinated ovaries and ovules is an alternative technique for the production of haploid plants. In apple, only Zhang and Lespinasse (1988) reported the induction of gynogenesis. Flower buds were collected at the balloon stage, when the embryo sac contained eight haploid nuclei. Three types of explants: ovaries, isolated ovules, or ovules excised but still attached to placental tissue were cultured. A callus induction rate of 32% mainly from the micropylar end and two embryo-like structures were obtained from the culture of ovules.

In our own experiments (Höfer, unpublished) flower buds of five apple genotypes, "Alkmene", "Clivia", "Pinova", "Remo" and "Reanda", were collected at the late balloon stage and at blossom. After sterilization of the buds, the ovaries and ovules were excised and incubated on MS basal medium supplemented with different combinations of growth regulators at 27 °C in the light or dark. The ovaries showed a strong secretion of phenols, and after two to four months, only some callus induction occurred. Two embryos were produced from isolated ovules on a medium containing 20 g/l sucrose,

Figure 3. Embryoid of the cv. "Remo" obtained 5 months after culture of unfertilized ovule.

0.1 mg/l indole-3-butyric acid (IBA) and 0.2 mg/l BA after five months of incubation (Fig. 3). However, these two structures were not capable of sustained growth.

2.7. *Induction of* in situ *parthenogenesis by irradiated pollen*

An *in vitro* haploidization technique was developed for *Petunia* by Raquin (1985) and then applied successfully to melon (Sauton and Dumas de Vaulx, 1987) and carrot (Rode and Dumas de Vaulx, 1987). It involves the induction of *in situ* parthenogenesis by pollination with irradiated pollen followed by the *in vitro* culture of the pollinated ovaries, immature seeds or embryos. For apple, haploid plants have been successfully induced in cv. Erovan after pollination with pollen irradiated at 500 to 1000 Gray (Gy), followed by *in vitro* culture of immature embryos on MS medium containing 0.1 mg/l NAA, 1 mg/l BA, and 1 mg/l gibberellic acid (GA_3) (Zhang *et al.*, 1988). The same technique was adapted to several other genotypes (Zhang and Lespinasse, 1991). The pollen parent used for these experiments was the same clone bearing a homozygous dominant marker gene *RR* encoding anthocyanin synthesis employed for *in vivo* gynogenesis (Lespinasse and Godicheau, 1980).

Pollen irradiation was performed by γ-rays from Cobalt 60 at different doses from 125 to 1500 Gy. Trees chosen as female parents were isolated in a cage before flowering, and three to four flowers per cluster were pollinated after manual emasculation at the balloon stage. Irradiation up to 1500 Gy only slightly affected germination of the pollen *in vitro*. However, the pollen tube length was reduced by radiation doses of 500 Gy and above (Zhang and Lespinasse, 1991). Studies of pollen tube growth *in situ* by fluorescence microscopy revealed that the pollen tubes from treatments at 125 to 1000 Gy were capable of passing through the style and reaching the embryo sac (Zhang, 1988). Fruit set and seed number, in all the cultivars tested, decreased with increasing doses of pollen irradiation. Fruit set was nearly suppressed when flowers were pollinated with the pollen irradiated at 1500 Gy. Our results using this same treatment indicated that the decreased fruit set and number of seeds were dependent on the genotype (Höfer, unpublished).

According to Zhang and Lespinasse (1991), three months after pollination is an appropriate stage for *in vitro* culture of immature embryos. About 70% of embryos thus obtained were able to germinate under *in vitro* conditions. If younger fruits were used, the frequency of survival of plants following embryo germination decreased considerably. If fruit samples were collected at the fully mature stage, some of the haploid embryos had degenerated. In our investigations, the optimal time for harvesting was four months after pollination. This variation in time could be correlated to the influence of different environmental conditions on fruit development. De Witte and Keulemans (1994) reported an increase in the amount of seeds with normally developed embryos which germinated better due to delayed picking.

Irradiation dose also has an important effect on this haploidization technique. Zhang *et al.* (1987) reported plants exhibiting a maternal phenotype (green colour) after using irradiation doses from 200 to 1000 Gy. Most of the plants obtained from 125 Gy irradiated pollen were hybrids (red colour). Haploid plants were obtained at doses from 200 to 1000 Gy, with 500 Gy as the most efficient dose for both embryo and haploid plant production.

In order to increase plant production efficiency, we tested alternative regeneration methods on the embryos. This included the above mentioned embryo culture, where *in vitro* embryos were placed in a chamber at 3 °C for two months to overcome their potential dormancy and thereafter transferred to the culture room for germination. Other alternative techniques including cotyledon culture according to Kouider *et al.* (1984) and embryo culture from mature fruits stored for six weeks at 3 °C (Barthe and Bulard, 1982) were employed in our laboratory. Preliminary results suggested that these regeneration methods are possible alternatives. All germinated and regenerated plants obtained from 250 and 500 Gy irradiated pollen expressed the green colour.

3. Conclusions and future prospects

Methods are now available to produce plants of *Malus* spp. via anther culture and parthenogenesis. Androgenesis has been successfully induced *in vitro* (Table 1), but the overall efficiency for embryogenesis is generally low. For the regeneration of androgenic embryos, two pathways of development are described: direct germination and adventitious shoot formation. Because of the higher plant regeneration rate reached by the latter process, this method is the most effective and reproducible protocol for plant regeneration from embryos in apple.

Embryo induction and development may be improved by the manipulation of physical, chemical, and physiological conditions. Unfortunately, in apple, no source of material from plants cultivated under defined conditions exists. Differentiation of flower buds occurs in the year prior to their maturation and depends strongly on the enviromental conditions. Therefore, anthers probably have different endogenous hormone levels from one season to another. The regeneration phase and the origin of the different ploidy levels also require further study.

In comparing *in vitro* androgenesis with *in situ* parthenogenesis induced by irradiated pollen, genotype has a much less marked effect on the latter and experiments using this technique are relatively easy to perform (Zhang and Lespinasse, 1992). Improvement will depend upon selection of an efficient radiation dose, optimization of the pollination method and seed harvest time. Optimal conditions should be deduced for alternative regeneration methods to rescue a maximum of haploid embryos. To realize the full potential of *in vitro* androgenesis and *in situ* parthenogenesis in apple, a greater understanding is still needed of factors that are involved in morphogenic competence.

4. References

Anonymous, 1976. A sharp increase of the frequency of pollen plant induction in wheat with potato medium. Acta Genet. Sin. 3: 25–31.

Barthe, Ph. and C. Bulard, 1982, Influence of agar and sucrose on the behaviour of dormant apple embryos cultured *in vitro*. New Phytol. 91 517–529.

Blaydes, D.F., 1966. Interaction of kinetin and various inhibitors in the growth of soybean tissues. Physiol. Plant. 8: 748–753.

Bouvier, L., 1993. Haploïdie chez le pommier (*Malus domestica* Borkh.) et le poirier (*Pyrus communis* L.). Ph.D. Thesis, Université de Paris.

Braniste, N. and I. Popescu, 1981. Preliminary data of obtaining pear and apple haploids by anther *in vitro*. Probleme de Genetica Teoretica si Aplicata 13: 409–414.

Chen, Z.G., 1986. Induction of androgenesis in woody plants. In: H. Hun and H. Yang (Eds.), Haploids in Higher Plants *In Vitro*, pp. 44–46. China Academic Publishers, Beijing.

Chu, C.C., C.C. Wang, C.S. Sun, C. Hsü, K.C. Yin, C.Y. Chu and F.Y. Bi, 1975. Establishment of an efficient medium for anther culture of rice through comparative experiments on the nitrogen sources. Sci. Sin. 18: 659–668.

D'Amato. F., 1985. Cytogenetics of plant cell and tissue cultures and their regenerates. CRC Crit. Rev. Plant Sci. 3: 73–112.

De Witte, K. and J. Keulemans, 1992. Androgenesis in apple: Importance of the plant material, Abstracts of EUCARPIA-Congress Reproductive Biology and Plant Breeding. Angers, July 6–11, pp. 157–158.

De Witte, K. & J. Keulemans, 1994. Restrictions of the efficiency of haploid plant production in apple cultivar Idared, through parthenogenesis *in situ*. In: H. Schmidt and M. Kellerhals (Eds.), Progress in Temperate Fruit Breeding, pp. 403–408. Kluwer Academic Publishers, Dordrecht.

Einset, J., 1945. The spontaneous origin of polyploid apples. Proc. Amer. Soc. Hort. Sci. 46: 91–93.

Fei, K.W. and G.R. Xue, 1981, Induction of haploid plantlets by anther culture *in vitro* in apple cv. "Delicious". Sci. Agric. Sin. 4: 41–44.

Hidano, Y., 1982. Callus and embryoid induction by anther culture in apple. Bull. Fac. Educ. Hirosaki Univ. 48: 69–74.

Höfer, M., 1994. *In vitro* androgenesis in apple: Induction, regeneration and ploidy level. In: H. Schmidt and M. Kellerhals (Eds.), Progress in Temperate Fruit Breeding, pp. 399–402. Kluwer Academic Publishers, Dordrecht.

Höfer, M., 1995. *In-vitro*-Androgenese bei Apfel. Gartenbauwissenschaft 60: 12–15.

Höfer, M. and V. Hanke, 1990. Induction of androgenesis *in vitro* in apple and sweet cherry. Acta Hort. 280: 333–336.

Höfer, M. and V. Hanke, 1994. Antherenkultur bei Apfel und Kirsche: Einfluß des Pollenentwicklungsstadiums – Korrelative Beziehungen zu morphologischen Blutenmerkmalen. Gartenbauwissenschaft 59: 225–228.

Johansson, L., 1983. Effects of activated charcoal in anther cultures. Physiol. Plant. 59: 397–403.

Johansson, L., A. Walin, A. Gedin, M. Nyman, E. Pettersson and M. Svensson, 1987. Anther and protoplast culture in apple and strawberry. Biennial Report 1986–1987, Div. of Fruit Breeding, Balsgard, Swedish Univ. Agr. Sci., pp. 83–92.

Kouider, M., S.R. Korban, R.M. Skirvin and M.C. Chu, 1984. Influence of embryonic dominance and polarity on adventitious shoot formation from apple cotyledons *in vitro*. J. Amer. Soc. Hort. Sci. 109: 381–384.

Kubicki, B., J. Telezynska and E. Milewska-Pawliczuk, 1975. Induction of embryoid development from apple pollen grains. Acta Soc. Bot. Pol. 44: 631–635.

Kuckuck, H., G. Kobabe and G. Wenzel, 1991. Fundamentals of Plant Breeding, Springer Verlag, Heidelberg.

Lespinasse, Y. and D. Noiton, 1986. Contribution á l'étude d'une plante haploïde de pommier (*Malus pumila* Mill.) Étude descriptive et comparaison avec des clones de ploïdie differente. I. Caractères végétatifs: entrenoeuds, feuilles et stomates. Agronomie 6: 659–664.

Lespinasse, Y. and E. Chevreau, 1987. Progressi nel miglioramento genetico del melo: cloni autocompatlbili e piante aploidi. Riv. Fruttic. Ortofloric. 69: 25–29.

Lespinasse, Y. and M. Godicheau, 1980. Création et description d'une plante haploïde de pommier (*Malus pumila* Mill.). Ann. Amélior. Plant 30: 39–44.

Milewska-Pawliczuk, E., 1978. Kultury pylnikov jabloni *Malus domestica*. Ph.D. Thesis, Warszawa SGGW, Akad. Rolnicza, Warsaw.

Milewska-Pawliczuk and B. Kubicki, 1977. Induction of androgenesis *in vitro* in *Malus domestica*. Acta Hort. 78: 271–276.

Murashige, T. and F. Skoog, 1962. A revised medium for rapid growth and biossays with tobacco tissue cultures. Physiol. Plant 15: 473–497.

Nakayama, R., K. Saito and R. Yamamoto, 1971. Studies on the hybridization in apple breeding II. Anther culture of the apple. Bull. Fac. Agric. Hirosaki Univ 17: 12–19.

Nakayama, R., K. Saito and R. Yamamoto, 1972. Studies on the hybridization in apple breeding. III. Callus formation and organ differentiation from anther culture. Bull. Fac. Agr. Hirosaki Univ. 19: 1–9.

Nitsch, J.P., 1972. Haploid plants from pollen. Z. Pflanzenzüchtg. 67: 3–18.

Patat-Ochatt, E.M., J. Boccon-Gibod, M. Duron and S.J. Ochatt, 1993. Organogenesis of stem and leaf protoplasts of a haploid golden delicious apple clone (*Malus* × *domestica* Borkh.). Plant Cell Rep. 12: 118–120.

Raquin, C., 1985. Induction of haploid plants by *in vitro* culture of *Petunia* ovaries pollinated with irradiated pollen. Z. Pflanzenzüchtg. 94: 166–169.

Rode, J.C. and R. Dumas de Vaulx, 1987. Obtention de plantes haploïdes de carotte (*Daucus carota* L) issues de parthénogenèse induite *in situ* par du pollen irradié et culture *in vitro* de graines immatures. C.R. Acad. Sci. Paris, Série III, 305: 225–229.
Sangwan, B.S. and R.S. Sangwan, 1987. Biochemical cytology of pollen embryogenesis. Intl. Rev. Cytol. 107: 221–272.
Sauton, A. and R. Dumas de Vaulx, 1987. Obtention de plantes haploïdes chez le melon (*Cucumis melo* L.) par gynogenèse induite par du pollen irradié. Agronomie 7: 141–148.
Sunderland, N., G.B. Collins and J.M. Dunwell, 1974. The role of nuclear fusion in pollen embryogenesis of *Datura innoxia* Mill. Planta 117: 227–241.
Van Harten, A.M. and C. Broertjes, 1989. Induced mutations in vegetatively propagated crops. Plant Breed. Rev. 6: 56–91.
White, P.R., 1963. The Cultivation of Animal and Plant Cells. Ronald Press, New York.
Wu, J.Y., 1981. Induction of haploid plants from anther culture of crabapple. J. Northeast. Agr. Coll. 3: 105–108.
Xue, G.R. and J.Z. Niu, 1984. A study on the induction of apple pollen plants. Acta Hort. Sin. 11: 161–164.
Xue, G.R., J.Z. Niu, Z.Y. Yang, Y.Z. Shi and K.W. Fei, 1990. Apple anther culture technique and the successful culture of pollen plantlets of 8 major apple cultivars. Sci. Agric. Sin. 23: 86.
Zarsky, V., J. Tupy and J. Vagera, 1986. *In-vitro*-culture of apple anther. In: Moet-Henessy Conference Fruit Tree Biotechnology, Moet Henessy, Paris, October, p. 75.
Zhang, Y.X., 1988. Recherche *in vitro* de plantes haploïdes chez le pommier cultivé (*Malus* × *domestica* Borkh.): androgenèse, gynogenèse, parthenogenèse *in situ* induite par du pollen irradié. Ph.D. Thesis, Université de Paris-Sud, Centre d'Orsay, Paris.
Zhang, Y.X. and Y. Lespinasse, 1988. Culture *in vitro* d'ovules non fécondés et d'embryons prélevés 8 jours après pollination chez le pommier cultivé (*Malus* × *domestica* Borkh.). Agronomie 8: 837–842.
Zhang, Y.X. and Y. Lespinasse, 1991. Pollination with gamma-irradiated pollen and development of fruits, seeds and parthenogenetic plants in apple. Euphytica 54: 101–109.
Zhang, Y.X. and Y. Lespinasse, 1992. Haploidy. In: F.A. Hammerschlag and R.E. Litz (Eds.), Biotechnology of Perennial Fruit Crops, pp. 57–75. CAB International, Wallingford.
Zhang, Y.X., J. Boccon-Gibod and Y. Lespinasse, 1987. Obtention d'embryons de pommier (*Malus* × *domestica* Borkh.) après culture d'anthères. C.R. Acad. Sci. Paris, Série III, 305: 443–448.
Zhang, Y.X., L. Bouvier and Y. Lespinasse, 1992. Microspore embryogenesis induced by low gamma dose irradiation in apple. Plant Breed. 108: 173–176.
Zhang, Y.X., Y. Lespinasse and E. Chevreau, 1988. Obtention de plantes haploïdes de pommier (*Malus* × *domestica* Borkh.) issues de parthénogenèse induite *in situ* par du pollen irradié et culture *in vitro* des pépins immatures. C.R. Acad. Sci. Paris, Série III 307: 451–457.
Zhang, Y.X., Y. Lespinasse and E. Chevreau, 1990. Induction of haploidy in fruit trees. Acta Hort. 280: 293–305.

15. Microspore and protoplast culture in *Ginkgo biloba*

J. TRÉMOUILLAUX-GUILLER, D. LAURAIN and J.C. CHÉNIEUX

Contents

1. Introduction

With respect to its origin, anatomy and reproduction, *Ginkgo biloba* shows unique characteristics in the plant kingdom. Morever, the specific secondary metabolites present in the leaves are highly valuable in medicine. Furthermore, this species can be considered as an interesting and original experimental model, because of a limited number of reports regarding *in vitro* cultures of *G. biloba*.

1.1. *Distribution and botanical traits*

The last remaining natural habitats of *G. biloba* are in the forests of the Chekiang province of Easten China (Michel, 1986). However due to its ornamental qualities and tolerance to air pollution (Christensen, 1972; Rohr, 1989), *G. biloba* (or *Ginkyo*) is widely cultivated in urban areas all over the world. It belongs to the family *Ginkgoaceae* and is an unique specimen of *Ginkgo*ales order, which appeared in the premian era, but reached its apogee in the jurassic era. Today *G. biloba* is considered as a prespermatophyte and the oldest tree in the world.

Despite its slow growth *G. biloba* is an elegant tree with an erect-trunk that can exceed 30 m in height (Rohr, 1989). The branches are dimorphous:

S.M. Jain, S.K. Sopory & R.E. Veilleux (eds.), In Vitro *Haploid Production in Higher Plants, Vol. 3,* 277–295.

Figure 1. One catkin with numerous microsporophylls in *Ginkgo biloba*. Magnified view (7.5×).

some (auxiblasts) are long with alternating leaves, while others are short (mesoblasts) with leaves and male or female flowers, because this species is dioecious. The bilobed leaves have a thick limb with dichotomal ribbing. The mode of reproduction is termed as oviparous (Favre-Duchartre, 1956), which is specific to prespermatophytes.

The male reproductive organs are on the mesoblasts. They are in the form of catkins with numerous microsporophylls (or stamens with a short filament) of some millimeters in length (Fig. 1). In temperate zones, towards the end of March, mciosis in the mother cells produces haploid microspores ($n = 12$ chromosomes), which develop pollen grains within 5 to 10 days. At maturation, each pollen grain surrounds itself with a double cell-wall (intine + exine). When anthesis takes place (mid-April), tetracellular pollen grains are formed from one reproductive cell, one vegetative cell, and two prothallus cells. The vegetative cell extrudes a pollen tube equipped with rhizoids and containing two ciliated spermatozoids (Favre-Duchartre, 1956).

The female organ consists of two erect ovules (orthotropous), one of which can develop completely. After meiosis, one tetrad cell, termed megaspore, evolves in prothallus (female gametophyte) (Favre-Duchartre, 1956). In the beginning, the prothallus is coenocytic, after which it divides and becomes surrounded with three tunica. At the beginning of September, when fecundation takes place, the spermatozoids become mobile and move from the pollen tube into the oosphere. After fecundation the proembryo keeps a coenocytic structure as far as eight divisions (Lee, 1955) and grows inside the female prothallus to lead to a mature zygotic embryo.

1.2. *Medicinal value*

For centuries, leaf extract of *Ginkgo biloba* has been used for therapeutic purposes (Kleijnen and Knipschild, 1992). Since time immemorial, the traditional Chinese pharmacopoeia mentioned the *Ginkgo* for the treatment of various inflammatory pathologies and cardiovascular disorders. Original compounds, flavonoids and unique diterpenes (Ginkgolides A, B, C, J) are analysed in *G. biloba* leaves (Van Beek *et al.*, 1990) and make this species useful in medicine. The leaf extract possesses a vasoregulating activity i.e. increased blood flow (Auguet *et al.*, 1986; Clostre, 1986; Kleijnen and Knipschild, 1992). Likewise the Ginkgolide B has powerful anti-Platelet Activating Factors (PAF) effect (Dupont *et al.*, 1986) warranting their efficiency in thrombosis, inflammatory reactions, cardiovascular disorders and cerebral insufficiency (Kleijnen and Knipschild, 1992).

1.3. *Previous haploid work* in vitro

It has previously been reported that in accordance with the conditions of *in vitro* culture, pollen development occurs in two pathways. Firstly, the pollen germination pathway of *G. biloba* was observed in the presence of complex and modified White (1943) and Tulecke (1953, 1957) media. Secondly, Tulecke (1953, 1957) reported that in the absence of germination, 0,1% of pollen led to the formation of a brittle white haploid tissue. At the same time, precocious embryogenic tissues were observed on the calli obtained from immature female gametophytes, cultured on either Knop's medium (Favre-Duchartre, 1956) or White's medium supplemented with coconut milk and 2,4–dichlorophenoxyacetic acid (2,4–D) (Tulecke, 1964). Also, similar tissue was initiated from mature female gametophytes cultured in the composite media amended with macroelements and organic compounds of Bourgin and Nistch (BN) (1967), microelements of Murashige and Tucker (MT) (1969) and 2,4–D and coconut milk (Rohr, 1980a). Recently, protoplasts isolated from immature prothalluses of *G. biloba* led to the formation of embryos in a modified liquid MT and BN media (Laurain *et al.*, 1993b).

2. New haploid works *in vitro*

For therapeutic purposes, it would be interesting to select haploid plants which synthesize new molecules or accumulate Ginkgolides or flavonoids at high levels. As previously reported, *in vitro* variability can allow optimal production of medicinal compounds (Deus and Zenk, 1982). Gametophyte embryos are useful tools for plant improvement. Indeed, isogenic diploids have been developed and from these plants it has been possible to isolate useful mutants (Bajaj *et al.*, 1988) and biochemical variants (De Paepe, 1986) notably variants with therapeutic properties (Sangwan-Norreel *et al.*, 1986). To attain these aims, embryogenic cultures were developed, either from immature microspores or from protoplasts isolated from male gametophyte suspensions. Therefore microspores of *G. biloba*, isolated at the uninucleate stage, were successively:
– cultured at different cell densities in the liquid Bourgin and Nitsch (1967)

Table 1. Culture media used for microspores, embryos, embryo clusters and suspensions of *Ginkgo biloba*

	Micropore culture		Embryo and embryogenic cluster culture			Embryogenic suspension		
	Liquid medium		Solid medium			Liquid medium		
Basal medium	BN	BN	BN	B5	B5	MT	MT	MT
Medium termed	BN	BNSC	A	B	C	D	E	F
Sucrose (M)	0.057	0.057	0.029	0.057	0.057	-	-	-
Glucose (M)	-	-	-	-	-	0.164	0.164	0.164
IAA (μM)	-	11.42	-	-	-	-	-	-
Kin (μM)	-	0.93	11.6	-	-	0.93	-	-
NAA (μM)	-	-	-	-	1.07	10.74	-	18.79
BA (μM)	-	-	-	-	8.87	-	-	-
Coconut milk (g/l)	120	120	120	-	-	-	-	-
Agar (g/l)	-	-	9	9	9	-	-	-

Mineral salts and organic additions to Bourgin &Nitsch(1967) medium = (BN) ; of Gamborg *et al.* (1976) = (B5) ; of Murashige & Tücker (1969) = (MT)
For all media, the pH was adjusted to 6 (before autoclaving).

Table 2. Various parameters were used to electropulse microspore suspension of *G. biloba*

Electric field ($V\ cm^{-1}$)	50	100	250	750	1000
Pulse duration (μs)	20 - 30 - 100	20 - 30 - 100	20 - 30 - 100	20 - 30 - 100	20 - 30 - 100
Pulse number	3 3 1	3 3 1	3 3 1	3 3 1	3 3 1

medium (Table 1) without growth factors (BN) (Exp. Ia) or supplemented with 11.42 μM indole-3-acetic acid (IAA) and 0.93 μM kinetin (KIN) BNSC (Exp. Ib) (Laurain *et al.*, 1993a);

- electrostimulated by short electrical pulses of various voltages (Exp. II) (Table 2) (Laurain *et al.*, 1993a);
- the microspore-derived embryogenic cell suspension was later used for protoplast isolation and culture (data unpublished).

Fig. 2 shows the sequence of different cultures started from isolated microspores of *G. biloba*.

2.1. *Embryogenesis from microspores*

2.1.1. *Protocols*

2.1.1.1. *Microspore isolation.* Mesoblasts bearing male inflorescences were

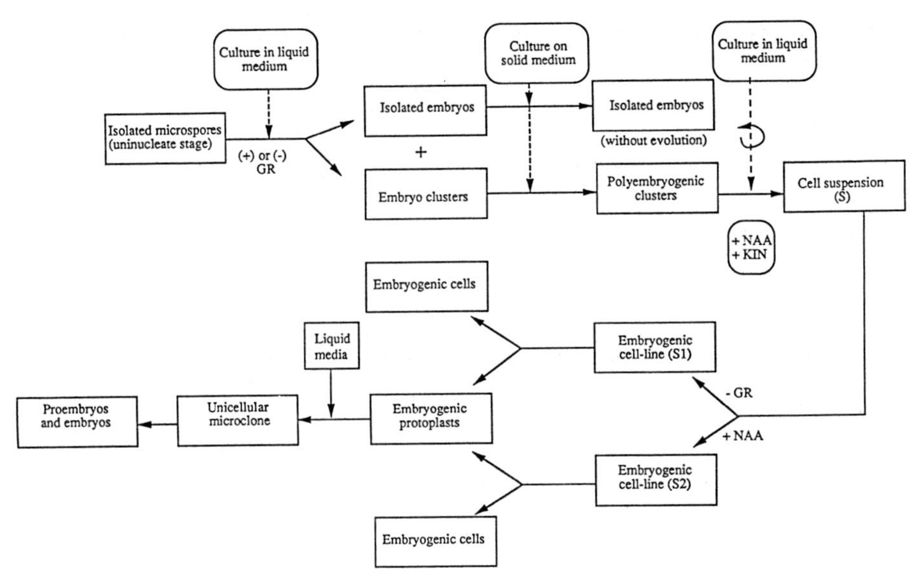

Figure 2. Sequence of the various cultures obtained from isolated microspores of *G. biloba*. GR – growth regulators; NAA – Naphtalene acetic acid; KIN – Kinetin; +: with, or −: without.

collected from a *G. biloba* tree in the Botanical Garden of Tours during a short period of time (from 23rd March to the 3rd April 1992) in order to obtain microsporophylls containing uninucleate microspores. The correct developmental stage was determined by nuclear staining with acetocarmine.

Fresh microsporophylls introduced into 7.5% (w/v) calcium hypochlorite solution, were vigourously shaken for 5 min to sterilize (2200 vibrations min^{-1}). After washing three times with sterilized distilled water, each microsporophyll was gently burst with forceps, liberating large quantities of microspores into 1 ml Bourgin and Nitsch (1967) liquid media either lacking or supplemented with growth regulators (Exp. Ia; Exps. Ib–II). The different liquid media used in all experiments are shown (Table 1). Five to eleven microsporophylls were necessary to achieve 1.5×10^4 to 5×10^4 microspores/ml.

2.1.1.2. *Microspore culture*. The basal BN (Bourgin and Nitsch 1967) medium was used with supplements as indicated in (Table 1). In our study, three experiments were conducted as follows: microspores were isolated at various densities (1.5×10^4 to 10^5/ml) and cultured either in free-hormones BN liquid medium (experiment Ia), or in BN liquid medium supplemented with 11.42 μM IAA and 0.93 μM KIN (experiment Ib-II). In experiment II, before being cultured, microspores were electrostimulated according to the conditions described (Table 2). Two milliliters of final suspension were plated in 55 mm Petri dishes (Exp. Ia,b) and 1 ml in 35 mm corning dishes (Exp. II). These microspore cultures were sealed with Parafilm, stored at 24 ± 1°C in the dark for 8 days and subsequently exposed to light (1000–1500 lux) at the same temperature. After two months, microspore cultures were routinely diluted by 1–2 ml (Exp. Ia,b) and 100–200 μl (Exp. II) of fresh liquid media at monthly intervals. After 4–5 months the embryos obtained from microspores were transferred onto various solid media with or without growth regulator supplements. Sucrose was added to some of the media at various concentrations. The BN (A) medium was found be most efficient (Table 1) (Laurain *et al.*, 1993a). All media were previously adjusted to pH 6 (before autoclaving).

2.1.1.3. *Electrostimulation conditions.* Isolated microspores of *G. biloba* were mixed (1.5 to 5 × 10^4 ml) in buffer solution containing 5 mM MES (N-morpholinoethane sulphonic acid), 3 mM $MgCl_2$ and 57.8 mM sucrose. From this microspore mixture samples of 100 μl were dispensed into corning dishes (35 mm diam) and were pulsed at various electric fields, pulse-durations and pulse-numbers (Table 2) using the Electropulsator ATEIM/CNRS (France). After the electric shock, 100 μl of the microspore suspension were diluted with 1 ml BN liquid medium supplemented with 11.42 μM IAA and 0.93 μM KIN. The electrostimulated microspore cultures were kept in the dark (as previously described) (Laurain *et al.*, 1993a).

2.1.2. *Early events in culture*

During the culture in liquid medium microspores at uninucleate stage, identified by nuclear staining with acetocarmine, were generally spherical in shape and displayed a dense cytoplasm with a central nucleus and small plastids (Fig. 3A). After 24 h, most microspores were 35 to 50 μm in diameter. At the same time, the viability of isolated microspores of *G. biloba*, was found to be respectively: 45% (Exp. Ia) without growth factors, 54% (Exp. Ib) with IAA and KIN, and 36 to 52% for electropulsed microspores (Exp. II) cultured in BNSC medium. Within 48–72 h the cellular volume of cultured microspores increased greatly and their diameter reached 102 to 125 μm.

The androgenic ability of microspores was influenced by various factors, such as microspore density, developmental stage, and growth regulators. Thus, the density of microspores in the culture medium was apparently a determining factor in achieving success of the culture (Keller and Stringam, 1978). Densities between 1.5 to 5 × 10^4/ml were ideal for *G. biloba* microspore culture. Likewise, in the case of *Brassica napus* microspore cultures, the optimal density was 10^4 to 4 × 10^4/ml.

The uninucleate microspores of *G. biloba* and many other plant species are ideal for the development of highly efficient microspore cultures (Keller and Stringam, 1978). For example, microspores of *Zea mays* (Coumans *et al.*, 1989), *Datura* (Sopory and Maheshwari, 1976), *Arachis hypogaea* (Willcox *et al.*, 1991), and some woody species such as *Citrus*, *Cupressus*, *Hevea* (Sangwan-Norreel and Duhoux, 1982) and *Aesculus hippocastanum* (Radojevic *et al.*, 1989) were most responsive. On the other hand, the most favourable

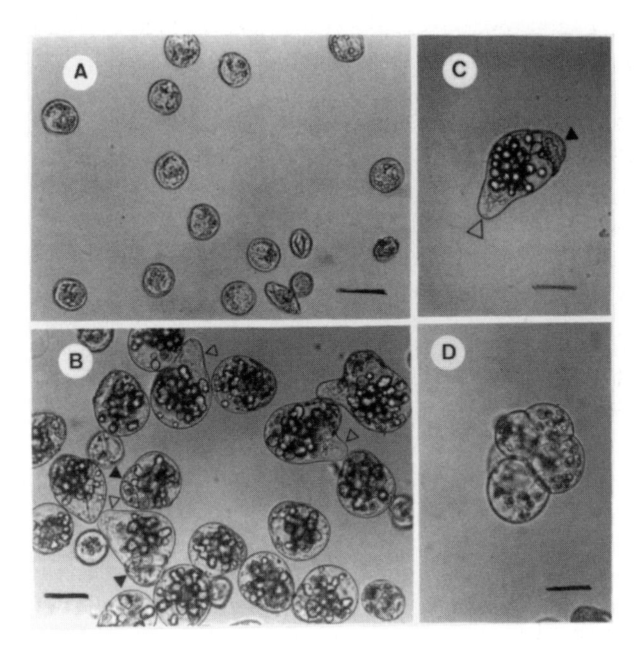

Figure 3. In vitro cultures of *Ginkgo biloba* microspores. (A) Freshly isolated microspores in BN liquid medium. (B-C) Microspores showing asymmetrical divisions ▲ with pollen-tube emergence △. (D) A small cluster formed by isodiametric cell divisions from one microspore. Bar = 100 μm.

time for the initiation of embryogenesis in microspores of *Nicotiana tabacum* was found to be just before or at the time of the first mitosis (Nitsch and Nitsch, 1969). Likewise, microspores of *B. napus* were reported to be highly embryogenic between the uninucleate stage and the early binucleate stage (Takahata and Keller, 1991). Similarly, the bicellular stage was favourable in *Prunus armeniaca* (Harn and Kim, 1972) and *Solanum carolinense* (Reynolds, 1986).

In the three experiments reported, *G. biloba* microspores exhibited a similar variation in their development. In some *G. biloba* microspores, both in the presence and the absence of growth factors, germination tended to begin with asymmetric divisions (Fig. 3B). After a week of incubation, 25 to 36% of microspores started dividing. In 11–24% of this microspore sample, pollen tubes developed from the vegetative cells after two to three weeks (Fig. 3B,C). However, other microspores showed atypical modes of division. Some cells (i.e., vegetative, generative or prothallial cells) released "enveloped nuclei" in the extracellular medium, by a phenomenon that we have named "inverted endocytosis". This process has been observed during our protoplast cultures in woody species such as *Choisya ternata*, *Rauwolfia*

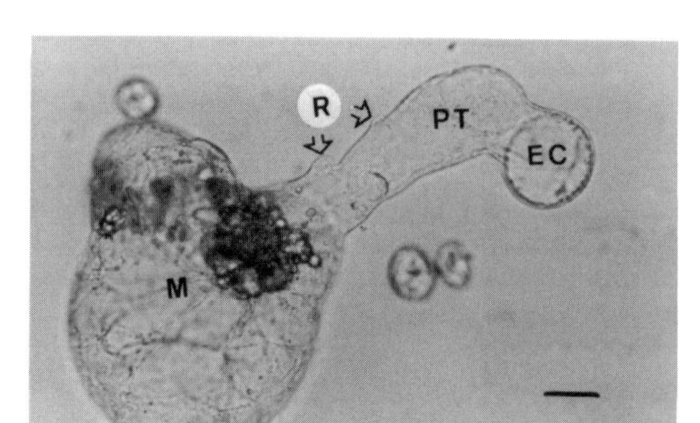

Figure 4. Growth of one pollen tube (PT) with rhizoid structures (R) and an embryogenic cell that was ejected by inverted endocytos (EC) during the culture of *Ginkgo biloba* culture. Bar = 100 μm.

vomitoria, *Ginkgo biloba*. In some microspores the inverted endocytosis (IE) could be observed with or without germination (Fig. 4). These nuclear expulsions generated egg cell-like structures, leading to unicellular microclones that directly produced globular embryos. On the other hand, without germination, some microspores directly evolved into microclones by internal cell division, others formed clusters by isodiametric cellular divisions, and all could generate embryos and androgenic clusters. *In vitro* culture of male gametophytes can take 2 pathways: a gametophytic (germination) or a sporophytic (embryogenesis) pathway (Sangwan-Norreel *et al.*, 1986). The isolated microspores of *G. biloba*, at the unicleate stage, cultured in the Bourgin and Nitsch (1967) medium, were able to complete their morphogenic programme and proceeded through embryogenesis.

2.1.3. *Embryo development in liquid medium*
After 6 weeks, proembryos led to the formation of globular, oblong, heart and torpedo shaped embryos (Fig. 5A–C) in the presence as well as absence of growth regulators and four weeks later, the embryo number was found to be 300 ml (Exp. Ia), 350 ml (Exp. Ib) and 3650 ml (standard Exp. II) (Table 3). The embryo formation occured earlier when the microspores were previously electrostimulated. Six to ten weeks later, after electrostimulation, the number of embryos obtained in experiment II was 250–600 ml and 2500–5300 ml, respectively (Table 3).

The efficiency of embryogenesis (defined as the number of embryos divided by the number of microspores originally plated) was either 1.8% (Exp. Ia) when the microspores were cultured without any growth regulators or 1.8% (Exp. Ib) in the presence of IAA and KIN. The highest frequency

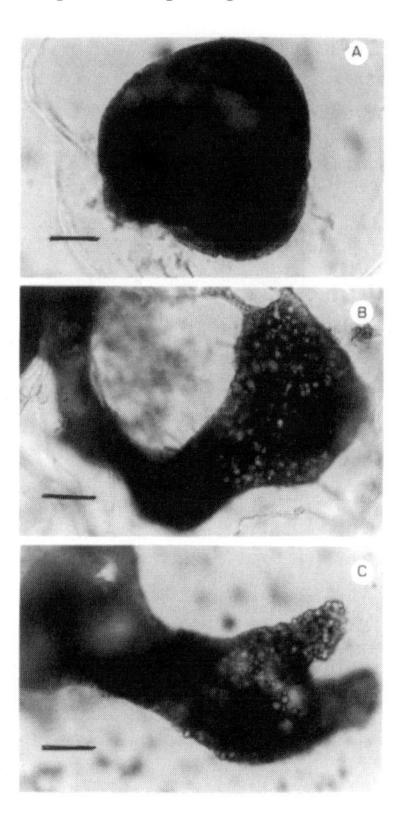

Figure 5. Various stages of embryo-development from microspores of *G. biloba*. (A) One embryo showed the heart-shape. (B-C) One embryo showed the torpedo-shape (B) and the young cotyledonary-shape (C). Bar = 100 μm.

of embryo formation occured (Exp. II) when the microspores were cultured in BN medium supplemented with IAA and KIN, and showed the following results: 21.70% for standard microspores and 16.70 to 23.30% for electrostimulated microspores (Table 3).

After four to five months in the liquid culture, the embryos had reached different stages of development. Slow growing embryos were transferred onto various solid media (Table 1) which was not detrimental to the embryos. However, only embryos in clusters continued to grow on the agar medium containing 11.6 μM KIN. The isolated embryos apparently did not develop any further. One month later, the embryo-clusters developed highly and from the latter a few embryos exhibited a club shape.

Our results show that *G. biloba* microspores did not require exogenous growth regulators. Similar results were observed during maize androgenesis (Mitchell and Petolino, 1991) and in other species where the endogenous hormonal level was sufficient to produce embryos (Coumans *et al.*, 1989).

Table 3. Embryo formation and embryo efficiency from microspores of *Ginkgo biloba* cultured for 4, 6 and 10 weeks

Experiments Efficiency of (Ia - Ib - II)	Formation of proembryos and embryos ; quantities of embryos (ml^{-1}) after 4.6 and 10 weeks			Densities of isolated microspores $x10^{-4}$ ml^{-1}	**Embryo** formation (%) after 10 weeks
	4	6	10		
Ia	*PE -	**E	300	1.60	1.80
Ib	PE -	E	350	1.9	1.8
II Microspores électro stimulated (volt cm^{-1})					
50	PE.E	400	4000	2.4	16.7
100	PE.E	250	3500	1.6	21.9
250	PE.E	600	2500	1.5	16.7
500	PE.E	260	5300	2.4	22.
750	PE.E	400	4900	2.1	23.3
1000	PE.E	380	4170	2.1	19.9
Standard microspores	PE -	150	3650	1.7	21.5

*PE (proembryos) ; ** E (embryos) - PE + E (non evaluated)

Microspores of *G. biloba* were cultured in BN basal medium (Exp Ia) and in BN medium supplemented with 2 mg/l IAA and 0.2 mg/l KIN (Exp. Ib - II).

However, coconut milk in Bourgin and Nitsch medium supplies cytokinins and is able to initiate androgenesis from *Ginkgo* microspores. As shown by Keller and Stringam (1978), cytokinins were essential for maximum microspore response in *Datura* and potato. On the other hand, we observed that the presence of IAA and KIN in the medium was not detrimental for embryo formation of *G. biloba*. Similarly, *Brassica campestris* microspores produced large numbers of embryos in the presence as well as absence of growth regulators (Sato *et al.*, 1989).

Our study demonstrated good resistance of microspores to high voltages. The electrical pulses seemed to be responsible for the early embryo development in electropulsed microspores. It is well-known that the application of electrical currents stimulates cell division and differentiation, especially in protoplasts of woody species regarded as recalcitrant to culture (Rech *et al.*, 1987; Chand *et al.*, 1988; Davey and Power, 1987; Montane and Teissié, 1989).

In experiment II (Table 3), the highest yields of embryos were obtained both from electrostimulated and control microspores of *G. biloba*. Both resulted in a large number of embryos which were increased by a factor of ten compared with experiments Ia–Ib carried out ten days earlier (with younger microspores). The increased yield of embryos cannot be due to the electrostimulation, but may be due to a particular phase of the uninucleate stage. Keller and Stringam (1978) reported that the uninucleate stage has been subdivided, and that the best response could be obtained at specific substages.

2.2. *Establishment of cell suspensions*

Embryogenic clusters from *G. biloba* microspores were first cultured in the liquid BNSC (Bourgin and Nitsch, 1967) medium with an addition of IAA and KIN (Table 1).

These polyembryogenic clusters (Fig. 6) were transferred onto various solid media: BN (A) (Bourgin and Nitsch, 1967) medium supplemented with 11.61 μM KIN, B5 (B) (Gamborg *et al.*, 1976) without growth regulators, B5 (C) with 1.07 μM NAA and 8.87 μM BA (Table 1) showed retarded growth. It took 6–8 weeks to achieve each subculture. The cellular growth increased greatly when the clusters were transferred into (Murashige and Tucker, 1969) MT(D) liquid medium shaken and supplemented with 10.74 μM NAA and 0.93 μM KIN. After one month of culture, primary cellular suspensions (S), composed of isolated cells or small aggregates were resuspended in two new liquid media [MT(E) without growth substances, and MT(F) supplemented with 18.79 μM NAA] (Table 1). This was done to initiate or induce a new embryogenesis. One month later the cellular lines obtained, termed respectively S_1 (Table 1) and S_2 (Fig. 2), exhibited embryogenic cells in the MT (E) and MT (F) media (data unpublished).

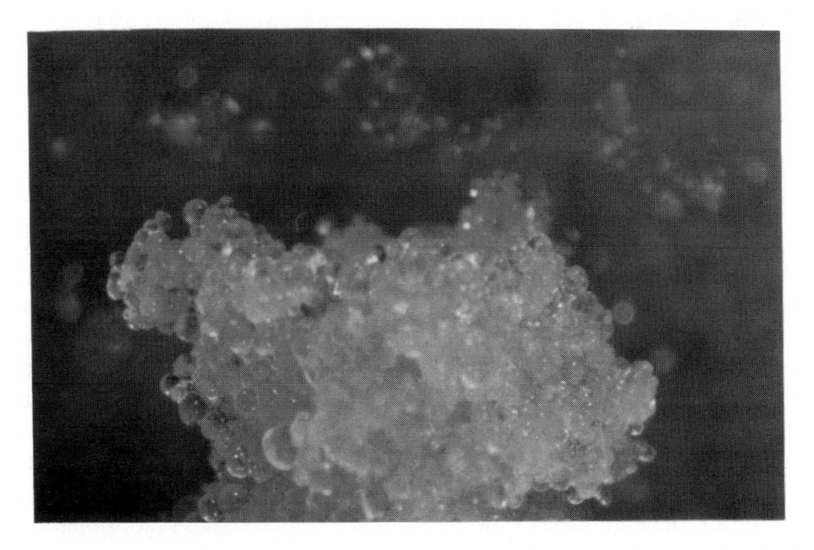

Figure 6. Embryogenic cluster of *G. biloba* on solid medium after five months of culture. Stereomicroscopic observation (60×).

2.3. *Embryogenesis from cell suspension protoplasts*

2.3.1. *Protoplast isolation and culture*

Protoplasts of *G. biloba* were isolated from a six-day-old subculture of cell suspension (S_1) and collected by filtration through a nylon sieve with a 100 μm mesh. In fact 650 ± 0.09 mg of fresh cells were dispersed into 4 ml of enzymatic mixture containing 0.50% (w/v) cellulase Onozuka R10 + 0.25% macerozyme R10 + 0.034 M $CaCl_2 \cdot 2H_2O$ + 0.5 M sorbitol (pH 5.5) and incubated in the dark at 25°C for 6–8 h. When debris and cellular aggregates were absent, the protoplast suspension was immediately centrifuged at 80 × g for 3 min. The supernatant was discarded and the protoplast pellet was resuspended in 4 ml of an isotonic enzyme-free medium (0.6 M sorbitol + 0.034 M $CaCl_2 \cdot H_2O$). The protoplasts were washed and centrifuged 3 times, and then resuspended in the basal culture media without growth regulators. After this, the protoplasts were diluted to 5×10^4/ml with each culture medium (Table 4). The various media under investigation received an addition of different combinations of growth factors (Table 4). For each modified medium of Bourgin and Nitsch, 1967 (BN); Murashige and Tucker, 1969 (MT), Gamborg *et al.*, 1976 (B5), replicates of 8, 4 and 8 were used, respectively.

2.3.2. *Results*

The procedure of protoplast isolation yielded 2×10^6 protoplasts per gram fresh weight of cell suspension S_1 (Fig. 7A). After purification test of viability by either fluorescein diacetate (FDA) or Evans blue staining showed that 96 to 98% of the protoplasts were viable (Fig. 7B). The spherical protoplasts had a diameter of 58.8 ± 13.8 μm and a clear cytoplasm containing small plastids. Some of them exhibited a dark, blurry-looking content which seemed to reflect a capacity for embryogenesis. The inoculation density was 5×10^4/ml in various Bourgin and Nitsch (1967), Murashige and Tucker (1969) and Gamborg *et al.* (1976) liquid media modified by adding 0.25 M sorbitol and 0.15 M glucose and various combinations of growth regulators (Table 4). After 24–48 h the viability of protoplasts was 75–90% in BN and in MT media. New cell wall formation, indicated indirectly by the change of shapes of the cells after 36–48 h in 19 ± 11% protoplasts and after 4–5 days, occurred in 49 ± 11% protoplasts. First atypic cellular divisions were observed after 36–48 h in 14 ± 2.5% protoplasts in BN medium. Microclones formed through repeated cytokinesis, either from multinucleate cells derived from protoplasts, or from "enveloped nuclei" released by the process called "inverted endocytosis", in all of tested media. After 5–7 days, the number of microclones were recorded as 1300 ± 700/ml and 700 ± 200 ml in BN and MT liquid media, respectively. In the beginning, some protoplasts produced long cytoplasmic extensions (100 to 200 μm), which looked like pollen tubes observed during the germination of microspores (Fig. 7C). The number of long cytoplasmic extensions increased to 450 ± 100/ml in BN liquid media

Table 4. Modified liquid media of Bourgin and Nitsch (1967), Murashige and Tücker (1969), Gamborg *et al.* (1976) used for the protoplast culture isolated from a cell line of *G. biloba*

	BN			MT						B5	
Medium termed	BN (0)	BN (1)	BN (2)	MT (0)	MT (1)	MT (2)	MT (3)	MT (4)	MT (5)	B5 (0)	B5 (1)
Basal medium	BN	BN	BN	MT	MT	MT	MT	MT	MT	B5	B5
Mineral medium	BN	BN	BN	*MT/2	MT/2	MT/2	MT/2	MT/2	MT/2	B5	B5
Glucose (M)	0.25	0.25	0.25	0.25	0.25	0.25	0.25	0.25	0.25	0.25	0.25
Mannitol (M)	0.35	0.35	0.35	0.35	0.35	0.35	0.35	0.35	0.35	0.35	0.35
µM											
NAA	-	-	-	-	10.74	10.74	2.69	-	-	-	-
2,4-D	-	-	-	-	-	-	-	-	-	-	0.45
IAA	-	11.42	17.13	-	-	-	-	-	11.42	-	-
KIN	-	0.46	0.46	-	-	0.93	-	-	0.93	-	
BA	-	-	-	-	-	-	8.87	8.87	-	-	-

BN : Bourgin & Nitsch (1967); MT : Murashige & Tucker (1969); B5 : Gamborg *et al.* (1976) ; Basal medium (mineral medium + vitamins) ; * MT/2 (half strength mineral MT medium). For all media, the pH was adjusted to 5.8

(with or without growth substances) within 6 days. After 2 weeks, microcolonies derived proembryos were obtained (Fig. 7D). Their numbers increased to 200–500/ml and 180–640/ml in BN, MT and B5 media in the presence or the absence of growth regulators (Table 5).

Two weeks later, some proembryos developed into globular embryos (Fig. 8A,B). In all media, the number of proembryos and embryos were found to be 550–2000/ml and 320–1500/ml, respectively, depending on the presence or the absence of growth regulators (Table 5). Embryogenesis was observed with an efficiency (the number of embryos divided by the number of protoplasts originally plated × 100) of 0.6 to 4% depending on the various media

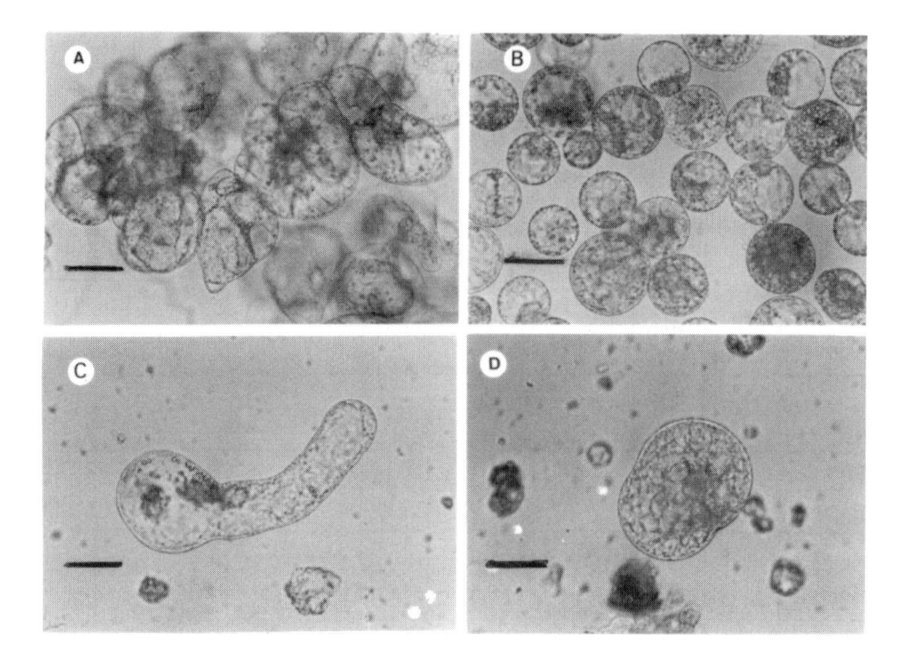

Figure 7. (A) Embryogenic cell suspension (S_1) of *G. biloba*. (B) Isolated protoplasts from the S_1 suspension. (C) A long cytoplasmic extension, like pollen-tube, produced by one protoplast. (D) A small cell-derived microclone by repeated nuclear divisions and cytokinesis. Bar = 100 μm.

tested (Table 5). After 2–3 months, the growth of embryos slowed down. When abscisic acid was added at various concentrations to the culture media, the development of embryos and embryo-clusters restarted.

Thus protoplasts isolated from a microspore-derived cell suspension of *G. biloba* were able to regenerate new cell walls and divide in growth regulator-free as well as growth regulator supplemented liquid media. It is well-known that most protoplasts absolutely require growth regulators at the beginning of culture. For example, protoplasts derived from gymnosperm cell lines required auxins in *Abies alba* (Lang and Kohlenbach, 1989), and required auxins and cytokinins in *Pinus taeda* (Gupta and Durzan, 1987) and *Larix* × *eurolepis* (Klimaszewska, 1989). Equally, in *Picea glauca* Attree *et al.* (1987) reported that protoplasts cultured without growth regulators, or with BA alone, died after 2 weeks. Moreover, protoplasts isolated from crown gall, excepting tumorous lines of *Zea mays* (Pedersen *et al.*, 1983), *Petunia* (Phillips and Darrell, 1988) and *Catharanthus roseus* (Trémouillaux-Guiller *et al.*, 1994) had an absolute requirement for growth regulators to initiate cell division (Scowcroft *et al.*, 1973). In contrast, protoplasts isolated from suspensions of *Citrus sinensis* (Harms and Potrykus, 1980) and from female prothalluses of *Ginkgo biloba* could regenerate without growth regulators

Table 5. Number of proembryos and embryos obtained from protoplasts derived from a cell line of *G. biloba* cultured in various media

Medium	Proembryos per ml after 2 weeks	Proembryos and embryos per ml after 4 weeks	* Efficiency of pro-embryo and embryo formation after 4 weeks
BN(0)	500 ± 190	2000 ± 400	4%
BN (1)	520 ± 250	1030 ± 1000	2%
BN (2)	640 ± 560	1500 ± 700	3%
MT (0)	200 ± 70	550 ± 300	1%
MT (1)	210 ± 90	580 ± 160	1.2%
MT (2)	190 ± 160	370 ± 110	1.2%
MT (3)	170 ± 80	600 ± 300	1.2%
MT (4)	160 ± 30	400 -	0.8%
MT (5)	180 ± 10	320 -	0.6%
B5 (0)	330 ± 80	700 ± 200	1.4%
B5 (1)	180 ± 50	730 ± 20	1.5%

BN: Bourgin & Nitsch (1967); MT: Murashige & Tucker (1969); B5:Gamborg *et al.*,(1976) modified liquid media
* Efficiency : the number of pro-embryos and embryos divided by the number of protoplasts originally plated
For each modification an average of 8, 4 and 8 replicates were made,respectively.

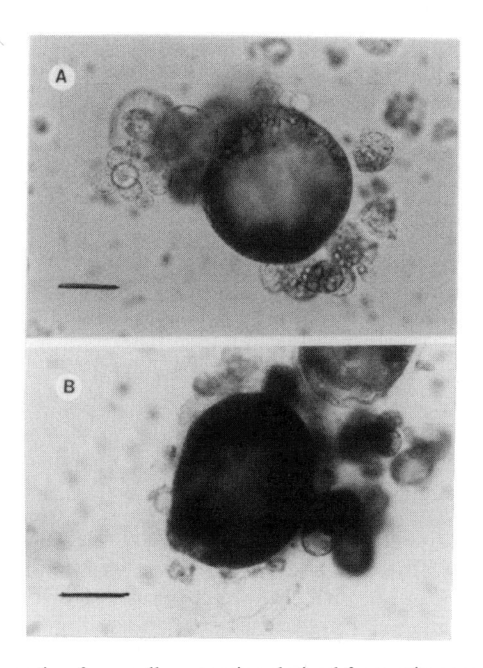

Figure 8. Embryo formation from cell suspension derived from microspores of *Ginkgo biloba*. (A,B) protoplast derived proembryo (A) and embryo (B). Bar = 100 μm.

(Laurain *et al.* 1993b). Moreover, no effect, either inhibiting or beneficial to growth, was observed in the presence of auxins and cytokinins during the present cultures of *Ginkgo biloba* protoplasts.

This rare ability of the *G. biloba* protoplasts to grow independent of growth regulators might be due to the species or genotype. Protoplasts from suspensions of *Ginkgo biloba*, cultured with or without growth factors, produced embryos directly. It should be emphasized that direct embryogenesis in tissues (Stolarz *et al.*, 1991), isolated cells (Conger *et al.*, 1983) and protoplasts is a rare phenomenon.

3. Conclusions and prospects

Our experiments have unequivocally shown that isolated microspores of *G. biloba* developed directly into androgenic embryos in the liquid Bourgin and Nitsch (1967) medium with or without addition of auxin. Embryogenic cell suspensions derived from microspores could be used for the isolation of viable protoplasts. These protoplasts demonstrated a great ability to regenerate cell walls, and to divide in the absence as well as in the presence of growth factors in various culture media. Microcolonies derived from protoplasts, comparable to egg-cells, developed into proembryos and later into embryos in all the investigated media. As far as we know, this is the first report on androgenesis from microspores and from microspore-derived protoplasts of *Ginkgo*. The resistance of microspores to electrostimulation, as well as the viability of the protoplasts isolated from embryogenic suspensions, allows us to envisage the production of transgenic embryos, after gene transfer with either electroporation or by co-culture with *Agrobacterium rhizogenes*. Thus methods of gene transfer by electroporation in microspores (Fennell and Hauptmann, 1992), and in intact cells (Klöti *et al.*, 1993) and by co-culture of microspores with bacteria (Pechan, 1989), have been recently reported. These studies have resulted in a need to further investigations in order to permit the complete development of androgenic embryos.

Furthermore, such studies would help elucidate the fundamental mechanisms of androgenesis. This would meet one of the most fundamental challenges posed by developmental biology.

4. References

Auguet, M., S. Delaflotte, A. Hellegouarch and F. Clostre, 1986. Bases pharmacologiques de l'impact vasculaire de l'extrait de *Ginkgo biloba*. Presse Méd. 15: 1524–1528.

Attree, S.M., D.I. Dunstan and L.C. Fowke, 1989. Plantlet regeneration from embryogenic protoplasts of white spruce (*Picea glauca*). Bio/Technology 7: 1060–1062.

Bajaj, Y.P.S., Furmanowa M. and O. Olszowska, 1988. Biotechnology of the micropropagation of medical and aromatic plants. In: Y.P.S. Bajaj (Ed.), Biotechnology in Agriculture and Foresty, Vol. 4. Medicinal and Aromatic Plants, pp. 58–94. Springer-Verlag, Berlin.

Bourgin, J.P. and J.P. Nitsch, 1967. Obtention de *Nicotiana* haploïdes à partir d'étamines cultivées *in vitro*. Ann. Physiol. Veg. 9: 377–382.

Chand, P.K., S.J. Ochatt, E.L. Rech, J.B. Power and M.R. Davey, 1988. Electroporation stimulates plant regeneration from protoplasts of the woody medicinal species *Solanum dulcamara* L. J. Exp. Bot. 39: 1267–1274.

Clostre, F., 1986. De l'organisme aux membranes cellulaires: les différents niveaux d'actions pharmacologiques de l'extrait de *Ginkgo biloba*. Presse Méd. 15: 1529–1538.

Conger, B.V., G.E. Hanning, D.J. Gray and J.K. McDaniel, 1983. Direct embryogenesis from mesophyll cells of orchard grass. Science 221: 850–851.

Coumans, M.P., S. Sohato and E.B. Swanson, 1989. Plant development from isolated microspores of *Zea mays* L. Plant Cell Rep. 7: 618–621.

Christensen, T.G., 1972. A study of the resistance of *Ginkgo biloba* L. to fungi: phytoalexin production induced by *Botrylis allii* Munn. Diss. Abstr. Int. b. 32: 4340.

Davey, M.R. and J.B. Power, 1987. Aspects of protoplasts and plant regeneration. In: K.J. Puite, J.J.M. Dons, H.J. Huizing, A.J. Kool, M. Koornneef and F.A. Krens (Eds.), Progress in Plant Protoplast Research, pp. 15–25. Kluwer Academic Publishers, Dordrecht.

De Paepe, R., 1986. Variation induite par l'androgénèse. Bull. Soc. Bot. Fr. 133, Actualités Bot. 4: 41–50.

Deus, B. and M.H. Zenk, 1982. Exploitation of plant cells for the production of natural compounds. Biotech. Bioengineer. 24: 1965–1974.

Dupont, L., O. Dideberg, G. Germain and P. Braquet, 1986. Structure of Ginkgolide B (BN 52021) monohydrate, a highly specific PAF/Acether Receptor Antagonist isolated from *Ginkgo biloba* L. Acta Cryst. 42: 1729–1762.

Favre-Duchartre, M., 1956. Contribution à l'étude de la reproduction chez le *Ginkgo biloba*. Rev. Cytol. Biol. Veg. XVII (1–2), pp. 214.

Fennel, A. and R. Hauptmann, 1992. Electroporation and PEG delivery DNA into maize microspores. Plant Cell Rep. 11, 567–570.

Gamborg, O.L., T. Murashige, T.A. Thorpe and I.K. Vasil, 1976. Plant tissue culture media. In Vitro 12: 473–478.

Gupta, P.K. and D.J. Durzan, 1987. Somatic embryos from protoplasts of lobolly pine proembryonal cells. Bio/Technology 5: 710–712.

Harms, C.T. and I. Potrykus, 1980. Hormone inhibition of *Citrus* protoplasts released by coculturing with *Nicotiana tabacum* protoplasts: its significance for somatic hybrid selection. Plant Sci. Lett. 19: 125–301.

Harn, C. and M.Z. Kim, 1972. Induction of callus from anthers of *Prunus armeniaca*. Korean J. Breed. 4: 49–53.

Keller, W.A. and G.R. Stringam, 1978. Production and utilization of microspore-derived haploïd plants. In: T.A. Thorpe (Ed.), Frontiers of Plant Cell Culture, pp. 113–122. IAPTC, Canada.

Kleijnen, J. and P. Knipschild, 1992. *Ginkgo biloba*. Lancet. 340: 1136–1139.

Klimaszewska, K., 1989. Recovery of somatic embryos and plantlets from protoplast cultures of *Larix* × *eurolepis*. Plant Cell Rep. 8: 440–444.

Klöti A., V.A. Iglesias, J. Wünn, P.K. Burkhardt, S.K. Datta and I. Potrykus, 1993. Gene transfer by electroporation into intact scutellum cells of wheat embryos. Plant Cell Rep. 12: 671–675.

Lang, H. and H. Kohlenbach, 1989. Cell differentiation in protoplast cultures from embryogenic callus of *Abies alba* L. Plant Cell Rep. 8: 120–123.

Laurain, D., J. Trémouillaux-Guiller, and J.C. Chénieux, 1993a. Embryogenesis from microspores of *Ginkgo biloba* L., a medicinal woody species. Plant Cell Rep. 12: 501–505.

Laurain, D., J.C. Chénieux, and J. Trémouillaux-Guiller, 1993b. Direct embryogenesis from female haploid protoplasts of *Ginkgo biloba* L., a medicinal woody species. Plant Cell Rep. 12: 656–660.

Lee, C.L., 1955. Fertilization in *Ginkgo biloba*. Bot. Gaz. 117: 79–100.

Michel, P.F., 1986. *Ginkgo biloba*, l'arbre qui a vaincu le temps. du félin, l'art du vivant, Paris, p. 108.

Mitchell, J.C. and J.F. Petolino, 1991. Plant regeneration from haploid suspension and protoplast cultures from isolated microspores of maize. J. Plant Physiol. 137: 530–536.
Montane, M.H. and J. Teissié, 1989. Electroactivation of plant protoplast division. Studia Biophysica 130: 223–226.
Murashige, T. and D.P.H. Tucker, 1969. Growth factor requirements of *Citrus* tissues cultures. In: J.V. Chapman (Ed.), Proc. First Int. Citrus Symp. Univ., Calif., Riverside, pp. 1155–1161.
Nitsch, J.P. and C. Nitsch, 1969. Haploid plants from pollen grains. Science 163: 85–87.
Ochatt, S.S. and J.B. Power, 1988a. An alternative approach to plant regeneration from protoplasts of sour cherry (*Prunus cerasis* L.). Plant Sci. 56: 75–79.
Pechan, P.M., 1989. Successful cocultivation of *Brassica napus* microspores and proembryos with *Agrobacterium*. Plant Cell Rep. 8: 387–390.
Pedersen, H.C., J. Christiansen J. and R. Wyndaele, 1983. Induction and *in vitro* culture of soybean crown gall tumors. Plant Cell Rep. 2: 201–204.
Phillips, R. and N.J. Darrel, 1988. A simple technique for single-cell cloning of crown gall tumour tissue; *Petunia* protoplast regeneration without exogenous hormones. J. Plant Physiol. 133: 447–451.
Pincemail, J. and C. Debry, 1986. Propriétés antiradicalaires de l'extrait de *Ginkgo biloba*. Presse Méd. 15: 1475–1479.
Radojevic, L., N. Djordjevic and B. Tucic, 1989. *In vitro* induction of pollen embryos and plantlets in *Aesculus carnea* Hayne through anther culture. Plant Cell. Tiss. Org. Cult. 17: 21–26.
Rech, E.L., S.J. Ochatt, P.K. Chand, J.B. Power and M.R. Davey, 1987. Electro-enhancement of division of plant protoplast-derived cells. Protoplasma 141: 169–176.
Reynolds, T.L., 1986. Pollen embryogenesis in anther cultures of *Solanum carolinense* L. Plant Cell Rep. 5: 273–275.
Rohr, R., 1980a. Développement *in vitro* du pollen de *Ginkgo biloba* L. Cytologia 45: 481–495.
Rohr, R., 1989. Maidenhair tree (*Ginkgo biloba* L.). In: Y.P.S. Bajaj (Ed.), Biotechnology in Agriculture and Forestry, Vol. 5, Trees II, pp. 251–267. Springer-Verlag, Berlin/Heidelberg.
Sangwan-Norreel, B.S. and E. Duhoux, 1982. Les conditions de la gynogénèse et de l'androgénèse expéimentales *in vitro* chez les arbres. Rev. Cytol. Biol. Végét. Bot. 5: 171–187.
Sangwan-Norreel B.S., R.S. Sangwan and J. Pare, 1986. Haploïdie et embryogénèse provoquée *in vitro*. Bull. Soc. Bot. France, 133, Actual. Bot. 4: 7–39.
Sato, T., T. Nishio and M. Harai, 1989. Plant regeneration from isolated microspore cultures of Chinese cabbage (*Brassica campestris* spp. *pekinensis*). Plant Cell Rep. 8: 486–488.
Scowcroft, W.R., M.R. Davey and J.B. Power, 1973. Crown gall protoplasts: isolation, culture and ultrastructure. Plant Sci. Lett. 1: 451–456.
Sopory, S.K. and S.C. Maheshwari, 1976. Development of pollen embryoids in anther cultures of *Datura innoxia*. J. Exp. Bot. 96: 49–57.
Stolarz, T., J. Macewicz and H. Lörz, 1991. Direct embryogenesis and plant regeneration from leaf explants of *Nicotiana tabacum* L. J. Plant Physiol. 1: 36–39.
Takahata, Y. and W.A. Keller, 1991. High frequency embryogenesis and plant regeneration in isolated microspore culture of *Brassica oleracea* L. Plant Sci. 74: 235–242.
Trémouillaux-Guiller J., H. Kodja and J.C. Chénieux, 1994. Single-cell cloning of a crown gall from protoplasts regenerated in hormone-free medium: establishment of pure transformed cell-lines of *Catharanthus roseus* G. Don. Plant Cell Tiss. Org. Cult. 37: 25–30.
Tulecke, W.R., 1953. A tissue derived from the pollen of *Ginkgo biloba*. Science 117: 599–600.
Tulecke, W.R., 1957. The pollen of *Ginkgo biloba*: *in vitro* culture and tissue formation. Amer. J. Bot. 44: 602–608.
Tulecke, W.R., 1964. A haploid tissue from the female gametophyte of *Ginkgo biloba*. Nature 203: 9–95.
Van Beek, T.A., H.A. Scheeren, T. Rantio, W.C.H. Melger and G.P. Lelyveld, 1990.

Determination of ginkgolides and bilobalide in *Ginkgo biloba* leaves and phytopharmaceuticals. J. Chromatogr. 543: 375–387.
White, P.R., 1943. A Handbook of Plant Tissue Culture. The Ronald Press Co., New York.
Willcox, M.C., S.M. Reed, J.A. Burns and J.C. Wynne, 1991. Effect of microspore stage and media on anther culture of peanut (*Arachis hypogaea* L.). Plant Cell. Tiss. Org. Cult. 24: 25–28.

16. Haploidy in forest trees

SNORRI BALDURSSON and M. RAJ AHUJA

Contents

1. Introduction

The main objective of this chapter is to review research on experimental induction of haploids in tree species since the pioneering efforts of LaRue and coworkers in the late fifties. Initially, however, we shall discuss some features of trees affecting haploid research, and give a brief outline of male and female gametophyte development *in vivo*, to set the stage for subsequent discussion on *in vitro* culture of these structures. The reader is also referred to Chen (1986, 1987), Rohr (1987), Bonga *et al.* (1988), Von Aderkas and

S.M. Jain, S.K. Sopory & R.E. Veilleux (eds.), In Vitro *Haploid Production in Higher Plants, Vol. 3*, 297–336.

Dawkins (1993) for recent reviews on haploidy in trees, and Singh (1978), Konar and Moitra (1980), Owens and Blake (1985), Pennell (1988), Sedgley and Griffin (1989), Blackmore and Knox (1990) for a fuller discussion of the development of reproductive structures of trees *in vivo*.

Cytogenetics and morphology of haploidy, and the potential applications of haploids and doubled haploids in forest tree genetics and improvement is discussed elsewhere (Baldursson and Ahuja, 1994).

2. Features of trees affecting haploid research

2.1. *Life history traits*

Development of all woody species from seed to adulthood is characterized by a transition from a strictly vegetative juvenile phase to a reproductive mature phase. Length of the juvenile period, under natural conditions, varies from a few years to several decades depending on species and environments, but is rarely shorter than 5 years in temperate forest trees (Hackett, 1985). Thus, most trees are large at reproductive maturity, which makes controlled growth in glasshouses impractical. Flower induction techniques developed over the last decades have shortened the juvenile period and lessened the space requirements significantly in many species (Ross and Pharish, 1985). As an extreme example, intensive management of birch (*Betula pendula*) in flower induction halls (constant light, CO_2 enrichment) has reduced the juvenile period to only 8 months (Viherä-Aarnio, 1991).

Temperate trees exhibit strict seasonality in flowering. Flowering in most species appears to be controlled more by gross seasonal changes rather than daylength alone, complicating attempts to induce flowering more than once a year (Sedgley and Griffin, 1989). However, grafts of *Picea glauca* can be made to flower twice a year by applying extra light and chilling periods in the greenhouse (Jan M. Bonga, personal communication). Production of sexual structures *in vitro* is another potential way of increasing the supply of haploid explants for experimentation. In *Sequoia sempervirens*, functional male and female cones were produced *in vitro* (Van Tran Thanh *et al.*, 1987), and successful *in vitro* flowering has been described in two species of bamboo (Nadgauda *et al.*, 1990).

Nevertheless, most researchers dealing with breeding or reproductive biology of trees must still make extensive use of field grown material. It follows that experiments must be restricted to the seasonal period of natural flowering. This period can be prolonged somewhat (two or three months at best) by early forcing of dormant flowerbuds and/or cold storage of flowering branches. However, the effects of such treatments on the response in e.g., anther culture (Section 5.3.3) are uncertain. Many temperate tree species, furthermore, flower irregularly or periodically (Sedgley and Griffin, 1989) making it uncertain that experimental material will be available every year (e.g., Baldursson *et al.*, 1993a). Even in the case of annual flowering, the dependency on field material invites fluctuations in response, due to climatic factors affecting physiological status and rate of microbial infection of donor trees. Finally, there may be logistic or practical difficulties in collecting the experimental material.

2.2. *Cytogenetics*

In plants, the term "haploid" is used for individuals containing the gametic chromosome number of the parent species. Depending on ploidy level of the parent, a distinction can be made between monohaploids and polyhaploids, arising from diploid or polyploid parents, respectively. Monohaploids contain only one set of chromosomes and are sexually sterile. Polyhaploids contain more than one set of chromosomes and may be sterile or fertile, depending on their genomic content (Kimber and Riley, 1963; Magoon and Khanna, 1963). Polyhaploids are generally more easily obtained than monohaploids. This is not surprising as the former may still contain some heterozygosity, masking unfavourable mutations, whereas in monohaploids, all lethal and semi-lethal recessive genes will be expressed.

Gymnosperms as a group are remarkably stable and conservative in their chromosome numbers (Sax and Sax, 1933; Khoshoo, 1961; Libby *et al.*, 1969), suggesting resistance to changes in chromosome numbers and ploidy levels. In conifers, a basic number of 12 is the general rule, varying from only 10 to 13 among eight of the nine families. In the remaining conifer family Podocarpaceae, chromosome numbers vary from 9 to 19. Usually the numbers are constant within a family. Only four truly polyploid conifers have been identified: two tetraploid species of *Juniperus*, a triploid *Larix* species, and the hexaploid *Sequoia sempervirens* (Libby *et al.*, 1969). It is interesting to note that *S. sempervirens* represents the only recorded example of a dihaploid gymnosperm growing in soil (Ball, 1987).

Chromosomes are usually smaller in the woody angiosperms with basic numbers ranging from 6 to 41. Furthermore, numbers often vary within families and genera and polyploidy is common (Libby *et al.*, 1969; Zhang *et al.*, 1990). *Hevea brasiliensis*, one of the most successful tree species in terms of haploid induction, is an allotetraploid (Chen, 1990).

3. Development of the male and female gametophytes *in vivo*

3.1. *The male gametophyte*

Male gametophyte ontogeny is basically similar in woody angiosperms and gymnosperms and will be dealt with in one section, while the large differences in megagametophyte development and structure merit separate sections for each of these groups.

Microsporogenesis encompasses the developmental events up to the first microspore division and includes formation of the pollen mother cell (PMC), meiosis, tetrad formation, and microspore maturation. "Microsporogenesis" and "pollen development" are often used interchangeably in the literature, and in that case both imply the whole set of developmental processes up to formation of mature pollen. However, the term "microspore" applies only to the uninucleate phase of the gametophyte and the term "pollen" is more correct when post-meiotic division has taken place (Pennell, 1988). PMCs differentiate from archaesporial tissue in the microsporangium (gymnosperms) or pollen sac (angiosperms). In most angiosperms the anther develops two lobes, each with two pollen sacs. In gymnosperms each microsporophyll bears two (Pinaceae) or more (Cupressaceae, Araucariaceae, Taxodiaceae) microsporangia on their abaxial surface (Sedgley and Griffin, 1989). Microsporogenesis is usually a continuous process once initiated in the spring (Blackmore and Knox, 1990), but can be interrupted at various stages by winter dormancy. Among the conifers, *Picea* and *Abies* overwinter at the PMC stage, whereas meiosis in *Larix*, *Pseudotsuga*, *Thuja* and *Tsuga* begins in the fall and becomes arrested at the pachytene or diplotene stage (Owens and Blake, 1985). In Betulaceae, meiosis occurs in the fall and the gametophytes overwinter as unicellular microspores. The microspores then divide and pollen development is completed when the temperature rises in the spring (Dunbar and Rowley, 1984).

In angiosperms, the first mitotic division of the haploid microspore gives rise to a bicellular pollen grain containing the vegetative and the generative cell. About two thirds of angiosperm species release pollen at this bicellular stage. In the remaining one third the generative nucleus undergoes a second mitotic division, while still in the anther, to form the two sperm nuclei (Sedgley and Griffin, 1989).

In *Taxus* spp., gametophytes are released from the mother plant as microspores (Pennell, 1988). Pollen development of Cupressaceae, Taxodiaceae and Taxaceae resembles that of angiosperms in that each microspore divides giving rise directly to a tube cell and a generative cell. Pollen is shed at this stage, except in *Taxus*. Pollen grains of other gymnosperms contain from two to a dozen or more cells at anthesis. In these the first microspore division is unequal, giving rise to a small prothallial cell and a large embryonal cell. The embryonal cell then divides again unequally forming a second prothallial cell and a larger antheridial initial which upon division gives rise to the tube cell and the generative cell. Pollen is shed at this four-celled stage or at the five-celled stage, after the generative cell has divided equally to form the stalk and body cells (Owens and Blake, 1985). In Araucariaceae the prothallial cells proliferate resulting in multicellular pollen grains (Singh, 1978).

Maturation of the male gametophyte is completed after the pollen grain has reached the stigma (angiosperms) or nucellus (gymnosperms) of the female reproductive structures of the same species. The various pollination mechanisms, as well as post-pollination development of male gametophytes in trees were reviewed by Owens and Blake (1985) and Sedgley and Griffin (1989).

In most angiosperms growth of the pollen tube is rapid and fertilization takes place within hours or days of pollination (Sedgley and Griffin, 1989). An exception is *Quercus velutina* with fertilization occurring 13 months after pollination (Mogensen, 1965).

Pollen germination is generally much slower in gymnosperms, taking from two weeks in *Picea engelmannii* (Owens and Blake, 1985), 3–5 weeks in *Picea sitchensis* (Owens and Molder, 1980; Baldursson *et al.*, 1993a), approximately 6 weeks in *Larix spp.* to several months, interrupted by winter dormancy, in *Pinus* (Singh, 1978). The reason for delayed pollen germination in gymnosperms may simply be to allow early pollination in spite of the slow megagametophyte development. Delayed fertilization as a reproductive strategy by the female to increase pollen competition has also been suggested (Owens and Blake, 1985).

3.2. *The female gametophyte*

3.2.1. *Gymnosperms*

Tissue differentiation within the ovule, and the early events of megasporogenesis up to the formation of the megaspore mother cell (MMC) are poorly known in gymnosperms (Pennell, 1988). The nucellus forms the bulk of the ovule, usually enveloped by one or rarely two or three (*Ephedra*, *Gnetum*) integuments (Konar and Moitra, 1980). The integuments eventually give rise to the micropyle.

MMCs differentiate from sporogenous tissue deep within the nucellus (Konar and Moitra, 1980). They generally undergo meiosis around the time of pollination (Owens and Blake, 1985). Only one megaspore of the tetrad, the functional megaspore, survives and develops further, the remaining meiotic products degenerate. Growth of the female gametophyte begins with free nuclear divisions of the functional megaspore giving rise to a coenocytic structure. The number of free nuclei is usually constant for a species (Singh, 1978), and varies from 256 in *Taxus* and *Torreya* to about 8.000 in *Ginkgo* (Konar and Moitra, 1980). Cell walls eventually develop between the nuclei, forming a multicellular gametophyte. In some Gymnosperms, e.g., *Taxus baccata* (Zenkteler and Guzowska, 1970; Rohr, 1987) and *Araucaria araucana* (Cardemil and Jordan, 1982), nuclear divisions are not always followed by complete wall formation so that the megagametophyte is partly composed of multinucleate cells.

One or more cells of the megagametophyte enlarge and function as archegonial initials (Konar and Moitra, 1980). These are terminal in the micropylar end of most genera, but occur laterally in the middle of the megagametophyte in *Sequoia* and *Sequoiadendron* (Sedgley and Griffin, 1989). Unequal divisions of the archegonial initials result in the formation of a large central cell and a small primary neck cell which upon further divisions produces the archegonial neck. The central cell divides to form a large egg cell and a smaller, often short lived, ventral canal cell. The cells immediately surrounding the archegonium divide actively acquiring dense cytoplasm and prominent nuclei to form the well-defined archegonial jacket. Thus the mature archegonium consists of an egg cell, ventral canal cell, neck cells and archegonial jacket. The number of archegonia in each megagametophyte usually lies between one and ten, although larger numbers are known in e.g., Podocarpaceae (Singh, 1978).

Following fertilization in Pinaceae, or at about the same time in incompatible crosses (Baldursson *et al.*, 1993a), the archegonia and central portion of the female gametophyte break down to form the corrosion cavity. The corrosion cavity is thought to nourish the embryo pushed into it by the elongation of the suspensor (Singh, 1978). At the time of fertilization the female gametophyte has attained more or less full size in Pinaceae although random cell divisions may continue (Singh, 1978). More active divisions in restricted areas of the gametophyte have been reported in several other gymnosperms. In some cases (*Athrotaxis*, *Cephalotaxus*, *Ephedra*, *Gnetum*, *Welwitshia*) large portions of the gametophyte develop after fertilization from these meristematic areas or layers (Singh, 1978).

Histochemical studies of the developing gymnosperm gametophyte are few and mainly restricted to the mature or germinating seed (Konar and Moitra, 1980). In recent years, accumulation of storage proteins in developing gametophytes and subsequent mobilization during germination have attracted much interest in relation to maturation of somatic embryos *in vitro* (reviewed in Nørgaard, 1992). During development of the female gametophyte, changes occur in the composition of carbohydrates, lipids and amino acids, as these are immobilized into food reserves in the form of starch, fats and proteins (Håkansson, 1956; Favre-Duchartre, 1958; Konar, 1958a,b; Johnson *et al.*, 1987; Gates and Greenwood, 1991).

In Pinaceae, the first appearance of food reserves coincides with the development of an embryo in the seed. Thus in *Picea abies*, the first starch grains appear in the gametophyte cells surrounding the embryo apex at the time when the suspensor cells of the embryo begin to elongate (Håkansson, 1956). A similar starch rich areas appear in *Larix decidua* just after fertilization (Von Aderkas and Bonga, 1988a). The mature gametophyte of *Picea abies* can be divided into three zones with regard to food reserves: the cells immediately surrounding the corrosion cavity have no starch, then follows a zone with simple starch grains which are later absorbed by the growing embryo, and finally the broad outer zone, rich in food reserves (Håkansson, 1956).

Favre-Duchartre (1958) studied the deposition of starch, lipids and lipoproteins in the gametophyte of *Ginkgo biloba*. Starch is the first reserve to appear during maturation of the archegonia, followed by lipoproteins and lipids. The mature gametophyte is layered in regard to food reserves. Lipids and lipoproteins are concentrated in the outermost layers and starch in the deeper layers. Hence, a distinct feature in *Ginkgo* and cycads, contra conifers, is the fact that food reserve accumulation has taken place prior to fertilization (Favre-Duchartre, 1958). This may well have a bearing on the different responses of gymnosperm megagametophytes cultured *in vitro*.

The presence of unknown growth promoters in megagametophytes of *Pseudotsuga menziesii* (Mapes and Zaerr, 1981) and *Ginkgo biloba* (Favre-Duchartre, 1958) has been suggested, based on observed promotion effects of gametophytes on embryo growth.

Since the female gametophyte develops from the functional megaspore it should in theory consist of only haploid cells. However, this is not always the case, diploid as well as polyploid cells, presumably derived from nuclear fusions in multinucleate cells (Bonga *et al.*, 1988), have been found scattered in mature gametophytes of *Taxus* spp. (Rohr, 1987), *Ginkgo biloba* (Avanzi and Cionini, 1971) and *Sequoia sempervirens* (Ball, 1987).

3.2.2. *Angiosperms*

Two major features distinguish female reproductive structures of angiosperms from those of gymnosperms. One is the enclosure of the ovule within an ovary. The other major difference is the reduction of the angiosperm female gametophyte to only seven cells in most cases, in contrast to the many hundreds or thousands of cells which comprise the gymnosperm megagametophyte.

Ovules are initiated from the placenta within the hollow basal part of the carpel (the ovary). The outermost layers of the ovule generally form two integuments partially enveloping the central mass, the nucellus. One of the nucellar cells differentiates into a megaspore mother cell (MMC). The MMC undergoes meiosis forming a linear tetrad, of which generally only one cell is functional. The surviving megaspore enlarges and undergoes three mitotic divisions to produce eight nuclei. Three of these migrate to the micropylar end of the embryo sac and form the egg apparatus consisting of the egg cell and two synergids. Another three nuclei form the antipodals at the chalazal end, and the two remaining polar nuclei occupy the shared cytoplasm of the central cell. Often the two polar nuclei fuse prior to fertilization. The mature female gametophyte thus consists of only seven cells (Sedgley and Griffin, 1989).

The mature embryo sac represents the end of the haploid generation in angiosperms. Double-fertilization results in the formation of a diploid zygote and a (generally) triploid primary endosperm nucleus, resulting from fusion between the two polar nuclei and a sperm nucleus.

The endosperm is a tissue unique to the angiosperms. In the great majority of families the endosperm is triploid but diploid, tetraploid, pentaploid and higher ploidy endosperms are also found (Lakshmi Sita, 1987). The endosperm is biparental and initiated at fertilization, in contrast to the strictly maternal haploid gametophytic tissue of gymnosperms. However, due to the common role as food reserve for the embryo the gymnosperm gametophyte is often wrongly referred to as an endosperm. For information on development of hardwood endosperms and culture *in vitro* the reader is referred to Lakshmi Sita (1987).

4. Methods of haploid induction

4.1. *Natural haploids*

In herbaceous angiosperms, naturally occurring haploids have been described in at least 71 plant species, representing 39 genera and 16 families (Kimber and Riley, 1963). Such natural haploids are frequently observed in polyembryonic seeds (Magoon and Khanna, 1963).

Only two reports have described the natural occurrence of mature haploid specimens of trees. One is an angiosperm, *Populus tremula* (Tralau, 1957), and the other a conifer, *Thuja plicata gracilis* (Polheim, 1968).

Among woody angiosperms, haploid twin embryos have been detected in *Robina pseudoacacia* (Kanezawa, 1948). Systematic searches for haploids among fruit tree seedlings commonly yield a few haploids. Pratassenja (1939) reported the first haploids in peach (*Prunus persica*). Hesse (1971) detected two spontaneous gynogenic peach haploids, recognized by two genetic markers. Later, Toyama (1974) found five haploids among 25 twin seedlings and a further 11 haploids from a population of non-twin seedlings in this species. Braniste *et al.* (1984) observed two haploid seedlings in pear (*Pyrus communis*). A survey of 50,000 apple seedlings, involving a marker gene from *Malus pumila* var. *Niedzwetzyana* coding for anthocyanin production, resulted in the identification of 13 haploids (Zhang *et al.*, 1990). Most of these haploid seedlings were weak and survived for only a short time.

Illies (1964) surveyed greenhouse sowings of *Picea abies* for polyembryonic or abnormal seedlings. Among 135 such seedlings she detected ten haploids, five of which arose from polyembryonic seeds and the remaining five from seeds with reversed embryos. No haploids were found among 400 normal looking seedlings. Later, Simak *et al.* (1968) detected a mosaic-aneuploid twin embryo in *Picea abies* with chromosome numbers ranging from 12–24. Isakov *et al.* (1981) obtained haploids at very low frequencies among seedlings of *Pinus silvestris*. Suspected haploids were selected using morphological criteria and subsequently cytologically analysed. Since most of the above mentioned haploid seedlings were sacrificed for cytological studies, their potential survival values are unknown. The studies showed, however, that spontaneous haploid seedlings do occur in trees and especially in connection with polyembryony or abnormal seeds. The weakest embryo in polyembryonic seed often aborts during seed development or during germination (Simak *et al.*, 1968). Hence, the possibility of early detection of such embryos with X-ray photography, and subsequent dissection and cultivation *in vitro* was discussed by several authors as a potentially useful method to obtain haploid trees (Toda and Sato, 1967; Simak *et al.*, 1968; Ching and Simak, 1971; Huhtinen, 1972; Brunkener, 1974). However, such attempts have been unsuccessful (Huhtinen, 1972). In general it may be concluded that the frequency of spontaneous haploids is far too low for any practical utilization.

4.2. *Haploids through experimentally induced parthenogenesis*

Different techniques have been employed to stimulate haploid parthenogenesis in herbaceous plants (reviewed in Kimber and Riley, 1963; Magoon and Khanna, 1963; Asker, 1980). These include delayed pollination, distant hybridization and pollination with genetically inert pollen, i.e., pollen inactivated with various chemical, physical or physiological treatments.

In woody angiosperms, experimentally induced haploid parthenogenesis was first described by Von Kopecky (1960) in *Populus* spp. He obtained six haploid seedlings of *P. alba*, following pollination with *P. tremula* pollen weakened by fermentation for one week in a stoppered test tube. Furthermore he obtained five haploid seedlings from a *P. alba* × *P. nigra* cross. Von Kopecky's results stimulated several studies aimed at inducing haploid parthenogenesis in poplars. Winton and Einspahr (1968) recovered four haploid/diploid chimaeras of *P. tremuloides* following pollination with compatible pollen, subjected to infrared heating at 50°C for 3–5 min, but not with *P. alba* pollen treated likewise. Stettler and Bawa (1971) pollinated thousands of female catkins of *P. trichocarpa* with a mixture of foreign pollen and irradiated (100 kR) compatible "mentor" pollen. They obtained haploids at the average rate of 0.1–0.3%, depending on female parent and treatment, with the most successful combination resulting in 1.7% haploid seedlings. Stettler and Bawa's use of pollen mixtures was based on the idea that inactivated compatible pollen would provide the necessary stimulus for acceptance and partial germination of the foreign pollen, and thereby stimulate divisions of the egg cell without fertilizing it. Such "mentor effects" were first described by Michurin who found that small doses of compatible pollen (mentor pollen) mixed with foreign pollen facilitated distant hybridization (Stettler and Ager, 1984). In *P. trichocarpa*, the pollen mixture also facilitated remote hybridization, whereas mentor pollen alone was found equally effective for haploid induction (Stettler, 1968; Stettler and Bawa, 1971).

Parthenogenetic haploid induction in poplars was perfected by Illies (1974a,b). She applied an aqueous solution of toluidine-blue to *P. tremula* catkins, at a critical time (12 h) after pollination with *P. alba* pollen, and obtained maternal seedlings at the high rate of 13–43% depending on the parent tree. The late application of toluidine-blue allowed for initial pollen germination and tube growth necessary for stimulation of the egg cell, while inactivating the male gametes and thus preventing fertilization. *P. alba* pollen, which is compatible with *P. tremula*, was used to introduce marker genes (pubescent leaves) to facilitate selection of maternal parthenogenic seedlings (glabrous leaves). Chromosome counts showed about two thirds of the maternal seedlings investigated to be haploid and the remaining aneuploid or rarely diploid.

In spite of Illies's promising results, haploid induction through induced parthenogenesis has evidently not been attempted in other hardwoods and this method seems to have been abandoned in favour of anther culture.

The only documented example of an attempted experimental parthenogenesis in a gymnosperm, is Livingston's (1972) systematic effort with *Pseudotsuga menziesii*. He applied irradiated pollen and different growth regulators to receptive ovulate cones but no haploid seedlings were obtained. Distant pollinations have been carried out in a number of gymnosperm species, frequently resulting in interspecific hybrids (e.g., Hagman, 1975). However, no haploid seedlings have been mentioned in these studies.

4.3. *Chromosome elimination*

A classic example of chromosome elimination is the production of haploid barley (*Hordeum vulgare*) after hybridization with *Hordeum bulbosum*. However, efficient anther/microspore culture techniques are gradually replacing the "bulbosum" method in this species (Kasha *et al.*, 1990). Some of the haploid tree seedlings obtained via induced parthenogenesis with the help of wide hybridization (see above), may have resulted from chromosome elimination but there is no example of deliberate use of this technique in trees.

4.4. *Induction of haploids* in vitro

With the introduction of *in vitro* techniques, a reliable production of haploid plants in large numbers became possible. Two methods have been used: (1) *in vitro* androgenesis through the culture of excised anthers, microsporophylls or isolated microspores, and (2) *in vitro* gynogenesis through the culture of excised ovaries, ovules or female gametophytes.

4.4.1. *Androgenesis* in vitro

The objective of experimental androgenesis is to trigger a transition from normal male gametophyte development towards the formation of haploid sporophyte. Androgenesis has found most widespread application in herbaceous crop plants and efficient methods have been developed for a number of species (reviews, e.g., Maheshwari *et al.*, 1982; Dunwell, 1985; Heberle-Bors, 1985; Prakash and Giles, 1987; Kasha *et al.*, 1990). Traditionally, androgenesis is induced in cultures of entire anthers. Isolated microspore cultures have been developed for some herbaceous plants, including the cereals (Kasha *et al.*, 1990), but not for any tree species so far.

Androgenesis *in vitro* can proceed via two quite dissimilar differentiation pathways. In one, the microspores behave like zygotes and proceed through

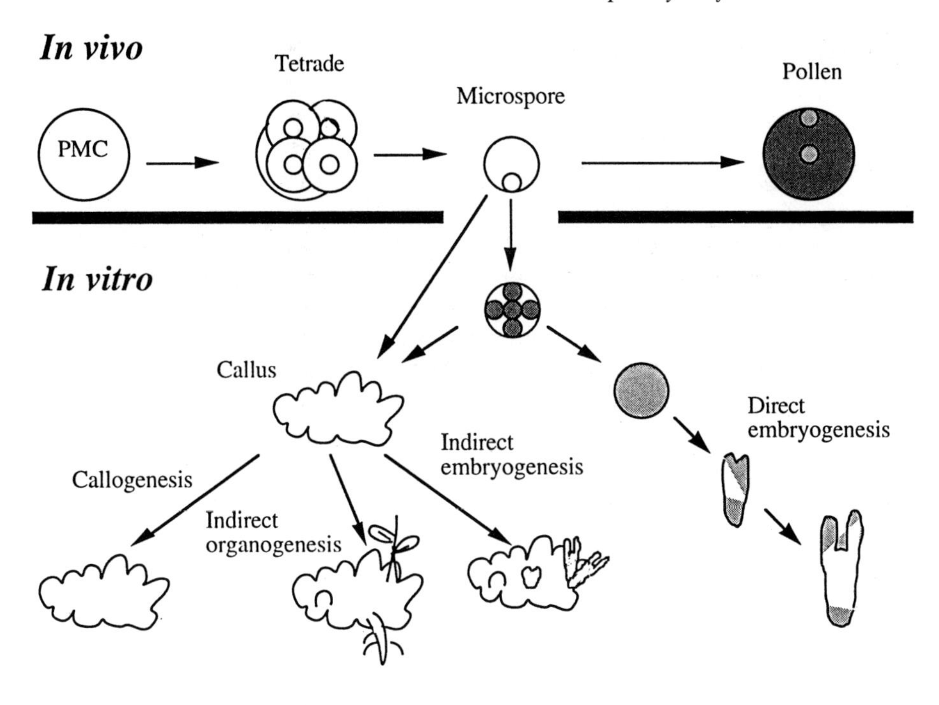

Figure 1. Schematic drawing showing angiosperm microsporogenesis *in vivo* and different developmental pathways of androgenesis *in vitro*.

characteristic stages of embryogenesis. In the other, the microspores initially form unorganized callus which upon subculture and appropriate stimuli undergoes indirect organogenesis, or embryogenesis, leading eventually to plantlet formation (Fig. 1).

4.4.2. *Gynogenesis* in vitro

The objective of gynogenesis is to trigger a transition from normal megagametophyte development towards the formation of haploid sporophyte. Gynogenesis has been extensively used for haploid tissue induction in gymnosperms. In angiosperms this method has been used as an alternative if the species in question does not respond to anther culture (Yang and Zhou, 1982). The disadvantage of gynogenesis is that, for practical reasons, only a limited number of genotypes can be isolated and tested, whereas androgenesis allows selection within a large population of genetically diverse cells.

As with androgenesis, there are several different developmental pathways of gynogenesis (Fig. 2).

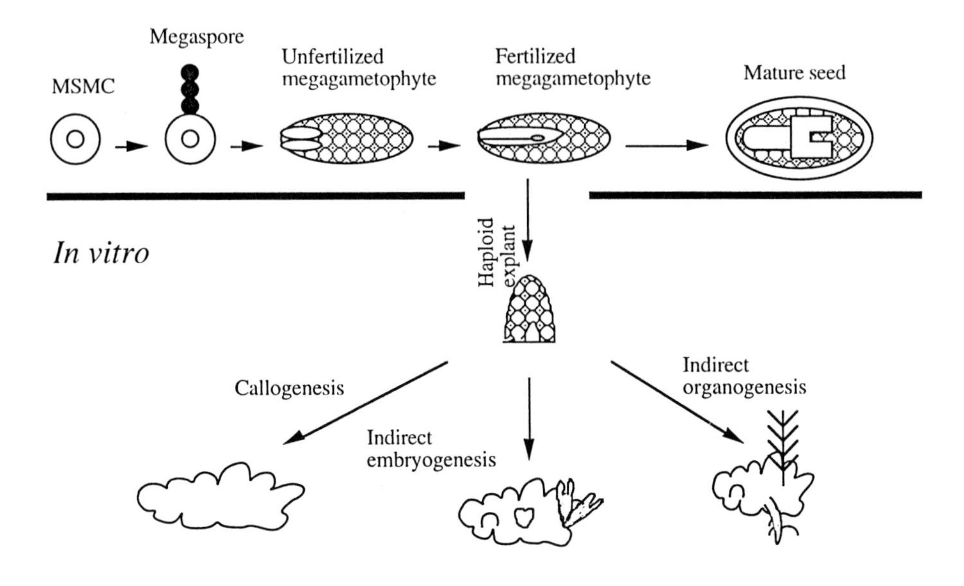

Figure 2. Schematic drawing showing conifer megagametophyte developement *in vivo* and different developmental pathways of gynogenesis *in vitro*.

5. Androgenesis in trees

5.1. *History and present status*

5.1.1. *Woody angiosperms*

Experimental androgenesis has been attempted in anther culture of at least 44 species of hardwoods, belonging to 22 genera (Table 1). Many authors did not describe any development further than early embryogenesis or organogenesis, leaving 22 species where plantlet formation has been successful to date.

Among forest trees, *Populus* species and hybrids have been by far the most responsive in anther culture. In fact, production of haploid plants from anther culture was first reported in this genus (Anonymous, 1975; Wang *et al.*, 1975). Subsequently, haploid plants have been obtained in seven poplar species and many interspecific hybrids (Table 1). The conventional method has been to induce callus growth on auxin rich media, followed by subculture on media with low auxin/cytokinin ratio for organogenesis (Wu and Nagarajan, 1990; Uddin *et al.*, 1988), or embryogenesis (Stoehr and Zsuffa, 1990). Using this approach up to 10% response (i.e., 10% of cultured anthers giving haploid plants) has been reported. Baldursson *et al.* (1993b) obtained direct

Table 1. Reports on attempted androgenesis with woody angiosperms through culture of anthers, containing immature microspores, *in vitro*

Species	Media and growth regulators (mg/l)[1]	Response observed	Referances
Aesculus x carnea	MS + 2,4-D (1), K (1)	plants via embryogenesis	Radojevic et al.,. 1989
A. hippocastaneum	MS + 2,4-D (1), K (1)	plants via embryogenesis	Radojevic, 1978
A. hippocastaneum	MS + 2,4-D (0-10), BA (0-20)	plants via embryogenesis	Jörgensen, 1991
A. parviflora	MS + 2,4-D (1), K (1)	callus	Radojevic, 1991
Albizia lebbek	MS + 2,4-D (0.5), NAA (2), K (2)	callus	De & Rao, 1983
Betula pendula	MS + 2,4-D (1-10), K (0.1-1)	callus, roots and shoots	Huhtinen, 1978
Carica papaya	1/2 MS + NAA (2), BA (1) + 6% S	plants via embryogenesis	Tsay & Su, 1985
Cassia fistula	MS + IAA (4), K (2)	callus	Bajaj & Dhanju, 1983
Chimonantus praecox	MS+ 2,4-D or GA_3 (1), K (1)	callus and globular embryoids	Radojevic & Kovor, 1986
Citrus limon	MSH + IAA, 2,4-D (1), BA (1)	callus	Drira & Benbadis, 1975
C. limon	MS+NAA (0.05), BA (0.5), K (0.5) + 4-6% S	callus and embryoids	Germana et al., 1991
C. deliciosa, C.limon, C.reticulata, C. sinensis	MT + IAA (0.1-1), BA (0.1-5) + 5% S	callus and a few shoots	Geraci & Starrantino, 1990
C. microcarpa	N_6 +2,4-D (0.5-2), K (2) + 6% S	plants via embryogenesis	Chen, 1987
Cocos nucifera	B + NAA (2), 2,4-D (0.25) + 15% CM + 6% S	embryoids	Thanh-Tuyen & De Guzman, 1983
Coffea arabica	LS + 2,4-D (0.1) + 5% CM	plants via embryogenic callus	Sondahl & Sharp, 1979
Fagus silvatica	MS + 2,4-D (5), BA (2.5)	one embryoid	Jörgensen, 1988
Hevea brasiliensis	RT + 2,4-D (1), NAA (1) K (1) + 5% CM + 7% S	plants via embryogenic callus	Chen et al., 1982; Chen, 1990

Table 1. Continued

Species	Media and growth regulators (mg/l)	Response observed	References
Litchi chinensis	MS + 2,4-D (2), NAA (0.5), K (1)	plants via embryogenic callus	Lianfang, 1990
Malus x domestica, several varieties	MS or M + IAA (1), K (1) and others	callus and embryoids	Milewska-Pawliczuk, 1990 and references therein,
Malus x domestica	MS+IBA (0.1), BA (0.5)	callus and embryoids	Zhang et al., 1990; 1992
M. prunifolia	MS + 2,4-D (2), K (2)	plants via organogenesis	Wu. 1981
M. pumila	MS + 2,4-D (0.4), IAA (4), K (0.2) +3-8% S	plants via embryogenesis	Fei & Xue, 1981
Peltophorum pterocarpum	MS + NAA (2), IAA (2), BA (2)	callus	De & Rao, 1983
Poinciana regia	MS + various auxins, K + 7% CM	callus and embryoids	Bajaj & Dhanju, 1983
Populus berolinensis	BN + 2,4-D (2), K (1)	plants via organogenesis	Anon, 1975
P x canescens	MS-1 + BA (1) + 6% M	embryoids	Baldursson et al., 1993
P. deltoides	MS + 2,4-D (2), K (1)	plants via organogenesis	Ho & Raj, 1985, Uddin et al., 1988
P. maximoviczii	MS + 2,4-D (0.5-2), K (0.1)	plants via embryogenic callus	Stoehr & Zsuffa, 1990
P. maximoviczii	MS-1+ BA (1) + 6% M	plants via embryogenesis	Baldursson et al., 1993
P. nigra	MS + 2,4-D (2) + K (1)	plants via organogenesis	Wang et al., 1975; Ho & Raj 1985
P. trichocarpa	MS-1 + BA (1) + 6% M	plants via embryogenesis	Baldursson et al., 1993
P. ussuriensis	BN + 2,4-D (2), K (1)	plants via organogenesis	Anon, 1975
Populus x various hybrids	BN or MS + 2,4-D (2), K (1)	plants via organogenesis	Wang et al., 1975; Anon, 1975; Ho & Raj 1985; Wu & Nagarajan, 1990
Prunus amygdalus	M + IAA (0.1), 2,4-D (1), K (0.2)	callus	Michellon et al., 1974
P. armeniaca	MS + NAA (2), K (2)	callus	Harn & Kim, 1972

Table 1. Continued

Species	Media and growth regulators (mg/l)	Response observed	References
P. cerasifera	M + 2,4-D (1), BA (5)	callus	Seirlis et al., 1979
P. cerasus	M+ BA (5-10)	plants via organogenesis	Seirlis et al., 1979
P. domestica	M + 2,4-D (1-10), BA (0-10)	callus	Seirlis et al., 1979
P. persica	M + IAA (0.1), 2,4-D (1), K (0.1)	callus	Michellon et al., 1974
P. persica	M + IAA (1), BA (1-5)	callus	Seirlis et al., 1979
P. persica	N + 2,4-D (1), Z (1)	callus	Todorovic et al., 1991
Quercus petres	MS or WPM + 2,4-D (2.5-5), BA (0-10)	plants via embryogenesis	Jörgensen, 1988, 1991
Ulmus americana	protoplast medium	isolated pollen protoplasts	Redenbaugh et al., 1980
U. americana	S + 2,4-D (6) + 4% S	callus	Redenbaugh et al., 1981
Vitis rupestris	NN + 2,4-D (1), BA (0.2)	plants via somatic embryogenesi and organogenesis	Cersosimo et al., 1990
V. thunbergii	N + NAA (2), BA (0.2)	callus, roots and shoots	Hirabayashi et al., 1976
V. vinifera	MS + NAA (0.1), K (0.3)	callus	Gresshoff & Doy, 1974
V. vinifera	B5 + 2,4-D (0.5), BA (2)	plants via embryogenic callus	Zou & Li, 1981
V. vinifera	1/2 MS + 2,4-D (1), BA (0.2)	plants via somatic emryogenesis	Mauro et al., 1986

[1)] Only induction media shown. Further development after imitation of callus or enbryoids commonly requires transfer to media lower in growth regulators. The concentration of carbohydrates is mentioned only when other than 2–3% sucrose. Abbreviations: Media: B (Blaydes, 1966), B_5 (Gamborg et al., 1968), BL (Brown & Lawrence, 1968), BMI-S1 (Krogstrup, 1986), BN (Bourgin & Nitsch, 1967), H (Heller, 1953), LM (Litvay et al., 1981), M (Miller, 1965), MS (Murashige & Skoog, 1962), MSH (MS with Heller's (1953) micronutrients, MS-1 (modified MS, see Baldursson et al., 1993), MT (Murashige & Tucker, 1969), N (Nitsch, 1972), NN (Nitsch & Nitsch, 1969), N_6 (Chu et al., 1975), R (Reinhart's medium as modified by Konar, 1963), RT (Chen, 1990), S (Sharpet al., 1972), W (White, 1943). Growth regulators: 2,4-D (2,4-dichlorophenoxyacetic acid), BA (benzyladenine), IBA (indolebutyric acid), IAA (indoleacetic acid), NAA (naphtaleneacetic acid), K (kinetin), Z (zeatin), G_3 (gibberellin) Carbohydrates: M (maltose), S (sucrose). Other: CH (casein hydrolysate), CM (cononut milk).

embryogenesis and subsequent plant formation in *Populus* spp. on media containing only cytokinin (BA) but the response was low (0.5–2%).

Other successful examples of haploid induction in hardwoods include *Hevea brasiliensis* (Chen, 1990). Culture techniques have been developed in this species yielding 100–140 embryos per 100 inoculated anthers, and with

10–40% of these developing into viable plantlets, depending on cultivar. Similar induction frequencies have been reported for *Aesculus* spp. (Radojevic, 1978, 1991; Radojevic *et al.*, 1989), but it is uncertain how many viable plants have been regenerated in this species.

Much effort has been invested in induction of haploids in fruit trees (Table 1). The European work (reviewed in Radojevic and Kovoor, 1986; Milewska-Pawliczuk, 1990; Zhang *et al.*, 1990), mostly emphasizing organogenesis from callus, has largely failed in producing haploid plants. Chinese workers, on the other hand, obtained plantlets via embryogenesis in *Citrus*, *Litchi*, *Malus*, *Prunus* and *Vitis* (Chen, 1987). Recently, haploid plants have been obtained via embryogenesis in *Carica papaya* (Tsay and Su, 1985) and haploid embryos in *Malus* × *domestica* (Zhang *et al.*, 1992).

5.1.2. *Gymnosperms*

Tulecke's (1953) and LaRue's (1954) work on *Ginkgo biloba* and *Taxus brevifolia*, respectively, demonstrated for the first time the possibility of obtaining subculturable tissue from pollen of flowering plants. This early work spurred interest and serious attempts in other gymnosperm species but the success has been sparse to date. Cell divisions and callus formation from pollen have been observed in 19 gymnosperm species (Table 2). In most cases these calli were composed of parenchymatous cells, with no apparent ability to differentiate organs. Root formation was observed in microspore callus of *Picea abies* (Simola, 1981, 1987), but ploidy of this callus was not determined. Tissue derived from pollen of *Taxus* spp. (LaRue, cited in Tulecke, 1959), was composed of long pollen tube-like cells which subsequently divided in polar fashion reminiscent of initial stages in gametogenetic embryogenesis in *Larix* (Bonga *et al.*, 1988). Similarly in *Zamia integrifolia* (LaRue, 1954), *Ephedra foliata* (Konar, 1963) and *Pinus resinosa* (Bonga, 1974) embryo-like structures have been described.

Although *in vitro* pollen or microsporophyll cultures have failed to produce haploid plants they have facilitated the study of later stages of microgametophyte development and spermatogenesis in gymnosperms (e.g., Ho and Rouse, 1970; Ho and Sziklai, 1971a,b; De Luca and Sabato, 1979; De Luca *et al.*, 1980; Rohr, 1980, 1987). Such studies have been difficult with fresh field material (Bonga *et al.*, 1988).

5.2. *Induction and ontogeny of androgenesis*

The exact mechanism triggering transition from gametophytic to sporophytic mode of development is still unknown. Evidence suggests that pollen dimorphism is a necessary condition for haploid induction in some species (see review by Sunderland and Huang, 1987), i.e., that the sporophytic potential is expressed only in morphologically anomalous pollen grains (p-grains). The timing of induction of the sporophytic potential is still not clear. Although this controversy will not be dealt with here, the question whether it happens

Table 2. Reports on attempted gymnosperm androgenesis *in vitro*

Species	Developmental stage of explant	Media and growth regulators (mg/l)[1]	Response observed	References
Cycadales				
Zamia floridana	immature microsporophylls	W + 2-20% S	pollen tube growth & multicellular structures	LaRue, 1954
Ceratozamia mexicana	mature pollen	W +2,4-D(0.6) + 20% CM	spermatogenesis and callus formation	De Luca et al., 1980
Ephedrales				
Ephedra foliata	mature pollen	R + 2,4-D (0.3) + 15% CM	callus and filamentous structures	Konar, 1963
Ginkgoales				
Ginkgo biloba	mature pollen	W + 20% CM	callus and septate structures	Tulecke, 1953, 1957
Coniferales				
Larix decidua	not mentioned	not mentioned	callus	Bonga, 1981
L. laricina	not mentioned	not mentioned	callus	Bonga, 1981
Picea abies	uni-nuclear microspores	N6 or MS + 0.1% CH	callus shoot initials and roots	Simola, 1981, 1987, Simola & Huhtinen 1986
Pinus mugo	not mentioned	not mentioned	callus	Bonga, 1981
P. nigra	not mentioned	not mentioned	callus	Bonga, 1981
P. resinosa	immature to mature microsporophylls	BL + 2,4-D (5)	callus and "embryo-like structures"	Bonga & Fowler, 1970; Bonga, 1974
Cupressus benthamii, C. macrocarpa, C. sempervirens	mature microsporophylls	WH	multicellular structures	Razmologov, 1973
C. arizonica	mature microsporophylls	BL,W or H + 2,4-D (1)	callus	Duhoux & Norreel, 1974
C. arizonica	mature pollen		isolated protoplasts	Duhoux, 1980.
Thuja orientalis	mature pollen	W +2,4-D (2) + 10% CM	callus	Rao & Mehta, 1968
Taxus brevifolia	mature pollen	W + 2,4-D (0.6) + 15% CM	pollen tubes and callus	Tulecke, 1959
Taxus baccata	microspores	not mentioned	pollen tubes and callus	Rohr, 1973
Torreya nucifera	mature microsporangia	W + 2,4-D + CM	callus	Tulecke & Sehgal, 1963

1) see footnote Table 1 for abbrevations

before or after meiosis (i.e., before or after inoculation of the anthers) is of utmost importance. If microspores acquire the sporophytic potential before or during meiosis, a view advocated by Heberle-Bors (1985) based on the discovery of pollen dimorphism in *Nicotiana tabacum*, manipulations of cultures of post-meiotic microspores are basically futile, and will never increase the yield beyond the number of pre-determined pollen grains. However, in other thoroughly studied plant species, e.g., *Datura innoxia* (Sunderland and Huang, 1987) and *Hyoscyamus niger* (Reynolds, 1990) there is ample evidence showing that induction takes place in post-meiotic microspores *in vitro*.

Different pathways of microspore embryogenesis have been described for herbaceous angiosperms, depending on the type of initial mitotic division in the microspores and the subsequent growth of proembryos from these cells (Raghavan, 1978; Sunderland and Evans, 1980; Maheshwari *et al.*, 1982; Sunderland and Huang, 1987). Initially, microspores may divide abnormally to form two equally sized vegetative cells, one or both of which multiply (route 1). Alternatively, microspores may follow the normal gametophytic pathway forming a large vegetative and a small generative cell. In this case non-polar cell divisions in the vegetative cell (route 2), the generative cell (route 3), or both may give rise to embryos. Variations on this basic scheme, including fusion of nuclei, have been seen, depending on species and developmental stage of microspores at culture initiation (Sunderland and Evans, 1980), but routes 1 and 2 are generally considered the most common ones.

5.2.1. *Woody angiosperms*

Radojevic and co-workers (Radojevic and Kovoor, 1986; Radojevic *et al.*, 1989; Radojevic, 1991) have described pollen dimorphism in *Aesculus* spp. based on size differences as well as differences in staining intensity with Schiff's reagent or fluorescein diacetate (FDA). Thanh-Tuyen and De Guzman (1983), working with *Cocos nucifera*, noted that embryogenic microspores were smooth, expanded and contained lightly staining cytoplasm, whereas darkly stained microspores were considered non-embryogenic. Except for these studies, no conclusive evidence of pollen dimorphism exists in hardwoods.

Low frequency of embryo induction in most hardwood species has prohibited systematic cytological studies concerning early stages of androgenesis. There is, however, no *a priori* reason to suspect that hardwood microspores behave differently from those of non-woody angiosperms. In *Hevea brasiliensis*, Chen (1990), reported that most microspores followed route 1 (see above) and a few microspores followed route 2. Still other microspores formed multinuclear pollen grains which failed to develop further. Similarly in *Cocos nucifera* (Thanh-Tuyen and De Guzman, 1983), route 1 was the dominant pathway, while Radojevic (1991) has described unequal initial divisions in *Aesculus hippocastanum* microspores, suggestive of route 2.

5.2.2. *Gymnosperms*
The only example of pollen dimorphism in gymnosperms, is Duhoux and Norreel's (1974) description of two size classes of pollen grains in *Juniperus*, related to their ability to form callus.

Cytology of gymnosperm androgenesis is largely unknown due to the fact that advanced stages of embryogenesis or organogenesis have never been observed. However, several authors have documented initial cellular events in culture. Tulecke (1957), described three abnormal developmental pathways of *Ginkgo* pollen, most of which were initiated in the tube cell. In one of these, elongated structures (septate tubes) were formed, giving rise later to subculturable callus, while the other pathways led to apparent developmental dead-ends in the form of intercalary structures or coenocytes (multinuclear cells). Coenocytes have also been detected in cultures of *Torreya nucifera* (Tulecke and Sehgal, 1963), *Taxus cuspida* (LaRue, 1954), *Pinus resinosa* (Bonga, 1974) and *P. mugo* (Bonga, 1981). Razmalogov (1973) observed unequal first divisions in *Cupressus* microspores resulting in the formation of large siphonogenic and a small generative cell, which upon separation from the large pollen grain, initiated cell divisions. In *Pinus resinosa*, callus growth was frequently observed to initiate in anomalous pollen grains in which the first mitosis had formed two equally sized and randomly distributed cells (Bonga, 1974; Bonga and McInnis, 1975), suggesting analogy to route 1 in the angiosperms (see above).

5.3. *Factors affecting androgenesis* in vitro

5.3.1. *Genotype of the donor tree*
Genotype of the donor plant has been considered one of the major determinants of yield in anther culture of herbaceous plants (Maheshwari *et al.*, 1982; Powell, 1990). Genotypic effects were also apparent in hardwoods when more than one donor tree was tested (Redenbaugh *et al.*, 1981; Tsay and Su, 1985; Chen, 1986; Uddin *et al.*, 1988; Milewska-Pawliczuk, 1990; Baldursson *et al.*, 1993b), although most genotypes responded poorly or not at all.

In gymnosperms, most researchers have used material from one or very few donor trees, thus precluding any conclusion on the effect of genotype.

5.3.2. *Developmental stage of pollen*
Mature pollen grains of angiosperms appear to be more or less permanently committed to the gametophytic pathway and without capacity for sporophytic divisions *in vitro* (Sunderland and Huang, 1987). Hence, immature microspores are used, with the optimal stage depending on species. In hardwoods, uninucleate microspores (i.e., between the tetrad stage and the first microspore mitosis) are the most responsive. Few investigators have attempted to define the developmental stage more closely. In *Aesculus* spp. (Radojevic, 1991) and *Malus domestica* (Kubicki *et al.*, 1976), anthers containing uninucleate microspores with centrally located nuclei responded best. In *Populus*

spp. (Wu and Nagarajan, 1990; Baldursson *et al.*, 1993b), *Carica papaya* (Tsay and Su, 1985), and *Cocos nucifera* (Thanh-Tuyen and De Guzman, 1983), the induction frequency was highest when the anthers contained late uninucleate microspores on the verge of mitosis. For convenience the responsive stage can usually be correlated with external features such as size and colour of flower buds or length of anthers (Chen, 1990; Radojevic and Kovoor, 1986).

Curiously, most attempts in gymnosperms have been made with mature microsporophylls or pollen (i.e., just prior to anthesis) (Table 2). The few systematic studies using microsporophylls at different stage of development appear to support the notion that relatively advanced pollen grains, i.e., after the first gametophytic mitosis, are the most responsive. When pre-meiotic microsporophylls of *Pinus resinosa* were cultured, meiosis and development up to the first gametophytic division proceeded normally in many microspores, but no or few calli were produced. Uninucleate microspores developed further *in vitro* and produced some calli and bipolar structures (see above), but material collected at the time of the first mitotic division was by far the most responsive (Bonga, 1974).

5.3.3. *Growth conditions and physiology of the donor tree*

In herbaceous species, growth conditions (temperature, photoperiod, light intensity) and the resulting physiological status of donor plants affects yield in anther culture in major ways (Maheshwari *et al.*, 1982; Dunwell, 1985; Olesen, 1987). Hence the practice of growing donor plants in controlled environments is considered necessary for reproducible results. Controlled growth of donor trees in growth chambers is usually not practical and it follows that no experiments have compared the response of material grown under different environmental regimes. Undoubtedly though, growth conditions of the donor will affect haploid induction in trees as in herbaceous plants. Several studies with coniferous forest trees have shown that environmental factors such as low (Ekberg and Eriksson, 1967; Ekberg *et al.*, 1968; Andersson *et al.*, 1969) and high temperatures (Owens and Blake, 1985; Chen, 1990), or drought (Chen, 1990), can cause severe meiotic irregularities *in vivo*. During early and hot summers, some cultivars of *Hevea brasiliensis* had anthers containing poorly differentiated and sterile microspores and as a rule no microspore embryos were obtained from such anthers (Chen, 1990). Similarly, annual variations in yield of haploid tissue were indicated in studies with *Populus trichocarpa* (Baldursson *et al.*, 1993b), and *Picea sitchensis* (Baldursson *et al.*, 1993a). Therefore, it is essential to investigate the condition and viability of the microspores prior to culture when field material is used.

5.3.4. *Culture media*

The basal media of Murashige and Skoog (MS) (1962) has been the most frequently used in anther culture of hardwoods, although several other media have been tried with success (Table 1). White's medium (White, 1943) and modifications thereof, on the other hand, has been most popular with gymnosperms (Table 2). Chen (1986) has suggested that hardwood anther cultures may need the relatively high levels of mineral salts provided by the MS medium. For microspore embryogenesis, however, the relative levels and ratios of nitrate and ammonium appear to be more important. In *Hevea brasiliensis* (Chen, 1986, 1990) it was possible to almost double the embryo induction frequency by reducing NO_3^- in relation to NH_4^+ in the MS medium. On the contrary, in *Populus trichocarpa* (Baldursson *et al.*, 1993b) and *Cocos nucifera* (Thanh-Tuyen and De Guzman, 1983) high levels of NO_3^- over NH_4^+ were beneficial for microspore embryogenesis.

Exogenous growth regulators have usually been considered an absolute requirement for androgenesis (Radojevic and Kovoor, 1986), although, again, the type(s) and optimal concentration(s) must be defined for each species. For hardwoods, relatively high levels of auxins and cytokinins, typically in one:one or two:one ratio, were used in the induction medium (Table 1). The initial combination of growth regulators may determine the differentiation pathway taken by the microspores. Exogenous auxins were necessary for callus induction in *Carica papaya* (Tsay and Su, 1985) and *Populus trichocarpa* (Baldursson *et al.*, 1993b), whereas media without growth regulators or media containing cytokinins only were sufficient for embryogenesis in these species. However, Radojevic (1978, 1991) found 2,4–D indispensable for microspore embryogenesis in *Aesculus hippocastanum*. For further development, i.e., organogenesis or embryo maturation and germination, subculture on media with lower levels of growth regulators was usually required. Gibberellin (GA_3) has been found useful for these later stages in some cases (Chen, 1986, 1987, 1990). In all species of gymnosperms tested, except *Ginkgo biloba* (Tulecke, 1957), exogenously applied auxins (mostly 2,4–D) at concentrations ranging from 0.25 μM (Razmologov, 1973) to 25 μM (Bonga and Fowler, 1970) appeared essential for callus induction and growth. Polyamines (putrescine, spermidine, spermine), in combination with different light sources, were found to enhance growth and root formation in microspore callus cultures of *Picea abies* (Simola and Huhtinen, 1986).

Carbohydrates exert a nutritional as well as osmotic effect in the culture medium, but little analytical work is available to distinguish the relative importance of these effects (Dunwell, 1985). Sucrose, being the major translocated carbohydrate in plant tissues, has been widely used in anther culture. Work with herbaceous crops has suggested that species shedding bicellular pollen grains (e.g., Solanaceae, Liliaceae) require low sucrose (2–4%) and those with tricellular pollen grains (e.g. Poaceae, Brassicaceae) benefit from high sucrose (high osmoticum) concentrations (6–15%), especially in the early phases of androgenesis (Dunwell, 1985; Powell, 1990). Whether this trend holds in woody angiosperms is difficult to asses based on available data. In species shedding tricellular pollen such as *Hevea brasiliensis* (Chen, 1986, 1990) and *Ulmus americana* (Redenbaugh *et al.*, 1981), high sucrose levels (7% and 4–10, respectively) were beneficial for embryogenesis and callus induction. The majority of hardwoods, however, have bicellular pollen grains (Brewbaker, 1967) and in some of these, e.g., *Cocos nucifera* (Thanh-Tuyen and De Guzman, 1983) and *Populus* × *simonigra* (Wu and Nagarajan, 1990), high sucrose levels (5–9%) were also found advantageous. Recently, maltose has been proven superior to sucrose in anther culture of some cereals (Last and Brettell, 1990; Powell, 1990; Roberts-Oehlschlager and Dunwell, 1990). This has been related to the rapid breakdown of sucrose into glucose and fructose and the inhibitory effects of the latter on embryogenesis (Powell, 1990; Roberts-Oehlschlager and Dunwell, 1990). Maltose has not been used as a carbon source in anther culture of tree species, except in *P. trichocarpa* (Baldursson *et al.*, 1993b). Sucrose at 2%, has been used exclusively as a carbon source in gymnosperms, but the experimental basis for that choice is lacking.

Natural extracts of undefined chemical composition were frequently employed in gymnosperm pollen culture media, including tomato juice (LaRue, 1954), yeast extract (Tulecke, 1957) and coconut milk (CM) (Tulecke, 1957; Tulecke and Sehgal, 1963; Konar, 1963; Rao and Mehta, 1969; Rohr, 1973; Duhoux, 1980). CM could be partially replaced by 0.1 mg/L naphtaleneacetic acid (NAA) and 1 mg/L Ca-pantothenate in *Ginkgo* pollen cultures (Tulecke, 1957) and Bonga (1974) found no enhancing effects of CM in *Pinus resinosa*. In hardwoods CM has been found beneficial in a few cases (e.g., Bajaj and Dhanju, 1983; Thanh-Tuyen and De Guzman, 1983).

Anther/microsporophyll cultures in woody species have been almost universally performed on media solidified with agar. However, there are indications that agar may contain compounds inhibitory to anther culture (Kohlenback and Wernicke, 1978) and for that reason agarose or the less expensive gellan-gums (e.g., Gelrite, Kelco) are often recommended as solidifying agents in anther cultures of herbaceous crops (Olesen, 1987). Gelrite solidified medium was successfully used for induction of embryogenesis in anther culture of *Populus trichocarpa* (Baldursson *et al.*, 1993).

5.3.5. *Pre-treatments*

Shock treatments of anthers or whole flowers, including high or low temperatures for various lengths of time prior to culture, have been recommended for many herbaceous species (Dunwell, 1985). Few reports have documented positive effects of such treatments in trees. Stoehr and Zsuffa (1990) indicated beneficial effects of four day cold (4°C) storage of catkins containing uninucleate microspores for callus production in *Populus maximowizcii*. Inoculation of *P. trichocarpa* anthers at 32°C for 24–72 h appeared beneficial (Baldursson, unpublished results). Six week cold storage of male strobili, harvested before the first microspore division, enhanced callus formation in pollen of *Pinus resinosa* (Bonga, 1974). This was presumably because the cold treatment stimulated abnormal cell divisions, resulting in pollen grains containing two to eight randomly distributed cells at the time of culture.

In *Pinus resinosa*, callus development from microsporophylls was significantly enhanced by centrifugation of the immature male strobili at 2000 rev/min for 30 min (Bonga and McInnis, 1975). The same treatment inhibited callusing of *Ulmus americana* anthers (Redenbaugh *et al.*, 1981).

6. Gynogenesis in trees

6.1. *Angiosperms*

Haploid plants have been obtained through ovule cultures in *Populus* (Wu and Nagarajan, 1990), and *Hevea* (Chen, 1990). In both genera, however, more efficient anther culture methods exist (see above). Three cultivars of apple, *Malus* × *domestica*, were tested in ovule culture and produced calli at a rate from 0 to 32% but plant regeneration was not obtained (Zhang *et al.*, 1990). In general, there are few reports on experimental gynogenesis in hardwoods and this section will, consequently, be devoted to the gymnosperms.

6.2. *Gymnosperms*

The multicellular megagametophytes of gymnosperms have long been considered an ideal source for experimental production of haploids. Their large size (from a few mm in many Pinaceae up to 3 cm in Araucariaceae) and situation in a sterile environment makes the megagametophytes amenable to isolation and manipulation in culture. Their homogeneous nature, being the product of only one cell (the megaspore), minimizes the possibility of chimaeric regenerants. And lastly, considering the close developmental proximity of the megagametophyte tissue to the egg cell and distinct meristematic phase, it may be expected to retain some morphogenic activity (Singh, 1978).

6.2.1. *History and present status*

Available reports on megagametophyte cultures in gymnosperms are summarized in Table 3. Duchartre (cited in LaRue, 1948) described, as early as 1888, putative root formation in isolated gametophytes of *Cycas thoursii*, forgotten in the germination bed. However, it was not until the mid-19th century that the first systematic attempts were made to regenerate plants *in vitro* from gymnosperm gametophytes (LaRue, 1948, 1950, 1954). LaRue's early work with the cycads demonstrated root and shoot formation in gametophyte cultures of *Cycas* and *Zamia* (Table 3). Later, small plantlets were obtained via gametophytic organogenesis and embryogenesis in *Zamia* (Norstog, 1965; Norstog and Rhamstine, 1967). De Luca *et al.* (1979) reported the formation of embryo-like structures in cultures of *Ceratozamia mexicana*, and choralloid roots of *Cycas revoluta* in culture (De Luca *et al.*, 1979; De Luca and Sabato, 1980). In a recent publication, megagametophyte organo-

Table 3. Reports on attempted gymnosperm gynogenesis *in vitro*

Species	Developmental stage of explant	Media and growth regulators (mg/l)[1]	Response observed	References
Cycadales				
Zamia floridana	around fertilization	White's minerals	roots and shoots	LaRue, 1948
Z. integrifolia	archegonial to mature	W + 2,4-D (1), K (5)	callus, roots, leaves, embryoids in 2- 6 months	Norstog, 1965,1982; Norstog & Rhamstine, 1967
Cycas circinalis	not mentioned	W + 2,4-D (10)	callus in 2-3 months	Norstog & Rhamstine, 1967
C. mexicana	mature	W + 2,4-D (1) K (5)	callus and embryoids in 7 months	De Luca et al., 1979
C. mexicana	approx. 1 month fertilized	B5 + 2,4-D (0.2-2), K (0-1)	roots and shoots	Chavez & Norstog, 1992
Cycas revoluta	mature	Not mentioned	roots in 1 year	LaRue, 1950
C. revoluta	mature	W + 2,4-D (1), K (1)	callus in 8 months "pseudobulbils"	De Luca et al., 1979
Encephalartos umbeluziensis	mature	MS + 2,4-D (10), K (1)	callus in 8 months	De Luca et al., 1979
Ephedrales				
Ephedra foliata	almost mature	W + 2,4-D (5) + 20% CM	callus within one week	Sankhala et al., 1967
Ephedra foliata	mature	MS + 2,4-D (2) + 20% CM	callus, roots and shoots in 24 days	Konar & Singh, 1979
Ephedra foliata	archegonial	MS + 2,4-D (2), K (2) + 20% CM	roots and shoots	Singh et al., 1981, Bhatnagar & Singh, 1984
Ginkgoales				
Ginkgo biloba	archegonial	W + 2,4-D (6) + 18% CM	callus in 2-3 months	Tulecke , 1964, 1967
Ginkgo biloba	4-6 months fertilized	MS + 2,4-D (1), IAA (0.1), 10% CM	callus	Rohr, 1977
Coniferales				
Araucaria araucana	mature	MS + NAA (10), K (1)	callus in 1-2 months	Cardemil & Jordan, 1982
Larix decidua	1-4 weeks fertilized	1/2 LM + BA (2)	plantlets via embryogenesis	Nagmani & Bonga, 1985
L. decidua	haploid embryonal masses	1/2 LM	plantlets from haploid protoplasts	von Aderkas, 1992
L. decidua	1-4 weeks fertilized	1/2 LM	plants in soil	von Aderkas & Bonga, 1993

[1] See footnote in Table 1 for abbreviations.

Table 3. Continued

Species	Developmental stage of explant	Media and growth regulators (mg/l)[1]	Response observed	References
L. leptolepis	1-4 weeks fertilized	1/2 LM	plantlets via embryogenesis	von Aderkas et al., 1990
L. x eurolepis	1-4 weeks fertilized	1/2 LM	plantlets via embryogenesis	von Aderkas et al., 1990
Picea abies	mature	not mentioned	callus and one plantlet	Bonga, 1977
P. abies	ca 1 month fertilized	N6 + 2,4-D (2), K (0.5)	callus, roots and shoots	Huhtinen et al., 1981, Simola & Honkanen, 1983
P. abies	ca 1 month fertilized	N6 + 2,4-D (2)+ K (0.5) with modified microelements	embryogenic callus	Simola & Santanen, 1990
P. abies	archegonial to fertilized	none	isolated protoplasts	Hakman et al., 1986
P. glauca	not mentioned	not mentioned	callus	Bonga, 1981
P. sitchensis	1-3 weeks fertilized	BMI-S1 + 2,4-D (2), K (1) BA (1)	callus in 1-3 months	Baldusson et al., 1993
Pinus nigra var. *austrica*	2-4 weeks fertilized	buffer + IAA	callus	Radforth & Bonga, 1960
P.lambertiana	mature	H + 2,4-D (1), K (0.1) + 15% CM	callus	Borchert, 1968
P. mugo	2-4 weeks fertilized	W + 2,4-D (0.1) or IAA (1) + 10% CM	callus	Bonga, 1974
P. mugo	2-4 weeks fertilized	BL + 2,4-D (5)	callus in 5-7 weeks	Bonga, 1974
P. resinosa	2-4 weeks fertilized	BL + 2,4-D (5)	callus	Bonga & Fowler, 1970
Pseudotsuga menziesii	around fertilization	MS + 2,4-D (7.5), K (1)	callus	Glock et al., 1988
Sequoia sempervirens	mature	MS + 2,4-D (1), K (1) + 15% CM	trees in soil	Ball ,1987
Taxus baccata	mature	W + 2,4-D (5)	callus in 3-4 months	Zenkteler & Guzowska, 1970
Taxus spp.	mature	BN + 2,4-D (2.5)	callus in 2 months	Rohr ,1982, 1987

[1)] See footnote in Table 1 for abbreviations.

genesis in *Ceratozamia mexicana* and *C. hildae* has been described, leading to shoot formation but rooting of adventitious shoots failed (Chavez and Norstog, 1992). Within the *Ephedrales*, root and shoot formation was reported in gametophyte cultures of *Ephedra foliata* (Konar and Singh, 1979; Singh *et al.*, 1981; Bhatnagar and Singh, 1984).

In the conifers, attempts to cultivate gametophytes *in vitro* have met with mixed success (Table 3). However, haploid plantlets were obtained via gametophytic embryogenesis in *Larix* spp. (Nagmani and Bonga, 1985; Von Aderkas *et al.*, 1990), and more recently plant regeneration was reported (Von Aderkas and Bonga, 1993). Von Aderkas (1992) has furthermore described embryogenesis, leading to plant regeneration, in haploid protoplasts of gametophytic origin in *Larix decidua*. Simola and Santanen (1990) observed formation of proembryos in megagametophyte callus of *Picea abies* (ploidy not mentioned). Earlier, Simola and Honkanen (1983) had obtained root and shoot production through organogenesis in haploid megagametophyte calli of this species. Ball (1987) produced shoots and plants from mature female gametophytes of *Sequoia sempervirens*. The plants had grown in the field for four years and grew as well as seedlings. Root smears from these plants gave exclusively diploid chromosome counts and the possibility of embryo origin could not be excluded. No morphogenesis has been demonstrated in other conifer species studied (Table 3).

6.3. *Induction and ontogeny of gynogenesis*

Gametophytic embryogenesis has been extensively studied in *Larix* (Von Aderkas and Bonga, 1988b) and compared to somatic embryogenesis in the same species (Von Aderkas *et al.*, 1991). During gametophytic embryogenesis, long cells were initially formed from the megagametophyte explant. These cells then divided in polar fashion forming a terminal cytoplasmically dense cell, which upon proliferation gave rise to embryonal mass. This mass consisted of axial embryos with dome-shaped embryo heads and suspensor. Once an embryonal mass had formed, gametic embryogenesis in *Larix* proceeded similarly as somatic embryogenesis in the same species (Von Aderkas *et al.*, 1991). Comparison of early stages was prohibited due to the fact that somatic embryogenesis was generally a form of cleavage polyembryony without early stages (Von Aderkas and Dawkins, 1993).

6.4. *Factors affecting gynogenesis*

6.4.1. *Developmental stage of the explant*

Calli exhibiting different degree of morphogenesis have been obtained from mature or near-mature megagametophytes in most gymnosperm genera studied (Table 3). However, when explants of different age were tested in culture an optimal stage for induction was found. Cultures of very young (pre-archegonial) megagametophytes have failed in all cases. In *Ginkgo biloba*, megagametophytes were most responsive for callus initiation during archegonial formation (approximately one month prior to fertilization) (Tulecke, 1967). Megagametophytes of *Zamia integrifolia* responded over a developmental period ranging from early archegonial (one to two months pre-fertilization) to approximately one month fertilized (Norstog, 1965).

Among the conifers, induction of haploid embryogenic tissue from *Larix* gametophytes was only possible during a few weeks period following fertilization (Nagmani and Bonga, 1985; Von Aderkas *et al.*, 1987; Von Aderkas and Bonga, 1988a). The responsive period coincided with the formation of the corrosion cavity, and accumulation of starch reserves in the gametophytes (von Aderkas and Bonga, 1988a). Formation of green non-embryogenic callus, also followed a phenocritical pattern which, interestingly, lagged about one month behind (Von Aderkas and Bonga, 1988a). Responsive periods, similar to that described for gametophytic embryogenesis in *Larix*, were also found for the induction of non-embryogenic haploid callus in *Pinus nigra* (Radforth and Bonga, 1960), *P. resinosa* (Bonga and Fowler, 1970), *P. mugo* (Bonga, 1974), *Picea glauca* (Von Aderkas and Bonga, 1988a) and *P. sitchensis* (Baldursson *et al.*, 1993b). Hakman *et al.* (1986) described an optimum period for protoplast isolation from *Picea abies* megagametophytes during archegonia formation.

6.4.2. *Genotype of donor tree*

Only a few reports mention differences in morphogenic response of gametophytes among donor trees. Callus induction was found dependent on the donor tree in *Picea abies* (Huhtinen *et al.*, 1981; Simola and Honkanen, 1983), *P. sitchensis* (Baldursson *et al.*, 1993a) and *Pseudotsuga menziesii* (Glock *et al.*, 1988). In contrast, no difference in induction (presumably initial gametophyte swelling and proliferation) rate was found among three trees of *Picea glauca* (Von Aderkas and Bonga, 1988), and five trees of *Larix* spp., tested for gametophytic embryogenesis, all showed a low response (Von Aderkas *et al.*, 1990).

6.4.3. *Culture media*

LaRue (1948, 1954) was able to maintain *Zamia* megagametophytes alive for years on just water, and obtained root and shoot differentiation on a simple mineral solution and sucrose. Similarly, megagametophytes of *Pinus nigra* var. *austriaca* (Radforth and Bonga, 1960) sustained growth and internal differentiation on a simple buffer with indoleacetic acid (IAA). Most investigators, however, have used complete media with macro- and microelements, vitamins, carbohydrates, growth regulators, and various organic supplements. The most commonly used basal salt formulations have been those of White (1943) and Murashige and Skoog (1962) (Table 3), although this choice often appears to have been based on convention rather than extensive experimentation. Simola and Santanen (1990) were able to improve growth of *Picea abies* megagametophyte callus lines up to 50%, by adding Ni and increasing the levels of several other microelements (B, Zn, I, Cu, Co) from those formulated by Murashige and Skoog (1962). In the same study they could not demonstrate any effects of changing the nitrogen (NH_4^+, NO_3) composition in the medium.

With the exception of cycads (LaRue, 1954; De Luca *et al.*, 1979; Chavez and Norstog, 1992), externally applied plant growth regulators appear essential for growth of gymnosperm megagametophytes *in vitro* (Table 3). For callus induction a combination of auxin (typically 2,4–D) and cytokinin (typically kinetin), or coconut milk, was generally used, but 2,4–D alone was sufficient in many cases (Bonga and Fowler, 1970; Zenkteler and Gusowska, 1970; Bonga, 1974; Rohr, 1982; Baldursson *et al.*, 1993b). Subculture of megagametophyte calli on media low in growth regulators or with an increased cytokinin/auxin ratio was sometimes required for organogenesis (*Ephedra foliata*: Konar and Singh, 1979; Singh *et al.*, 1981; Bhatnagar and Singh, 1984), or embryogenesis (*Larix decidua*: Nagmani and Bonga, 1985). In other cases organogenesis has been obtained on relatively auxin-rich induction media (*Zamia integrifolia*: Norstog, 1965; *Ceratozamia mexicana*: De Luca *et al.*, 1979; Chaves and Norstog, 1992; *Picea abies*: Simola and Honkanen, 1983).

Addition of organic nitrogen (casein hydrolysate, glutamine, aspargine and other amino acids) to the culture media has been common, and was beneficial in some cases. Thus, in *Zamia integrifolia* addition of glutamine and aspargine to a medium containing auxin and cytokinin, doubled the number of gametophytes exhibiting a morphogenic response (Norstog, 1982). In *Picea abies*, organic nitrogen was favourable for growth of an embryogenic megagametophyte callus line whereas a non-embryogenic callus line was unaffected (Simola and Santanen, 1990). Earlier, Simola and Honkanen (1983) had found a combination of three polyamines to favour development of roots in megagametophyte callus of *Picea abies*.

Chaves and Norstog (1992) used 6% sucrose as a carbon source in megagametophyte cultures of *Ceratozamia* spp., other investigators have used sucrose at 2–3%.

6.4.4. *Pre-treatments and explant types*

Simola and Honkanen (1983) observed that cold storage (4°C) of cones for a minimum of 12 days enhanced callus induction frequencies in *Picea abies* megagametophyte cultures. In *Picea sitchensis* cold storage for up to 3 weeks was ineffective, whereas incubation of megagametophytes at 33°C for 2–4 days appeared beneficial for callus induction (Baldursson, 1994).

A significant enhancement of haploid callus induction was obtained by culturing *Picea sitchensis* megagametophytes in a low oxygen environment (4% O_2) for the first seven weeks in culture. Through such partial simulation of the *in ovulo* environment, up to 90% of megagametophytes from the most responsive donor tree could be induced to form haploid callus (Baldursson, 1994).

To secure purely haploid cultures, the fertilized gametophytes must be rid of the diploid embryo before inoculation. Transverse cuts of the gametophytes result in two different types of explants, i.e., a proximal one containing the archegonia (archegonial or micropylar end) and a distal more or less uniform one (chalazal end). Longitudinal cuts, on the other hand, are bilaterally symmetrical. Whole embryo-free explants may also be obtained through induced parthenocarpy (the formation of fruit without fertilization) by incompatible (mentor) pollination (Von Aderkas and Bonga, 1988a; Baldursson *et al.*, 1993b). Bonga (1974) found none of the chalazal explants of *Pinus mugo* and *P. nigra* to be responsive, whereas those cut longitudinally produced callus. In *Ephedra foliata* transversely and longitudinally cut gametophytes produced callus and roots in equal frequencies, but the longitudinally cut ones were more productive in both accounts (Bhatnagar and Singh, 1984). These variations may simply have been due to volume differences of the explant, or differences in the size of the area of injured surface in contact with the medium. Alternatively, it may be speculated that the archegonia and/or the maturing embryo(s) contributed to the better response of longitudinally cut explants. However, Von Aderkas *et al.* (1987) found chalazal ends of *Larix decidua* significantly more responsive than archegonial ends for callus induction. Baldursson (1994) compared four types of mentor-pollinated gametophyte explants of *Picea sitchensis*: whole, longitudinal halves, archegonial and chalazal halves, and found whole explants to respond best.

Explants of mature gametophytes of *Araucaria araucana* only formed callus when co-cultivated with proliferating embryonic tissue (Cardemil and Jordan, 1982). In *Taxus baccata*, activation and proliferation of mature endosperm also depended on the presence of the embryo (Zenkteler and Guzowska, 1970). These reports suggested the presence of a growth-promoting factor, being conveyed from the embryo to the gametophyte. The reverse, i.e., promoting effect of the gametophyte on growth of zygotic embryos *in vitro*, has been reported in *Pseudotsuga menziesii* (Mapes and Zaerr, 1981).

7. Conclusions and recommendations

7.1. *Woody angiosperms*

In light of the considerable effort invested in anther culture of woody angiosperms (Table 1) the results appear disappointing. Presently, haploid techniques have been developed for only a few genera (*Aesculus*, *Hevea*, *Populus*) up to a point approaching usefulness in tree improvement (see Baldursson and Ahuja, 1996). However, for any given tree species, the efforts have been negligible compared to that for most herbaceous crops. In the cereals (e.g., wheat and barley) progress was also slow until highly responsive "model" genotypes were identified, enabling reproducible systematic experiments for methodical improvements (Andersen *et al.*, 1991).

Identification of responsive genotypes emphasises the genetic aspect of the ability to respond in culture. This has not been appreciated enough in trees, since only one or very few donor genotypes have been tested in most studies. A more fruitful approach calls for testing a large number of donors.

Responsive material may also be obtained through hybridization of responsive genotypes, followed by selection of recombinants among the offsprings. An F_1 hybrid (Haikin 2) of *Hevea brasiliensis*, was found to surpass both parents in androgenic capacity (Chen, 1986). This approach has also been used successfully in the grasses (Halberg *et al.*, 1990).

When a new species is tested in anther culture, flowering branches of the donor trees should, if possible, be grafted on rootstocks and maintained indoors in a defined environment to avoid physiological fluctuations. If field material must be used, the general vigour and viability of the microspores should be checked by methods such as fluorescein diacetate (FDA) staining (Widholm, 1972). Anthers containing mid to late-uninucleate microspores should be used as explants. The MS basal medium has been widely used and is probably the best bet during screening of donor genotypes. Although auxins are generally sufficient for the induction of microspore callusing, embryogenesis may require the addition of cytokinins and lowering or exclusion of auxins. Hence it is wise to include two regimes of growth regulator combinations in the screening process, i.e., high and low auxin:cytokinin ratios. Similarly, media containing high (6–9%) and low (3%) sucrose concentrations should be included in the screening process. Finally, it is probably safer to use agarose or Gelrite as a solidifying agent rather than agar.

In most tree species the yield of macroscopic structures (embryos, calli) in anther culture will be very low in the initial screening process. For a more sensitive screening test, cytological observations on early events in culture are recommended. The viability and early divisions of the microspores can be monitored by FDA staining and microscopic observations. Alternatively, it is possible to search for multicellular pollen grains or proembryos by fixing anthers after 10–14 days in culture, followed by staining with nuclear stains such as Feulgen. By this method Olesen (1987) was able to identify responsive genotypes in anther culture of *Lolium perennne*.

7.2. *Gymnosperms*

Experimental androgenesis in gymnosperms has met with only marginal success compared with angiosperms. In that light it is rather ironic that gymnosperms were the first group where androgenic callus induction was demonstrated (*Ginkgo biloba*: Tulecke, 1953). Another curiosity is that gymnosperm pollen appears to have the ability to respond later in the developmental process (i.e., after the first microspore division) than is generally the case for angiosperms. The low response of gymnosperm microspores *in vitro* compared to those of angiosperms is indeed difficult to explain. Microsporogenesis and pollen ontogeny are basically very similar in both groups, except for the presence of the usually sterile and ephemeral prothallial cells in the former (Section 3.2). In this regard gymnosperm pollen is less specialized and might *a priori* be more amenable to culture.

Morphogenesis is more common in megagametophyte cultures, leading to plantlet formation in a few cases (cf. Tables 2 and 3). The reason why female gametophytes respond better than male gametophytes *in vitro* may simply be their larger size and abundant nutrient reserves, providing buffering to suboptimal culture conditions. Although mature megagametophytes of gymnosperms have exhibited organ differentiation *in vitro*, an optimal developmental stage for culture initiation appears to exists. In the conifers this is a two to four week period following fertilization. In terms of cultural practices and the necessity of testing several donor genotypes, the same general comments apply as for woody angiosperms.

Some authors (Simola and Honkanen, 1983; Von Aderkas *et al.*, 1987; Bonga *et al.*, 1988) have attributed the limited success with haploid induction in conifers to the high genetic load of lethal and semi-lethal recessive genes carried by these organisms. However, the frequency of callus induction could be improved dramatically by optimization of culture methods (Baldursson, 1994), suggesting that other factors may be also be important. The generally low effort per species with forest trees has been discussed above. Furthermore, most studies on conifers, with the notable exception of *Larix* (see above), have attempted to obtain haploids via indirect organogenesis. The reason for this is mostly historical since experimental embryogenesis as a phenomenon was only recently "discovered" in conifers (Hakman and Von Arnold, 1985; Nagmani and Bonga 1985). Indirect organogenesis from subcultured diploid somatic callus has rarely been achieved in conifers. Hence, the problem may partly lie in the low organogenic capacity of conifer callus rather than in haploidy, or the expression of deleterious genes, *per se*. Therefore, gametophytic embryogenesis should be emphasised in future studies.

Finally, the scarcity of natural haploids, the low vigour of experimentally produced haploid plants, and the conservatism exhibited by conifers in chromosome numbers and ploidy, indicate that conifers are in general intolerant to genome manipulations. Artificial doubling of chromosomes in haploid organogenic calli or embryonal masses prior to plant regeneration might be a fruitful approach. Indeed, the plant obtained by Von Aderkas and Bonga (1993) from haploid embryogenic culture of *Larix decidua* was spontaneously diploid. Presently, there is no report of artificial chromosome doubling in conifers and it is uncertain if conventional methods such as colchicine treatment will work. Alternatively in *Larix* spp. with a functional protoplast regeneration system (Von Aderkas, 1992), chromosome doubling prior to regeneration might be achieved through somatic hybridization.

8. References

Andersen, S.B., A. Olesen, N. Halberg and S. Madsen, 1991. Haploids in grasses based on knowledge from cereals. In: A.P.M. Nijs and E. Elgersma (Eds.), Fodder Crop Breeding: Achievements, Novel Strategies and Biotechnology, pp. 129–134. Pudoc, Wageningen.

Andersson, E., I. Ekberg and G. Eriksson, 1969. A summary of meiotic investigations in conifers. Stud. Forest. Suec. 70: 19.

Anonymous, 1975. Induction of haploid poplar plants from anther culture *in vitro*. Sci. Sin. 18: 771–777.

Asker, S., 1980. Gametophytic apomixis: Elements and genetic regulation. Heriditas 93: 277–293.

Avanzi, S. and P.G. Cionini, 1971. A DNA cytophotometric investigation on the development of the female gametophyte of *Ginkgo biloba*. Caryologia 24: 105–116.

Bajaj, Y.P.S. and M.S. Dhanju, 1983. Pollen embryogenesis in three ornamental trees – *Cassia fistula*, *Jacaranda acutifolia*, and *Poinciana regia*. J. Tree Sci. 2: 16–19.

Baldursson, S., J.V. Nørgaard and P. Krogstrup, 1993a. Factors influencing haploid callus initiation and proliferation in megagametophyte cultures of sitka spruce (*Picea sitchensis*). Silvae Genetica 42: 79–86.

Baldursson, S., J.V. Nørgaard, P. Krogstrup and S.B. Andersen 1993b. Microspore embryogenesis in anther culture of three species of *Populus*, and regeneration of dihaploid plants of *Populus trichocarpa*. Can. J. For. Res. 23: 1812–1825.

Baldursson, S., 1994. Promotion of haploid callus induction in cultured mega-gametophytes of *Picea sitchensis*. In: Proc. 5th Int. IUFRO Work. Party S2.04–07 "Biotechnology of Trees", ICONA-CENEAN, Balsain, Spain, 18–22 Oct., 1993 (in press).

Baldursson, S. and R.M. Ahuja, 1996. Cytogenetics and potential of haploidy in forest tree genetics and improvement. In: S.M. Jain, S.K. Sopory and R.E. Veilleux (Eds.), *In Vitro* Haploid Production in Higher Plants, Vol. 1, pp. 49–66. Kluwer Academic Publishers, Dordrecht.

Ball, E.A., 1987. Tissue culture multiplication of *Sequoia*. In: J.M. Bonga and D.J. Durzan (Eds.), Cell and Tissue Culture in Forestry, Vol. 3, pp. 146–158. Martinus Nijhoff Publishers, Dordrecht.

Bhatnagar, S.P. and M.N. Singh, 1984. Organogenesis in the cultured female gametophyte of *Ephedra foliata*. J. Exp. Bot. 35: 268–278.

Blackmore, S. and R.B. Knox, 1990. Microsporogenesis: the male programme of development. In: S. Blackmore and R.B. Knox (Eds.), Microspores: Evolution and Ontogeny, pp. 1–10. Academic Press, London.

Blaydes, D.F., 1966. Interaction of kinetin and various inhibitors in the growth of soybean tissue. Physiol. Plant. 19: 748–753.

Bonga, J.M., 1974. *In vitro* culture of microsporophylls and megagametophyte tissue of *Pinus*. In Vitro 9: 270–277.

Bonga, J.M., 1977. Applications of tissue culture in forestry. In: J. Reinert and Y.P.S. Bajaj (Eds.), Applied and Fundamental Aspects of Plant Cell Tissue and Organ Culture, pp. 93–108. Springer-Verlag, Berlin.

Bonga, J.M., 1981. Haploid tissue culture and cytology of conifers. In: Colloque International Sur La Culture *In Vitro* des Espèces Forestières, IUFRO Section S2 01 5, Fountainbleau, 31 August – 4 September 1981, pp. 283–294.

Bonga, J.M. and D.P. Fowler, 1970. Growth and differentiation in gametophytes of *Pinus resinosa* cultured *in vitro*. Can. J. Bot. 48: 2205–2207.

Bonga, J.M. and A.H. McInnis, 1975. Stimulation of callus development from immature pollen of *Pinus resinosa* by centrifugation. Plant Sci. Lett. 4: 199–203.

Bonga, J.M., P. von Aderkas and D. James, 1988. Potential application of haploid cultures of tree species. In: J.W. Hanover and D.E. Keathley (Eds.), Genetic Manipulation of Woody Plants, pp. 57–78. Plenum Publishers, New York.

Borchert, R. 1968. Spontane Diploidisierung in Gewebekulturen des Megagametophyten von *Pinus lambertiana*. Z. Pflanzenphysiol. 59: 389–392.

Bourgin, J.P. and J.P. Nitsch, 1967. Production of haploid *Nicotiana* from excised stamens. Ann. Physiol. Vég. 9: 377–382.

Braniste, N., A. Popescu and T. Coman, 1984. Production and multiplication of *Pyrus communis* haploid plants. Acta Hort. 161: 147–150.

Brewbaker, J.L., 1967. The distribution and phylogenetic significance of binucleate and trinucleate pollen grains in the angiosperms. Am. J. Bot. 54: 1069–1083.
Brown, C.L. and R.H. Lawrence, 1968. Culture of pine callus on a defined medium. Forest Sci. 14: 62–64.
Brunkener, L., 1974. A review of methods for the production of haploids in seed plants. Research Notes, Royal College of Forestry, Stockholm, Nr. 13.
Cardemil, L. and M. Jordan, 1982. Light and electron microscopic study *in vitro* cultured female gametophyte of *Araucaria araucana* (Mol.) Koch. Z. Pflanzenphysiol. 107: 329–338.
Cersosimo, A., M. Crespan, G. Paludetti and A.A. Altamura, 1990. Embryogenesis, organogenesis and plant regeneration from anther culture in *Vitis*. Acta Hort. 280: 307–314.
Chavez, V.M. and K. Norstog, 1992. *In vitro* morphogenesis of *Ceratozamia hildae* and *C. mexicana* from megagametophytes and zygotic embryos. Plant Cell Tiss. Org. Cult. 30: 93–98.
Chen, Z., 1986. Induction of androgenesis in woody plants. In: H. Hu and H. Yang (Eds.), Haploids of Higher Plants *In Vitro*, pp. 42–66. Springer-Verlag, Berlin.
Chen, Z., 1987. Induction of androgenesis in hardwood trees. In: J.M. Bonga and D.J. Durzan (Eds.), Cell and Tissue Culture in Forestry, Vol. 2, pp. 247–268. Martinus Nijhoff Publishers, Dordrecht.
Chen, Z., 1990. Rubber (*Hevea brasiliensis* Muell. Arg.): *In vitro* production of haploids. In: Y.P.S. Bajaj (Ed.), Biotechnology in Agriculture and Forestry, Vol. 12. Haploids in Crop Improvement I, pp. 215–236. Springer-Verlag, Berlin.
Chen, Z., C. Qian, M. Qin, X. Xu and Y. Xiao, 1982. Recent advances in anther culture of *Hevea brasiliensis* (Muell.-Arg.). Theor. Appl. Genet. 62: 103–108.
Ching, K. and M. Simak, 1971. Competition among embryos in polyembryonic seeds of *Pinus silvestris* L. and *Picea abies* (l.) Karst. Royal Collage of Forestry, Dept. of Forestry and Genetics, Stockholm. Research Notes 30.
Chu, C.C., C.C. Wang, C.S. Sun, C. Hsu, K.C Yin, C.Y. Chu and C.Y. Bi, 1975. Establishment of an efficient medium for anther culture of rice, through comparative experiments on the nitrogen sources. Sci. Sin. 18: 659–668.
De, D.N. and P.V.L. Rao, 1983. Androgenetic haploid callus of tropical leguminous trees. In: S.K. Sen and K.L. Giles (Eds.), Plant Cell Culture in Crop Improvement, pp. 469–474. Plenum Press, New York.
De Luca, P. and S. Sabato, 1979. *In vitro* spermatogenesis of *Encephalartos* Lehm. Caryologia 32: 241–245.
De Luca, P. and S. Sabato, 1980. Regeneration of coralloid roots on cycad megagametophytes. Plant Sci. Lett. 18: 27–31.
De Luca, P., V. La Valva and S. Sabato, 1980. Spermatogenesis and tissue formation in cycad pollen grains. Caryologia 33: 261–265.
De Luca, P., A. Moretti and S. Sabato, 1979. Regeneration in megagametophytes of Cycads. Giorn. Bot. Ital. 113: 129–143.
Drira, N. and A. Benbadis, 1975. Analyse, par culture d'anthères *in vitro*, des potentialités androgénétiques de deux espéces de *Citrus* (*Citrus medica* L. et *Citrus limon* L. Burm.). C.R. Acad. Sci. Paris, 281: 1321–1324.
Duhoux, E., 1980. Protoplast isolation of gymnosperm pollen. Z. Pflanzenphysiol. 99: 207–214.
Duhoux, E. and B. Norreel, 1974. Sur l'isolement de colonies tissulaires d'origine pollinique à partir de cones males du *Juniperus chinensis* L., du *Juniperus communis* L., et du *Cupressus arizonica* G., cultivés *in vitro*. C.R. Acad. Sci. (Paris) 279: 651–654.
Dunbar, A. and J.R. Rowley, 1984. *Betula* pollen development before and after dormancy: exine and intine. Pollen et Spores 26: 299–338.
Dunwell, J.M., 1985. Haploid cell cultures. In: R.A. Dixon (Ed.), Plant Cell Culture – A Practical Approach, pp. 21–37. IRL Press, Oxford, Washington DC.
Ekberg, I. and G. Eriksson, 1967. Development and fertility of pollen in three species of *Larix*. Hereditas 57: 303–311.

Ekberg, I., G. Eriksson, and Z. Sulikova, 1968. Meiosis and pollen formation in *Larix*. Hereditas 59: 427–438.

Favre-Duchartre, M., 1958. *Ginkgo*, an oviparous plant. Phytomorphology 8: 377–390.

Fei, K.W. and G.R. Xue, 1981. Induction of haploid plantlets by anther culture *in vitro* in apple cv. "Delicious". Chinese Agr. Sci. 4: 41–44 (in Chinese, English abstract).

Gamborg, O.L., R.A. Miller and K. Ojima, 1968. Nutrient requirements of suspension cultures of soybean root cells. Exp. Cell Res. 50: 151–158.

Gates, J.C. and M.S. Greenwood, 1991. The physical and chemical environment of the developing embryo of *Pinus resinosa*. Am. J. Bot. 78: 1002–1009.

Geraci, G. and A. Starrantino, 1990. Attempts to regenerate haploid plants from *in vitro* cultures of *Citrus* anthers. Acta Hort. 280: 315–320.

Germana, M.A., F.G. Crescimanno, F. De Pasquale and W.Y. Ying, 1991. Androgenesis in 5 cultivars of *Citrus limon* L. Burm. f. Acta Hort. 300: 315–324.

Glock, H., H.H. Hattemer and A. Steinhauer, 1988. Untersuchungen über die Überlebensfähigkeit von Endospermkulturen de Douglasie (*Pseudotsuga menziesii*) *in vitro*. Botanica Acta 101: 240–245.

Gresshoff, P.M. and C.H. Doy, 1974. Derivation of haploid cell line from *Vitis vinifera* and the importance of the stage of meiotic development of anthers for haploid culture of this and other genera. Z. Pflanzenphysiol. 73: 132–141.

Hackett, W.P., 1985. Juvenility, maturation and rejuvenation in woody plants. Hortic. Rev. 7: 109–155.

Hagman, M., 1975. Incompatability in forest trees. Proc. R. Soc. London, Ser. B. 188: 167–182.

Håkansson, A., 1956. Seed development of *Picea abies* and *Pinus silvestris*. Medd. Stat. Skogforskningsinst. (Stockholm) 46: 1–23.

Hakman, I., S. von Arnold and H. Fellner-Feldegg, 1986. Isolation and DNA analysis of protoplasts from developing female gametophytes of *Picea abies* (Norway spruce). Can. J. Bot. 64: 108–112.

Hakman, I. and S. Von Arnold, 1985. Plantlet regeneration through somatic embryogenesis in *Picea abies* (Norway spruce). J. Plant. Physiol. 121: 149–158.

Halberg, N., A. Olesen, I.K.D. Tuveson and S.B. Andersen, 1990. Genotypes of perennial ryegrass (*Lolium perenne* L.) with high anther-culture response through hybridization. Plant Breed. 105: 89–94.

Harn, C. and M.Z. Kim, 1972. Induction of callus from anthers of *Prunus armeniaca*. Korean J. Breed. 4: 49–53 (in Korean, English abstract).

Heberle-Bors, E., 1985. *In vitro* haploid formation from pollen: a critical review. Theor. Appl. Genet. 71: 361–374.

Heller, R., 1953. Research on the mineral nutrition of plant tissues. Ann. Sci. Nat. Bot. Biol. Vég. 11th Ser. 14: 1–223.

Hesse, C.O., 1971. Monoploid peaches, *Prunus persica* Batsch: description and meiotic analysis. J. Amer. Soc. Hort. Sci. 96: 326–330.

Hirabayashi, T., I. Kozaki and T. Akihama, 1976. *In vitro* differentiation of shoots from anther callus in *Vitis*. Hortic. Sci. 11: 511.

Ho, R.H. and Y. Raj, 1985. Haploid plant production through anther culture in poplars. For. Ecol. Managem. 13: 133–142.

Ho, R.H. and G.E. Rouse, 1970. Pollen germination of *Larix sibirica* (Siberian larch) *in vitro*. Can. J. Bot. 48: 213–215.

Ho, R.H. and O. Sziklai, 1971a. Germination and development of lodgepole pine pollen *in vitro*. Can. J. For. Res. 1: 12–19.

Ho, R.H. and O. Sziklai, 1971b. Pollen germination of *Tsuga heterophylla in vitro*. Can. J. Bot. 49: 117–119.

Huhtinen, O., 1972. Production and use of haploids in breeding conifers. IUFRO Genetics, Sabrao Joint Symposia, Tokyo, D-3 (I): 1–8.

Huhtinen, O., 1978. Callus and plantlet regeneration from anther cultures of *Betula pendula* Roth. In: IV Int. Congr. Plant Cell Tissue Cult., Calgary, p. 169 (Abstr.).
Huhtinen, O., J. Honkanen and L.K. Simola, 1981. Effects of genotype and nutrient media on callus production and differentiation of Norway spruce endosperms cultured *in vitro*. In: Colloque International Sur La Culture *In Vitro* des Especès Forestières, IUFRO Section S2 01 5, Fountainbleau, 31 August – 4 September 1981, pp. 307–311.
Illies, Z.M., 1964. Auftreten haploider Keimlinge bei *Picea abies*. Naturwissenschaften 51: 442.
Illies, Z.M., 1974a. Experimentally induced haploid parthenogenesis in the poplar section Leuce after late inactivation of the male gamete with toluidine-blue. In: H.F. Linskens (Ed.), Fertilization in Higher Plants, pp. 335–340. North-Holland Publishing Company, Amsterdam.
Illies, Z.M., 1974b. Induction of haploid parthenogenesis in aspen by post-pollination treatment with toluidine-blue. Silvae Genet. 23: 221–226.
Isakov, Y.N., A.K. Butorina and L.S. Muraya, 1981. Isolation of spontaneous haploids in scots pine and the prospect of their utilization in forest genetics and breeding. Sov. Genet. 17: 485–489.
Johnson, M.A., J.A. Carlson, J.H. Conkey and T.L. Noland, 1987. Biochemical changes associated with zygotic pine embryo development. J. Exp. Bot. 38: 518–524.
Jörgensen, J., 1988. Embryogenesis in *Quercus petraea* and *Fagus silvatica*. J. Plant Physiol. 132: 638–640.
Jörgensen, J., 1991. Androgenesis in *Quercus petraea*, *Fagus silvatica* and *Aesculus hippocastanum*. In: R.M. Ahuja (Ed.), Woody Plant Biotechnology, pp. 353–354. Plenum Press, New York (Abstr.).
Kanezawa, R., 1948. Haploid and tetraploid in locust trees (*Robina pseudacacia* L.). Bull. Tokyo Univ. Forest. 36: 11–18.
Kasha, K.J., A. Ziauddin and U.H. Cho, 1990. Haploids in cereal improvement: Anther and microspore culture. In: J.P. Gustafson (Ed.), Gene Manipulation in Plant Improvement II, pp. 213–235. Plenum Press, New York.
Khoshoo, T.N., 1961. Chromosome numbers in gymnosperms. Silvae Genet. 10: 1–9.
Kimber, G. and R. Riley, 1963. Haploid angiosperms. Bot. Rev. 29: 480–531.
Kohlenbach, H.W. and W. Wernicke, 1978. Investigations on the inhibitory effect of agar and the function of active carbon in anther culture. Z. Pflanzenphysiol. 86: 463–472.
Konar, R.N., 1958a. A qualitative survey of the free amino acids and sugars in the developing female gametophyte and embryo of *Pinus roxburghii* Sar. Phytomorphology 8: 168–173.
Konar, R.N., 1958b. A quantitative survey of some nitrogenous substances and fats in the developing female gametophyte and embryo of *Pinus roxburghii* Sar. Phytomorphology 8: 1174–176.
Konar, R.N., 1963. A haploid tissue from the pollen of *Ephedra foliata* Boiss. Phytomorphology 13: 170–174.
Konar, R.N. and A. Moitra, 1980. Ultrastructure, cyto- and histochemistry of female gametophyte of gymnosperms. Gamete Res. 3: 67–97.
Konar, R.N. and M.N. Singh, 1979. Production of plantlets from the female gametophytes of *Ephedra foliata* Boiss. Z. Pflanzenphysiol. 95: 87–90.
Krogstrup, P. 1986. Embryo-like structures from cotyledons and ripe embryos of Norway spruce (*Picea abies*). Can. J. For. Res. 16: 664–668.
Kubicki, B., J. Telezynska and E. Milewska-Pawliczuk, 1976. Induction of embryoid development from apple pollen grains. Acta Soc. Bot. Pol. 44: 631–635.
Lakshmi Sita, G., 1987. Triploids. In: J.M. Bonga and D.J. Durzan (Eds.), Cell and Tissue Culture in Forestry, Vol. 2, pp. 269–284. Martinus Nijhoff Publishers, Dordrecht.
LaRue, C.D., 1948. Regeneration in the megagametophyte of *Zamia floridana*. Bull. Torrey Bot. Club 75: 597–603.
LaRue, C.D., 1950. Regeneration in Cycas (abstract). Am. J. Bot. 37: 664.
LaRue, C.D., 1954. Studies on growth and regeneration in gametophytes and sporophytes of gymnosperms. Brookhaven Symp. Biol. 6: 187–208.

Last, D.I. and R.I.S. Brettell, 1990. Embryo yield in wheat anther culture is influenced by the choice of sugar in the culture medium. Plant Cell Rep. 9: 14–16.

Libby, W.J., R.F. Stettler and F.W. Seitz, 1969. Forest genetics and forest-tree breeding. Annu. Rev. Genet. 3: 469–494.

Lianfang, F., 1990. Litchi (*Litchi chinensis* Sonn.): *In vitro* production of haploid plants. In: Y.P.S. Bajaj (Ed.), Biotechnology in Agriculture and Forestry, Vol. 12. Haploids in Crop Improvement I, pp. 264–274. Springer-Verlag, Berlin.

Litvay, J.D., M.A. Johnson, D. Verma, D. Einspahr and K. Weyerauch, 1981. Conifer suspension culture medium development using analytical data from developing seeds. IPC Technical Paper Series, No. 115: 1–17.

Livingston G.K., 1972. Experimental studies on the induction of haploid parthenogenesis in Douglas-fir and the effects of radiation on the germination and growth of Douglas-fir pollen. Dissert. Abstr. Int. B 32: 4331–4332.

Magoon, M.L. and K.R. Khanna, 1963. Haploids. Caryologia 16: 191–235.

Maheshwari, S.C., A. Rashid and A.K. Tyagi, 1982. Haploids from pollen grains: Retrospect and prospect. Am. J. Bot. 69: 865–879.

Mapes, M.O. and J.B. Zaerr, 1981. The effect of the female gametophyte on the growth of cultured Douglas-fir embryos. Ann. Bot. 48: 577–582.

Mauro, M.Cl., C. Nef and J. Fallot, 1986. Stimulation of somatic embryogenesis and plant regeneration from anther culture of *Vitis vinifera* cv. Cabernet-Sauvignon. Plant Cell Rep. 5: 377–380.

Michellon, R., J. Hugard and R. Jonard, 1974. Sur l'isolement de colonies tissulaires de pecher (*Prunus persica* Batsch, cultivars Dixired et Nectared IV) et d'amandier (*Prunus amygdalus* Stokes, cultivar Ai) à partir d'anthères cultivées *in vitro*. C.R. Acad. Sc. Paris, 278: 1719–1722.

Milewska-Pawliczuk, E., 1990. Apple (*Malus domestica* Borkh.): *In vitro* induction of androgenesis. In: Y.P.S. Bajaj (Ed.), Biotechnology in Agriculture and Forestry, Vol. 12. Haploids in Crop Improvement I, pp. 250–263. Springer-Verlag, Berlin.

Miller, C.O. 1965. Evidence for the natural occurrence of zeatin and derivatives: compounds from maize which promote cell division. Proc. Natl. Acad. Sci. USA: 1052.

Mogensen, H.L., 1965. Ovule abortion in *Quercus* (Fagaceae). Am. J. Bot. 62: 160–165.

Murashige, T. and F. Skoog, 1962. A revised medium for rapid growth and bio-assays with tobacco tissue cultures. Physiol. Plant. 15: 473–497.

Murashige, T. and D.P.H. Tucker, 1969. Growth factor requirements of *Citrus* tissue culture. In: H.D. Chapman (Ed.), Proc. 1st Citrus Symp. Vol. 3, pp. 1155–1161. Univ. Calif., Riverside Publication.

Nagmani, R. and J.M. Bonga, 1985. Embryogenesis in subcultured callus of *Larix decidua*. Can. J. For. Res. 15: 1088–1091.

Nadgauda, R.S., V.A. Parasharami and A.F. Mascarenhas, 1990. Precocious flowering and seeding behaviour in tissue-cultured bamboos. Nature 344(6264): 335–336.

Nitsch, J.P., 1972. Haploid plants from pollen. Z. Pflanzenzüchtg. 67: 3–18.

Nitsch, J.P. and C. Nitsch, 1969. Haploid plants from pollen grains. Science 163: 83–87.

Norstog, K., 1965. Induction of apogamy in megagametophytes of *Zamia integrifolia*. Am. J. Bot. 52: 993–999.

Norstog, K., 1982. Experimental embryology of gymnosperms. In: B.M. Johri (Ed.), Experimental Embryology of Vascular Plants, pp. 25–51. Springer-Verlag, Berlin.

Norstog, K. and E. Rhamstine, 1967. Isolation and culture of haploid and diploid cycad tissues. Phytomorphology 17: 374–381.

Nørgaard, J.V., 1992. Somatic embryogenesis in *Abies nordmanniana* LK. Ph.D. Thesis, The Royal Veterinary and Agricultural University, Copenhagen.

Olesen, A., 1987. Anther culture of perennial ryegrass (*Lolium perenne* L.): The production of haploid plants and their potential in relation to traditional breeding strategies. Ph.D. Thesis, The Royal Veterinary and Agricultural University, Copenhagen.

Owens, J.N. and M.D. Blake, 1985. Forest tree seed production: A review of the literature

and recommendations for future research. Petawawa Nat. For. Inst., Can. For. Serv., Inf. Rep. PI-X-53.
Owens, J.N. and M. Molder, 1980. Sexual reproduction of sitka spruce (*Picea sitchensis*). Can. J. Bot. 58: 886–901.
Pennell, R.I., 1988. Sporogenesis in conifers. Adv. Bot. Res. 15: 179–196.
Polheim, F., 1968. *Thuja gigantea gracilis* Beissn. – ein Haplont unter den Gymnospermen. Biol. Runschau 6: 84–86.
Powell, W., 1990. Environmental and genetic aspects of pollen embryogenesis. In: Y.P.S. Bajaj (Ed.), Biotechnology in Agriculture and Forestry, Vol. 12. Haploids in Crop Improvement I, pp. 45–65. Springer-Verlag, Berlin.
Prakash, J. and K.L. Giles, 1987. Induction and growth of androgenic haploids. Int. Rev. Cytol. 107: 273–293.
Pratassenja, G.D., 1939. Production of polyploid plants. Haploid and triploids in *Prunus persica*. Plant Breed. Abstr. 9: 1602 (Abstr.).
Radforth, N.W. and J.M. Bonga, 1960. Differentiation induced as season advances in the embryo-gametophyte complex of *Pinus nigra* var. *austriaca*, using indoleacetic acid. Nature 185: 332.
Radojevic, L., 1978. *In vitro* induction of androgenic plantlets in *Aesculus hippocastanum*. Protoplasma 96: 369–374.
Radojevic, L., 1991. Horse chestnut (*Aesculus* spp.). In: Y.P.S. Bajaj (Ed.), Biotechnology in Agriculture and Forestry, Vol. 16. Trees III, pp. 111–141. Springer-Verlag, Berlin.
Radojevic, L. and A. Kovoor, 1986. Induction of haploids. In: Y.P.S. Bajaj (Ed.), Biotechnology in Agriculture and Forestry, Vol. 1. Trees, pp. 65–86. Springer-Verlag, Berlin.
Radojevic, L., N. Djordjevic and B. Tucic, 1989. *In vitro* induction of pollen embryos and plantlets in *Aesculus carnea* Hayne through anther culture. Plant Cell Tiss. Org. Cult. 17: 21–26.
Raghavan, V., 1978. Origin and development of pollen embryoids and pollen calluses in cultured anther segments of *Hyoscyamus niger* (henbane). Am. J. Bot. 65: 984–1002.
Rao, N.M. and A.R. Mehta, 1969. Callus tissue from the pollen of *Thuja orientalis* L. Indian J. Exp. Biol. 7: 132–133.
Razmologov, V.P., 1973. Tissue culture from the generative cell of the pollen grain of *Cupressus* spp. Bull. Tor. Bot. Club. 100: 18–22.
Redenbaugh, M.K., R.D. Westfall and D.F. Karnosky, 1980. Protoplast isolation from *Ulmus americana* L. pollen mother cells, tetrads and microspores. Can. J. For. Res. 10: 284–289.
Redenbaugh, M.K., R.D. Westfall and D.F. Karnosky, 1981. Dihaploid callus production from *Ulmus americana* anthers. Bot. Gaz. 142: 19–26.
Reynolds, T.L., 1990. Ultrastructure of pollen embryogenesis. In: Y.P.S. Bajaj (Ed.), Biotechnology in Agriculture and Forestry, Vol. 12. Haploids in Crop Improvement I, pp. 66–82. Springer-Verlag, Berlin.
Roberts-Oehlschlager, S. and J.M. Dunwell, 1990. Barley anther culture: Pre-treatment on mannitol stimulates production of microspore-derived embryos. Plant Cell Tiss. Org. Cult. 20: 235–240.
Rohr, R., 1973. Ultrastructure des spermatozoides de *Taxus baccata* L. obtenus à partir de cultures aseptiques de microspores sur un milieu artificiel. C.R. Acad. Sci. (Paris) 277: 1869–1871.
Rohr, R. 1977. Évolution en culture *in vitro* des prothalles fémelles âgés chez le *Ginkgo biloba* L. Z. Pflanzenphysiol. 85: 61–69.
Rohr, R., 1980. Développement *in vitro* du pollen *Ginkgo biloba* L. Cytologia 45: 481–495.
Rohr, R., 1982. Activation et proliferation des cellules du megagametophyte de *Taxus* cultive *in vitro*. Can. J. Bot. 60: 1583–1589.
Rohr. R., 1987. Haploids (Gymnosperms). In: J.P. Bonga and D.J. Durzan (Eds.), Cell and Tissue Culture in Forestry, Vol. 2, pp. 230–246. Martinus Nijhoff Publishers, Dordrecht.
Ross, S.D. and R.P. Pharish, 1985. Promotion of flowering in tree crops: Different mechanisms and techniques, with special reference to conifers. In: M.G.R. Cannell and J.E. Jackson

(Eds.), Attributes of Trees as Crop Plants, pp. 383–397. Institute of Terrestrial Ecology, Huntingdon.

Sankhala, N., D. Sankhala and U.N. Chatterji, 1967. *In vitro* induction of proliferation in female gametophytic tissue of *Ephedra foliata* Boiss. Naturwissenschaften 54: 203.

Sax, K. and H.J. Sax, 1933. Chromosome number and morphology in the conifers. J. Arnold Arboretum 14: 356–375.

Sedgley, M. and A.R. Griffin, 1989. Sexual Reproduction of Tree Crops. Academic Press, London.

Seirlis, G., A. Mouras and G. Salesses, 1979. Tentatives de culture *in vitro* d'anthères et de fragments d'oranges chez les *Prunus*. Ann. Amélior. Plant. 29: 145–161.

Sharp, W.R., R.S. Raskin and H.E. Sommer, 1972. The use of nurse culture in the development of haploid clones of tomato. Planta 104: 357–361.

Simak, M., A. Gustafsson, and K. Ching, 1968. Occurrence of mosaic aneuploid in polyembryonic Norway spruce seed. Stud. Forest. Suec. 67: 8.

Simola, L.K., 1981. Ultrastructure of callus cultures from trees. In: Colloque International Sur La Culture *In Vitro* des Espèces Forestières, IUFRO Section S2 01 5, Fountainbleau, 31 August – 4 September 1981, pp. 201–210.

Simola, L.K., 1987. Structure of cell organelles and cell wall in tissue cultures of trees. In: J.M. Bonga and J.J. Durzan (Eds.), Cell and Tissue Culture in Forestry, Vol. 1, pp. 389–419. Martinus Nijhoff Publishers, Dordrecht.

Simola, L.K. and J. Honkanen, 1983. Organogenesis and fine structure in megagametophyte callus lines of *Picea abies*. Physiol. Plant. 59: 551–561.

Simola, L.K. and O. Huhtinen, 1986. Growth, differentiation, and ultrastructure of microspore callus of *Picea abies* as affected by nitrogenous supplements and light. N. Z. J. For. Sci. 16: 357–368.

Simola, L.K. and A. Santanen, 1990. Improvement of nutrient medium for growth and embryogenesis of megagametophyte and embryo callus lines of *Picea abies*. Physiol. Plant. 80: 27–35.

Singh, H., 1978. Embryology of Gymnosperms. Gebrüder Borntraeger, Berlin.

Singh, M.N., R.N. Konar and S.P. Bhatnagar, 1981. Haploid plantlet formation from female gametophytes of *Ephedra foliata* Boiss. *in vitro*. Ann. Bot. 48: 215–220.

Sondahl, M.R. and W.R. Sharp, 1979. Research in *Coffea* spp. and applications of tissue culture methods. In: W.R. Sharp, P.O. Larsen, E.F. Paddock and V. Raghaven (Eds.), Plant Cell and Tissue Culture, Principles and Applications, pp. 527–584. Ohio University Press, Columbus, Ohio.

Stettler, R.F., 1968. Irradiated mentor pollen: its use in remote hybridization of black cottonwood. Nature 219: 746–747.

Stettler, R.F. and K.S. Bawa, 1971. Experimental induction of haploid parthenogenesis in black cottonwood. Silvae Genet. 20: 15–25.

Stettler R.F. and A.A. Ager, 1984. Mentor effects in pollen interactions. In: H.F. Linskens and J. Heslop-Harrison (Eds.), Cellular Interactions, (Encyclopedia Plant Physiology), pp. 609–623. Springer-Verlag, Berlin.

Stoehr, M.U. and L. Zsuffa, 1990. Induction of haploids in *Populus maximowiczii* via embryogenic callus. Plant Cell Tiss. Org. Cult. 23: 49–58.

Sunderland, N. and L.J. Evans, 1980. Multicellular pollen formation in cultured barley anthers. II: The A, B and C pathways. J. Exp. Bot. 31: 501–514.

Sunderland, N. and B. Huang, 1987. Ultrastructural aspects of pollen dimorphism. Int. Rev. Cytol. 107: 175–219.

Thanh-Tuyen, N.T. and E.V. de Guzman, 1983. Formation of pollen embryos in cultured anthers of coconut (*Cocos nucifera* L.). Plant Sci. Lett. 29: 81–88.

Toda, R. and T. Sato, 1967. The detection of abnormal embryo in pine seed by soft X-ray photography, and frequency of the abnormal embryo seed. J. Jpn. For. Soc. 49: 429–436.

Todorovic, R.R., P.D. Misic, D.M. Petovic and M.A. Mirkovic, 1991. Anther culture of peach cultivars "Cresthaven" and "Vesna". Acta Hort. 300: 331–333.

Toyama, T.K., 1974. Haploidy in peach. HortScience 9: 187–188.
Tralau, H., 1957. Über eine Haploide Form von *Populus tremula* aus Uppland. Botaniska Notiser, Lund.
Tsay, H.S. and C.Y. Su, 1985. Anther culture of papaya (*Carica papaya* L.). Plant Cell Rep. 4: 28–30.
Tulecke, W., 1953. A tissue derived from the pollen of *Ginkgo biloba*. Science 117: 599–600.
Tulecke, W., 1957. The pollen of *Ginkgo biloba*: *In vitro* culture and tissue formation. Am. J. Bot. 44: 602–608.
Tulecke, W., 1959. The pollen cultures of C.D. LaRue: A tissue from the pollen of *Taxus*. Bull. Torrey Bot. Club 86: 283–289.
Tulecke, W., 1964. A haploid tissue culture from the female gametophyte of *Ginkgo biloba* L. Nature 203: 94–95.
Tulecke, W., 1967. Studies on tissue cultures derived from *Ginkgo biloba* L. Phytomorphology 17: 381–386.
Tulecke, W. and N. Sehgal, 1963. Cell proliferation from the the pollen of *Torreya nucifera*. Contr. Boyce Thompson Inst. 22: 153–163.
Uddin, M.R., M.M. Meyer and J.J. Jokela, 1988. Plantlet production from anthers of Eastern cottonwood (*Populus deltoides*). Can. J. For. Res. 18: 937–941.
Van Tran Thanh, K., D. Yilmaz-Lentz and T.R. Trinh, 1987. *In vitro* control of morphogenesis in conifers. In: J.P. Bonga and D.J. Durzan (Eds.), Cell and Tissue Culture in Forestry, Vol. 2, pp. 168–182. Martinus Nijhoff Publishers, Dordrecht.
Viherä-Aarnio, A., 1991. History of birch (*Betula* spp. L.) breeding in Finland. In: Rep. Found. Forest Tree Breed. 1, Helsinki, pp. 49–57.
Von Aderkas, P., 1992. Embryogenesis from protoplasts of haploid European larch. Can. J. For. Res. 22: 397–402.
Von Aderkas, P., J.M. Bonga, and R. Nagmani, 1987. Promotion of embryogenesis in cultured megagametophytes of *Larix decidua*. Can. J. For. Res. 17: 1293–1296.
Von Aderkas, P. and J.P. Bonga, 1988a. Morphological definition of phenocritical period for initiation of haploid embryogenic tissue from explants of *Larix decidua*. In: R.M. Ahuja (Ed.), Somatic Cell Genetics of Woody Plants, pp. 29–38. Kluwer Academic Publishers, Dordrecht.
Von Aderkas, P. and Bonga, J.M. 1988b. Formation of haploid embryoids of *Larix decidua*: early embryogenesis. Am. J. Bot. 75: 690–700.
Von Aderkas, P., K. Klimaszewska, and J.M. Bonga, 1990. Diploid and haploid embryogenesis in *Larix leptolepis*, *L. decidua*, and their hybrids. Can. J. For. Res. 20: 9–14.
Von Aderkas, P., K. Klimaszewska, J.N. Owens and J.M. Bonga, 1991. Comparison of larch embryogeny *in vivo* and *in vitro*. In: M.R. Ahuja (Ed.), Woody Plant Biotechnology, pp. 139–155. Plenum Press, New York.
Von Aderkas, P. and J.M. Bonga, 1993. Plants from haploid tissue of *Larix decidua*. Theor. Appl. Genet. 87: 225–228.
Von Aderkas, P. and M.D. Dawkins, 1993. Haploid embryogenesis in trees. In: M.R. Ahuja (Ed.), Micropropagation of Woody Plants, pp. 58–65. Kluwer Academic Publishers, Dordrecht.
Von Kopecky, F., 1960. Experimentelle Erzeugung von haploiden Weißpappeln (*Populus alba* L.). Silvae Genet. 9: 102–105.
Wang, C., Z. Chu and C. Sun, 1975. The induction of *Populus* pollen-plants. Acta Bot. Sin. 18: 56–62 (in Chinese, English abstract).
Widholm, J.M., 1972. The use of fluorescein diacetate and phenosafranine for determining viability of cultured plant cells. Stain Technol. 47: 189–194.
White, P.R., 1943. Nutrient deficiency studies and an improved inorganic nutrient for cultivation of excised tomato roots. Growth 7: 53–65.
Winton, L.L. and D.W. Einspahr, 1968. The use of heat-treated pollen for aspen haploid production. Forest Sci. 14: 406–407.

Wu, J.Y., 1981. Induction of haploid plants from anther culture of crabapple. J. Northeast Agr. Coll. 3: 105–108 (in Chinese, English abstract).

Wu, K. and P. Nagarajan, 1990. Poplars (*Populus* spp.): *In vitro* production of haploids. In: Y.P.S. Bajaj (Ed.), Biotechnology in Agriculture and Forestry, Vol. 12. Haploids in Crop Improvement I, pp. 237–249. Springer-Verlag, Berlin.

Yang, H.Y. and C. Zhou, 1982. *In vitro* induction of haploid plants from unpollinated ovaries and ovules. Theor. Appl. Genet. 63: 97–104.

Zenkteler, M.A. and I. Guzowska, 1970. Cytological studies on the regenerating mature female gametophyte of *Taxus baccata* L., and mature endosperm of *Tilia platyphyllos* Scop. *in vitro* culture. Acta Soc. Bot. Pol. 39: 161–173.

Zhang, Y.X., Y. Lespinasse and E. Chevreu, 1990. Induction of haploidy in fruit trees. Acta Hort. 280: 293–305.

Zhang, Y.X., L. Bouvier and Y. Lespinasse, 1992. Microspore embryogenesis induced by low gamma dose irradiation in apple. Plant Breed. 108: 173–176.

Zou, C. and P. Li, 1981. Induction of pollen plants of grape (*Vitis vinifera* L.) Acta Bot. Sin. 23: 79–81.

17. Haploidy in bamboo (*Sinocalamus latiflora*) by anther culture

HSIN-SHENG TSAY, JIA-YAN HSU and CHANG-CHING YEH

Contents

1. Introduction

Bamboo is a monocotyledonous evergreen plant belonging to the family Gramineae and subfamily Bambusoideae. Its distribution covers tropical, subtropical and temperate zones of the world. *Sinocalamus latiflora* Munro (Syn. *Bambusa oldhamii* Munro) is the major bamboo species growing in Taiwan. It can be utilized as vegetable, ornamental plant and paper pulp for various purposes. Bamboo flowers only once during the end of its lifetime, which occurs at the end of its first fruiting season (Nadgauda *et al.*, 1990). The improvement of this species is not possible through conventional breeding methods due to irregular flowering habit, extremely low seed viability, and dependence of seed production on unpredictably long intervals (6–50 years) (Mehta *et al.*, 1982). Propagation through seeds is also undependable on account of the long interval needed to seed (50–60 years in *S. latiflora*). Vegetative propagation through conventionally used offset is beset with problems such as bulkiness and non-availability of the propagules and difficulties in transport over long distances (Mehta *et al.*, 1982; Rao *et al.*, 1985).

Haploid production is an effective method for the production of homozygous lines, which are used to develop hybrid cultivars in cross-pollinated crops. This chapter describes the successful anther culture of *S. latiflora* (Munro) McClure resulting in pollen embryo formation and subsequent plant development. The techniques described should be of value for future cultivar improvement of this economically important species.

S.M. Jain, S.K. Sopory & R.E. Veilleux (eds.), In Vitro *Haploid Production in Higher Plants, Vol. 3,* 337–347.

Table 1. Influence of microspore developmental stage on the formation of embryos and callus in bamboo anther culture

Microspore developmental stage	No. of anthers cultured	Anthers producing embryos and callus	
		Number	%
Tetrads	135	0	0.0 (0–3)
Early-uninucleate	1070	9	0.8 (0–2)
Mid-uninucleate	1540	57	3.7 (3–5)
Late-uninucleate	1090	47	4.3 (3–5)
Early-binucleate	520	31	6.0 (4–9)

Basal medium contained N_6 inorganic salts and MS organic substances supplemented with 1 mg/l BA, 1 mg/l 2,4–D, 6% sucrose, 0.2% charcoal and 0.8% Sigma agar. Data in parentheses are the 95% confidence limits.

2. Procedure

2.1. *Anther stage and bud selection*

The developmental stage of pollen at the time of anther excision and culture is an important factor affecting androgenesis. It has been shown that the production of embryos or callus in cultured anthers, generally, occurs only within a short period of microsporogenesis and the optimal response stage varies depending on the species (Sunderland and Dunwell, 1977; Sunderland, 1980). In bamboo, anthers in late-uninucleate to early-binucleate stage are highly responsive in culture (Table 1). Accurate developmental stage can be determined by fixing anthers in Carnoy's solution (ethanol:chloroform:acetic acid = 6:3:1) and followed by strong acid (15% CrO_3:10% HNO_3:5% HCl = 2:1:1) treatment (Tsay, 1980) and staining with aceto-carmine.

Inflorescences or spikelets were collected from field-grown bamboo plants in early summer (May to June of 1989–1992). The panicles of bamboo with microspores at different developmental stage are shown in Fig. 1. Bamboo spikelets with anthers suitable for anther culture are shown in Fig. 2. Florets can be exposed after 3–4 layers of glumes have been removed (Fig. 3). Spikelet with 5–6 florets can be used as culture materials. Each floret contains 6 anthers (Figs. 4–5).

2.2. *Bud pretreatment*

Pretreatment of bamboo in florescences at 10°C for 7 days prior to anther culture was found to enhance microspore embryogenesis (Table 2). In the early work with experimental androgenesis, both the induction for sporophytic development and the continued division of the microspores to form callus or embryos were usually carried out under the same conditions by inoculating the excised anthers on a specified culture medium. However,

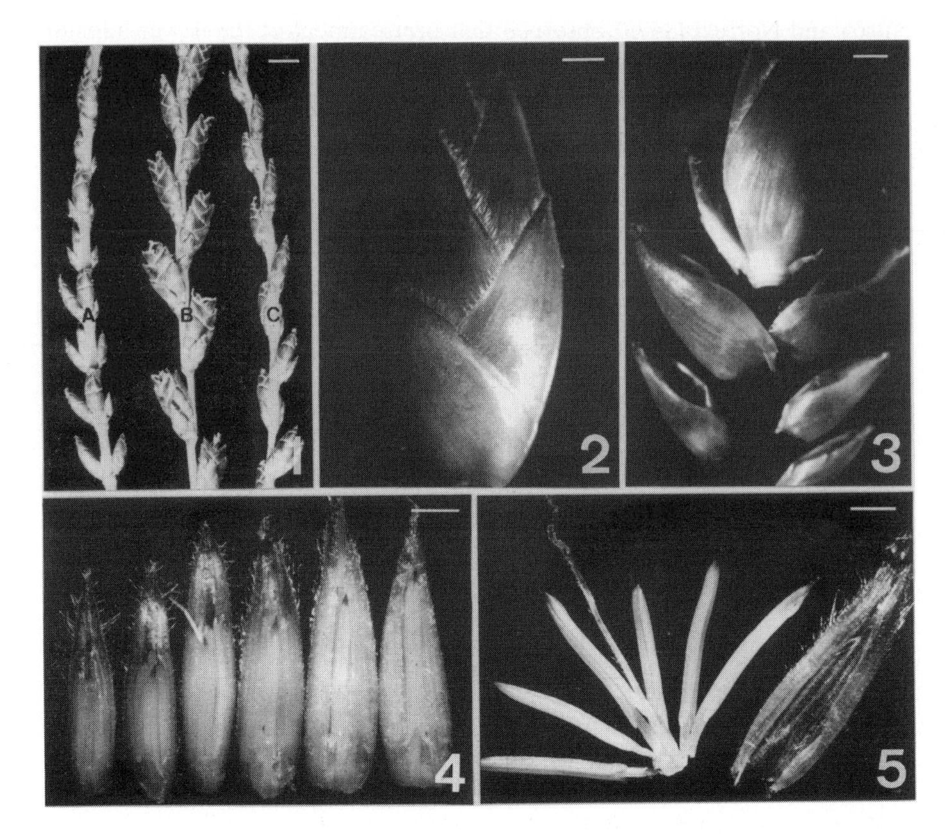

Figure 1. 1. The panicles of bamboo with microspores at different developmental stages (bar = 10 mm). Inflorescence with microspores before uninucleate stage (A), at post-binucleate stage (B), and between uninucleate to early-binucleate stage (C). 2. Bamboo spikelets with anthers suitable for anther culture (bar = 1 mm). 3. Floret exposed after 3–4 layers of glumes were removed (bar = 1 mm). 4. Spikelet with 5–6 florets can be used as culture materials (bar = 1 mm). 5. Each floret contains 6 anthers (bar = 1 mm).

Table 2. Influence of cold (10°C) pretreatment of inflorescence before anther inoculation on the formation of embryos and callus in bamboo anther culture

Duration of cold treatment (days)	No. of anthers cultured	Anthers producing embryos and callus	
		Number	%
0	1338	49	3.7 (3–5)
7	631	36	5.7 (4–9)

Anthers with microspores at early-uninucleate to early-binucleate stages were used. Basal medium was the same as shown in Table 1, except 9% sucrose was used. Data in parentheses are the 95% confidence limits.

Nitsch and Norreel (1973) observed that pretreatment of the flower buds of *Datura innoxia* to low temperature for a short period of time prior to culture yielded more pollen embryos and plants. The beneficial effects of low-temperature pretreatment have since been confirmed in anther culture of other species (Bajaj, 1977, 1983, 1990; Malhotra and Maheshwari, 1977; Sunderland, 1978; Genovesi and Magill, 1979; Sunderland and Roberts, 1979; Sunderland and Wildon, 1979; Huang and Sunderland, 1980; Lin and Tsay, 1984; Tsay and Chen, 1984).

In addition to cold shock, other pretreatments such as high temperature (Keller and Armstrong, 1978, 1979; Yeh and Tsay, 1988), placing the detached tillers in water at room temperature (Wilson *et al.*, 1978) and putting the excised anthers in a water-saturated atmosphere (Dunwell, 1981) have also been found beneficial to embryo and plant yields. These discoveries have led to the formulation of a concept that the induction phase is of great importance in androgenesis (Sunderland, 1980), and this phase is probably nutrient-independent (Dunwell, 1981).

2.3. *Sterilization and anther dissecting*

For sterilization, the spikelets were first immersed in 70% alcohol for 30 s followed by 1% sodium hypochlorite solution for 10 min. After rinsing with sterile water 3–4 times, flower buds were separated and anthers dissected out with forceps.

2.4. *Culture medium*

Although the induction phase of androgenesis may be nutrient-independent, continued division of the induced microspores to the formation of embryos or callus does require the presence of appropriate nutrients in the culture medium (Nitsch, 1969; Wang *et al.*, 1974). Clapham (1973) first discovered that the high concentration of ammonium ions in LS medium (Linsmaier and Skoog, 1965) was inhibitory to callus formation from barley microspores. Subsequently, Chu *et al.* (1975) developed the N_6 medium which is characterized by having a low concentration of $(NH_4)_2SO_4$ and a high concentration of KNO_3. For bamboo anther culture, medium is derived from MS (Murashige and Skoog, 1962) or N_6 basic salts with the following modifications: 1 mg/l 2,4–D, 1 mg/l BA, 2 g/l charcoal, 9% sucrose, 0.8% agar and pH 5.7 ± 0.1. The results also showed that N_6 was more efficient than MS basic salts (Table 3). Similar phenomena have been reported in other cereal crops (Chu, 1978; Miao *et al.*, 1978; Genovesi and Magill, 1979; Chen *et al.*, 1982; Nitsch *et al.*, 1982; Tsay *et al.*, 1982).

High sucrose concentrations have been found to be beneficial for plant production from anther culture of many plant species, especially for the Gramineae (Clapham, 1973; Ouyang *et al.*, 1973; Ono and Larter, 1976; Chen *et al.*, 1978; *Miao et al.*, 1978; Tsay *et al.*, 1986). In bamboo, evaluation

Table 3. Effect of mineral salts and plant growth regulators on the formation of embryos and callus from bamboo anthers cultured at the late uninucleate pollen stage

Medium composition			No. of anthers cultured	Anthers producing embryos and callus	
Salts	2,4-D (mg/l)	NAA (mg/l)		Number	%
N_6	1		113	5	4.4 (1–10)
N_6	3		90	2	2.2 (0–7)
N_6	5		145	2	1.4 (0–5)
N_6		3	125	4	3.2 (1–8)
N_6		5	115	1	0.9 (0–5)
MS	3		175	2	1.1 (0–5)
MS	5		70	0	0.0 (0–6)

Basal medium: N_6 (Chu *et al.*, 1975) or MS (Murashige and Skoog, 1962) salts with 1 mg/l BA, 2 g/l charcoal, 6% sucrose and 0.8% Sigma agar. Anthers with microspores at early-uninucleate to early-binucleate stages were used. Data in parentheses are the 95% confidence limits.

Table 4. Influence of sucrose concentration on the formation of embryos and callus in bamboo anther culture

Sucrose concentration (%)	No. of anthers cultured	Anthers producing embryos and callus	
		Number	%
3	970	25	2.6 (1–3)
6	1060	21	2.0 (1–3)
9	905	41	4.5 (3–5)
12	1420	57	4.0 (3–5)

Anthers with microspores at early uninucleate to early binucleate stages were used. Basal medium was the same as shown in Table 1. Data in parentheses are the 95% confidence limits.

of medium composition indicated that 9% sucrose concentration was optimal (Table 4). The reason for a high sucrose concentration requirement is not known. Wang *et al.* (1974) and Chaleff and Stolarz (1981) pointed out that osmotic pressure of the nutrient medium is at least partly responsible. Chen (1978) reported that high sucrose concentrations show differential promotive effects on rice anthers cultured at different developmental stages. Thus, the optimal sucrose concentration may change with the developmental stage of the microspores.

2.5. *Incubation of anthers*

Temperature plays important roles in androgenesis. The temperatures used for bamboo anther culture generally fall between 25 and 30°C. Table 5 shows that culture of anthers at 25°C for one month followed by a temperature of 30°C resulted in higher rate of embryo and callus initiation compared with those cultured constantly at 25°C. The same response was observed in rice

Table 5. Influence of high temperature during culture on the formation of embryos and callus in bamboo anther culture

Anther source of donor plant	Treatment	No. of anthers cultured	Anthers producing embryos and callus	
			Number	%
A	CK	1435	49	3.4 (2–4)
	LT	1158	78	6.7 (6–9)
B	CK	1359	4	0.3 (0–1)
	LT	1282	32	2.6 (2–4)
	AT	1195	33	2.8 (2–4)

Microspore developmental stage of anthers cultured and basal medium composition were the same as in Table 4. Anther donor plant A and B representing cultured anthers were collected from two individual plants. CK – cultured anthers were incubated at 25 ± 1°C; LT – cultured anthers were incubated at 25 ± 1°C for 1 month and transferred to 30 ± 1°C; AT – cultured anthers were incubated at 30 ± 1°C. Data in parentheses are the 95% confidence limits.

anther culture. However, concomitant with the increase in anther response is an increase in the number of albino plants (Wang *et al.*, 1978). In bamboo, the influence of temperature on the albino plant formation needs more study.

2.6. *Callus formation and plant regeneration*

Anthers turned brown within 2–3 weeks of culture. Continued culture on the same medium under the same environment for about two months resulted in callus proliferation (Fig. 6). The percentage of anthers producing callus was highest on medium containing 1 mg/l 2,4-D (Table 3). Pollen-derived callus was observed to be of two types. The embryogenic callus which later gave rise to plantlets was compact, opaque and white in appearance. Shiny globular bodies, comparable to normal zygotic embryos were observed on this type of callus (Fig. 7). The other type, a non-embryogenic callus (Fig. 8) appeared yellowish and friable. No plantlets could be induced from this callus.

The compact, organized callus maintained totipotency for more than 15 months through several subcultures on the same callus initiation and embryo induction medium. The embryos formed *in vitro* showed features characteristic of grass embryos including a well-developed scutellum and coleoptile (Fig. 9). Continued culture of these embryogenic calli on the anther culture medium or transfer of the callus to an auxin-free medium and culture in the same environment for about one month resulted in embryo germination and establishment of rooted plantlets (Fig. 10).

Microscopic examination of the root tip cells from anther-derived plantlets revealed their true microspore origin as evidenced by the haploid number of chromosomes ($n = 36$). Cytohistological studies of androgenic anthers of bamboo found that the embryos were derived from microspores (Figs. 11–

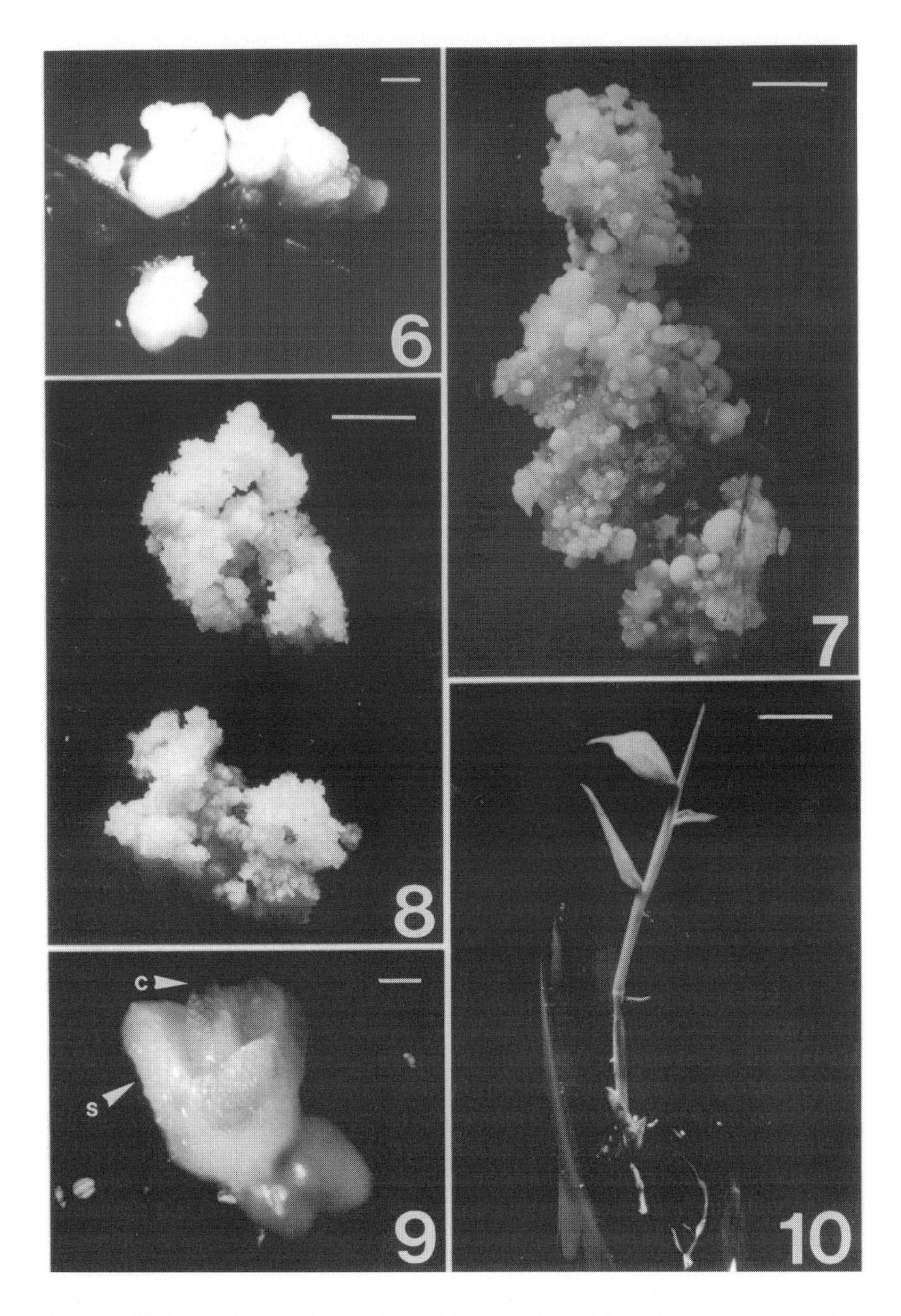

Figure 6. Embryogenic callus from anthers of bamboo after 60 days of culture on N_6 medium supplemented with 1 mg/l 2,4–D, 1 mg/l BA, 2 g/l charcoal, 9% sucrose and 0.8% Sigma agar (bar = 1 mm). 7. Clusters of embryos protruding from embryogenic callus (bar = 5 mm). 8. Non-embryogenic callus (bar = 5 mm). 9. Germinating embryo showing scutellum (S) and coleoptile (C) (bar = 1 mm). 10. Haploid plant induced through embryo formation of bamboo anther culture (bar = 10 mm).

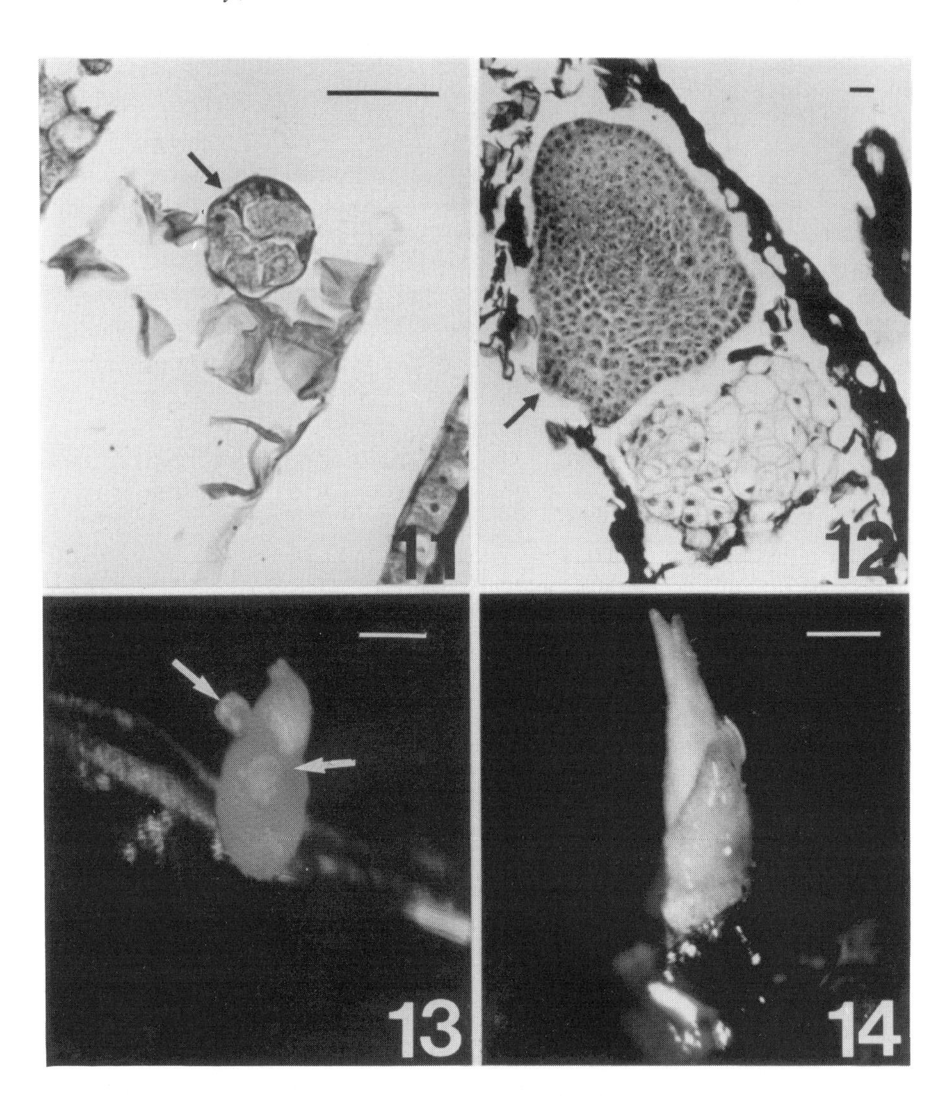

Figure 11. A multinucleate microspore (arrow) in a 6-week-old cultured anther. 12. A growing giant embryo (arrow) in a 10-week-old cultured anther. 13. Secondary embryos (arrow) at the surface of a primary torpedo shaped embryo in a 12–week-old cultured anther. 14. Haploid plant directly induced from a microspore-derived embryo. (Figs. 11–12, bar = 0.1 mm; Figs. 13–14, bar = 2 mm).

12). Fig. 11 showed a multinucleate microspore in a 6–week-old culture. A growing giant embryo was observed in a 10-week-old culture (Fig. 12). Secondary embryos were regenerated at the surface of a primary torpedo shaped embryo in a 12–week-old culture in the same medium (Fig. 13).

Haploid plants were directly induced from these embryos (Fig. 14). Six plantlets were successfully transplanted to pots and grew for 6 months. These transplanted haploid plants died eventually due to a lack of vigor.

As a member of the grass family, bamboo has been considered difficult to culture *in vitro*. Nadgauda *et al.* (1990) reported an *in vitro* inflorescence culture system with which viable seeds were produced from culture of panicles with spikelets in *Bambusa arundinacea* Wild and *Dendrocalamus brandisii* Kurz. Other reports included the aseptic culture of embryos from mature seeds (Alexander and Rao, 1968), the release of protoplasts from *Bambusa* leaf tissue (Tseng *et al.*, 1975), and callus initiation from leaves and shoot tips of three bamboo species (Huang and Murashige, 1983). There have also been several reports claiming *in vitro* regeneration of bamboo plants from a variety of explant sources. For example, callus cultures derived from mature seeds of *Bambusa arundinacea* (Mehta *et al.*, 1982) and *Dendrocalamus strictus* (Rao *et al.*, 1985), mature zygotic embryos of *Sinocala muslatiflora* (Yeh and Chang, 1987), leaves of *Phyllostachys viridis* (Hassan and Debergh, 1987) and inflorescences of *B. oldhamii* and *B. beecheyana* (Yeh and Chang, 1986a,b) have regenerated intact plants via somatic embryogenesis. The requirement for plant regeneration through somatic embryogenesis of *B. oldhamii*, *B. beecheyana* and *S. latiflora* was MS salts enriched with 3 mg/l 2,4–D and 2 mg/l kinetin (Yeh and Chang, 1986a,b, 1987).

Haploid plant regeneration from bamboo anther culture was not reported until 1990 (Tsay *et al.*, 1990; Yeh *et al.*, 1991). Although some haploid plants were obtained from anther-derived callus subcultured on an auxin free medium, solidified N_6 medium containing 1 mg/l 2,4–D and 1 mg/l BA was also satisfactory for both inducing embryos and subsequent plantlet development, a result similar to those described by Yeh and Chang (1986a,b, 1987).

3. Conclusion

Although anther culture has been used extensively on many Gramineae species, including important cereal crops, it has remained virtually untried with bamboos. The protocols and conditions reported here could lead to the initiation of breeding programmes for bamboo improvement. By chromosome doubling of the haploid plants, completely homozygous diploid plants can be obtained in one generation. The procedure is thus able to shorten the breeding time by 5–6 generations. Since the vegetative growth phase of some bamboo species lasts for over 60 years, the time saved by anther culture makes cultivar improvement possible. Further studies are needed to determine the method for chromosome doubling from these haploid plants.

4. References

Alexander, M.P. and T.C. Rao, 1986. *In vitro* culture of bamboo embryos. Curr. Sci. 37: 415.

Bajaj, Y.P.S., 1977. *In vitro* induction of haploids in wheat (*Triticum aestivum* L.). Crop Improv. 4: 54–64.

Bajaj, Y.P.S., 1983. *In vitro* production of haploids. In: D.A. Evans, W.R. Sharp, P.V. Ammirato, and Y. Yamada (Eds.), Handbook of Plant Cell Culture. Vol. 1. Techniques for Propagation and Breeding, pp. 228–287. MacMillan Press, New York.

Bajaj, Y.P.S., 1990. Biotechnology in Agriculture and Forestry. Vol. 12. Haploids in Crop Improvement 1. Springer-Verlag, Berlin.

Chaleff, R.S. and A. Stolarz, 1981. Factors influencing the frequency of callus formation among cultured rice (*Oryza sativa*) anthers. Physiol. Plant. 51: 201–206.

Chen, C.C., 1978. Effects of sucrose concentration on plant production in anther culture of rice. Crop Sci. 18: 905–906.

Chen, L.J., P.C. Lai, C.H. Liao and H.S. Tsay, 1982. Medium evaluation for rice anther culture. In: A. Fujiwara (Ed.), Plant Tissue Culture, pp. 551–552. Maruzen, Tokyo.

Chen, Y., C.H. Tso, J.F. Wang and K.H. Chang, 1978. On screening of anther culture media for hybrid *Oryza sativa* L. subsp. Keng X *O. sativa* subsp. Shien by orthogonal test. In: H. Hu (Ed.), Proc. Symp. Anther Culture, pp. 40–49. Science Press, Beijing.

Chu, C.C., 1978. The N_6 medium and its applications to anther culture of cereal crops. In: H. Hu (Ed.), Proc. Symp. Plant Tissue Culture, pp. 45–50. Science Press, Beijing.

Chu, C.C., C.S. Wang, C.S. Sun, C. Hsu, K.C. Yin, C.Y. Chu and F.Y. Bi, 1975. Establishment of an efficient medium for anther culture of rice through comparative experiments on the nitrogen sources. Sci. Sin. 17: 657–668.

Clapham, D., 1973. Haploid *Hordeum* plants from anthers *in vitro*. Z. Pflanzenzüchtg. 69: 142–155.

Dunwell, J.M., 1981. Stimulation of pollen embryo induction in tobacco by pretreatment of excised anthers in a water-saturated atmosphere. Plant Sci. Lett. 21:9–13.

Genovesi, A.D. and C.W. Magill, 1979. Improved rate of callus and green plant production from rice anther culture following cold shock. Crop Sci. 19: 662–664.

Hassan, A.E. and P. Debergh, 1987. Embryogenesis and plantlet development in the bamboo *Phyllostachys viridis* (Young) McClure. Plant Cell Tiss. Org. Cult. 10: 73–77.

Huang, B. and N. Sunderland, 1980. Stress pretreatment in barley anther culture. John Innes Inst. Annu. Rep. No. 71: 63–64.

Huang, L.C. and T. Murashige, 1983. Tissue culture investigation of bamboo I. Callus culture of *Bambusa*, *Phyllostachys* and *Sasa*. Bot. Bull. Acad. Sin. 24: 31–52.

Keller, W.A. and K.C. Armstrong, 1978. High frequency production of microspore-derived plants from *Brassica napus* anther cultures. Z. Pflanzenzüchtg. 80: 100–108.

Keller, W.A. and K.C. Armstrong, 1979. Stimulation of embryogenesis and haploid production in *Brassica campestris* anther cultures. Theor. Appl. Genet. 55: 65–67.

Lin, S.L. and H.S. Tsay, 1984. Effect of cold treatment on rice anther culture. J. Agric. Res. China 28: 45–49.

Linsmaier, E.M. and F. Skoog, 1965. Organic growth factor requirements of tobacco tissue culture. Physiol. Plant. 18: 100–127.

Malhotra, K. and S.C. Maheshwari, 1977. Enhancement by cold treatment of pollen embryoid development in *Petunia hybrida*. Z. Pflanzenphysiol. 85: 177–180.

Mehta, U., I.V.R. Rao and H.Y.M. Ram, 1982. Somatic embryogenesis in bamboo. In: A. Fujiwara (Ed.), Proc. 5th Intl. Cong. Plant Tissue and Cell Culture, pp. 109–110. The Jpn. Assoc. Plant Tissue Culture, Tokyo.

Miao, S.H., C.S. Kuo, Y.L. Kwei, A.T. Sun, S.Y. Ku, W.L. Lu and Y.Y. Wang, 1978. Induction of pollen plants of maize and observations on their progeny. In: H. Hu (Ed.), Proc. Symp. Plant Tissue Culture, pp. 23–33. Science Press, Beijing.

Murashige, T. and F. Skoog, 1962. A revised medium for rapid growth and bioassays with tobacco tissue cultures. Physiol. Plant. 15: 473–497.

Nadgauda, R.S., V.A. Parasharami and A.F. Mascarenhas, 1990. Precocious flowering and seeding behavior in tissue-cultured bamboo. Nature (London) 344: 335–336.

Nitsch, C. and B. Norreel, 1973. Effet d'un choc thermique sur le pouvoir embryogène du pollen de *Datura innoxia* cultivé dans l'anthère ou isolé de l'anthère. C.R. Acad. Sci. Paris Sér. D. 276: 303–306.

Nitsch, C., S. Anderson, M. Godard, M.G. Neuffer and W.P. Sheridan, 1982. Production of haploid plants of *Zea mays* and *Pennisetum* through androgenesis. In: E.D. Earle and Y. Demarly (Eds.), Variability in Plants Regenerated from Tissue Culture, pp. 69–91. Praeger, New York.
Nitsch, J.P., 1969. Experimental androgenesis in *Nicotiana*. Phytomorphology 19: 389–404.
Ono, J. and E.N. Larter, 1976. Anther culture of *Triticale*. Crop Sci. 16: 120–122.
Ouyang, J.W., H. Ju, C.C. Chuang and C.C. Tseng, 1973. Induction of pollen plants from anthers of *Triticum aestivum* L. cultured *in vitro*. Sci. Sin. 16: 79–95.
Rao, I.U., I.V.R. Rao and V. Narang, 1985. Somatic embryogenesis and regeneration of plants in the bamboo *Dendrocalamus strictus*. Plant Cell Rep. 4: 191–194.
Sunderland, N., 1978. Strategies in the improvement of yields in anther culture. In: H. Hu (Ed.), Proc. Symp. Plant Tissue Culture, pp. 65–86. Science Press, Beijing.
Sunderland, N., 1980. Anther and pollen culture. In: D.R. Davies and D.A. Hopwood (Eds.), The Plant Genome, pp. 171–183. Innes Charity, Norwich.
Sunderland, N. and J.M. Dunwell, 1977. Anther and pollen culture. In: H.E. Street (Ed.), Plant Tissue and Cell Culture, pp. 223–265. Blackwell, Oxford.
Sunderland, N. and M. Roberts, 1979. Cold-pretreatment of excised flower buds in float culture of tobacco anthers. Ann. Bot. 405–414.
Sunderland, N. and D.C. Wildon, 1979. A note on the pretreatment of excised flower buds in float culture of *Hyoscyamus* anthers. Plant Sci. Lett. 15: 169–175.
Tsay, H.S., 1980. Strong acid treatment for making smears of tobacco microspores. J. Agric. Res. China 29: 103–106.
Tsay, H.S. and L.J. Chen, 1984. The effects of cold shock and liquid medium on callus formation in rice anther culture. J. Agric. Res. China 33: 24–29.
Tsay, H.S., L.J. Chen, T.H. Tseng and P.C. Lai, 1982. The culture of rice anthers of Japonica × Indica crosses. In: A. Fujiwara (Ed.), Plant Tissue Culture, pp. 561–562. Maruzen, Tokyo.
Tsay, H.S., C.C. Yeh and J.Y. Hsu, 1990. Embryogenesis and plant regeneration from anther culture of bamboo (*Sinocalamus latiflora* (Munro) McClure). Plant Cell Rep. 9: 135–143.
Tsay, S.S., H.S. Tsay and C.Y. Chao, 1986. Cytochemical studies of callus development from microspore in cultured anther of rice. Plant Cell Rep. 5: 119–123.
Tseng, T.C., D.F. Liu and S.Y. Shaio, 1975. Isolation of protoplasts from crop plants. Bot. Bull. Acad. Sin. 16: 55–60.
Wang, C.C., C.S. Sun and Z. Chu, 1974. On the conditions for the induction of rice pollen plantlets and certain factors affecting the frequency of induction. Acta. Bot. Sin. 16: 43–53.
Wang, C.C., C.S. Sun, C.C. Chu and S.C. Wu, 1978. Studies on the albino pollen plantlets of rice. In: H. Hu (Ed.), Proc. Symp. Plant Tissue Culture, pp. 149–160. Science Press, Beijing.
Wilson, H.M., G. Mix and B. Foroughi-Wehr, 1978. Early microspore divisions and subsequent formation of microspore calluses at high frequency in anthers of *Hordeum vulgare* L. J. Exp. Bot. 29: 227–238.
Yeh, C.C. and H.S. Tsay, 1988. The effect of temperature treatment and medium composition on rice anther culture. J. Agric. Res. China 37: 250–256.
Yeh, C.C., J.Y. Hsu and H.S. Tsay, 1991. Effects of microspore developmental stage, sucrose concentration and temperature treatment on embryoid and callus formation of bamboo (*Sinocalamus latiflora* (Munro) McClure) anther culture. Chinese Agron. J. 1: 47–56.
Yeh, M.L. and W.C. Chang, 1986a. Plant regeneration through somatic embryogenesis in callus culture of green bamboo (*Bambusa oldhamii* Munro). Theor. Appl. Genet. 73: 161–163.
Yeh, M.L. and W.C. Chang, 1986b. Somatic embryogenesis and subsequent plant regeneration from inflorescence callus of *Bambusa beecheyana* Munro var. Beecheyana. Plant Cell Rep. 5: 409–411.
Yeh, M.L. and W.C. Chang, 1987. Plant regeneration via somatic embryogenesis in mature embryo-derived callus culture of *Sinocalamus latiflora* (Munro) McClure. Plant Sci. 51: 93–96.

18. Haploidy in coffee

Y. RAGHURAMULU and N.S. PRAKASH

Contents

1. Introduction

Coffee is one of the most important commercial crops grown in many countries. Each day all over the world, millions of coffee lovers enjoy drinking coffee. Judged by total volume, coffee is one of the leading commodities in the international trade. Currently, it provides revenue over $ US 10 billion annually to the producing countries, and work for an estimated 20 million people in the areas of cultivation, processing and marketing throughout the world (Wrigley, 1988). During the year 1991–92, world's green coffee production was 61.2 million tonnes (Anonymous, 1993).

The genus *Coffea* is the most economically important member of the family Rubiaceae, which is reported to comprise 60 species (Purseglove, 1968). All the important species of coffee have originated in tropical Africa. Out of a total of 60 species, only three viz., arabica (*Coffea arabica* L.), robusta (*C. canephora*, Pierre ex Froehner) and liberica coffee (*C. liberica* Bull ex. Hiern) are commercially cultivated. Arabica coffee is preferred over all other species because of its superior quality and it accounts for nearly 80 per cent of the global coffee trade. Robusta coffee contributes to the remaining 20 per cent, and has become important with the increase in sale of instant coffee since second world war. Liberica coffee stands the poor third with a share of less than one per cent of world coffee trade (Bridson, 1987).

The first commercial plantations of coffee were believed to be established in Yemen, which dominated the world coffee trade during 17th century. These commercial plantations were forbidden to visitors. In spite of such

S.M. Jain, S.K. Sopory & R.E. Veilleux (eds.), In Vitro *Haploid Production in Higher Plants, Vol. 3,* 349–363.

stringent precautions, seven seeds were reported to have been brought into India by a Muslim pilgrim named Baba Budan during the early 17th century. Likewise the spread of coffee to other countries was also based on a series of unauthorised introductions with a very small number of seeds/plants. This indicates that the majority of the commercial arabica plantations around the world are derived from a very narrow genetic base. Systematic surveys for the collection of wild coffees in Ethiopia were started by Cramer in 1928, and since then most of the wild coffee growing areas have been explored except the South-East Africa region (Berthaud and Charrier, 1988).

2. Genetics and improvement

Arabica coffee (*Coffea arabica* L.) is the only self fertile, allotetraploid ($2n = 4x = 44$) species of the genus *Coffea* and all the remaining species are cross pollinating diploids ($2n = 22$). Arabica produces superior quality coffee (Vishveshwara, 1971), but is suscesptible to major diseases like coffee leaf rust (*Hemileia vastatrix* Berk and Br.), coffee berry disease (*Colletotrichum coffeanum* Noack) (Bettencourt and Rodrigues, 1985; Kushalappa, 1989) and pests like stem borer (*Xylotrechus quadripes* Chevrolat) (D'Souza *et al.*, 1983), nematodes [*Pratylenchus coffeae* (Zimm) Filip and Stek] (Kumar, 1979), etc. Robusta on the other hand is far more tolerant to most of these diseases/pests, but it produces inferior quality coffee. Hence, most of the world's coffee improvement programmes are centered around *C. arabica*. Conventional breeding programmes have been successful in evolving desirable high yielding selections/intervarietal hybrids of arabica coffee. However, subsequently many of these selections/hybrids are found to be susceptible to the main diseases and pests of coffee. The transfer of resistance genes present in the wild coffee species to the selected arabica cultivars offers good promise. The ploidy variation observed between *C. arabica* and other diploid species is a major bottleneck for transferring the desirable traits of diploids into the tetraploid arabica coffee (Sondahl and Lauritis, 1992). However despite such limitations, some interspecific hybrids like "Arabusta" from Ivory Coast and "Icatu" from Brazil have been developed by crossing *C. arabica* with spontaneous/artificial tetraploids of *C. canephora*. In India, another interspecific hybrid between *C. arabica* × *C. canephora* was developed by adopting triploid breeding strategy. Besides these, spontaneous tetraploid interspecific hybrids like Kalimas, Kawisari, Devamachy and Hybrido-de-Timor have been isolated from among the population where arabica and robusta coffees are cultivated together. These natural hybrids have been successfully utilized in coffee breeding programmes of India (Sreenivasan, 1989; Sreenivasan *et al.*, 1993) and elsewhere (Bettencourt and Rodrigues, 1985; Cambrony, 1985; Wrigley, 1988; Carvalho *et al.*, 1989; Castillo-Z, 1989; Echeverri and Fernandez, 1989).

Interest in the exploitation of these tetraploid interspecific hybrids has constantly been increasing owing to their better adaptability, productivity and durable resistance to coffee leaf rust disease (*H. vastatrix* B. and Br.) (Capot, 1972; Sreenivasan *et al.*, 1993).

3. Utility of haploids in coffee improvement programmes

Induction of haploids by *in vitro* methods was first reported from anther cultures of *Datura innoxia* by Guha and Maheshwari (1966). *In vitro* androgenesis by anther culture or isolated microspore culture has been reported in 247 species and hybrids belonging to 34 families (Maheshwari *et al.*, 1983). Haploid production methods have received considerable interest in recent years (Aslam *et al.*, 1991; Draz *et al.*, 1991; Ma *et al.*, 1991; Burnett *et al.*, 1992). But the cultivated varieties resulting from haplomethods are few and mostly obtained in tobacco, rice, wheat, barley and maize (See Bollon and Raquin, 1987). Unlike herbaceous angiosperms, there have been few *in vitro* haploidization studies in tropical and sub-tropical fruit trees and the success has been rather limited (Zhang and Lespinasse, 1992).

In a perennial crop like coffee, acceptable level of homozygosity is achieved only after 15–18 years of conventional breeding. Production of haploid lines would greatly help in reducing the long pre-breeding cycles. In case of self-incompatible diploid species like *C. canephora* it is impossible to develop inbred lines through conventional methods. In such cases, *in vitro* haploid production techniques offer a unique scope in obtaining inbred lines. Such homozygous inbred lines could be used in the production of F_1 hybrids. In self-fertile tetraploid *C. arabica* breeding programmes, culture of anthers/microspores is of special interest with regard to the major diseases like coffee leaf rust. Production of haploids from complex hybrids having multiple desirable genes, and doubling of their chromosome complement followed by regeneration will result in quick propagation of desirable genetic information (Jansen, 1992). Furthermore the resulting homozygous diploids will have a double dose of genes for resistance and may have an advantage over their parent lines because of a larger spectrum of resitance (Monaco *et al.*, 1977). Such encouraging results have already been reported in a multiple disease resistant tobacco hybrid (Burk *et al.*, 1979).

Haploids could also be used in mutation breeding. The mutagenic treatment of haploid plants followed by doubling will enable all the mutations including the recessive ones to be expressed and to contribute to genetic improvement (Zhang and Lespinasse, 1992; Szarejko *et al.*, 1991; Kasha *et al.*, 1991).

In vitro haploid production approaches are also a source of gametoclonal variation (Baenziger *et al.*, 1991; Yung *et al.*, 1991). Such variation sometimes confers new characters which could be difficult to select via classical breeding methods. It could be an additional tool to increase variability of species like *C. arabica* which have a narrow genetic base. Isolated haploid cells and protoplasts could be an ideal material for genetic transformation through which transgenic homozygous diploid plants could be obtained, after doubling their chromosome number. Haploid protoplasts could also have potential utilization in somatic hybridization (Bollon and Raquin, 1987).

3.1. *Haploids in tetraploid* Coffea arabica

In natural populations of tetraploid arabica species, plants with reduced chromosome number ($n = 2x = 22$) have been observed occasionally and these plants are referred as dihaploids. Dihaploid coffee plants have been known to occur for a long time in the tetraploid species, *Coffea arabica* (Mendes and Bacchi, 1940; Carvalho, 1952; Cramer, 1957; Vishveshwara, 1960; Sreenivasan *et al.*, 1981; Prakash *et al.*, 1993a). They have been grouped earlier under the varietal name "monosperma" due to their low seed set compared to tetraploid arabicas and high frequency of single seeded fruits, which resemble peaberries (Sybenga, 1960). Dihaploids were also reported in the progenies of several cultivars of *C. arabica* viz., Typica, Bourbon, Maragogype, Semperflorens, Laurina, Erecta, Caturra and San Ramon with a frequency of one in 10,000 seedlings (Carvalho, 1952). In India too, Sreenivasan *et al.* (1981) reported a similar frequency of dihaploids. Although polyembryony is considered to be a possible reason for spontaneous occurrence of dihaploids, Mendes (1944) could not establish a correlation between dihaploid origin and polyembryony.

3.2. *Morphology of dihaploids in* C. arabica

In general, the dihaploid plants are morphologically quite distinct from tetraploids, with compact stature, reduced bush spread and number of primaries (Table 1, Fig. 1a). The branches are relatively slender and the leaves are small and narrow (Fig. 1b). The flowering behaviour is normal but the flowers are small in size with low fertility which may probably be due to abnormal meiosis. It was observed that the fruit set was very low as a result of open pollination or by hand pollination. Irrespective of the mode of pollination, the few fruits produced were found to be single-seeded (Sybenga, 1960; Prakash *et al.*, 1993a).

3.3. *Anatomical features of dihaploids*

Franco (1939) established a negative correlation between chromosome number and stomatal frequency. Later, several reports gave a detailed record on the increase in the number of stomata and decrease in stomatal plastids in dihaploids compared with tetraploids of arabica (Rao and Vishveshwara, 1960; Orozco and Cassalett, 1974; Boaventura *et al.*, 1981; Mishra *et al.*, 1991; Sreenivasan *et al.*, 1992). Studies on the transection of leaf lamina of dihaploids revealed a structure similar to that of most woody perennials

Table 1. Comparative morphometrics of dihaploid and tetraploid plants (Prakash *et al.*, 1993a)

Character	Cauvery/Catimor S.4345		Cauvery/Catimor S.4349	
	Dihaploid	Tetraploid	Dihaploid	Tetraploid
Plant height (cm)	68.7	99.4	77.5	90.8
Bush spread (cm)	77.5	104.4	66.2	98.6
Stem girth (mm)	39.1	48.7	33.4	48.4
No. of primaries	16.0	23.2	20.0	25.6
Mean length of primaries (cm)	56.4	71.8	51.6	76.2
No. of nodes/primary	18.6	23.1	16.4	23.3
Leaf area (cm^2)	28.6	59.0	27.5	50.0
Stomatal frequency/mm^2	271.5	190.0	297.5	179.5
Stomatal length (μ)	23.6	29.7	23.8	27.8
Stomatal plastid number	#M8.8	20.8	#M9.3	19.6
Pollen grain diameter (μ)	22.9	28.5	23.9	30.4
Pollen fertility	57.96	88.5	61.34	90.4

having dorsiventral symmetry. The distinct features included uniseriate upper and lower epidermal layers covered with cuticle, well-differentiated mesophyll tissue into compact palisade and loosely arranged spongy layers; vascular bundles with normal orientation of xylem and phloem tissues (Fig. 1c). However, the thickness of lamina, epidermis, palisade tissue and spongy tissue and diameter of spongy cells were found to be less in dihaploids compared with higher ploidy levels of *Coffea* (Prakash *et al.*, 1993b).

3.4. *Leaf rust resistance in dihaploids*

Under field conditions, dihaploids of arabica are less infected with leaf rust pathogen (*Hemileia vastatrix*) which may probably be due to the absence of fruiting in these plants (Eskes, 1989). Resistance of coffee cultivars and species of different ploidy level were also compared under laboratory condition (Mazzafera *et al.*, 1987) and indicated that leaf discs of dihaploids and tetraploids of "Catuai" showed similar susceptibile characters whereas dihaploids and hexaploids of "Bourbon Vermelho" showed relatively prolonged latent period than their respective tetraploid counterparts.

3.5. *Cytological behavior of dihaploids*

The meiotic behaviour of dihaploids was studied in detail (Mendes and Bacchi, 1940; Berthaud, 1976; Kammacher, 1981; Prakash *et al.*, 1993a) and many irregularities characteristic of haploids have been reported. At diakinesis/metaphase I, several univalents were commonly found. Mendes and Bacchi (1940) noticed 1 to 6 bivalents in 71% of the cells. According to Prakash *et al.* (1993a), the organization of metaphase I plates was not clear and that resulted in random distribution of univalents throughout the cyto-

Figure 1. (a) A spontaneous dihaploid plant of *C. arabica*. (b) Branches of a dihaploid (left) and a tetraploid (right) *C. arabica*. (c) A transection of leaf lamina of a dihaploid *C. arabica*. (d) PMC of a dihaploid *C. arabica* showing random distribution of chromosomes [4 II (Marked) + 14 I] × 1600. (e) Callusing in anthers of × hybrid. (f) Regenerated plantlet from anthers of × hybrid through somatic embryogenesis.

plasm (Fig. 1d). The frequency of univalents and bivalents per cell varied between 10–22 and 0–6, respectively. The presence of bivalents indicated that some homology exists within one genome of *C. arabica* (Mendes and Bacchi, 1940), although the importance of non-homologous pairing is not known (Sybenga, 1960). Berthaud (1976) reported 19% of aneuploid pollen mother cells due to cytomixis phenomena at metaphase I and numerous bivalents (2–8) in the cells with 22 chromosomes. At anaphase I, separation was equal (11:11) in some of the cells while unequal segregations, laggards, precocious cleavage of the chromosomes resulting in the division of univalents were common anomalies observed in all dihaploids. The abnormalities in the early stages were carried forward and second meiotic divisions were also reported to be irregular. The tetrad stage manifested many variations such as micronuclei, and more than four nuclear groups. The pollen fertility was considerably reduced due to the disturbed meiosis. However, the sterility was not complete despite several irregularities which may be attributed to the restoration of euploid condition in some cells at the beginning of the meiotic process (Kammacher, 1981).

During megasporogenesis too, irregularities such as degeneration of ovules and abnormal divisions of the embryo sac were noticed (Bacchi, 1941).

These detailed studies on meiotic behaviour confirmed the dihaploid status and supported the concept that *C. arabica* is an allotetraploid of ancient origin. Kammacher (1981) opined that because of allotetraploid orgin of *C. arabica*, the dihaploids of this species are segmentary dihaploids formed by the combination of two sets of partially homologous chromosomes. Among them, one reproduces the structure of genome "C" the basic genome of all the known *Coffea* diploids. The second genome of *C. arabica* had its origin from a genus related to *Coffea* but not *Coffea* and more possibly this origin goes back to a Rubiaceae taxon without any survivors in nature at present (Kammacher, 1981).

4. Haploids in diploid species of *Coffea*

In diploid species of *Coffea* ($2n = 2x = 22$), the frequency of naturally occurring haploids ($n = x = 11$) is very low compared with dihaploid frequencies in the tetraploid species *C. arabica*. Dublin and Parvais (1975) reported for the first time the occurrence of spontaneous haploids in *C. canephora*, which were obtained from monoembryonic seeds. The frequency was three haploids in 20,000 plantlets. These haploid plantlets were small with short internodes and slightly crinkled narrow leaves. The chromosome number was found to be $n = 11$.

5. Production of haploids in coffee

In coffee, two pathways have been identified for the production of haploid plants, viz. *in vivo* and *in vitro* approaches.

5.1. In vivo *approach*

In this method, efforts are mainly directed towards utilization of spontaneous haploids to produce doubled haploids. Dublin and Parvais (1975) reported the spontaneous development of haploid plants in *Coffea canephora*. Couturon (1982) first proposed a method of recovery of haploid plants of *C. canephora* by grafting embryos obtained from polyembryonic fruits. Mature embryos were grafted onto freshly germinated seedlings at hypocotyl region by side grafting method. He obtained 82 haploid plants from 623 grafted polyembryos, a frequency of 13.16 per cent. After chromosome duplication by treating the caulinary apices with colchicine, the first thirty diploidised haploids obtained were reported to have satisfactory vigour compared with that of heterozygous diploid plants of *C. canephora*. Adopting the same technique of grafting polyembryos, Lashermes *et al.* (1993) developed more than 750 doubled haploids of *C. canephora*. They reported inbreeding depression in general vigour and reproductive aspects. The performance of hybrids involving these doubled haploids demonstrated an improvement in heterosis. The number of recoverable haploids with the grafting method is mainly dependent on the number of haploid embryos produced in association with polyembryony. Since their frequency of occurrence is limited, it is not possible to obtain large numbers of desirable haploids.

5.2. In vitro *approach*

In vitro induction of haploids is a versatile technique for the production of haploids and dihaploids, either with whole anther or isolated microspore culture (Maheshwari *et al.*, 1983).

In the first report, Sharp *et al.* (1973) successfully cultured anthers of *C. arabica* cv. Bourbon Amarelo on Linsmaier and Skoog's (1965) medium. Rapidly growing, friable, white callus cells with a dihaploid ($2x = 22$) and haploid ($x = 11$) chromosome numbers were obtained by culturing anthers at the uninucleate stage of microspore development in the dark at 25°C. They observed pro-embryo formation only in dihaploid callus cells which, however, failed to undergo subsequent development.

Monaco *et al.* (1977) cultured individual isolated microspores of *C. arabica* and *C. liberica* for the first time by using the nurse culture technique, as described by Sharp *et al.* (1972) for tomato explants. In this technique, microspores were nursed on filter paper discs placed on top of intact anthers of the same species or on fast-growing *C. arabica* callus. However, the growth of such individually cultured microspores was reported to be unsatisfactory, and only one of 100 nurse cultures of *C. arabica* proliferated into callus. These workers also observed a correlation between callusing vs. floral bud size, anther size, and stage of microsporogenesis in *C. liberica*. The anthers were cultured at the pollen mother cell, leptotene I and binucleate stages of development. Maximum fresh weight of callus (0.993 g after 90 days) was obtained in anthers at the binucleate pollen stage.

Regeneration of plantlets from anthers of some of the diploid species of *Coffea* through callusing or through direct embryogenesis in *C. arabica* has been reported but nothing is known about their chromosome number (Ramos, 1987 cf. Eskes, 1989). At CENICAFE, Colombia, more than 1000 androgenic embryos, which regenerated into plantlets, are reported to have been obtained from anthers at the uninucleate pollen stage (Londono and Marulanda cf. Villalobos, 1989).

At Centro de Investigacao das Ferrugens do Cafeeiro, Portugal, Carneiro (1993) obtained embryogenesis and plant regeneration in Catuai and Catimor cultivars by culturing isolated microspores and whole anthers in MS and Nitsch liquid and solidified media supplemented with different combinations of growth regulators, vitamins and amino acids.

Neuenschwander *et al.* (1993) described a technique of mechanical isolation of microspores from anthers of *C. arabica* cv. Catuai and Catimor. This method, in their opinion was superior to the whole bud maceration technique, and was the first step towards an efficient single haploid cell regeneration system.

Ascanio and Arcia (1994) reported the influence of anther development stage and heat shock on androgenesis in *C. arabica* cv. Garnica. Callusing was observed only in anthers at the uninucleate pollen grain stage when cultured on Murashige and Skoog (1962) medium containing IBA (2 mg/l) and BA (8 mg/l). Subsequent androgenesis occurred in calli of the anthers that received pre-treatment of 5°C for 24 or 48 h and cultured on MS medium with IAA (0.5 mg/l) and 2–ip (1 mg/l). The embryos after subculturing developed into plants of which 10% were normal ($4x = 44$) and 74% had small leaves while others had purple, albino or narrow leaves with dihaploid chromosome number ($2x = 22$).

6. Coffee anther culture in India

In India, work on different tissue culture techniques was initiated at Central Coffee Research Institute, Balehonnur, Karnataka, during 1985 with the main objectives of developing suitable protocols on coffee for rapid multiplication, inducing somaclonal variation, embryo rescue, haploid production, somatic hybridization and cryo-preservation as complements to the conventional breeding programme for crop improvement. Anther culture was mainly aimed at recovering haploids from commercially important cultivars of arabica coffee for crossing with other diploid species of *Coffea*, and for producing stable doubled haploids in cross-pollinating robusta and other interspecific hybrids like Sln.6 and Congensis × Robusta. As an initial step, Raghuramulu (1989) tried to identify the most suitable stage of anther development and culture medium for coffee anthers. Anthers from three cultivars of *C. arabica* and one cultivar of *C. canephora* at different stages of development were cultured on a modified MS medium (Murashige and Skoog, 1962) supplemented with different growth regulator combinations. Irrespective of composition of the medium, the anthers from green buds at microsporogenesis stage responded well. The contamination levels were also less when anthers were excised from green buds three days after blossom showers. Moderate to profuse callusing was obtained with anthers of *C. arabica* cvs. Sln.7.3 and S.795, when the basal MS medium was supplied with a growth regulator combination of BA:IAA (0.2–0.4 mg/l) or Kin (1–2 mg/l):2 4–D (0.1–0.2 mg/l). The callus obtained was creamy and non-friable. The anthers from grown up candle stage buds (at pollen grain stage) or from opened flowers (anthesis stage) were found to be very difficult to culture owing to very high contamination levels. Even the one or two uncontaminated cultures failed to produce callus. The callus from anthers of Sln.7.3 and S.795 did not respond to further subcultures and they became dark brown to black after 8 months (Table 2).

Recently plantlet regeneration has been achieved by Muniswamy and Sreenath (unpublished, personal communication) from cultured anthers of Congensis × Robusta interspecific hybrid. Anthers at the uninucleate microspore stage were pretreated initially at different temperatures and cultured on a modified MS medium (Murashige and Skoog, 1962) supplemented with different growth regulator combinations. The cultures were incubated at 25 ± 2°C in dark initially for 2 months and then in alternate diffused light. The response of anthers was dependent on the concentrations of growth regulators in the medium and pretreatment given before culturing. The most encouraging response was obtained with BA and IBA (1 mg/l each) in anthers pretreated initially at 5°C for 24 h and extended to 10°C for another 24 h. Very little to moderate creamy or white calli (Fig. 1e) were observed in these cultures, which turned greenish immideately after transferring to diffused light and then to brownish colour. This brown callus was subcultured on a MS medium containing NAA (0.05 mg/l) and Kin (0.25 mg/l). Androgenesis was observed in 4.2% of these cultures after 9 months, and the resultant embryos were further multiplied and regenerated into plantlets successfully (Fig. 1f). More than 1000 androgenic embryos were regenerated into plantlets and transferred to potted vermiculite for *ex vitro* hardening. However, the chromosome complement of these plantlets is yet to be ascertained.

These results indicate that factors like genotype, developmental stage of anther or pollen, pre-treatment and composition of culture medium greatly influence the response of cultured coffee anthers.

7. Conclusion

Overall the success achieved so far in production of haploid/dihaploid plants in *Coffea* spp. is quite encouraging. In other perennial tree crops, the success

Table 2. In vitro response of coffee anthers at different sages of development (Raghuramulu, 1989)

Material	Source/Stage of anthers at inoculation	Growth regulators (mg/l)	No. of cultures initiated	Contamination (%)	Response
C. arabica					
cv. Sln.7.3	Green buds (Microsporogenesis stage)	BA:IAA(0.4:0.2)	21	0.00	Expansion of anthers followed by profuse callusing
		BA:IAA(0.2:0.2)	21	28.60	
		KIN:2,4-D(1.0:0.1)	42	9.50	
cv. S. 795	''	BA:2,4-D(1.0:0.2)	14	21.40	Expansion of anthers followed by moderate callusing
		KIN:2,4-D(1.0:0.2)	14	42.90	
		KIN:2,4-D(2.0:0.1)	5	100.00	
Portugal Introduction	Candle stage buds (Pollen grain stage)	KIN:IAA (1.0:0.2)	18	39.00	No callusing
		IAA:KIN (1.0:0.2)	17	67.70	
		KIN:IAA (2.0:1.0)	11	100.00	
C. canephora					
cv. BR. 11	''	KIN:IAA (1.0:0.2)	7	57.10	No callusing
		IAA:KIN (1.0:0.2)	5	40.00	
		KIN:IAA (2.0:1.0)	14	28.60	
		IAA:KIN (2.0:1.0)	14	100.00	
''	Flowers (Anthesis stage)	KIN:IAA (2.0:1.0)	7	100.00	Nil
		IAA:KIN (2.0:1.0)	7	100.00	

* In MS basal medium (Murashige and Skoog, 1962).

of anther culture has been rather limited. Androgenic haploid plants have been obtained only in 13 of 29 fruit crop species studied (Zhang and Lespinasse, 1992). In coffee, most of the reports on regeneration of androgenic plants have appeared only during the past 7–8 years. Perhaps the day may not be too far when these *in vitro* haploids/dihaploid plants find a place in the regular breeding programmes of coffee for its improvement.

8. Acknowledgements

We are grateful to Dr. R. Naidu, Director of Research and Dr. M.S. Sreenivasan, Head, Division of Botany, Central Coffee Research Institute for their interest and encouragement. The help rendered by Dr. H.L. Sreenath and Mr. B. Muniswamy, Division of Tissue Culture, CCRI, in providing their unpublished information and relevant photographs, is duly acknowledged.

9. References

Anonymous, 1993. World coffee production. Café Cacao Thé 37(1): 86.

Ascanio, E.C.E. and M.M.A. Arcia, 1994. Effect of anther development stage and a heat shock on androgenesis in *Coffea arabica* L. var. Garnica. Café Cacao Thé 38: 75–80.

Aslam, F.N., M.V. MacDonald and D.S. Ingram, 1991. Haploid production in rapid cycling *Brassica campestris* and *B. napus*. Cruciferae Newsl. 14–15: 98–99.

Bacchi, O., 1941. Observances citologicas em *Coffea*, VII. A macrosporogense na variedade "monosperma". Bragantia 1: 483–490.

Baenziger, P.S., V.D. Keppenne, M.R. Morris, C.J. Peterson and P.J. Mattern, 1991. Quantifying gametoclonal variation in wheat doubled haploids. Cereal Res. Comm. 19: 33–42.

Berthaud, J., 1976. Étude cytogénétique d'un Haploïde de *Coffea arabica* L. Café Cacao Thé 20: 91–96.

Berthaud, J. and A. Charrier, 1988. Genetic resources of *Coffea*. In: R.J. Clarke and R. Macrae (Eds.), Coffee, Vol. 4. Agronomy, pp. 1–42. Elsevier Science Publishing Co., Inc., New York.

Bettencourt, A.J. and C.J. Rodrigues, Jr., 1985. Principles and practice of breeding for resistance to rust and other diseases. In: R.J. Clarke and R. Macrae (Eds.), Coffee, Vol. 4. Agronomy, pp. 199–234. Elsevier Science Publishing Co., Inc., New York.

Boaventura, Y.M.S., D.M. Medina, M.J.F.R. Vieira and H.V. de Arruda, 1981. Numero de cloroplastos e nivel de ploidia em especies de *Coffea* L. Revta Brasil Bot. 4: 15–21.

Bollon, H. and C. Raquin, 1987. Haplomethods: A tool for crop improvement. In: Nestle Research News 1986–87, pp. 81–90. Nestec Ltd., Switzerland.

Bridson, D.M., 1987. Nomenclature notes on *Psilanthus* (including Coffee section. *Paracoffea* Miquel) (Rubiaceae tribe Coffeae) Kew Bull. 42: 453–460.

Burk, L.G., J.F. Chaplin, G.V. Gooding and N.T. Powell, 1979. Quantity production of anther derived haploids from a multiple disease resistant tobacco hybrid. I. Frequency of plants with resistance or susceptibility to tobacco mosaic virus (TMV), potato virus Y (PVY) and root knot (RK). Euphytica 28: 201–208.

Burnett, L., S. Yarrow and B. Huang, 1992. Embryogenesis and plant regeneration from isolated microspores of *Brassica rapa* L. ssp. Oleifera. Plant Cell Rep. 11: 215–218.

Cambrony, H.R., 1985. Arabusta and other interspecific fertile hybrids. In: R.J. Clarke and R.

Macrae (Eds.), Coffee, Vol. 4. Agronomy, pp. 263–291. Elsevier Science Publishing Co., Inc., New York.

Capot, J., 1972. L'amélioration du cafeier en côte d'Ivoire les hybrides Arabusta. Café Cacao Thé 16: 3–16.

Carneiro, M.F., 1993. Induction of double haploids in *Coffea arabica* cultivars via anther or isolated microspores culture. In: Proc. 15th Intl. Sci. Coll. on Coffee, Vol. I, p. 133. ASIC, Montpellier (Abstr.).

Carvalho, A., 1952. Taxonomia de *Coffea arabica* L. VI. Caracters morphologicas des haploides. Bragantia 12: 201–212.

Carvalho, A., A.B. Eskes and L.C. Fazuoli, 1989. I. Breeding for rust resistance in Brazil. In: A.C. Kushalappa and A.B. Eskes (Eds.), Coffee Rust: Epidiomology, Resistance and Management, pp. 295–307. CRC Press, Inc., Boca Raton, FL.

Castillo-Z, J., 1989. II. Breeding for rust resistance in Colombia. In: A.C. Kushalappa and A.B. Eskes (Eds.), Coffee Rust: Epidiomology, Resistance and Management, pp. 307–316. CRC Press, Inc., Bota Racon, FL.

Couturon, E., 1982. Obtention d'haploïdes spontanes de *Coffea canephora* pierre par l'utilisation du greffage d'embryons. Café Cacao Thé 26: 155–160.

Cramer, P.J.S., 1957. A review of literature of coffee research in Indonesia. STC Editorial, IICA, Turrialba, Costa Rica.

Draz, A.E., J. Zepata and G.S. Khush, 1991. Development of dihaploid rice lines through anther culture (I). Intl. Rice Res. Newsl. 16(5): 6.

D'Souza, G.I.D., P. Krishnamoorthy Bhat, G.H. Venkataramaiah and M.J. Chacko, 1983. Pests of coffee. In: All India Scientific Writer's Society (Eds.), Agricultural Entomology, Vol. II, pp. 277–293. All India Scientific Writer's Society, New Delhi.

Dublin, P. and J.P. Parvais, 1975. Note sur les premiers haploïdes spontanés de couverts chez le *Coffea canephora* var. robusta. Café Cacao Thé 19: 191–196.

Echeverri, J.H. and C.F. Fernandez, 1989. IV. The promecafe program for Central America. In: A.C. Kushalappa and A.B. Eskes (Eds.), Coffee Rust: Epidiomology, Resistance and Management, pp. 323–329. CRC Press, Inc., Bota Racon, FL.

Eskes, A.B., 1989. Resistance. In: A.C. Kushalappa and A.B. Eskes (Eds.), Coffee Rust: Epidemiology, Resistance and Management, pp. 171–292. CRC Press, Inc., Bota Racon, FL.

Franco, C.M., 1939. Relation between chromosome number and stomata in *Coffea*. Bot. Gaz. 100: 817–827.

Guha, S. and S.C. Maheshwari, 1966. Cell division and differentiation of embryos in the pollen grain of *Datura in vitro*. Nature 212: 97–98.

Jansen, R.C., 1992. On the selection for specific genes in doubled haploids. Heredity 69: 92–95.

Kammacher, P., 1981. On the meiotic behaviour of *Coffea arabica* dihaploids. In: Proc. 9th Intl. Sci. Coll. on Coffee, pp. 717–724. ASIC, Paris.

Kasha, K.J., A. Ziauddin., E. Reinbergs and D.E. Falk, 1991. Use of haploids in induced mutation in barley and wheat. Cereal Res. Comm. 19: 101–108.

Kumar, A.C., 1979. Relative tolerance or susceptibility of Arabica, Robusta and Excelsa coffees to *Pratylenchus coffeae*. In: C.S. Venkatram (Ed.), Proc. Second Annual Symp. on Plantation crops, PLACROSYM II, 1979, pp. 20–26.

Kushalappa, A.C., 1989. Introduction In: A.C. Kushalapp and A.B. Eskes (Eds.), Coffee Rust: Epidemiology, Resistance and Management, pp. 1–11. CRC Press, Inc., Bota Racon, FL.

Lashermes, P., A. Charrier and E. Couturon, 1993. On the use of double haploids in genetics and breeding of *Coffea canephora* P. In: Proc. 15th Intl. Sci. Coll. on Coffee, Vol. I, pp. 218–225. ASIC, Paris, 1993.

Linsmaier, E.M. and F. Skoog, 1965. Organic growth factor requirements of tobacco tissue cultures. Physiol. Plant. 18: 100–127.

Ma, H.C., C.E. Wassom and G.H. Liang, 1991. Direct regeneration of maize haploids via anther culture. Cytologia 56: 103–106.

Maheshwari, S.C., A. Rashid and A.K. Tyagi, 1983. Anther/pollen culture for production of haploids and their utility. Newsl. Intl. Assoc. Plant Tiss. Cult. 41: 2–9.
Mazzafera, P., A. Carvalho, L.C. Fazuoli and F.A. Levy, 1987. Infeccao por *Hemileia vastatrix* em especies e variedades de cafe com diferentes niveis de ploidia. In: 14° Cong. Bras. Pesq. Caf., 1987, p. 99.
Mendes, A.J.T., 1944. Observacoes Citologicas em *Coffea* VII. Poliembrionia. Bragantia 4: 693–708.
Mendes, A.J.T. and O. Bacchi, 1940. Observacoes citologicas em *Coffea* V. Uma variedade haploide (di-haploide)de *C. arabica* L. Instituto Agronmico do Estado, em Campinas Boletin Tecnico 77: 1–26.
Mishra, M.K., N.S. Prakash and M.S. Sreenivasan, 1991. Relationship of stomatal length and frequency to ploidy level in *Coffea* L. J. Coffee Res. 21: 32–41.
Monaco, L.C., M.R. Sondahl, A. Carvalho, O.J. Crocomo and W.R. Sharp, 1977. Applications of tissue culture in the improvement of coffee. In: J. Reinert and Y.P.S. Bajaj (Eds.), Applied and Fundamental Aspects of Plant Cell Tissue and Organ Culture, pp. 109–129. Springer-Verlag, Berlin.
Murashige, T. and F. Skoog, 1962. A revised medium for rapid growth and bioassays with tobacco tissue cultures. Physiol. Plantarum 15: 473–497.
Neuenschwander, B., M. Dofour and T.W. Bauman, 1993. Haploid cell colony formation from mechanically isolated microspores of *Coffea arabica*. In: Proc. 15th Intl. Sci. Coll. on Coffee, Vol. 11, pp. 760–761. ASIC, Paris.
Orozco, C.F.J. and D.C. Cassalett, 1974. Relacion entre las caracteristicas estomaticas y el numero cromosomica de un hibrido interspecifico en cafe. Cenicafe 25: 33–49.
Prakash, N.S., M.K. Mishra, D. Padmajyothi and M.S. Sreenivasan, 1993a. Cytomorphological studies on two polyhaploid forms of *Coffea arabica* L. In: M.K. Nair, R.D. Iyer, V. Rajagopal and P.S.P.V. Vidyasagar (Eds.), Proc. of Tenth Plantation Crops Symp., PLACROSYM X, 1992, pp. 253–257.
Prakash, N.S., D. Padmajyothi, M.K. Mishra, A.S. Ram and M.S. Sreenivasan, 1993b. Ploidy level – its influence on leaf anatomical features of *Coffea* L. J. Coffee Res. 23: 75–83.
Purseglove, J.W., 1968. Tropical Crops, Dicotyledons II. Longmans, London.
Raghuramulu, Y., 1989. Anther and endosperm culture of coffee. J. Coffee Res. 19: 71–81.
Rao, Y.R.A. and S. Vishveshwara, 1960. Studies in polyploids of *Coffea*. Indian Coffee 24: 303–306.
Sharp, W.R., R.S. Raskin and H.E. Sommer, 1972. The use of nurse culture in the development of haploid clones in tomato. Planta 104: 357–361.
Sharp, W.R., L.S. Caldas, O.J. Crocomo, L.C. Monaco and A. Carvalho, 1973. Production of *Coffea arabica* callus of three ploidy levels and subsequent morphogenesis. Phyton 31: 67–74.
Sondahl, M.R. and J.A. Lauritis, 1992. Coffee. In: F.A. Hammerschlag, and R.E. Litz (Eds.), Biotechnology in Agriculture, No. 8. Biotechnology of Perennial Crops, pp. 401–420. CAB International, UK.
Sreenivasan, M.S., 1989. III. Breeding coffee for leaf rust resistance in India. In: A.C. Kushalappa and A.B. Eskes (Eds.), Coffee Rust: Epidemiology, Resistance and Management, pp. 316–335. CRC Press, Inc., Bota Racon, FL.
Sreenivasan, M.S., A.S. Ram and N.S. Prakash, 1993. Tetraploid interspecific hybrids in coffee breeding in India. In: Proc. 15th Intl. Sci. Coll. on Coffee, Vol. I, pp. 226–233. ASIC, Paris.
Sreenivasan, M.S., N.S. Prakash and M.K. Mishra, 1992. Evaluation of some indirect ploidy indicators in *Coffea* L. Café Cacao Thé 36: 199–205.
Sreenivasan, M.S., M. Ramachandran and K.R. Sundar, 1981. Frequency of polyploids in *Coffea arabica* L. In: S. Vishveshwara (Ed.), Proc. Fourth Annual Symp. on Plantation Crops. PLACROSYM IV, 1981, pp. 23–28.
Sybenga, J., 1960. Genetics and cytology of coffee. A literature review – Bibliograph. Genet. 11: 217–316.
Szarejko, I., M. Maluszynski., K. Polok and A. Kilian, 1991. Doubled haploids in the mutation

breeding of selected crops. In: P.H. Kitto (Ed.), Plant Mutation Breeding for Crop Improvement, Vol. 2, pp. 355–378. IAEA, Vienna.
Villalobos, V.M., 1989. Advances in tissue culture methods applied to coffee and cocoa. In: Proc. Intl. Symp. on Plant Biotechnologies for Developing Countries organised by CTA and FAO, 26–30 June 1989, Luxembourg, pp. 247–250.
Vishveshwara, S., 1960. Occurrence of haploid *Coffea arabica* L. cultivar Kents. Indian Coffee 24: 123–124.
Vishveshwara, S. 1971. Breeding for quality in coffees. Indian Coffee 35: 509.
Wrigley, G., 1988. Coffee. Longman Scientific and Technical, Essex, England.
Yung, C.H., E.A. Wernsman and G.V. Gooding Jr., 1991. Characterization of potato virus Y resistance from gametoclonal variation in flue-cured tobacco. Phytopathology 81: 887–891.
Zhang, Y.X. and Y. Lespinasse, 1992. Haploidy. In: F.A. Hammerschlag and R.E. Litz (Eds.), Biotechnology in Agriculture, No. 8, Biotechnology of Perennial Fruit Crops, pp. 57–75. CAB International, UK.

19. Haploidy in alfalfa

DANIEL Z. SKINNER and GEORGE H. LIANG

Contents

1. Introduction

Plants having a single genome are known as haploids. Alternatively, haploid plants also are defined as those that have the gametic number of chromosomes. However, in the case of a polyploid species like alfalfa (*Medicago sativa* L.), a tetraploid with $2n = 4x = 32$, haploid plants actually are true diploids with two homologous sets of chromosomes ($2n = 2x = 16$). As far as we know, true haploid (or "monoploid"; $2n = 1x = 8$) forms of alfalfa have not yet been reported. We will refer to "haploid" alfalfa plants derived directly from a tetraploid plant through halving the number of genomes, resulting in $2n = 2x = 16$, as dihaploids. The term "diploid" will be used to refer to naturally occurring *M. sativa* with 16 chromosomes.

Alfalfa is one of the most important forage crops and is produced on about 40 million hectares worldwide. Germplasm resources can be traced to nine distinct sources, and the majority of the alfalfa is grown in temperate zones. Improvement of alfalfa is a complex problem. It is an autotetraploid and the host of a large number of insects and diseases; it requires insect pollination; a *Rhizobium* symbiotic relationship for nitrogen fixation and adequate growth; and optimum conditions for seed production and forage production are different. Even with all of these complicating factors, breeding improved alfalfa has been largely successful. The availability of dihaploid alfalfa plants can facilitate both basic and applied research on this crop species. The reduced chromosome number could: (1) simplify genetic and cytogenetic analysis because of disomic segregation and exclusively bivalent formation; (2) facilitate crosses with available diploid germplasm and later transfer of the improved diploids back to the tetraploid level via $2n$ gametes, colchicine doubling, somatic cell fusion, or bridging crosses using triploids; (3) produce tetraallelic plants for maximum heterozygosity; (4) help in detecting recessive mutants in mutation research; and (5) provide dihaploid calli, cell suspensions, and protoplasts for other basic research.

Although the cultivated alfalfa is tetraploid, diploid plants also have been known for many years (Bolton and Greenshields, 1950; Lesins, 1957; Stan-

S.M. Jain, S.K. Sopory & R.E. Veilleux (eds.), In Vitro *Haploid Production in Higher Plants, Vol. 3,* 365–375.

ford and Clement, 1958; Clement and Lehman, 1962). Naturally occurring populations of $2n = 16$ *M. sativa* plants likewise have been found (Clement, 1962). In fact, we now know that collections of plants formerly referred to as the separate species *M. coerulea* are merely diploid forms of *M. sativa* (reviewed by Lesins and Gillies, 1972). This diploid form has been found growing in nature and distributed over rather large areas (Lesins and Lesins, 1964) and, undoubtedly, represents the ancestral form of the modern-day tetraploid forms. The naturally occurring diploid *M. falcata* plant types, although morphologically distinct, interbreed freely with diploid *M. sativa* and give rise to hybrid swarms with intergrading morphology and unimpaired fertility. *M. falcata* is also now considered by some researchers to be a subspecific form of *M. sativa*. Hence, diploid alfalfa is not at all uncommon. This natural occurrence of $2n = 2x = 16$ plants has been utilized in alfalfa improvement programmes. Diploid *M. falcata* reportedly was used in the development of eight tetraploid cultivars grown in the US, including widely-grown cultivars Ladak, Narragansett, and Vernal (Bolton, 1962). These cultivars have been used as parent material for numerous derivative germplasms and cultivars; hence, the diploid form of the plant figures prominently in the historical development of alfalfa as a crop.

A diploid line of alfalfa has been developed through the crossing of dihaploid plants with diploid *M. falcata* types, followed by selection for fertility and reproductive stability, to yield populations of 16–chromosome plants that behave essentially as cultivated alfalfa (Bingham and McCoy, 1979). These populations, collectively known as CADL [cultivated alfalfa at the diploid level (Bingham and McCoy, 1979)], provide a valuable resource for research purposes and are being widely used in studies relating crop performance to ploidy level and gene dosage.

With the natural occurrence of diploids, there is no need to breed *M. sativa* plants having 16 chromosomes. However, it is conceivable that a dihaploid form of a particular tetraploid alfalfa plant might be needed for research relating to gene dosage or other ploidy effects. Our objective for this chapter is to discuss methods by which dihaploid alfalfa can be developed from a tetraploid cultivar.

Bingham (1971) described a method of obtaining dihaploids from tetraploid alfalfa stocks through $4x$-$2x$ crosses using either suction-emasculated male fertile or cytoplasmic male sterile plants as female. Twenty eight of 34 seed parents produced at least one dihaploid offspring (Bingham, 1971), presumably from parthenocarpic egg development. Although surprisingly large differences occurred among the efficiencies of pollen parents in stimulating dihaploid production, a total of 178 dihaploids was recovered (Bingham, 1971). A similar approach was used later to scale the genomic complement of a selected alfalfa plant to seven different ploidy levels (Bingham and Saunders, 1974). Their study also found that $3x$-$2x$ crosses often gave rise to $2n = 16$ plants (63% of progeny), suggesting that scaling a given tetraploid first to the triploid and then to the dihaploid level might be possible

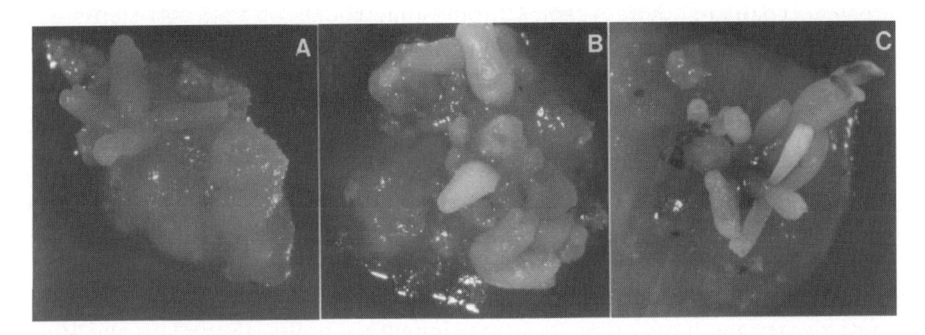

Figure 1. Embryogenesis on alfalfa callus derived from anther culture. Typical embryoid formation is shown at (A) four, (B) six, and (C) eight weeks after culture initiation.

(Binek and Bingham, 1970). Thus, the interploidy cross method is relatively efficient in producing dihaploids from most seed parent genotypes.

Although in most of the alfalfa genotypes ploidy can be reduced through breeding methods, inducing genome reduction in certain plants can be extremely difficult. In those cases, more elaborate, tissue culture-based methods may be successful.

Haploids or polyhaploids can be obtained efficiently through anther and microspore culture in certain species, such as rice (*Oryza sativa* L.) (Chen, 1986a,b; Reddy *et al.*, 1985); *Brassica* (Chuong and Beversdorf, 1985; Duijs *et al.*, 1992; Epps *et al.*, 1992; Hansen and Svinnset, 1993; Lichter, 1982; Swanson *et al.*, 1987; Takahata and Keller, 1991); *Vicia faba* L. (Zhou, 1988); tobacco (*Nicotiana tabacum* L.) (Kyo and Harada, 1985); sugarbeet (*Beta vulgaris* L.) (Geyt *et al.*, 1985); barley (*Hordeum vulgare*) (Xu and Sunderland, 1981; Wei *et al.*, 1986); wheat (*Triticum aestivum*) (Datta and Wenzel, 1987); corn (*Zea mays*) (Dupuis *et al.*, 1987; Wan and Widholm, 1992); *Lilium* and *Tulipa* (Tanaka and Ito, 1980; Tanaka *et al.*, 1987); *Datura* (Sangwan-Norreel, 1977); and linseed (*Linum usitatissimum*) (Nichterlein and Friedt, 1993). The technique has not been successful in our laboratory with sorghum [*Sorghum bicolor* (L.) Moench].

Several attempts have been made in our laboratory to recover alfalfa dihaploids through culture of mid- to late-uninucleate stage microspores. Anthers from a diverse array of alfalfa cultivars; Anik, Buffalo, Cuf 101, and Lancota; two germplasm resources, CADL and KS10; and several clones of Ladak were cultured in an effort to develop dihaploids and to determine factors influencing callus formation, embryogenesis, and plant regeneration (Song, 1990). Calli formed from explants from all plant sources. Microscopic examination showed that calli were initiated predominantly from broken filaments, and occasionally from anther walls. Embryogenesis, followed by plantlet formation (Fig. 1), occurred in the calli from one Lahonton, two KS10, and two Ladak clones, indicating that embryogenesis and regeneration are genotype-dependent events. Chromosome counts on root meristem cells of more than 30 regenerants revealed that all regenerated plants were tetraploids, presumably derived from maternal tissue. Extensive study of cultured anthers from a variety of sources revealed that one week after placing the anthers on medium, microspores inside the anthers had degenerated and rarely could be found (Song, 1990).

In another study (Saunders and Bingham, 1972), callus was initiated from immature anthers, immature ovaries, cotyledons, internode sections, and seedling hypocotyls. Histological studies of the anthers as they developed callus tissue indicated that most proliferation was of maternal tissue, although some tissue may have originated from immature pollen (Saunders and Bingham, 1972). Of 226 plants originating from potentially diploid callus (166 from anthers and 60 from ovaries), most were tetraploid, at least nine were octoploid, and none were dihaploid (Saunders and Bingham, 1972).

Successful recovery of alfalfa dihaploids has been reported from a single study (Zagorska *et al.*, 1990). Flower buds were treated with various combinations of temperature regimes and gamma radiation prior to culturing anthers with microspores at the mid- to late-uninucleate stage. All eight cultivars responded differently for callus formation from cultured anthers on Blaydes medium supplemented with 2,4–D, NAA, or kinetin. For plant regeneration, Blaydes medium supplemented with (2–isopentenyl)adenine (2-ip), myo-inositol, or yeast extract was reported to be effective in inducing organogenesis and subsequent plant regeneration. Chromosome counts of the regenerated plants showed that 50% were tetraploid ($2n = 32$), others were reported to be "mixoploid" (although the discussion presented suggests that the authors were actually referring to heteroploidy), and two plants were dihaploid (Zagorska *et al.*, 1990). Further refinement of the techniques should improve the efficiency of dihaploid recovery. We review here the available information and our unpublished data on microspore and anther culture, and briefly discuss how dihaploid alfalfa plantlets may be produced and utilized in the future.

2. General procedures

For culturing microspores at the uninucleate stage, collect flower buds 1 mm in length and sterilize with 75% EtOH for 15 s and then with 1.3% sodium hypochlorite (25% commercial bleach) for 10 min, followed by four rinses in sterile water (Song, 1990). Keep the flower buds in a humid environment until you are ready to grind them.

Grind the flower buds in a sterilized mortar and pestle containing 10 ml liquid medium (PT4–15, Table 1), filter through a 52 μm mesh, and centrifuge 10 min at 250 × g in a sterilized tube. Collect the pelleted microspores and resuspend in PT4–15, then spread them on PT4–15 medium solidified with agar or Gelrite.

Table 1. Composition of PT4–15, modified SHAP, and SPG-2 media (mg/l)

Component	PT4–15	Modified SHAP	SPG-2
KNO_3	2,500	2,500	1,000
$CaCl_2$ H_2O	250	200	400
$MgSO_4$	170.8	170.8	146.4
KH_2PO_4	–	–	300
$NH_4H_2PO_4$	300	300	–
NH_4NO_3	–	–	1,000
$FeSO_4$ $7H_20$	27.8	15	27
Na_2-EDTA	37.3	20	37
$MnSO_4$ H_2O	12	10	10
$ZnSO_4$ $7H_2O$	6	1	6
H_3BO_3	6	5	3
KI	1	1	0.8
$CuSO_4$ $5H_2O$	0.1	0.2	0.1
$CoCl_2$ $6H_2O$	0.1	0.1	0.1
Na_2MoO_4 $2H_2O$	0.2	0.1	0.2
Glycine	2	–	–
Thiamine HCl	3	5	1
Pyridoxine HCl	0.5	0.5	1
Nicotinic acid	3	5	1
Biotin	1.5		
Glutamic acid	2	–	–
Ascorbic acid	10	–	–
Lactoalbumin hydrolysate	100	–	–
Myoinositol	1	100	100
Sucrose	90,000	30,000	30,000
Glucose	100	–	–
2,4,5–T	1	–	–
Kinetin	1	0.1	–
2,4–D	–	0.5	–
pCPA	–	2.0	–
Proline	5.76 g	3	3
Alanine	2.67 g	1.5	1.5
Agar	6 g	6 g	6 g

Incubate cultured microspores in the dark at 25°C. Callus formation normally initiates within one week of culture. Three to 4 weeks later, transfer the calli to a regeneration medium (modified SHAP medium or SPG-2 medium, Table 1) for plantlet formation in an incubation chamber at 25°C under a light intensity of about 50 μmol s^{-1} m^{-2}. Plantlet formation often is apparent 4 weeks after initiation (Fig. 1).

After the regenerated plantlets have grown at least one leaf and a well-developed root system, transplant them into pots containing sterilized vermiculite. Irrigate with liquid SHAP medium without sucrose, alanine, or proline. To acclimate the regenerated plantlets and prevent desiccation, place them in a transparent plastic box within a growth chamber and gradually remove the cover over a period of a week or more as new leaves appear. Transfer acclimated plants to a greenhouse where they may be transplanted into pots of sterilized soil.

Identify dihaploids by chromosome counts from root tips. If necessary, make cuttings from well-established regenerants and allow them to proliferate new roots and use for chromosome counts. Ten to 20 days later, collect actively dividing root tips and treat with an enzyme solution (5% pectinase and 2.5% cellulase) for 60–90 min at 37°C in a water bath (Song *et al.*, 1988). Remove the root tips from the enzyme solution. Place three to five root tips on a slide, add 1 drop of acetocarmine (1%), and tap gently with the tip of a forceps to disperse the cells. After 1 to 2 min, place a cover slip over the cells, apply pressure on the cover slip with your thumb, and heat the slide over a burner for several short periods to clear the background. Examine the slide under the microscope for counting the chromosome numbers.

3. Discussion and prospects

The difficulties in producing dihaploids of alfalfa via anther culture may be the result of callusing from cells of anther walls or filaments, which is detrimental to the developmental process of immature microspores. The degenerating somatic cells of the anther walls could release substances toxic to dihaploid cells (Geyt *et al.*, 1985). Therefore, microspore culture conditions may be adapted to avoid the effects of degeneration of somatic cells.

To block the normal pathway of microspore development and stimulate microspores to form calli and undergo embryogenesis, various stimuli have been applied to isolated microspores in several other species. High or low temperature shock (Keller and Armstrong, 1979; Datta and Wenzel, 1987); split temperature regime – high temperature treatment for a period of time and then culture at normal temperatures (Chuong and Beversdorf, 1985); sugar starvation (Aruga and Nakajima, 1985; Wei *et al.*, 1986); other nutrient starvation (Kyo and Harada, 1985); and preculturing of microspores inside the anther before extraction (Datta and Wenzel, 1987), have been attempted with different degrees of success. Gamma radiation apparently stimulated alfalfa microspores to divide and, ultimately, regenerate into dihaploid plants, but with very poor efficiency (Zagorska *et al.*, 1990). Applications of the other stimulation methods described above to alfalfa may improve the efficiency of dihaploid production in this crop.

Pollination with pollen of distantly-related species has induced dihaploid formation in a few cases. For example, wide crosses using maize pollen have been successful in producing dihaploids of wheat (Laurie and Bennett, 1988a, 1989, 1990) and oats (Rines and Dahleen, 1990). Cross-fertilization between wheat and sorghum also might have taken place (Laurie and Bennett, 1988b). Pollination of alfalfa with pollen from a distantly related species might stimulate dihaploid formation; however, we are unaware of any research along these lines.

From a breeding standpoint, if doubled, tissue culture-derived dihaploid plants or lines are to be used in alfalfa improvement, they should be compared with doubled dihaploids obtained by traditional methods in order to determine their usefulness. For agronomic traits, several factors could affect the means and the variances of doubled dihaploid lines for any genetically encoded trait including: (1) reduced chances to undergo recombination in the presence of linkages; (2) gametophytic selection in microspore induction and seedling development; and (3) culture-induced variation, such as somaclonal variation or aneuploidy. However, with increased population size and use of F_2 or backcross plants, chances for recombination could be increased. Improvement of the nutrient medium and culture conditions and better timing for collecting the microspores could reduce the occurrence of undesirable off-types.

Field trials have produced conflicting reports on the performance of doubled dihaploids relative to that of lines from single-seed descent. In spring wheat, acceptable levels of performance were observed among doubled dihaploid lines compared to single-seed descent lines (Mitchell *et al.*, 1992). On the other hand, lower-yielding doubled dihaploid lines also have been reported in wheat (Baenziger *et al.*, 1989). In barley, no differences were found in doubled dihaploid lines and lines from single-seed descent for yield, heading date, or height (Choo *et al.*, 1982). Picard *et al.* (1988) reported a large number of high-yielding wheat lines were obtained from anther culture. In tobacco, dihaploid lines derived from anther culture were reported to have lower yield (12 to 21%) and were agronomically inferior to selfed progenies of the parental lines (Brown and Wernsman, 1982). Furthermore, they reported that each cycle of anther culture of the doubled dihaploid lines resulted in vigor reduction approximately equivalent to the magnitude of selfed progenies of their respective parents. These different results may be due to the different crops and plant materials used. Large-scale tests with materials from a broad genetic base should be performed before definite conclusions are drawn.

Somaclonal variation, or spontaneous genetic alteration in cells grown *in vitro*, is commonly reported. The plausible causes of somaclonal or protoclonal variation are many, including mitotic segregation from chromosome nondisjunction, chromosome loss or deletion, somatic crossing over, somatic meiosis, repair of DNA lesions, mutation, genome reorganization, activity of transposable genetic elements, gene amplification and deletion, or cryptic virus elimination. The exact cause is not known, and controlled experiments are necessary to determine the cause in each case.

In the case of alfalfa, it has been suggested that mobile genetic elements are activated by the tissue culture process. Bingham and Clement (1989) described six independent alleles of mutable loci. One of the mutable loci, designated *c2–m4*, was much more active in tissue culture (active frequency about 0.23), than in whole plants (active frequency <0.001). A genetic study of *c2–m4* showed that the level of mutability increased as the dosage of the allele increased, indicating that each copy of the allele could function independently (Ray and Bingham, 1991).

Chromosome number variation of alfalfa tissue culture regenerants also is common. In one study, 58% of mesophyll protoplast regenerants from a tetraploid plant were octoploid, others were various heteroploids, and only 30% were tetraploid (Johnson *et al.*, 1984). An earlier study (McCoy and Bingham, 1977) recovered both tetraploids and octoploids from diploid cell suspensions. This persistent tendency to spontaneous polyploidization of alfalfa from various cell sources suggests that recovery of true polyhaploids from microspore culture may be difficult; even if embryogenic callus from microspores is obtained. Therefore, it is possible that the reported lack of polyhaploid production from gametic cells of alfalfa was actually the result of karyotypic instability of the regenerated plants, rather than regeneration from maternal tissues.

Stability of the organellar and nuclear genomes of regenerated plants can be analyzed on a molecular level. Recently, restriction patterns of mtDNA from regenerated sugarcane plants derived from cell suspensions or from the parental plants were compared (Chowdhury and Vasil, 1993). The results demonstrated that sugarcane mtDNA was stable through the regeneration process. On the other hand, rearrangement of the mt-genome following regeneration is known in alfalfa (He and Liu, 1994), sorghum (Kane *et al.*, 1992), and wheat (Morère-Le Paven *et al.*, 1992) and in faba bean cell suspension cultures (Negruk *et al.*, 1986). In our laboratory so far, chloroplast DNA probes have detected no changes in the chloroplast genomes of alfalfa plants regenerated after 90 or 140 days in callus culture (unpublished). Likewise, Rose *et al.* (1986) did not observe mtDNA variations in alfalfa protoclones. Genomic stability of the regenerated plants may be organ/tissue or species dependent and perhaps is a time-related event.

Methods of production of alfalfa dihaploids are available, but may not function for a particular plant genotype. As we have reviewed here, several avenues of research are being pursued, and progress toward methods that will permit the extraction of dihaploids from any alfalfa plant likely will become available within a relatively short time. These dihaploids undoubtedly will find application in basic and applied genetic research of alfalfa, and will enhance the developent of useful cultivars.

4. References

Aruga, K. and T. Nakajima, 1985. Factors affecting the process of embryo formation from pollen grains in tobacco. Jpn. J. Breed. 35: 127–135.

Baenziger, P.S., D.M. Wesenberg, V.M. Smail, W.L. Alexander and G.W. Schaeffer, 1989.

Agronomic performance of wheat doubled-haploid lines derived from cultivars by anther culture. Plant Breed. 103: 101–109.
Binek, A. and E.T. Bingham, 1970. Cytology and crossing behavior of triploid alfalfa. Crop Sci. 10: 303–306.
Bingham, E.T., 1971. Isolations of haploids of tetraploid alfalfa. Crop Sci. 11: 433–435.
Bingham, E.T. and W.M. Clement Jr., 1989. Alfalfa transposable elements and variegation. Dev. Genet. 10: 552–560.
Bingham, E.T. and J.M Saunders, 1974. Chromosome manipulations in alfalfa: scaling the cultivated tetraploid to seven ploidy levels. Crop Sci. 14: 474–477.
Bingham, E.T. and T.J. McCoy, 1979. Cultivated alfalfa at the diploid level: Origin, reproductive stability, and yield of seed and forage. Crop Sci. 19: 97–100.
Bolton, J.L. and J.E.R. Greenshields, 1950. A diploid form of *Medicago sativa* L. Science 112: 275–277.
Bolton, J.L., 1962. Alfalfa; Botany, Cultivation, and Utilization. Interscience Publishers, New York.
Brown, J.S. and E.A. Wernsman, 1982. Nature of reduced productivity of anther-derived dihaploid lines of flue-cured tobacco. Crop Sci. 22: 1–5.
Chen, Y., 1986a. Anthers and pollen culture of rice. In H. Hu and H. Yang (Eds.), Haploids of Higher Plants *in Vitro*, Chapter 1, pp. 1–25. Springer-Verlag, New York.
Chen, Y., 1986b. The inheritance of rice pollen plant and its application in crop improvement. In: H. Hu and H. Yang (Eds.), Haploids of Higher Plants *in Vitro*, Chapter 7, pp. 110–135. Springer-Verlag, New York.
Choo, T.M., E. Reinbers and S.J. Park, 1982. Comparison of frequency distributions of doubled haploids and single seed descent lines in barley. Theor. Appl. Genet. 61: 215–218.
Chowdhury, M.K.U. and I.K. Vasil, 1993. Molecular analysis of plants regenerated from embryogenic cultures of hybrid sugarcane cultivars (*Saccharum* spp.). Theor. Appl. Genet. 86: 181–188.
Chuong, P.V. and W.D. Beversdorf, 1985. High frequency embryogenesis through isolated microspore culture in *Brassica napus* L. and *B. carinata* Braun. Plant Sci. 39: 219–226.
Clement, W.M. Jr., 1962. Chromosome numbers and taxonomic relationships in *Medicago*. Crop Sci. 2: 25–28.
Clement, W.M. Jr. and W.F. Lehman, 1962. Fertility and cytological studies of a dihaploid plant of alfalfa, *Medicago sativa*. Crop Sci. 2: 451–453.
Datta, S.K. and G. Wenzel, 1987. Isolated microspore derived plant formation via embryogenesis in *Triticum aestivum* L. Plant Sci. 48: 49–54.
Duijs, J.G., R.E. Voorrips, D.L. Visser and J.B.M. Custers, 1992. Microspore culture is successful in most crop types of *Brassica oleracea* L. Euphytica 60: 45–55.
Dupuis, I., P. Roeckel, E. Matthys-Rochon and C. Dumas, 1987. Procedure to isolate viable sperms cells from corn (*Zea mays* L.) pollen grains. Plant Physiol. 85: 876–878.
Epps, D.J., D. Hutchinson and W.A. Keller, 1992. *In vitro* culture of isolated microspores and regeneration of plants in *Brassica campestris*. Plant Cell Rep. 11: 234–237.
Geyt, J.V., K. Dhalluin and M. Jacobs, 1985. Induction of nuclear and cell divisions in microspores of sugar beet (*Beta vulgaris* L.). Z. Pflanzenzüchtg. 95: 325–335.
Hansen, M. and K. Svinnset, 1993. Microspore culture of swede (*Brassica napus* ssp. *rupifera*) and the effects of fresh and conditioned media. Plant Cell Rep. 12: 496–500.
He, Q.Q., and K.Z. Liu, 1994. The variation of chloroplast DNA and mitochondrial DNA in tissue culture of alfalfa. Acta Agron. Sin. 20: 33–38.
Johnson, L.B., D.L. Stuteville, S.E. Schlarbaum and D.Z. Skinner, 1984. Variation in phenotype and chromosome number in alfalfa protoclones regenerated from nonmutagenized calli. Crop Sci. 24: 948–951.
Kane, E.J., A.J. Wilson and P.S. Chourey, 1992. Mitochondrial genome variability in sorghum cell culture protoclones. Theor. Appl. Genet. 83: 799–806.
Keller, W.A. and K.C. Armstrong, 1979. Stimulation of embryogenesis and haploid production

in *Brassica campestris* anther cultures by elevated temperature treatment. Theor. Appl. Genet. 55: 65–67.

Kyo, M. and H. Harada, 1985. Studies on conditions for cell division and embryogenesis in isolated pollen culture of *Nicotiana rustica*. Plant Physiol. 79: 90–94.

Laurie, D.A. and M.D. Bennett, 1988a. The production of haploid wheat plants from wheat × - maize crosses. Theor. Appl. Genet. 76: 393–397.

Laurie, D.A. and M.D. Bennett, 1988b. Cytological evidence for fertilization in hexaploid wheat × sorghum crosses. Plant Breed. 100: 73–82.

Laurie, D.A. and M.D. Bennett, 1989. The timing of chromosome elimination in hexaploid wheat × maize crosses. Genome 32: 953–961.

Laurie, D.A. and M.D. Bennett, 1990. Early post-pollination events in hexaploid wheat × maize crosses. Sex. Plant Reprod. 3: 70–76.

Lesins, K., 1957. Cytogenetic study on a tetraploid plant at the diploid chromosomal level. Can. J. Bot. 35: 181–196.

Lesins, K. and I. Lesins, 1964. Diploid *Medicago falcata* L. Can. J. Genet. Cytol. 66: 152–163.

Lesins, K. and C.B. Gillies, 1972. Taxonomy and cytogenetics of *Medicago*. In: C.H. Hanson (Ed.), Alfalfa Science and Technology, pp. 53–86. Am. Soc. Agron., Madison, WI.

Lichter, R., 1982. Induction of haploid plants from isolated pollen of *Brassica napus*. Z. Pflanzenphysiol. 105: 427–434.

McCoy, T.J. and E.T. Bingham, 1977. Regeneration of diploid alfalfa plants from cells grown in suspension culture. Plant Sci. Lett. 10: 59–66.

Mitchell, M.J., R.H. Busch and H.W. Rines, 1992. Comparison of lines derived by anther culture and single-seed descent in a spring wheat cross. Crop Sci. 32: 1446–1451.

Morère-Le Paven, M.C., J. De Buyser, Y. Henry, F. Corre, C. Hartmann and A. Rode, 1992. Multiple patterns of mtDNA reorganization in plants regenerated from different *in vitro* cultured explants of a single wheat variety. Theor. Appl. Genet. 85: 9–14.

Negruk, V.I., G.I. Eisner, T.D. Redichkina, N.N. Dumanskaya, D.I. Cherny, A.A. Alexandrov, M.F. Shemyakin and R.G. Butenko, 1986. Diversity of *Vicia faba* circular mtDNA in whole plants and suspension cultures. Theor. Appl. Genet. 72: 541–547.

Nichterlein, K. and W. Friedt, 1993. Plant regeneration from isolated microspores of linseed (*Linum usitatissimum* L.). Plant Cell Rep. 12: 426–430.

Picard, E., C. Parisot, M. Causse, P.H. Barbant, G. Doussinault, M. Trottet and M. Rousset, 1988. Comparison of doubled haploid method with other breeding procedures in wheat (*Triticum aestivum*) when applied to populations. In: T.E. Miller and R.M.D. Koebner (Eds.), Proc. Int. Wheat Genet. Symp., 7th. Inst. Plant Sci. Res., Cambridge, pp. 1155–1159.

Ray, I.M. and E.T. Bingham, 1991. Inheritance of a mutable phenotype that is activated in alfalfa tissue culture. Genome 34: 35–40.

Reddy, V.S., S. Leelavathi and S.K. Sen, 1985. Influence of genotype and culture medium on microspore callus induction and green plant regeneration in anthers of *Oryza sativa*. Physiol. Plant. 63: 309–314.

Rines, H.W. and L.S. Dahleen, 1990. Haploid oat plants produced by application of maize pollen to emasculated oat florets. Crop Sci. 30: 1073–1078.

Rose, R.J., L.B. Johnson and R.J. Kemble, 1986. Restriction endonuclease studies on the chloroplast and mitochondrial DNA of alfalfa (*Medicago sativa* L.) protoclones. Plant Mol. Biol. 6: 331–338.

Sangwan-Norreel, B.S., 1977. Androgenic stimulating factors in the anther and isolated pollen grain culture of *Datura innoxia* Mill. J. Exp. Bot. 28: 843–852.

Saunders, J.W. and E.T. Bingham, 1972. Production of alfalfa plants from callus tissue. Crop Sci. 12: 804–808.

Song, J.S., 1990. Microspore culture in alfalfa (*Medicago sativa* L.). Ph.D. Dissertation, Part V. Kansas State Univ., Manhattan, KS.

Song, J.S., E.L. Sorensen and G.H. Liang, 1988. A new method to prepare root tip chromosomes in alfalfa. Cytologia 53: 641–645.

Stanford, E.H. and W.M. Clement Jr., 1958. Cytology and crossing behavior of a haploid alfalfa plant. Agron. J. 50: 589–592.

Swanson, E.B., M.P. Coumans, S.C. Wu, T.L. Barsby and W.D. Beversdorf, 1987. Efficient isolation of microspores and the production of microspore-derived embryos from *Brassica napus*. Plant Cell Rep. 6: 94–97.

Takahata, Y. and W.A. Keller, 1991. High frequency embryogenesis and plant regeneration in isolated microspore culture of *Brassica oleracea* L. Plant Sci. 74: 235–242.

Tanaka, I. and M. Ito, 1980. Induction of typical cell division in isolated microspores of *Lilium longiflorum* and *Tulipa gesneriana*. Plant Sci. Lett. 17: 279–285.

Tanaka, I., C. Kitazume and M. Ito, 1987. The isolation and culture of lily pollen protoplasts. Plant Sci. 50: 205–211.

Wan, Y. and J.M. Widholm, 1992. Formation of multiple embryo-like structures from single microspores during maize anther culture. Plant Cell Rep. 11: 529–531.

Wei, Z.M., M. Kyo and H. Harada, 1986. Callus formation and plant regeneration through direct culture of isolated pollen of *Hordeum vulgare* cv. "Sabarlis". Theor. Appl. Genet. 72: 252–255.

Xu, Z.H. and N. Sunderland, 1981. Glutamine, inositol, and conditioning factors in the production of barley pollen callus *in vitro*. Plant Sci. Lett. 23: 161–168.

Zagorska, N., R. Stereva and P. Robeva, 1990. Alfalfa (*Medicago* spp.): *In vitro* production of haploids. In: Y.P.S. Bajaj (Ed.), Biotechnology in Agriculture and Forestry 12: Haploids in Crop Improvement I, pp. 458–471. Springer-Verlag, New York.

Zhou, C., 1988. Isolation and purification of generative cells from fresh pollen of *Vicia faba* L. Plant Cell Rep. 7: 107–110.

20. Haploid of soybean

CHING-YEH HU, GUANG-CHU YIN* and MARIA HELENA BODANESE ZANETTINI

Contents

1. Introduction

Soybean (*Glycine max*, L. Merr.; $2n = 40$) is an annual seed legume for the production of oil and protein. It is the world's most important plant protein source for human consumption and animal feeds, in addition to being one of the most important sources of a high quality, low cholesterol edible oil.

The domestication of soybean started in northeastern China more than 3,000 years ago and it was introduced to the U.S. about 230 years ago. The shift toward a meat-based diet in developed countries after the Second World War stimulated soybean cultivation for feed in the U.S., Brazil and Argentina. Today North and South America produce more than three-quarters of all the soybeans grown in the world. In the Orient, China remains the major producer. A map of world soybean production, trade, and per capita consumption was published by the National Geographic Society (Hapgood, 1987) based on USDA sources.

With the extremely narrow genetic base of soybean cultivars of the major Western producing nations (Delannay *et al.*, 1983; Vello *et al.*, 1988), hybrid breeding provides an attractive means to introduce novel traits into the soybean gene pool. Traits, such as stress tolerance (Marshall and Broue, 1981; Newell and Hymowitz, 1982), disease resistance (Mignucci and Chamberlain, 1978; Singh *et al.*, 1974) and high protein content (Lin and

*This review article is dedicated to late Prof. G.C. Yin (1936–1994) in honor of his contributions to the advance in soybean research. Prof. Yin passed away, suddenly, after the completion of the first draft of this manuscript.

S.M. Jain, S.K. Sopory & R.E. Veilleux (eds.), In Vitro *Haploid Production in Higher Plants, Vol. 3*, 377–395.

Yin, 1983) are known to exist in wild soybean relatives. Using conventional breeding to produce new cultivars after hybridization is very time consuming. With anther culture, on the other hand, target genes can be introduced into soybean and released in pure lines in just one generation following hybridization. Thus, it would provide a quick means of breeding new cultivars for the ever-changing market.

Soybean is a recalcitrant species for *in vitro* regeneration. Although callus can be induced readily from tissues at various developmental stages, the totipotency for organogenesis diminishes soon after seed germination (Saka *et al.*, 1980), and somatic embryogenesis is confined to early embryonic development only (Ranch *et al.*, 1985).

The recalcitrant nature of this species makes progress in haploid plant induction sluggish. Only scant published information on this subject is available. Matsubayashi and Kuranuki (1975) tested 13 species belonging to 4 different families for their ability to respond to anther culture. They classified soybean as a non-responsive species. Callus induction from soybean anther culture, nevertheless, was first reported by Tang *et al.* (1973). One year later, Ivers *et al.* (1974) produced diploid calli from soybean anther culture. Liu and Zhao (1986) were successful in haploid callus formation from isolated microspore culture. Yin *et al.* (1982) obtained sporadic haploid plantlets from Chinese soybean cultivars through organogenesis of microspore-derived calli. Similar results were reported by another Chinese group (Jian *et al.*, 1986). Furthermore, anther culture of U.S. soybean cultivars produced embryos (Zhuang *et al.*, 1991). Seven embryo-like structures were also reported recently in anther culture of Chinese cultivars (Ye *et al.*, 1994). Sporadic plantlets have also been obtained from androgenic embryos of Brazilian soybean cultivars (E. Kaltchuk-Santos and M.H.B. Zanettini, unpublished data). In spite of the above efforts, there seems no reproducible, efficient system yet available for the production of soybean androgenic plantlets.

In this chapter, we will review and summarize the current status of soybean anther and microspore culture. Hopefully, we may also suggest a way for future workers for the eventual success in routine dihaploid soybean plant production.

2. The nature of soybean pollen grains

Based on fluorescent microscopy by Corriveau and Coleman (1988), the mature soybean pollen grain was classified as a bicellular type. A detailed study, employing similar procedure, was made by X.J. Zhuang and C.Y. Hu (unpublished observation). They stained living pollen grains with a DNA-specific fluorochrome, mithramycin (100 μg/ml in 10 mM $MgCl_2$ aqueous solution) and examined them under a Zeiss 48 7709 combination of excitation (435) and emission filters. The work revealed that approximately 0.4% of the mature soybean pollen grains of the U.S. cultivars Williams 82 and

Asgrow A1929 were of tricellular type (see media sucrose section for possible significance).

Dimorphism in soybean pollen grains has been clearly demonstrated by Kaltchuk-Santos *et al.* (1993) with four Brazilian cultivars (Fig. 1A,B). From the binucleate stage onward, interspersed among the starch-filled normal grains is a second class of smaller (ca. 3/4 the normal size) grains which lacked starch and stained lightly (Fig. 1B). Such atypical grains have been named "E pollen" (Sunderland and Wicks, 1969), "S pollen" (Horner and Street, 1978), or "P pollen" (Heberle-Bors and Reinert, 1979). The frequency of such atypical pollen varied significantly among cultivars. Of 7,500 pollen grains examined in each cultivar, the percentage of P pollen in cv. Decada, IAS5, BR4, and Ipagro21 was 1.86 ± 1.84, 0.04 ± 0.11, 0.09 ± 0.18, and 0.17 ± 0.47, respectively. The percentage of small, lightly-stained, uninucleate putative precursors (Fig. 1A) was 3.69 ± 3.55, 0.25 ± 0.24, 0.94 ± 1.19, and 1.32 ± 4.13, respectively.

The atypical pollen grains observed by Kaltchuk-Santos *et al.* (1993) resembled the products of the initial division of one of the four established routes of *in vitro* androgenesis (Sunderland, 1974). In this route the first microspore division is symmetric, resulting in two equal-sized cells with diffuse vegetative-type nuclei. The putative uninucleate precursors of the P pollen may have already diverged and were predestined to follow such route in culture. This route has been considered one of the principal routes in the formation of haploid callus and embryos (Rashid and Street, 1974; Wilson *et al.*, 1978) and was also one of the routes identified in soybean microspore callus formation (Yin *et al.*, 1980). The significance of pollen dimorphism in relation to androgenic tissue/plant production, nevertheless, was based primarily on indirect statistical correlations and is by no means conclusive.

3. Floral bud selection

A soybean floral bud is covered by five sepals which fuse at the base into a calyx. There is a pair of needle-like bracts located on the outside of the calyx (Fig. 1C). The rate of growth of the bracts is slower than that of the calyx-covered floral buds. The correspondence of microsporogenesis stages to external floral bud features was first established by Ivers *et al.* (1974) for soybean cv. Hark. They selected and cultured anthers of stage 3 buds, having 2.5 mm bud length with 30% of the microspores at late tetrad and 70% at individualized developmental stages. The correlation between the floral bud morphological size index to the microspore developmental stages for Chinese soybean cultivars was established by Yin *et al.* (1982). They reported that when the bract length was the same as the bud length, the microspores were at the tetrad stage. When the bracts were equal to 2/3–3/4 of the bud length (bud length = 2.5 to 3.5 mm), the microspores were at the uninucleate stage. In this size range, the microspores ranged from early uninucleate stage with

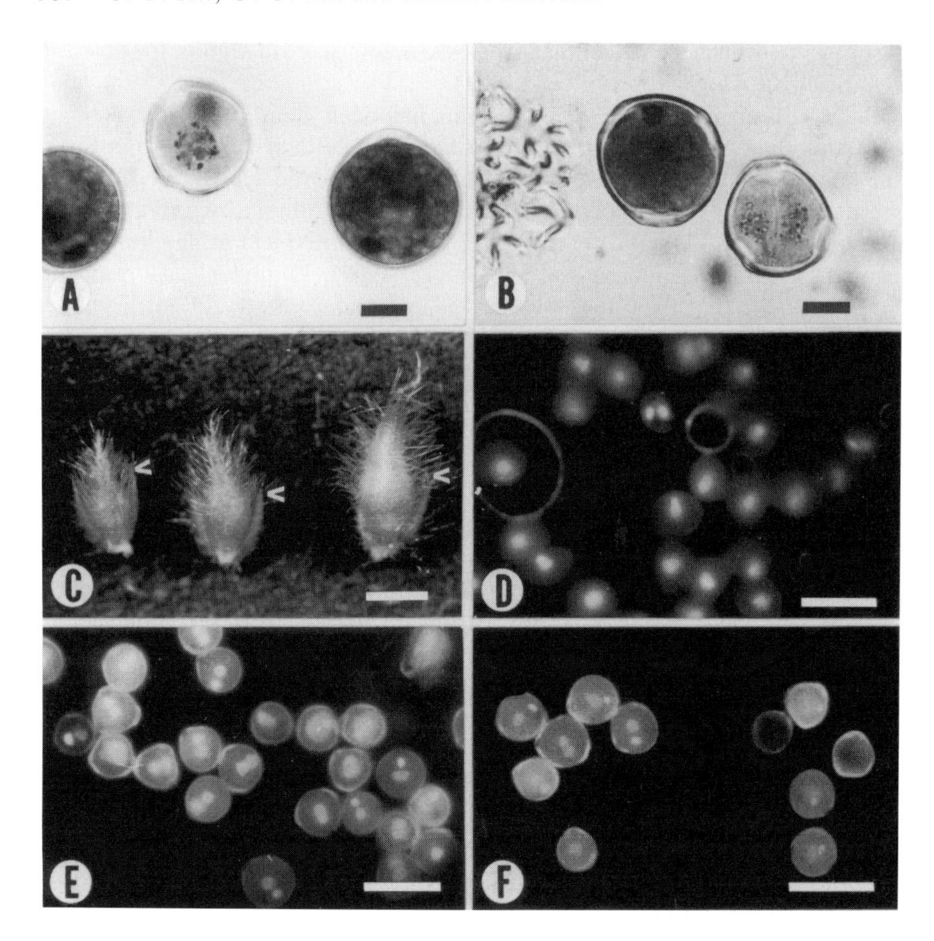

Figure 1. (A,B) Pollen dimorphism in soybean. Bars = 10 um. (A) Two normal binucleate pollen grains containing starch grains and deeply stained cytoplasm which obscures the nuclei and one uninucleate putative precursor of P pollen, which is smaller in size and exhibits a weaker staining reaction. (B) Normal binucleate pollen and P pollen containing two equal-sized cells with diffuse vegetative-type nuclei. Note the wall separating the nuclei of P pollen. (C-F) Correlation between floral bud morphology and microspore development in soybean. (C) Floral bud stages. Left: bract length = 2/3 of bud length; Center: bract length = 1/2 of bud length; Right: bract length = 1/3 of bud length. Arrow heads pointing to the tips of the bracts. Bar = 2 mm. (D-F) Corresponding microspores stained with mithramycin fluorochrome to show the nuclei. They were examined with Zeiss 48 7709 combination of excitation (435) and emission filters. Bars = 50 μm. (D) Uninucleate and early binucleate pollen from buds with bract length = 2/3 of bud length. (E) Middle binucleate pollen from buds with bract length = 1/2 of bud length. (F) Late binucleate pollen from buds with bract length = 1/3 of bud length.

Table 1. The correlation of microspore developmental stages with the external floral bud features of soybean cv. Williams 82 and Asgrow A1929 and callus production after 4 months *in vitro* anther culture

BUD LENGTH (mm)	BRACT/BUD RATIO	ANTHER COLOR	POLLEN STAGES (%) UNI	BI	EMPTY	CALLUS TOTAL#	SMOOTH %
-	1 - 3/4	TW	-	-	-	-	-
3.9 ± 0.63	2/3	TY	57	42	1	129	59
4.4 ± 0.71	1/2	OY	10	79	12	93	18
5.0 ± 0.67	1/3	OYG	<1	67	33	55	20

UNI – uninucleate; BI – Bi-nucleate; – data not available; TW – translucent whitish; TY – translucent, pale yellow; OY – semi-opaque, light yellow. OYG – opaque, light yellow with slight green shade.

a centrally located nucleus, to the mid-uninucleate stage with the nucleus away from the center, to a late-uninucleate stage with the nucleus against pollen wall. When the bracts length become 1/2 the bud length, the microspores are at the binucleate stage. The above correlation was based on field-grown plants. When plants are grown in different seasons or in the greenhouse, this correlation may be somewhat off and would need to be re-standardized.

Larger buds were tested by X.J. Zhuang and C.Y. Hu (unpublished data) using U.S. cultivars Williams 82 and Asgrow A1929. Essentially the same correlations between microspore stages to bud length and bract/bud ratio as reported by Yin *et al.* (1982) were found. See Table 1 for bud length, bract/bud ratio, anther color, pollen stages and *in vitro* responses of different bud size groups. The microspores from the anthers of different bud stages are shown in Fig. 1C – F. They chose to culture floral buds with the bract/bud ratios from 1/3 to 2/3.

The developmental stage of microspores in the same bud might not necessarily be identical. Based on the coloration, the degree of yellow shade, of the anther wall, the ten anthers in a given bud, especially the younger buds, seemed at somewhat different developmental stages (C.Y. Hu, unpublished observation). This observation, however, has not been substantiated with quantitative data.

G.C. Yin (unpublished data) found that anthers derived from field-grown plants responded to culture conditions better than those from greenhouse-grown plants. Their group had also experienced that anthers harvested from the full bloom stage of the donor plants responded better than those harvested from plants approaching the end of their blooming period. Again, there was no statistical data available to confirm such observations.

For anther culture of Chinese cultivars, the ideal microspore developmental stages were determined to be early to mid-uninucleate (Yin *et al.*, 1982).

They found that anthers at the tetrad stage did not respond to the culture conditions and would stay translucent white on the medium with no obvious change. When, on the other hand, anthers with microspores at the late-uninucleate and binucleate stages were cultured, Yin *et al.* (1982) observed a gradual blackening of the anther wall and the disintegration of the microspores. This resulted in a quiescent appearance of the black-colored anthers on culture medium. Only anthers containing microspores at early and mid-uninucleate stages responded to culture conditions by gradually turning brown and producing calli. With U.S. cultivars, X.J Zhuang and C.Y Hu (unpublished data; Table 1; Fig. 2) observed anthers dissected from floral buds with a bract/bud ratio >2/3, containing uninucleate microspores that remained quiescent, translucent-white on the medium after 2.5 months of *in vitro* culture. The cytoplasm of their microspores shrunk and detached from the pollen wall (Fig. 2A,C). Auto-fluorescence using Zeiss 48 7702 combination of excitation (370) and emission filters indicated that starch grains had not yet developed in the cytoplasm of these microspores (Fig. 2D). They, nevertheless, obtained large amounts of calli with anthers dissected from floral buds with a bract/bud ratio between 1/3 and 2/3, containing late-uninucleate and binucleate microspores; such anthers turned brown in culture. Most of the microspores in these anthers had dense cytoplasm, which contained many auto-fluorescent starch granules, after 2.5 months of *in vitro* culture (Fig. 2A,B,E,F). Progressively fewer calli were produced as anthers containing microspores at more advanced binucleate stages were cultured. A certain percentage of anthers at a more advanced stage (dissected from

Figure 2. Age of anthers and their *in vitro* responses (photo taken 2.5 months after *in vitro* culture). (A) Cultured anthers. Left: anthers dissected from floral buds with bract length >2/3 of the bud length containing microspores prior to the desirable stage. They stayed quiescent, translucent-white on the medium with no obvious change. Middle: anthers dissected from floral buds with bract length between 1/3 to 2/3 of the bud length containing microspores at the desirable stage. They swelled, turned brown and were capable of producing callus *in vitro*. Right: anthers dissected from floral buds with bract length <1/3 of the bud length containing microspores past the desirable stage. They turned black and stayed quiescent on the medium. Each ruler div. = 1 mm. (B) Four calli developed from a single anther (arrow head) of the desirable stage. The two calli at right and bottom are larger, rough-surfaced, light-yellow colored, non-embryogenic ones and the two at upper left, next to the ruler are smaller, smooth-surfaced, yellowish-green colored, embryogenic calli. Each ruler div. = 1 mm. (C-H) Light (C, E, G) and auto-fluorescent (D, F, H) microscopic images of microspores from anthers of the above three age groups. Auto-fluorescence was examined with Zeiss 48 7702 combination of excitation (370) and emission filters. All photos with the same magnification. Bar (on C) = 20 μm. (C,D) From the left side anther group of A. Note the inactive, shrunken cytoplasm which detached from pollen walls. Auto-fluorescence indicates that starch grains had not yet developed in the cytoplasm. (E,F) From the middle anther group of A. Note the dense cytoplasm containing granular, auto-fluorescent starch grains. (G,H) From the right side anther group of A. Most pollen disintegrated with only a few pollen grains left in the anthers. Note that the disintegrating cytoplasm demonstrated reduced auto-fluorescence.

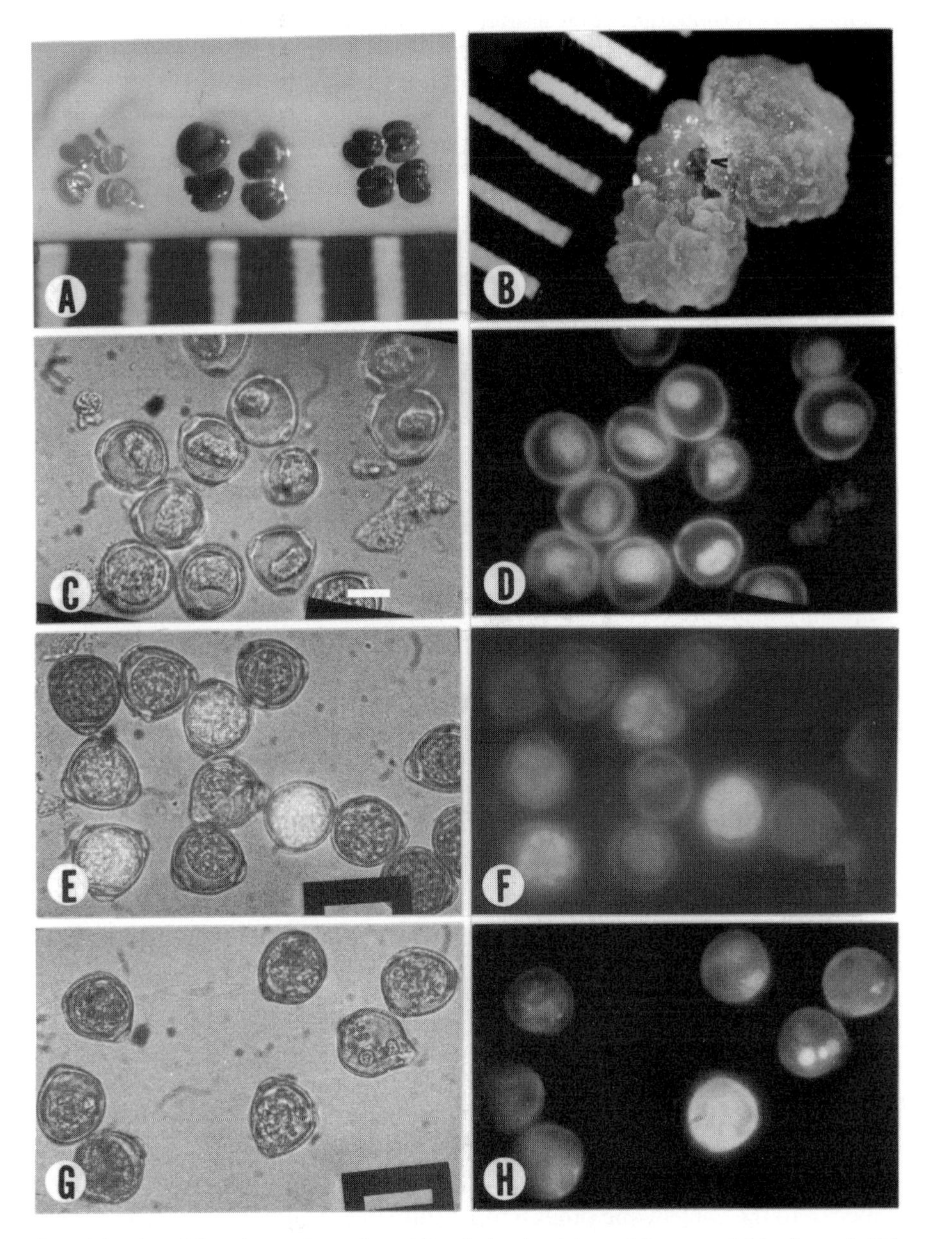

floral buds with a bract length <1/3 of the bud length) turned black and did not respond on culture medium. There were very few microspores left in these black, quiescent anthers after 2.5 months of culture (Fig. 2A,G,H). The disintegrating cytoplasm exhibited reduced auto-fluorescence. The two

Table 2. The effects of pre-culture temperature incubations of floral buds on callus production and pollen survival of soybean cvs. Williams 82 and Asgrow A1929 after *in vitro* anther culture. The control anthers were cultured without pre-culture incubation. Pollen viability was determined with fluorescein diacetate fluorochrome 2 months after *in vitro* anther culture. Calli types were counted 3 months after *in vitro* anther culture

CV.	TEMP-DAYS (°C)	# OF ANTHER CULTURED	SMOOTH CALLUS (#) TOTAL	WITH EMB	ANTHERS WITH ALIVE POLLENS (%)
Wms. 82	Control	360	16	0	80
	4 - 8	180	59	5	73
	8 - 4	480	61	5	100
	8 - 8	360	21	5	85
	37 - 1	300	0	0	18
A1929	Control	360	13	0	82
	4 - 8	180	44	3	73
	8 - 4	120	11	0	75
	8 - 8	360	17	0	70
	37 - 1	300	1	0	8

studies above differed on androgenic capacities of microspores at the late uninucleate and binucleate stages. Cultivar, the growth conditions of donor plants, and media differences might be some of the possible causes for such discrepancy. The non-uniformity of microspore stages to bud morphology and human error during indexing could not be ruled out as other possible causes.

4. Pre-treatment of floral buds

To test the effects of pre-culture temperature treatments, the floral bud clusters of cvs. Williams 82 and Asgrow A1929 were treated at 4°C for 8 days, 8°C for 4 and 8 days, and 37°C for 1 day, along with a non-treated control (X.J. Zhuang and C.Y. Hu, unpublished data; Table 2). With Williams 82, the number of calli, and calli with embryo-like structures produced under 4°C – 8 day and 8°C – 4 day pre-treatments were similar and superior to the control and other temperature pre-treatments. Cultivar A1929 responded better when 4°C for 8 days treatment was given. The 37°C – 1 day pre-culture heat shock consistently produced few or no callus and a very high percentage of pollen degeneration in culture. With Chinese cultivars, Ye *et al.* (1994) pretreated floral buds for 3 to 5 days at 0–1, 4–5, and 7–

8°C and obtained 10.6, 18.2 and 24.4% callus induction. G.C. Yin (unpublished data) found that pretreating floral buds at a temperature range of 2–4°C for 3 to 12 days was effective in pollen callus induction for Chinese soybean cultivars. Liu and Zhao (1986) reported a stimulating effect on microspore division when they pre-treated the floral bud clusters of Chinese soybean cultivars with low temperature (7–8°C) for 5 or 8 days. They found that anthers exposed to 9.0 μM 2.4–D during pre-treatment exhibited a faster rate of microspore division.

The above studies generally agreed that low temperature, 4–8°C, pre-treatment of the floral buds for a few days was beneficial for callus induction. Both G.C. Yin (unpublished data) and Liu and Zhao (1986) detected equal division of the microspores during low temperature pre-treatment. Statistical data, however, were not provided.

5. Surface disinfestation and dissection

Dense trichomes are present on the surface of soybean floral buds (Fig. 1C), making disinfestation difficult due to the trapping of air bubbles. To wet the trichome surface completely, buds should first be dipped in 70–75% ethanol with stirring for 1 min. Then, they should be immersed in a disinfectant solution of 1% NaOCl (20% Clorox) or 0.1% HgCl containing a trace amount of detergent such as Tween under ultrasonic vibration for about 20 min. After they have been rinsed with 3–4 changes of sterile distilled water, the buds are ready for dissection.

During dissection, buds are opened with two sterile dissecting needles. The short filaments of the ten stamens are attached to the lower portion of the corolla. The anthers can be excised from the filaments and transferred, with or without partial filament, onto the culture medium. The needles should be re-sterilized after dissecting each bud by dipping them in 70–75% ethanol. The ethanol can be removed by wiping the tool on a dry sterile paper towel. The optimum density of anthers plated on culture medium has not yet been determined. However, Ye *et al.* (1994) found that the callus induction was 20.5% when three or more anthers were inoculated in a group, whereas only 6.4% anthers produced callus when they were inoculated singly.

6. Culture media

6.1. *Basal medium*

Yin *et al.* (1982) tested several basal media including MS (Murashige and Skoog, 1962), Nitsch (Nitsch and Nitsch, 1969), N_6 (Chu, 1978) and B5 (Gamborg *et al.*, 1968) for soybean anther culture and obtained the best callus induction with B5 formulation (data were not shown). Jian *et al.* (1986)

Table 3. B5 Long medium. This medium is an enriched formulation of B5 medium (Gamborg *et al.*, 1968) with 16 organic compounds added to the B5 salt base. It is manufactured and marketed in a dehydrated form by the Carolina Biological Supply Co., 2700 York Road, Burlington, NC 27215, U.S.A.

COMPONENTS	CONCENTRATION (mg/l)
B5 Salt Base	3,081
p-Aminobenzoic Acid	0.2
Ascorbic Acid	0.4
d-Biotin	0.00025
Choline Chloride	0.2
Folic Acid	0.015
Niacin	0.5
D-Pantothenic Acid	0.4
Pyridoxine HCl	0.5
Riboflavin	0.015
Thiamine HCl	0.5
L-Arginine F.B.	40.0
L-Asparagine (Andy)	40.0
Glycine	20.0
L-Glutamine	60.0
L-Phenylalanine	20.0
L-Tryptophan	40.0

tested MS, White (1963) and B5 basal media for soybean anther culture and determined that B5 provided the highest frequency of callus induction (no quantitative data given). They made further improvements to the B5 formulation by changing the concentrations of $MgSO_4 \cdot 7H_2O$, $CaCl_2 \cdot 2H_2O$, and $NaH_2PO_4 \cdot 5H_2O$ to 185, 250 and 50 mg/l, respectively. With this modification, they were able to get up to 36.4% callus formation. Liu and Zhao (1986) used enriched B5 medium (Kao, 1982) for their soybean microspore culture work. Zhuang *et al.* (1991) and E. Kaltchuk-Santos and M.H.B. Zanettini (unpublished data) used B5 salts enriched with 16 organic compounds ("B5 long medium"; Carolina Biological Supply Co., Burlington, NC, U.S.A.; Table 3) and gelled with 0.3% agarose as the basal medium in their anther culture work.

6.2. *Growth regulators*

Plant species have been grouped into those that require growth regulators (Gramineae, Cruciferae) and those that do not (Solanaceae) with regard to anther culture. Soybean seems to belong to the former group. The key growth regulator 2,4-D has been essential for soybean microspore callus induction. IAA and NAA were not as affective as 2,4-D (Yin *et al.*, 1982, data not shown; Ye *et al.*, 1994, see below). In anther culture of Chinese cultivars, Yin *et al.* (1982) showed that 9.0 μM 2,4-D supported 22.4% callus induction. The percentage of callus induction dropped drastically when higher or lower concentrations were used (1.40 and 1.05% callus induced with 4.5 and 13.6 μM 2,4–D, respectively). Data obtained by Ye *et al.* (1994) were not as drastically different among different 2,4-D treatments: 23.9, 36.9, 25.5 and 18.6% callus were induced by 4.5, 9.0, 13.6 and 18.1 μM 2,4-D, respectively. Both groups found that 9.0 μM 2,4–D was optimum for soybean anther callus induction. Ye *et al.* (1994) also tested different concentrations of NAA on soybean anther callus induction. Callus induction percentages were 12.8, 16.7, 23.3, 27.2 and 25.0% for 5.4, 10.8, 16.1, 21.5, 26.9 μM NAA, respectively. This data indicated that NAA was not as effective as 2,4–D in soybean anther callus induction.

The concentration of 9.0 μM 2,4-D was employed by Zhuang *et al.* (1991), E. Kaltchuk-Santos and M.H.B. Zanettini (unpublished data) and Jian *et al.* (1986). In addition, 2.2 μM BA was supplemented by Zhuang *et al.* (1991) and E. Kaltchuk-Santos and M.H.B. Zanettini (unpublished data) and 8.9 μM BA, and 2.3 μM Kin by Jian *et al.* (1986). A much lower level of 2,4-D (0.23–0.90 μM) was used with the addition of 0.46–2.3 μM Zea and 2.7–5.4 μM NAA when isolated soybean microspores were cultured (Liu and Zhao, 1986).

6.3. *Sucrose*

Sucrose concentration in the medium is another important factor in soybean androgenesis. Plant species may be separated, empirically, into low (2–4%) and high (8–12%) sucrose requirement for better anther culture results. This division of species seems to be related to separation according to mature pollen which can be either bicellular (Solanaceae, Liliaceae) or tricellular (Gramineae, Cruciferae). The former group requires low osmotic conditions and the latter, high osmotic conditions for anther culture. The pollen grains of soybean are primarily bicellular (Corriveau and Coleman, 1988); however, a small percentage (0.4% for cv. Williams 82 and Asgrow A1929) of pollen grains are tricellular (X.J. Zhuang and C.Y. Hu, unpublished observation).

Yin *et al.* (1982) tested 6, 9, 12, 18, and 24% sucrose in anther culture of four Chinese soybean cultivars. The average percentage of callus formation was 3.2, 12.7, 25.7, 1.6 and 1.6, respectively. However, the differences in the percentage of callus formation were not as drastic when different concentrations of sucrose were tested by Ye *et al.* (1994). They obtained 33.3, 36.3, 39.8, 46.4 and 39.5% callus formation from medium containing 3, 6, 9, 12, and 15% sucrose. The data from both studies indicated that sucrose concentration around 12% was optimum for callus induction for soybean anther culture. With three Brazilian soybean cultivars, E. Kaltchuk-

Santos and M.H.B. Zanettini (unpublished data) tested the effects of sucrose concentrations on androgenic callus induction by removing the diploid callus which formed during the first month of culture. With 12% sucrose, the percentages of callus formation for cvs. IAS5, RS7 and Decada were 9.8, 23.2, and 6.9, respectively; with 9% sucrose, the percentages were 28.5, 42.6, and 46.6, respectively.

Some authors believe that, although sucrose at high concentration acts effectively as a trigger for the initiation of pollen division, it is not required for the further development. A significant increase in androgenic embryo yield has been obtained by sequential reduction in sucrose concentration in *Brassicas napus* anther culture medium during the first two weeks of culture (Chen and Chen, 1983). C.Y. Hu (unpublished data) tested the effect of reduced sucrose concentrations in soybean after an initial 2 week incubation on high sucrose medium. After cultured anthers (420 anthers/treatment) were transferred from 12% sucrose to 12, 9, 6, and 3% sucrose in two experiments, an average of 22.5, 8.0, 5.0, and 0.0 smooth, potentially embryogenic, calli were obtained, respectively. Similar results were recorded when anthers were transferred from the medium containing 9% initial sucrose concentration to media supplemented with 9, 6, and 3% sucrose concentrations. The above data indicate that a high concentration of sucrose was required both to initiate soybean microspore division as well as maintain callus development.

Data obtained by Ye *et al.* (1994) suggested that higher sucrose concentrations in medium were capable of reducing the percentage of diploid callus formation. Using 3, 6, 9, 12, and 15% sucrose in medium, they found that 30, 60.9, 66.2, 64.7, and 84.3%, respectively, of the callus obtained was haploid. Jian *et al.* (1986) also stated "When the concentration of sucrose was 12%, there was an enhancement of pollen-callus formation and an inhibition of somatic callus formation."

6.4. *Recommended soybean anther culture medium*

To suggest a reasonable starting point, we recommend the use of the following formulation as the initial soybean anther culture medium: B5 Long medium (Carolina Biological Supply Co., Burlington, NC, U.S.A.; Table 3) supplemented with 9.0 μM 2,4–D, 2.2 μM BA, and 9–12% sucrose. To avoid the possible inhibitory impurities in agar, 0.3% agarose should be used to gel the medium. This medium composition is by no means ideal for successful soybean anther culture. It can be improved depending on the genotype and culture conditions.

7. Cell division and callus formation

When anthers at the proper developmental stage are cultured, the anther wall will turn brown in about ten days and remain dark brown on culture medium. Callus formation occurs rapidly on the brown anthers. The highest frequency of callus formation is obtained after 20–30 days in culture. Based on the site of origin (Yin *et al.*, 1982), and chromosome count (Ye *et al.*, 1994) these early calli are, primarily, derived from diploid tissues such as anther wall, connective tissue and filaments with 40 chromosomes per nucleus. They are light-yellow, rough-surfaced, and relatively fast-grown tissues. We frequently excised and discarded these calli at the end of the first month of incubation. The remaining brown anthers were transferred onto fresh medium with the same composition for androgenic calli production.

Yin *et al.* (1982) reported that the content of more than half of the microspores disintegrated after 10 days of incubation. At this time, about 10% microspores started or had completed the first mitosis. The first mitotic division, in most cases, is unequal type of division and produces a generative and a vegetative nucleus. There was a small fraction of microspores engaged in equal division that resulted in two vegetative nuclei. In the hetero-nucleate grains, the division capacity of the generative nucleus was limited. Most of the nuclei in the multinucleate or multicellular grains seemed to be derived from the vegetative nucleus. After a 15–20 day incubation, the majority of the cultured microspores degenerated. Those remaining were mostly multinucleate grains and a small fraction of grains had more than three cells. After a 25 day incubation, the contents of the multinucleate grains disintegrated. At the same time multicellular grains with more than 100 cells appeared which made up about 2% of the total grains. The contents of some of those grains had already been released from the broken pollen walls and formed multicellular androgenic tissue clusters.

Jian *et al.* (1986) reported their seven years' work on soybean anther culture. Concerning the development of pollen *in vitro*, they simply stated, "After 5 days of culture, each of the uninucleate microspores divided into two equal nuclei, and after 10 days, the uninucleate pollen developed into pollen embryos, which eventually developed into globular embryos and embryoids." This statement contradicts all the other reports on anther culture of soybean. The "globular embryos" and "pollen embryos" in this article may have referred to the multicellular pollen grains which, they stated, "further developed into callus or embryoid."

Although the mitotic activities of the responding microspores were observed as early as at the end of low temperature pre-treatment period (Yin *et al.*, unpublished observation; Liu and Zhao, 1986), their rate of division apparently was slow compared to that of the somatic cells. Macroscopic androgenic calli started to appear after the first month of cultivation. Yin *et al.* (1982), by cytological examination, found that the calli that appeared 35 days after inoculation were composed of approximately 40% haploid, 11% diploid, and 49% aneuploid cells. Chromosome counts performed by Ye *et al.* (1994) on the calli produced after the first month of incubation were mostly from 14 to 26. Jian *et al.* (1986) stated that 50% of the anther calli were of microspore origin after 1–2 months of *in vitro* culture. The above studies agreed that these late-appearing calli were largely of haploid origin.

They were mostly light-green, either rough or smooth-surfaced, slow-growing tissues.

Upon subculture, the somatic callus grew fast and aged quickly, exhibiting no capacity for regeneration other than rooting. While the androgenic calli developed slowly on fresh medium, they aged slowly, and occasionally demonstrated shoot regeneration and/or embryogenic potential.

Liu and Zhao (1986) used an enriched liquid B5 medium (originally developed for soybean protoplast culture by Kao, 1982) supplemented with 2,4–D, Zea, and NAA to culture isolated soybean microspores. They obtained macroscopic milky-white calli from three-week-old dark-incubated cultures. These calli were successfully multiplied by subculture on a solid B5 medium supplemented with 2.2 μM BA, 9.9 μM IBA, 5.7 μM IAA and 2.3 μM Kin.

8. Shoot organogenesis and plantlet production

Yin *et al.* (1982) transferred light-green and rough-surfaced androgenic calli, 3 mm diam, onto B5-based differentiation media containing various cytokinin/auxin combinations to induce organogenesis about one week after the calli appeared on anthers. The organogenic capacity of these calli diminished if the tissues were permitted to stay too long on the callus induction medium. Roots were frequently produced from the transferred calli with sporadic appearance of shoot buds. Over several years, the frequency of bud production was always too low to recommend an appropriate bud differentiation medium (G.C. Yin, unpublished data). Once a bud was produced, it was excised and placed on a rooting medium with reduced cytokinin, 2.7–10.8 μM NAA and 10–100 mg/l Amanica (a rooting compound, the active ingredient could not be obtained) for plantlet production. Root-tip squashes confirmed the haploid nature of the regenerated plantlets.

Ye *et al.* (1994) obtained 14.6% root and 0.46% bud regeneration on various types of complex callus differentiation media. They also emphasized the importance of early transfer of the calli onto differentiation medium, ideally 5 to 7 days after their appearance on the cultured anthers. The capability of bud differentiation disappeared if the calli were subcultured onto the callus induction medium.

Apparently, sporadic shoot buds and plantlets were also obtained by Jian *et al.* (1986). Gradually improving the formulation of regeneration media, they obtained up to 1% shoot formation and 19 plantlets in one year.

Liu and Zhao (1986) transferred microspore derived calli, as they reached about 7 mm diam, onto regeneration media. Thirty-three regeneration media containing various combinations of growth regulators were tested. Some combinations resulted in the induction of roots and/or bud primordia but none supported bud growth to lead to plantlet production.

9. Embryogenesis and plantlet production

From smooth calli of U.S. cultivars, Zhuang *et al.* (1991) obtained (without a change in medium composition) embryo-like structures (Fig. 3A – C) although most of these structures were swollen and expressed various degrees of abnormalities. A small fraction of them resembled normal globular, heart or early cotyledon stage zygotic embryos. One callus produced more than ten well-defined but abnormal embryos. The ploidy levels of these embryos were not determined. Since the somatic tissues of mature soybean are known to be highly recalcitrant, it is unlikely that these embryos were derived from the anther wall/connective/filament tissues. Embryo-like structures were also obtained from cultured anthers of Chinese (Jian *et al.*, 1986; Ye *et al.*, 1994) and Brazilian (E. Kaltchuk-Santos and M.H.B. Zanettini, unpublished data) cultivars.

Numerous embryo producing calli were transferred onto Ranch *et al.* (1986) soybean somatic embryo "maturation" and/or "multiplication" media. However, no further embryonic development or multiplication took place following such subculture (Zhuang and Hu, unpublished data).

On the other hand, E. Kaltchuk-Santos and M.H.B. Zanettini (unpublished data) transferred the embryo-containing calli of Brazilian soybean cultivars onto media without growth regulators. On this new medium, 39 embryonic structures developed leaf-like structures, but only a few appeared normal. All 39 embryonic structures were again transferred onto medium with reduced (1%) sucrose. New leaves developed from many of the embryonic structures, one of which elongated and resulted in a complete plantlet (Fig. 3D). An attempt to obtain mitotic figures from root tips of this plantlet was unsuccessful.

10. Concluding remarks

Due to the recalcitrant nature of soybean, haploid induction techniques have only progressed slowly over the last two decades. Only fragments of basic information are presently available. Nevertheless, a solid base has been laid down by these pioneering studies. Major accomplishments are listed below:

a) Anthers with microspores at the uninucleate and early binucleate stages are more responsive than those at tetrad and late binucleate stages. The morphological indications of floral buds with microspores at the desirable stages for anther culture are when bract length equals 1/3–3/4 of the bud length (bud length = 3–5 mm) and the anthers are turning from translucent to opaque, showing light yellow color, but before the appearance of slight greenish-yellow shade.
b) Low temperature, 4–8°C, pre-treatment of the floral buds for 3 to 8 days before anther culture is beneficial.
c) B5 medium (Long formula [Table 3], if available) with 9–12% sucrose,

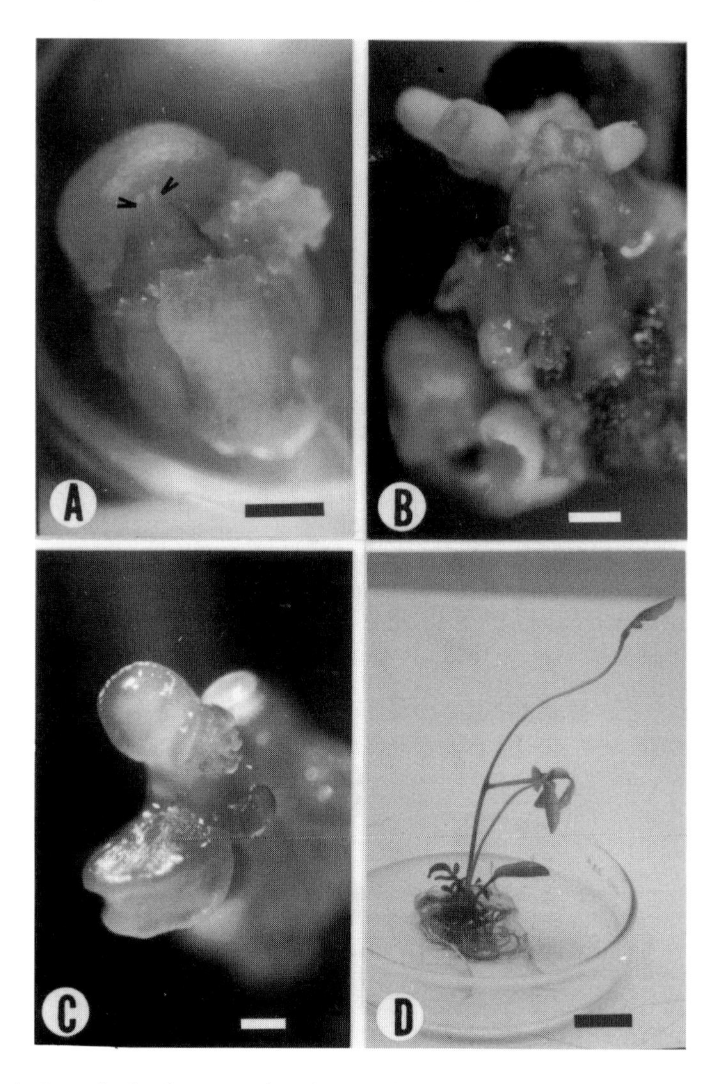

Figure 3. Androgenic development of soybean anthers cultured *in vitro*. (A-C) Abnormal embryoids developed from the smooth-surfaced calli. (A) A cluster of swelled embryoids, from a 5–months-old culture, tightly assembled together. Arrow heads on the hypocotyl of one embryoid pointing to the cotyledons of another embryoid. Bar = 2 mm. (B) Twelve well-defined, mostly trumpet-shaped, embryoids, from a 5-months-old anther culture, ranging from globular to early cotyledon stages. Bar = 1 mm. (C) Embryoids, from a 2.5–months-old anther culture, at cotyledon stage. The largest one with its two cotyledons fused together. Bar = 0.2 mm. (D) A plantlet developed from an androgenic embryoid. Bar = 20 mm.

9.0 μM 2,4-D, 2.2 μM BA, and 0.3% agarose is a good starting soybean anther culture medium.

d) Calli formed after the first month of culture are mostly of somatic origin but those formed during the second month are more likely of androgenic origin.

e) For bud induction, androgenic callus should be transferred onto the regeneration medium within one week after its appearance.

f) Androgenic embryos appear directly on the initial anther culture medium.

Statistically verified data are lacking in all aspects of soybean anther culture. To assist in the search of techniques leading to routine haploid plant production, such data need to be obtained. Thus, diverse areas of research activities are wide open and awaiting investigation.

To respond to changing needs, breeding new soybean cultivars is constantly necessary. Anther culture enables the development of completely homozygous lines from heterozygous F_1 plants in just a single generation. Homozygosity takes years to achieve by traditional breeding and it is an essential feature in self-pollinated soybean for new cultivar production. The potential benefits for saving time and labor costs in breeding justify our efforts toward development of soybean anther culture technology.

11. Acknowledgements

This work was supported in part by an USDA/OICD grant #58-319R-0-008, grant #90-240360-9 from NJ Comm. Sci. and Tech., NJ Governor's Challenge Grant to WPC, the Assigned Research Time Program from WPC, Fundacao de Amparo a Pesqisa do Rio Grande do Sul (FAPERGS) and Conselho Nacional de Desenvolvimento Cientifico e Tecnologico (CNPq). Thanks is due to Dr. Robert F. Callahan and Dr. Jack M. Widholm who have critically read this manuscript.

12. References

Chen, Z. and Z. Chen, 1983. High frequency induction of pollen-derived embryoids from anther cultures of rape (*Brassica napus* L.) Kexue Tongbao 28: 1690–1694.

Chu, C., 1978. The N_6 medium and its application to anther culture of cereal crops. In: Proc. Symp. on Plant Tissue Culture, pp. 43–50. Sci. Press, Peking.

Corriveau, J.L. and A.W. Coleman, 1988. Rapid screening method to detect potential biparental inheritance of plastid DNA and results for over 200 angiosperm species. Amer. J. Bot. 75: 1443–1458.

Delannay, X., D.M. Rodgers and R.G. Palmer, 1983. Relative genetic contributions among ancestral lines to North American soybean cultivars *Glycine max*. Crop Sci. 23: 944–949.

Gamborg O.L., R.A. Miller and K. Ojima, 1968. Nutrient requirements of suspension cultures of soybean root cells. Exp. Cell Res. 50: 151–158.

Hapgood, F., 1987. Soybean. Nat. Geographic 172(1): 66–91.

Heberle-Bors, E. and J. Reinert, 1979. Androgenesis in isolated pollen cultures of *Nicotiana tabacum*: dependence upon pollen development. Protoplasma 99: 237–245.

Horner, M. and H.E. Street, 1978. Pollen dimorphism – origin and significance in pollen plant formation by anther culture. Ann. Bot. 42: 763–777.

Ivers, D.R., R.R. Palmer and W.R. Fehr, 1974. Anther culture in soybean. Crop Sci. 14: 891–893.

Jian, Y.Y., D.P. Liu, X.M. Luo and G.L. Zhao, 1986. Studies on induction of pollen plants in *Glycine max* (L.) Merr. Chiang-su-Nung-Yeh-Hseuh-Pao J. Agri. Sci. (Nanjing) 2 (Suppl): 26–30.

Kaltchuk-Santos, E., M.H.B. Zanettini and E. Mundstock, 1993. Pollen dimorphism in soybean. Protoplasma 174: 74–78.

Kao, K.N. 1982. Plant protoplast fusion and isolation of heterokaryocytes. In: L.R. Wetter and F. Constabel (Eds.), Plant Tissue Culture Methods, pp. 49–56. Paririe Regional Lab. N.R.C. of Canada, Saskatchewan.

Lin Z.P. and G.C. Yin, 1983. The study on storage protein of soybean. Soybean Sci. 2: 232–238 (in Chinese).

Liu, D.P. and G.L. Zhao, 1986. Callus formation from pollen culture *in vitro* of soybean. Soybean Sci. 5: 17–20.

Marshall, D.R. and P. Broue, 1981. The wild relatives of crop plants indigenous to Australia and their use in plant breeding. J. Aust. Inst. Agric. Sci. 47: 149–154.

Matsubayashi, M. and K. Kuranuki, 1975. Embryogenic responses of the pollen to varied sucrose concentrations in anther culture. Sci. Rep. Fac. Agri., Kobe Uni., Japan. 11: 215–230.

Mignucci, J.S., and D.W. Chamberlain, 1978. Interactions of *Microphaera diffusa* with soybeans and other legumes. Phytopathology 68: 169–173.

Murashige, T. and F. Skoog, 1962. A revised medium for rapid growth and bioassays with tobacco tissue cultures. Physiol. Plant. 15: 473–497.

Newell, C.A. and T. Hymowitz, 1982. Successful wide hybridization between the soybean and a wild perennial relative, *G. tomentella* Hayata. Crop Sci. 22: 1062–1065.

Nitsch, J.P. and C. Nitsch, 1969. Haploid plants from pollen grains. Science 163: 85–87.

Ranch J.P., L. Oglesby and A.C. Zielinski, 1985. Plant regeneration from embryo-derived tissue cultures of soybeans. In Vitro Cell. Dev. Biol. 21: 653–658.

Ranch J.P., L. Oglesby and A.C. Zielinski, 1986. Plant regeneration from tissue cultures of soybean by somatic embryogenesis. In: I. Vasil (Ed.), Plant Regeneration and Genetic Variability, Vol. 4, Cell Culture and Somatic Cell Genetics of Plants, pp. 97–110. Academic Press, New York.

Rashid, A. and H.E. Street, 1974. Segmentations in microspores of *Nicotiana tabacum* which lead to embryoid formation in anther cultures. Protoplasma 80: 323–324.

Saka, H., T.H. Voqui-Dinh and T.Y. Cheng, 1980. Stimulation of multiple shoot formation on soybean stem nodes in culture. Plant Sci. Lett. 19: 193–201.

Singh, B.B., S.C. Gupta and B.D. Singh, 1974. Sources of field resistance to rust and yellow mosaic diseases of soybean. Indian J. Genet. Plant Breed. 34: 400–404.

Sunderland, N., 1974. Anther culture as a means of haploid induction. In: K. Kasha (Ed.), Haploids in Higher Plants: Advances and Potential, pp. 91–122. Univ. of Guelph Press, Guelph.

Sunderland, N. and F.M. Wicks, 1969. Cultivation of haploid plants from tobacco pollen. Nature 224: 1227–1229.

Tang, W.T., T.S. Ling and C.S. Chang, 1973. Effects of kinetin and auxin on callus formation in anther tissue cultures of soya bean. J. Agri. Asso. China (Chung-hua Nung-hsueh Hui Pao) #83: 1–7.

Vello, N.A., D.M. Hiromoto and A.J.B.V. Azevedo Filho, 1988. Coefficient of parentage and breeding of Brazilian soybean germplasm. Rev. Bras. Genet. 11: 679–697.

White, P.R., 1963. The Culture of Animal and Plant Cells. 2nd edn. Ronald Press Co., New York.

Wilson H.M., G. Mix and B. Foroughi-Wehr, 1978. Early microspore divisions and subsequent formation of microspore callus at high frequency in anthers of *Hordeum vulgare*. J. Exp. Bot. 29: 227–238.

Ye, X.G., Y.Q. Fu and L.Z. Wang, 1994. Study on several problems of soybean anther culture. Soybean Sci. 13: 193–199.

Yin, G.C., X.Z. Li, Z. Xu, L. Chen, Z.Y. Zhu and F.Y. Bi, 1980. A study of anther culture of *Glycine max*. Kexue Tongboa (English edn.) 25: 976 (Abstr.).

Yin, G.C., Z.Y. Zhu, Z. Xu, L. Chen, X.Z. Li and F.Y. Bi, 1982. Studies on induction of pollen plant and their androgenesis in *Glycine max* (L.) Merr. Soybean Sci. 1: 69–76 (in Chinese with English abstract).

Zhuang X.J., C.Y. Hu, Y. Chen and G.C. Yin, 1991. Embryoids from soybean anther culture. In Vitro Cell. Dev. Biol. 27(3, Part II): 145A (Abstr. # 432).

In Vitro Haploid Production in Higher Plants
Volume 3 – Important Selected Plants

M.R. Ahuja, Institute of Forest Genetics, Sieker Landstrasse 2, 22927 Grosshansdorf, Germany. Fax: +49 4102 696 200.

P.G. Arnison, FAAR Biotechnology Group Inc., P.O. Box 58012, 1516 Orleans Blvd., Orleans, Ontario, Canada K1C 7H4. Fax: 613-830-1086.

S. Baldursson, Iceland Forest Research Station, Mogilsa, IS 270, Mosfellsbaer, Iceland. Fax: +354 1 667750.

J.C.Chénieux, EA 1370 DRED-Plant Biotechnology Laboratory, Faculty of Pharmacy, 21 Avenue Monge, 37200 Tours, France.

A. Cersosimo, Istituto Sperimentale per Il Tobacco, 35071 Bovolone (VR) 8 Febbraio 1994, Via Canton, 14, Italy. Fax: 45 710 1097.

M.M. Fitch, USDA, ARS, Experiment Station HSPA, 99–193 Aiea Heights Drive, Aiea, HI 96701, USA. Fax: 808-486-5020.

M.J. Hennerty, University College Dublin, Department of Crop Science, Horticulture & Forestry, Faculty of General Agriculture, Belfield Road, Dublin 4, Ireland. Fax: +353-1-706 1104.

Monica Höfer, Bundesanstalt fur Zuchtungsforschung an Kulturpflanzen, Institut fur Obstzuchtung, Pilinitzer Platz 2, D-8054, Dresden, Germany. Fax: 351-39366.

J.Y. Hsu, Department of Agronomy, Taiwan Agricultural Research Institute, Weufeng, Taichung 41301, Taiwan.

C.-Y. Hu, The William Paterson College of New Jersey, Department of Biology, Wayne, NJ 07470, USA. Fax: 201-595-3414.

C.S. Hunter, Department of Biological Sciences, U.W.E., Bristol, Frenchay Campus, Bristol BS16 1QY, UK. Fax: +44 272 763 871.

B. Keimer, Maribo Seed, Hojbygardsvej 14, DK-4960 Holeby, Denmark.

E.R.J. Keller, Institute of Plant Genetics and Crop Plant Research, Corrensstr. 3, D-06466 Gatersleben, Germany. Fax: +49 161 251 9832.

W.A. Keller, National Research Council of Canada, 110 Gymnasium Road, Saskatoon, Saskatchewan, Canada S7N 0W9.

L. Korzun, Institute of Plant Genetics and Crop Plant Research, Corrensstr. 3, D-06466 Gatersleben, Germany.

G.H. Liang, USDA-ARS and Department of Agronomy, Kansas State University, Manhattan, KS 66506-56501, USA.

D. Laurain, EA 1370 DRED-Plant Biotechnology Laboratory, Faculty of Pharmacy, 31 Avenue Monge, 37200 Tours, France.

Y. Lespinasse, INRA, Station d'Amélioration des Éspeces Fruitières et Ornamentales, Beaucouze, 49000 Angers, France.

P.H. Moore, U.S. Department of Agriculture, Agricultural Research Service, P.O. Box 1057, Aiea, HI 96701, USA.

S.J. Ochatt, INRA, Centre de Recherches d'Angers, Station d'Amélioration des Éspeces Fruitières et Ornamentales, BP 57-49071, Beaucouze Cedex, France. Fax: 33 41 73 61 02.

C.E. Palmer, Department of Plant Science, University of Manitoba, Winnipeg, Manitoba, Canada R3T 2N2.

H.C. Pedersen, Danisco Seed, Hojbygardsvej 14, DK-4960 Holeby, Denmark.

M.C. Pedroso, Universidade de Lisboa, Faculdade de Ciencias, Department Biologia Vegetal, Bloco, Piso 1, Campo Grande, 1700 Lisboa, Portugal. Fax: 31 1 759 77 16.

N.S. Prakash, Central Coffee Research Institute, Coffee Research Station, Pin 577117, Chikmagalur, District Karnataka, India.

J.A. Przyborowski, Department of Plant Breeding and Seed Production,

University of Agriculture and Technology, ul. Pl. Lódzki 3, PL-10-724 Olsztyn, Poland.

Y. Raghuramulu, Central Coffee Research Institute, Coffee Research Station, Pin 577117, Chikmagalur District, Karnataka, India.

J. Regner, Höhere Bundeslehr- und Versuchsanstalt Klosterneuburg, Rehgraben 2, A-2103 Langenzersdorf, Austria.

G.L. Rotino, Istituto Sperimentale per l'Orticoltura Sezionedi Via Paullese, 28 20075 Montanaso Lombardo (MI), Italy. Fax: +39 371 681 72.

A.J. Sayegh, Faculty of Agriculture, University College Dublin, Belfield, Dublin 4, Ireland.

D.Z. Skinner, USDA-ARS and Department of Agronomy, Kansas State University, Manhattan, KS 66506-5501, USA.

R. Theiler-Hedtrich, Swiss Federal Research Station, CH-8820 Wadenswil, Switzerland.

J. Trémouillaux-Guiller, Université François Rabelais-Tours, Faculté des Sciences Pharmaceutiques, 31 Avenue Monge-37200 Tours, France. Fax: +33 47 27 66 60.

H.S. Tsay, Department of Agronomy, Taiwan Agricultural Research Institute, 289 Chang-cheng Road, Wu-feng 41301, Taichung, Republic of China.

R.E. Veilleux, Department of Horticulture, Virginia Polytechnic Institute and State University, College of Agriculture and Life Science, Blacksburg, VA 24061-0383, USA. Fax: 540-231-3083.

C.-C. Yeh, Department of Agronomy, Taiwan Agricultural Research Institute, Wufeng, Taichung 41301, Taiwan.

G.-C. Yin, Soybean Institute, Heilongiang Academy of Agricultural Science, Harbin, P.R.O. China.

M.H.B. Zanettini, Departamento de Genetica, Universidade Federal do Rio G. Sol, Porto Alegre, RS Brasil.

Y.X. Zhang, Laboratoire de Biochimie, INRA-GEVES, La Magneraud, B.P. 52, 17700 Surgères, France.

Species and subject index

Current Plant Science and Biotechnology in Agriculture

1. H.J. Evans, P.J. Bottomley and W.E. Newton (eds.): *Nitrogen Fixation Research Progress.* Proceedings of the 6th International Symposium on Nitrogen Fixation (Corvallis, Oregon, 1985). 1985 ISBN 90-247-3255-7
2. R.H. Zimmerman, R.J. Griesbach, F.A. Hammerschlag and R.H. Lawson (eds.): *Tissue Culture as a Plant Production System for Horticultural Crops.* Proceedings of a Conference (Beltsville, Maryland, 1985). 1986 ISBN 90-247-3378-2
3. D.P.S. Verma and N. Brisson (eds.): *Molecular Genetics of Plant-microbe Interactions.* Proceedings of the 3rd International Symposium on this subject (Montréal, Québec, 1986). 1987 ISBN 90-247-3426-6
4. E.L. Civerolo, A. Collmer, R.E. Davis and A.G. Gillaspie (eds.): *Plant Pathogenic Bacteria.* Proceedings of the 6th International Conference on this subject (College Park, Maryland, 1985). 1987 ISBN 90-247-3476-2
5. R.J. Summerfield (ed.): *World Crops: Cool Season Food Legumes.* A Global Perspective of the Problems and Prospects for Crop Improvement in Pea, Lentil, Faba Bean and Chickpea. Proceedings of the International Food Legume Research Conference (Spokane, Washington, 1986). 1988 ISBN 90-247-3641-2
6. P. Gepts (ed.): *Genetic Resources of* Phaseolus *Beans.* Their Maintenance, Domestication, Evolution, and Utilization. 1988 ISBN 90-247-3685-4
7. K.J. Puite, J.J.M. Dons, H.J. Huizing, A.J. Kool, M. Koorneef and F.A. Krens (eds.): *Progress in Plant Protoplast Research.* Proceedings of the 7th International Protoplast Symposium (Wageningen, The Netherlands, 1987). 1988 ISBN 90-247-3688-9
8. R.S. Sangwan and B.S. Sangwan-Norreel (eds.): *The Impact of Biotechnology in Agriculture.* Proceedings of the International Conference The Meeting Point between Fundamental and Applied in vitro Culture Research (Amiens, France, 1989). 1990. ISBN 0-7923-0741-0
9. H.J.J. Nijkamp, L.H.W. van der Plas and J. van Aartrijk (eds.): *Progress in Plant Cellular and Molecular Biology.* Proceedings of the 8th International Congress on Plant Tissue and Cell Culture (Amsterdam, The Netherlands, 1990). 1990 ISBN 0-7923-0873-5
10. H. Hennecke and D.P.S. Verma (eds.): *Advances in Molecular Genetics of Plant–Microbe Interactions.* Volume 1. 1991 ISBN 0-7923-1082-9
11. J. Harding, F. Singh and J.N.M. Mol (eds.): *Genetics and Breeding of Ornamental Species.* 1991 ISBN 0-7923-1094-2
12. J. Prakash and R.L.M. Pierik (eds.): *Horticulture – New Technologies and Applications.* Proceedings of the International Seminar on New Frontiers in Horticulture (Bangalore, India, 1990). 1991 ISBN 0-7923-1279-1
13. C.M. Karssen, L.C. van Loon and D. Vreugdenhil (eds.): *Progress in Plant Growth Regulation.* Proceedings of the 14th International Conference on Plant Growth Substances (Amsterdam, The Netherlands, 1991). 1992 ISBN 0-7923-1617-7
14. E.W. Nester and D.P.S. Verma (eds.): *Advances in Molecular Genetics of Plant–Microbe Interactions.* Volume 2. 1993 ISBN 0-7923-2045-X
15. C.B. You, Z.L. Chen and Y. Ding (eds.): *Biotechnology in Agriculture.* Proceedings of the First Asia-Pacific Conference on Agricultural Biotechnology (Beijing, China, 1992). 1993 ISBN 0-7923-2168-5

Current Plant Science and Biotechnology in Agriculture

16. J.C. Pech, A. Latché and C. Balagué (eds.): *Cellular and Molecular Aspects of the Plant Hormone Ethylene.* 1993 ISBN 0-7923-2169-3
17. R. Palacios, J. Mora and W.E. Newton (eds.): *New Horizons in Nitrogen Fixation.* Proceedings of the 9th International Congress on Nitrogen Fixation (Cancún, Mexico, 1992). 1993 ISBN 0-7923-2207-X
18. Th. Jacobs and J.E. Parlevliet (eds.): *Durability of Disease Resistance.* 1993 ISBN 0-7923-2314-9
19. F.J. Muehlbauer and W.J. Kaiser (eds.): *Expanding the Production and Use of Cool Season Food Legumes.* A Global Perspective of Peristent Constraints and of Opportunities and Strategies for Further Increasing the Productivity and Use of Pea, Lentil, Faba Bean, Chickpea, and Grasspea in Different Farming Systems. Proceedings of the Second International Food Legume Research Conference (Cairo, Egypt, 1992). 1994 ISBN 0-7923-2535-4
20. T.A. Thorpe (ed.): In Vitro *Embryogenesis in Plants.* 1995 ISBN 0-7923-3149-4
21. M.J. Daniels, J.A. Downie and A.E. Osbourn (eds.): *Advances in Molecular Genetics of Plant-Microbe Interactions.* Volume 3. 1994 ISBN 0-7923-3207-5
22. M. Terzi, R. Cella and A. Falavigna (eds.): *Current Issues in Plant Molecular and Cellular Biology.* Proceedings of the VIIIth International Congress on Plant Tissue and Cell Culture (Florence, Italy, 1994). 1995 ISBN 0-7923-3322-5
23. S.M. Jain, S.K. Sopory and R.E. Veilleux (eds.): In Vitro *Haploid Production in Higher Plants.* Volume 1: Fundamental Aspects and Methods. 1996 ISBN 0-7923-3577-5
24. S.M. Jain, S.K. Sopory and R.E. Veilleux (eds.): In Vitro *Haploid Production in Higher Plants.* Volume 2: Applications. 1996 ISBN 0-7923-3578-3
25. S.M. Jain, S.K. Sopory and R.E. Veilleux (eds.): In Vitro *Haploid Production in Higher Plants.* Volume 3: Important Selected Plants. 1996 ISBN 0-7923-3579-1
26. S.M. Jain, S.K. Sopory and R.E. Veilleux (eds.): In Vitro *Haploid Production in Higher Plants.* Volume 4: Cereals. 1996 ISBN 0-7923-3978-9
27. I.A. Tikhonovich, N.A. Provorov, V.I. Romanov and W.E. Newton (eds.): *Nitrogen Fixation: Fundamentals and Applications.* 1995 ISBN 0-7923-3707-7
28. N.L. Taylor and K.H. Quesenberry: *Red Clover Science.* 1996 ISBN 0-7923-3887-1
29. S.M. Jain, S.K. Sopory and R.E. Veilleux (eds.): In Vitro *Haploid Production in Higher Plants.* Volume 5: Oil, Ornamental and Miscellaneous Plants. 1996 ISBN 0-7923-3979-7

KLUWER ACADEMIC PUBLISHERS – DORDRECHT / BOSTON / LONDON